# RECENT TRENDS IN AEROELASTICITY, STRUCTURES, AND STRUCTURAL DYNAMICS

Raymond L. Bisplinghoff

# Recent Trends in Aeroelasticity, Structures, and Structural Dynamics

*Prabhat Hajela, editor*

Papers from the R. L. Bisplinghoff Memorial Symposium
University of Florida
February 6–7, 1986

UNIVERSITY PRESSES OF FLORIDA
University of Florida Press
Gainesville

UNIVERSITY PRESSES OF FLORIDA is the central agency for scholarly publishing of the State of Florida's university system, producing books selected for publication by the faculty editorial committees of Florida's nine public universities: Florida A&M University (Tallahassee), Florida Atlantic University (Boca Raton), Florida International University (Miami), Florida State University (Tallahassee), University of Central Florida (Orlando), University of Florida (Gainesville), University of North Florida (Jacksonville), University of South Florida (Tampa), University of West Florida (Pensacola).

ORDERS for books published by all member presses should be addressed to University Presses of Florida, 15 NW 15th Street, Gainesville, FL 32603

Copyright © 1987 by the Board of Regents of the State of Florida

Printed in the U.S.A.

Library of Congress Cataloging-in-Publication Data

R.L. Bisplinghoff Memorial Symposium (1986: University of Florida)
   Recent trends in aeroelasticity, structures, and structural dynamics.

  "Papers from the R.L. Bisplinghoff Memorial Symposium,
University of Florida, February 6–7, 1986."
  1. Aeroelasticity.  2. Structures, Theory of.
3. Airframes.  I. Hajela, Prabhat, 1956–
II. Bisplinghoff, Raymond L.  III. Title.
TL574.A37R58  1986    629.132′362    87–8120
ISBN 0–8130–0871–9

To the memory of
Raymond L. Bisplinghoff
1917–1985

CONTENTS

Preface ix

Foreword

Some Memories and Some Notes from a Recent Survey of Aeroelasticity    xi
   H. Ashley

### I. Aeroelastic Problems in Helicopters, Propellers and Turbomachinery

Recent Trends in Rotary-Wing Aeroelasticity    3
   P. P. Friedmann
A Survey of Current Problems in Turbomachine Aeroelasticity    48
   F. Sisto
Experimental Classical Flutter Results of a Composite Advanced Turboprop Model    63
   O. Mehmed and K. R. V. Kaza
Whirl Flutter of Swept Tip Propfans    75
   M. I. Young and C. A. Harper
Recent Developments in Flutter Suppression Techniques for Turbomachinery Rotors    87
   O. O. Bendiksen
Review of Floquet Theory in Stability and Response Analyses of Dynamic Systems with Periodic Coefficients    101
   G. H. Gaonkar and D. A. Peters
Identification of Structural Dynamic Systems    120
   S. Hanagud, M. Meyyappa, and J. I. Craig

### II. Aeroelasticity and Unsteady Aerodynamics

Nonlinear Aeroelasticity: An Overview    144
   E. Dowell
Experimental Studies in Aeroelasticity of Unswept and Forward Swept Graphite/Epoxy Wings    149
   J. Dugundji
Low-Speed Aeroelasticity of Bridge Decks    161
   R. H. Scanlan
Unsteady Supersonic Aerodynamics of Planar Lifting Surfaces Accounting for Arbitrary Time-Dependent Motion    173
   L. Librescu
Aeroelasticity of Very Light Aircraft    187
   I. Kroo
Problems and Progress in Aeroelasticity for Interdisciplinary Design    203
   E. Carson Yates, Jr.

## III. Structures, Structural Dynamics and Active Control

| | |
|---|---|
| An Overview of Finite Element Formulations by Generalized Variational Principles<br>*T. H. H. Pian* | 222 |
| Catastrophic Failure of Laminated Cylinders Under Internal Pressure<br>*J. W. Mar* | 236 |
| Natural Frequencies of Triangular Plates Using Characteristic Orthogonal Polynomials in Rayleigh-Ritz Method<br>*R. B. Bhat* | 251 |
| Structural Stability in Turbulent Flow<br>*Y. K. Lin* | 259 |
| On the Transient Dynamics During the Orbiter Based Deployment of Flexible Members<br>*V. J. Modi* | 271 |
| Interactive Structural and Controller Synthesis for Large Spacecraft<br>*S. C. McIntosh, Jr. and M. A. Floyd* | 283 |
| Damage in Composite Laminates from Central Impacts at Subperforation Speeds<br>*L. E. Malvern, C. T. Sun, and D. Liu* | 298 |
| Dynamic Responses of Orthotropic Plates under Moving Masses<br>*O. P. Agrawal and S. Saigal* | 313 |

## IV. Optimal Design

| | |
|---|---|
| Structural Tailoring for Aircraft Performance<br>*T. A. Weisshaar* | 336 |
| Structural Optimization Methods for Industrial Application<br>*G. N. Vanderplaats* | 353 |
| Constrained Optimization Techniques for Active Control of Aeroelastic Response<br>*V. Mukhopadhyay* | 371 |
| First and Second-Order Sensitivity Analyses of Linear and Nonlinear Structures<br>*R. T. Haftka and Zenon Mróz* | 384 |
| A Survey of Methods and Problems in Aeroelastic Optimization<br>*P. Hajela* | 401 |

# PREFACE

The collection of papers assembled in this volume represents a lasting tribute to the memory of the late Professor R. L. Bisplinghoff, an eminent scientist and scholar who made significant contributions to the fields of aeroelasticity and aeronautical structures. An even more noteworthy contribution was the inspiration that Professor Bisplinghoff provided to his students and coworkers: this sentiment was repeatedly expressed by the various speakers who participated in the symposium and who are also the principal contributors to this volume.

Digital computing machinery has revolutionized the approach to the analysis and design of aerospace structural systems. This has nowhere been more evident than in the development of general structural analysis capabilities such as the finite element method. Professor Bisplinghoff actively participated in this development and is credited with early contributions to the analysis of wing structures by such discrete methods. Almost thirty years have elapsed since the computer was introduced into the work environment, bringing in its wake a host of new problems and solution strategies.

Although these developments have been periodically reported in various conferences and publications, this symposium brought together a group of recognized experts to assess the present state of the art and to unify trends for future research efforts. The papers compiled here are therefore either comprehensive surveys of well-developed fields of study or new contributions in emerging areas of technology. The subject areas covered include fixed and rotary wing aeroelasticity; aeroelastic considerations in rotating machinery; aeroelastic problems in bridge design; structural analysis and structural dynamics in aerospace applications; aeroservoelastic considerations; and the emerging discipline of optimal structural design. The book is organized in four sections to reflect this distribution.

My association with Professor Bisplinghoff was at the University of Florida, where he taught for eight years as an adjunct professor in the Department of Engineering Sciences. His untimely death made this association short but has left me with pleasant memories. As an assistant professor in my first position, I was impressed by his concern for my professional well-being. He always had the time to discuss any aspect of my research plans and was full of constructive suggestions. I and my colleagues at the University of Florida felt it was only natural to organize and host a befitting memorial for a truly talented and gracious individual.

I would like to thank Dr. A. Ezra for the symposium's support received from the National Science Foundation under Grant No. NSF-ECE-8602170. I also wish to acknowledge the support of other sponsors of the symposium, including the Scientific Advisory Board of the U.S. Air Force and the Massachusetts Institute of Technology. A very special note of gratitude is extended to Professor K. T. Millsaps for his numerous suggestions and enthusiastic support at every stage of the planning process. Finally, I would like to thank Ms. Pamela Herring and Ms. Gail Luparello for providing invaluable assistance in organizational matters.

Prabhat Hajela

# FOREWORD

## SOME MEMORIES AND SOME NOTES FROM A RECENT SURVEY OF AEROELASTICITY*

Professor Holt Ashley
Department of Aeronautics and Astronautics
Stanford University
Stanford, CA 94305

## INTRODUCTION AND APPRECIATION

Ladies and Gentlemen. You see before you a "Junior Author," and I am deeply proud to be one. Let me begin by thanking the organizers of this Memorial Symposium - and especially my former student Professor Prabhat Hajela - for giving me the opportunity to be part of the program and to share some personal reminiscences of the wonderful man to whom this Symposium is dedicated.

It is easy to make an impression on an audience by formally listing the achievements and honors of Raymond L. Bisplinghoff. He was a member of both National Academy of Sciences and National Academy of Engineering; he held directorships in nine corporations; he was a Fellow of four professional engineering societies and Honorary Fellow of two (Royal Aeronautical Society and AIAA); and one could go on and on. But there are also subjective and non-quantifiable dimensions of a human being that only those can attempt to describe who had the privilege of many years of personal association. Therefore, let me use the first part of this talk to relate to you some "RLB" stories, some random recollections.

• I first encountered Ray when I walked into the M.I.T. course XVI Introduction to Aeronautical Structures subject (16.20) in June 1946. Here was this small man with a high-pitched voice, just hired out of the U.S. Navy by Professor Jerome C. Hunsaker, the Department Head. It soon became apparent, however, that he knew exactly what he was talking about and could communicate it to his students brilliantly, In fact, he was one of the few who had the following ability: if you copied into your notes everything he wrote on the blackboard, you had available nearly a complete course text.

• Soon I started to work for him and others in the new M.I.T. Aeroelastic & Structures Research Laboratory, whose Directorship he took over in 1950 from Professor Manfred Rauscher. He was not only a legendary leader, but a skillful shortstop on our softball team.

---

*Revision of an oral presentation

- After our M.I.T. group had presented in 1952 its first Summer Course on "Aeroelasticity", Ray took Bob Halfman and me aside, saying "I think it's time for us to write a book". It was an extraordinarily friendly, pleasant collaboration. The result [1] appeared in 1955 and is still in print, having sold nearly 10,000 copies. The 1952 course notes were the first draft. The only hitch in the process was when Cornell Laboratory reprinted the whole thing within their own covers (we had forgotten to insert the phrase "copyright by the authors"). Subsequently Ray, Bob and I participated in several other books with several co-authors, but I believe we were all proudest of the <u>first</u>.

- At the Ninth International Congress of Applied Mechanics in Brussels, 1956, we shared a couple of adventures. While crossing the street with me, he was struck on the left leg by a mad hit-and-run driver. Although his shoe was thrown 30 feet, there were fortunately no breaks - just a very uncomfortable bruise. The most frightening portion of the incident was our ride to the hospital in a Belgian ambulance. Also at this Congress he told us his plan to seek the Ph.D. at E.T.H., Zurich. He said, "Holt, I'm nearly 40 years old and still don't have a doctor's degree!" He earned it in record time, producing a fascinating thesis on tests and theory for wings subjected to aerodynamic heating. This was also the subject of his memorable Wright Brothers Lecture [2].

- At M.I.T. Ray's door was always open. But generally his colleagues took advantage of his hospitality only when they had a minor triumph to share. I recall that Zartarian and I burst in on him in 1956 to show how we could use piston theory to predict supersonic flutter speeds with simple algebra. He really seemed to take as much pleasure as we did from such discoveries.

- Each of us, I am sure, has a list of what we believe to be Ray's greatest research contributions. There were many, but my list includes: a tensor/dyadic approach to developing equations of motion for flexible structures which permitted large excursions in rigid-body degrees of freedom (<u>decades</u> ahead of its time at University of Cincinnati); one of the first finite-element analyses of built-up box beams; his work on high temperature structures; and full development of means for predicting aircraft response to dynamic loads, especially nuclear blast waves.

- My later associations with him included eighteen months at National Science Foundation and since 1982, as a colleague in a little company called RANN, Inc., Alfred J. Eggers is the head of RANN and we talked about Ray recently. In Al's view, Ray's outstanding characteristic (among many) was the way he combined an aspiration for the <u>highest</u> quality with deep concern for what is best for the United States. His espousal of the RANN program at N.S.F. in the early 1970's exemplified his "patriotic perfectionism".

• I last saw Ray at the Cosmos Club, D.C. in February 1984. Although already he had lost a lot of weight, he was still full of humor and enthusiasm. It is in that spirit that I shall recall him - and those memories will remain with me warmly as long as I am able to remember.

THE AMR REVIEW [3]

Bisplinghoff first reviewed aeroelasticity for <u>Applied Mechanics Reviews</u> [4] in 1958. My follow-on appeared in February 1970 [5]. Last year Professors Charles Steele and George Springer persuaded 25 of us to prepare up-dates on earlier reviews on various topics for a special-issue publication in late 1986. Its preparation is still on schedule.

While it would be inappropriate here to duplicate or give details of Ref. [3], let me mention some highlights. The review attempts to build on Ref. [5], noting that progress in the field since 1970 has been generally evolutionary. 127 carefully-selected references are added to the original 125. One of the first subjects is the "people of aeroelasticity," with the sad news of the passing of the three "giants": Bisplinghoff in March 1985, H.G. Kussner in 1984, and I.E. Garrick in 1981.

The organizations are listed which, in the author's opinion, have fostered particularly significant contributions to aeroelasticity in the post-war era. By design, this list does not contain names of individual universities. If it did, however, I would place M.I.T. at the top of the roster.

Finally, the review selects and summarized four areas for more detailed discussion:

• Active control of aeroelastic phenomena.

• Rotating machinery (axial-flow gas turbines, propellers and helicopter rotors).

• Transonic flutter and associated tools of computational fluid dynamics.

• Problems involving separated flow.

The reader is referred to Ref. [3] for further details.

REFERENCES

1. Bisplinghoff, R.L., Ashley, H., and Halfman, R.L.,

*Aeroelasticity*, Addison-Wesley Publishing Co., Reading, MA., 1955 (second printing in 1957; translated into Russian 1958 by V.P. Shidlovsky et al).

2. Bisplinghoff, R.L., "Some Structural and Aeroelastic Considerations of High Speed Flight- The Nineteenth Wright Brothers Lecture," *Journal of the Aeronautical Sciences*, Vol. 23, pp. 289-321, 1956.

3. Ashley, H., "Aeroelasticity," to appear in a special issue of *Applied Mechanics Reviews*, October 1986 (tentative date).

4. Bisplinghoff, R.L., "Aeroelasticity," *Applied Mechanics Reviews*, Vol. II, pp. 99-103, 1958.

5. Ashley, H., "Aeroelasticity," *Applied Mechanics Reviews*, Vol. 23, pp. 119-129, 1970.

I. Aeroelastic Problems in Helicopters, Propellers and Turbomachinery

# RECENT TRENDS IN ROTARY-WING AEROELASTICITY

Peretz P. Friedmann
Mechanical, Aerospace and Nuclear Engineering Department
University of California
Los Angeles, California 90024, U.S.A.

## Abstract

The purpose of this paper is to survey the principal developments which have occurred in the field of rotary-wing aeroelastcity during the past five year period. This period has been one of considerable activity and approximately one hundred papers have been published on this topic. To facilitate this review the field has been divided into a number of areas in which concentrated research activity has taken place. The main areas in which recent research is reviewed are: (1) structural modeling; (2) aerodynamic modeling; (3) aeroelastic problem formulation using automated or computerized methods; (4) aeroelastic analyses in foward flight; (5) coupled rotor/fuselage analyses; (6) active controls and their application to aeroelastic response and stability; (7) application of structural optimization to vibration reduction; and (8) aeroelastic analysis and testing of special configurations. These areas are reviewed with different levels of detail and some useful observations regarding potentially rewarding areas of future research are made.

## Nomenclature

| | |
|---|---|
| $\bar{a}$ | = elastic axis aerodynamic center offset, nondimensionalized by semichord |
| b | = semichord nondimensionalized with respect to blade radius |
| $C(\bar{s})$ | = generalized Theodorsen lift deficiency function |
| $C_1$ | = inflow parameter |
| c | = chord |
| $C_T$ | = thrust coefficient |
| $C_{DU}$ | = unsteady drag coefficient |
| $C_W$ | = weight coefficient, approximately equal to $C_T$ |
| $C_{\theta CT}$ | = control system stiffness |
| $c_\zeta$ | = damping coefficient for lag damper |
| $D_\zeta$ | = structural damping in lag |
| EA | = axial stiffness |
| EI | = bending stiffness |
| E | = expected value |
| e | = blade root offset |
| GJ | = torsional stiffness |
| h | = plunge displacement, also hub height above CG |
| k | = reduced frequency |
| $\ell_T$ | = length of swept tip portion of blade |
| $\ell_i$ | = length of finite element |
| $L_C$ | = circulatory lift |
| $M_C$ | = circulatory moment |
| M | = Mach number |
| $M_1$ | = dynamic inflow parameter |

$n$ = number of blades
$Q(t)$ = downwash velocity at ¾ chord
$R_c$ = elastic coupling parameter
$R$ = blade radius
$R_x, R_y$ = fuselage translations in x and y directions, respectively
$T$ = transfer or control matrix
$u, u(\dot{x}, \dot{y}), u(x, y, \dot{x}, \dot{y})$ = control, reduced state control of hub rate and rate and position
$U_{T0}$ = constant part of velocity of approach in airfoil theory
$U_T(t)$ = velocity of approach in airfoil theory
$u, v, w$ = axial, lag and flap elastic blade displacements
$W_Z, W_\theta, W_{\Delta\theta}$ = weighting matrices for vibration levels, control angles and rates control angle, respectively
$X_1, X_2$ = augmented states
$Z_0, Z_i$ = vector of uncontrolled and controlled vibration levels, respectively
$\alpha$ = angle of attack
$\alpha_0$ = mean angle of attack
$\bar{\alpha}$ = oscillatory part of $\alpha$
$\beta_R, \beta_P$ = regressing and progressing flap modes
$\beta_P$ = precone angles
$\beta_{BB}$ = built in blade to beam angle
$\beta_k, \zeta_k$ = flap and lag deflections of the $k^{th}$ blade, respectively
$\gamma$ = lock number
$\delta_3$ = pitch-flap coupling
$\epsilon$ = basis for ordering schemem, magnitude of blade slopes
$\zeta, \eta$ = principal axes of cross section
$\zeta_R, \zeta_P$ = regressing and progressing lag mode
$\theta_0$ = collective pitch
$\theta_x, \theta_y$ = roll and pitch of fuselage, respectively
$\theta$ = pitch mode for coupled rotor/body analysis
$\theta$ = pitch angle for airfoil in ONERA model
$\theta_{HHn}$ = higher harmonic control pitch angle
$\theta_{HHC}, \theta_{HHS}, \theta_{HHO}$ = various components of higher harmonic control
$\theta_{AC(\psi)}, \theta_{AS(\psi)}$ = components of active control
$\theta_{Ak}$ = actively controlled pitch angle of $k^{th}$ blade
$\lambda$ = inflow mode
$\Lambda_V, \Lambda_H$ = vertical and horizontal ply angle orientation
$\Lambda$ = sweep angle of blade tip
$\mu$ = advance ratio
$\rho_A$ = density of air
$\sigma$ = blade solidity
$\phi$ = roll mode, coupled rotor/fuselage model
$\phi_c, \phi_s$ = phase angles for HHC
$\hat{\phi}, \phi$ = torsional deformation and torsional quasicoordinate, respectively
$\psi$ = Azimuth angle
$\bar{\omega}_{F1}, \bar{\omega}_{L1}$ = fundamental rotating flap and lag frequencies nondimensionalized with respect to $\Omega$
$\bar{\omega}_\phi, \bar{\omega}_{T1}$ = fundamental torsional, rotating flap and lag frequencies nondimensionalized with respect to $\Omega$
$\Omega$ = speed of rotation
$(\cdot)$ = derivative with respect to $\psi$ or $t$

## 1. Introduction and Objectives

During the last twenty years there has been a tremendous proliferation in the literature dealing with rotary-wing aeroelastcity which indicates that understanding of rotor and coupled rotor/fuselage dynamics plays a central role in the design of successful rotorcraft. This vigorous activity has also resulted in a considerable number of survey papers which have dealt with various aspects of rotary-wing aeroelastic stability and response problem. These surveys are listed in chronological orders in Refs. [1-10]. One of the first significant reviews of rotary-wing dynamic and aeroelastic problems was provided by Loewy [1] where a wide range of dynamic problems were reviewed in considerable detail. A more restricted survey emphasizing the role of unsteady aerodynamics and vibration problems in forward flight was presented by Dat [2]. Flight dynamics problems of hingeless rotorcraft including experimental results was treated by Hohenemser [3]. Blade stability was also discussed in Ref. 3, since it is considered to be part of the broader flight dynamics problem. Two comprehensive reviews of rotary-wing aeroelasticity were presented by Friedmann [4,5]. In Ref. 4 a detailed chronological discussion of the flap-lag and coupled flap-lag-torsion problems in hover and forward flight was presented emphasizing the inherently nonlinear nature of the hingeless blade aeroelastic stability problem. The nonlinearities considered were geometrical nonlinearities due to moderate blade deflections. In Ref. 5 the role of unsteady aerodynamics, including dynamic stall, was examined together with the treatment the nonlinear aeroelastic problem in forward flight. Finite element solutions to rotary-wing aeroelastic problems were also considered together with the treatment of coupled rotor/fuselage problems. Another detailed survey by Ormiston [6] discussed the aeroelasticity of hingeless and bearingless rotors, in hover, from an experimental and theoretical point of view.

In addition to these papers which have emphasized primarily the aeroelastic stability problem two other surveys have dealt exclusively with the vibration problem and its control in rotorcraft [7,8]. One could therefore classify these papers as related to the aeroelastic response of the rotor, the vibrations caused by this aeroelastic response and the study of various passive, semi-active and active devices for controlling such vibrations. Finally it should be mentioned that in a very recent comprehensive review paper by Johnson [9,10] both the aeroelastic stability and the rotorcraft vibration problems were reviewed in the context of dynamics of advanced rotor systems.

The purpose of this paper is to survey the principal developments which have occurred in the field of rotary-wing aeroelasticity during the past five year period and thus it represents an extension of the previous two papers written by the author [4,5]. This period has been very productive and over one hundred papers were published on this topic. To facilitate this review the subject matter has been subdivided into a number of areas in which a concentrated research activity had taken place. Each area is reviewed as a separate topic and a list of these topics, including a brief description, is provided below.

(1) <u>Structural Modeling</u>; In this area there was continued interest in geometrically nonlinear structural models for hingeless and bearingless rotor configurations. Finite element models for bearingless rotors have been developed. Structural models for composite blades and curved or swept blades were also introduced.

(2) <u>Aerodynamic Modeling</u>; Previously developed dynamic stall models have been refined and extended. There have been a number of additional studies aimed at the understanding of dynamic inflow. Arbitrary motion aerodynamics for rotary-wing applications were developed and applied to a number of simple problems.

(3) <u>Aeroelastic Problem Formulation</u>; Automated generation of complicated equations of motion using numerical methods or computerized symbolic manipulation was the primary activity.

(4) <u>Aeroelastic Analyses in Forward Flight</u>; A number of studies dealing with aeroelastic stability and response of hingeless and bearingless rotors were performed. There was continued interest in numerical treatment of equations with periodic coefficients.

(5) <u>Coupled Rotor/Fuselage Aeromechanical Analyses</u>; A number of coupled rotor/fuselage analyses have been developed and correlated with experimental data.

(6) <u>Active Controls and their Application to Vibration Alleviation and Blade Stability Augmentation</u>; This area was by far the most vigorous. Many studies, primarily experimental, have been aimed at vibration reduction by higher harmonic control. A few studies have also considered blade stability augmentation.

(7) <u>Application of Structural Optimization to Vibration Reduction</u>; In this area modern structural optimization was used to tailor fundamental blade frequencies such that vibration levels in forward flight were significantly reduced.

(8) <u>Aeroelastic Analysis and Testing of Special Configurations</u>; Such as circulation in controlled rotors, constant lift rotors, hybrid heavy lift helicopters, and bearingless/hingeless configurations which were tested in wind tunnels.

Based on the review of these research areas a number of observations regarding potentially rewarding areas of future research are made. Finally it should be noted that the author apologizes for papers which were inadvertently omitted in this survey.

## 2. Structural Modeling

Previous research during the last fifteen years [4-6] has established the importance of geometrically nonlinear terms in the analysis of hingeless and bearingless rotors. These geometrically nonlinear terms are associated with the assumption of moderate rotations (or blade slopes) and small strains and require the use of nonlinear beam kinematics in the development of the structural, inertia and aerodynamic operators associated with the rotary-wing aeroelastic problems. This kinematical nonlinearity produces, in many cases, coupling between the bending and torsional motions of the blade. This important coupling effect can not be obtained in an accurate and consistent manner without incorporation of the geometrical nonlinearities. Therefore a considerable number of recent studies have been aimed at providing improved capabilities for dealing with this particular class of problems.

Alkire [11] extended the analysis presented in Ref. 12 to obtain a better understanding of the role of built-in pretwist and elastic twist in the derivation of transformation matrices which relate the position vectors of the undeformed and deformed states of the blade. Two distinct approaches were considered, one in which the built-in twist is applied initially before the elastic deformations occur and a second approach where the pretwist is combined with the twist due to deformation. It was shown that these two approaches could be related to each other. Furthermore a procedure was developed for evaluating transformation matrices which remain unaffected by rotation sequences or the treatment of pretwist. Hodges [13] in a recent study has developed a nonlinear beam element for the analysis of rotating blades in which the assumption of moderate deflections has been abandoned. His analysis, which is intended to capture large rotations, is based on the systematic simplification of the kinematic relations using a less restrictive assumption whereby extensional strain is neglected compared to unity. Furthermore the transformations used in this study utilized Tait-Bryan orientation angles and Rodrigues parameters instead of Euler angles, which have been used in many previous studies. The final equations are based on the assumptions of isotropic stress-strain relations. This study served as the theoretical basis of the beam element used in the GRASP computer program [14]. This beam element [13] represents an important contribution since it is based on a minimal number of assumptions restricting the magnitude of the deflections experienced by the rotor blade. Associated with this model one finds both mathematical elegance and complexity. Thus the cost effectiveness of this model for rotary-wing aeroelastic analyses remains to be demonstrated. It is quite possible that for most applications the previous models, based on the assumption of moderate deflections, could prove themselves adequate.

Equations of blade equilibrium which were based on moderate deflection beam theories [5] frequently utilize ordering schemes to neglect higher order nonlinear terms. In such ordering schemes the slopes of the blade are assigned an order of magnitude $\epsilon$ and terms of order of magnitude $\epsilon^2$ are neglected compared to terms of order one. By assigning orders of magnitude to the various parameters in the problem this approach leads to equations which contain second order nonlinear terms. In a study by Crespo DaSilva and Hodges [15,16] the influence of retaining the next level of terms in the equations of motion was considered, this approach yields more exact equations, which include third order nonlinear terms. In the second part of this study [16] the influence of these third orders on blade response and stability was considered, using a global Galerkin method to solve the equations of motion [5]. The results indicated that at relatively high collective pitch values ($\theta_0 > 0.2$) and for a blade which was very soft in torsion ($\bar{\omega}_\phi = 2.5$) the third order terms can influence both, the equilibrium position and the stability of the blade. This influence which is more pronounced for stiff-in-plane hingeless blades, is mild at practical values of the collective pitch setting for soft-in-plane blades.

Many previous studies of structural models of rotor blades [4-6] were restricted to initially straight blades. To remedy this situation Rosen and Rand [17,18] developed a structural nonlinear model for the behavior of curved helicopter blades. The model is very general and it allows for complicated geometries, boundary conditions and structural property distributions. In this model large deformations of the blade are accounted for. A somewhat restrictive assumption is that the undeformed rod lies initially in a plane. This assumption, combined with the pretwist of the blade and the retention of the curvature terms causes these equations to be cumbersome. Large deflections are treated using Euler angles and it is assumed that strains are small and negligible compared to unity. This model appears to be an improvement on a previous model which was used by the same authors [19].

Finite element modeling of hingeless and bearingless blade configuration was another area where a fair amount of activity occurred. Sivaneri and Chopra [20] developed a finite element model for hingeless blades which was very similar to an earlier model developed by Friedmann and Straub [21]. Subsequently Sivaneri and Chopra [22] have extended this finite element model to bearingless rotors. The flexbeam type of bearingless rotor is modeled by using regular beam finite elements for the outboard portion, a rigid clevis, and multiple beams to represent the flexbeam and the torque tube, as shown in Fig. 1. Special displacement compatibility conditions are enforced at the clevis. The fifteen degree of freedom finite element model used for modeling the outboard portion of the beam, shown in Fig. 2, is based on a cubic interpolation for the bending degrees of freedom, v, w and the axial degree of freedom u, and a quadratic interpolation for torsion $\phi$. This method consists essentially of developing a special redundant root element for the flexbeam.

Finite elements for bearingless rotor modeling have been also developed by Hodges et al. [14] as part of a general rotorcraft aeromechanical program called GRASP. The structural modeling capability in GRASP is quite general enabling one to model any bearingless rotor configuration. Reference 14 utilizes higher order finite elements as opposed to the conventional finite elements used in Refs. 20-22. Another, sixteen degree of freedom, finite element model has been used in Ref. 23 to study the influence of a compressible lifting surface theory on the coupled flap-lag-torsional aeroelastic stability of a hingeless rotor blade in hover.

All the structural models discussed above were restricted to isotropic blades. One of the more important recent developments was the emergence of structural models suitable for the analysis of composite rotor blades, which are widely used on modern helicopters. Mansfield and Sobey [24] made a pioneering attempt to develop the stiffness properties of graphite fiber composite rotor blades and they also tried to explore the potential of this model for aeroelastic tailoring. Despite its innovative nature this study fell short of it stated objectives.

A comprehensive and important study by Hong and Chopra [25] presented, for the first time, an aeroelastic model for a composite rotor blade in hover. In this study a moderate deflection, coupled flap-lag-torsional analysis of a laminated box beam was developed in which terms up to the second order in blade slopes were retained. The nonlinear strain displacement relations were taken from Hodges and Dowell [26]. Each laminate wall of the box beam, representing the blade spar shown in Fig. 3, was assumed to consist of a number of composite plies at arbitrary orientation of the ply angles. Constitutive relations were obtained assuming that each lamina of the laminates is <u>orthotropic</u> and there is no shear stiffness through the thickness distribution. The equations of blade motion were obtained using Hamilton's principle. In these equations the axial stiffness EA, the bending stiffness EI and the torsional stiffness GJ are effective section stiffnesses which depend on ply lay-up and orientation.

Identification of coupling effects due to the composite structure of the blade was facilitated by the introduction of six constants $\kappa_{\beta 1},...\kappa_{\beta 6}$ which are unique to the composite blade and depend on laminate orientation and layup. The

$\kappa_{p1}$ resembles a pitch-lag coupling term, $\kappa_{p2}$ resembles a pitch-flap coupling terms and $\kappa_{p3}$ is due to nonlinear extension-torsion coupling terms. The other three constants were less significant. Using quasisteady aerodynamics, and a finite element model, illustrated by Fig. 2, the dynamic equations of motion were solved in a conventional manner [5] to obtain stability boundaries, which were presented as root locus plots. Results were calculated for a hingeless rotor with the following properties: $\gamma = 5.0$; $\sigma = 0.10$; $c/R = 0.08$; $\beta_p = 0$; $C_T/\sigma = 0.10$. Some typical results are shown in Figs. 4 and 5. The root locus plot for the lag mode eigenvalue for a symmetric case is shown in Fig. 4. For this case the side flanges of the box in Fig. 3 have non zero-ply angles. The full lines show the results for positive ply angles and the heavy broken lines correspond to the negative ply angels. The influence of the ply angle ($\Lambda_V$) variation is considerable. A positive ply angle destabilizies the lag mode and a negative angle stabilizes the lag mode. It turns out that this effect is primarily due to the $\kappa_{p1}$ coupling term which represents pitch-lag coupling. The other coupling terms are either zero or play a negligible role. The light broken line in Fig. 4 represents the results with the $\kappa_{p1}$ coupling term neglected and thus is shows only the influence of play angle variation on the stiffness terms corresponding to an equivalent isotropic blade analysis. Figure 5 shows similar results for the lag degree of freedom for antisymmetric ply angles on the side flanges and zero ply angles on the top and bottom flanges. In this case the major coupling term is $\kappa_{p3}$ which is due to the extension-torsion coupling. This is a nonlinear coupling term and it indicates the importance of a nonlinear analysis for this class of structures.

Another important practical and complicated theoretical problem is the structural modeling of the aeroelastic behavior of rotor blades with swept tips. An analytical study which illustrates the effect of blade sweep on rotor vibratory hub, blade and control system loads was conducted by Tarzanin and Vlaminck [27]. The portion of the blade tip which was swept was located at 0.87R and two sweep angles, 10° and 20°, were considered. Sweep introduces powerful inertia, aerodynamic and structural coupling effects. The analytical model used in Ref. 27 could not represent in a consistent manner all the effects due to sweep therefore a technique called simulated sweep was used in which local inertia, elastic and aerodynamic axes were adjusted in an approximate manner. The authors concluded that tip sweep influences both blade vibrations and stability and recommended the development of improved analytical methods needed for a better fundamental understanding of dynamics of blades with swept tips.

A hingeless rotor with a swept tip is shown in Fig. 6. An important study capable of simulating such a hingeless rotor blade configuration was recently completed by Celi [28]. The model developed in this study is based on the dynamic equations of equilibrium presented in Ref. [29]. The blade is modeled using the Galerkin finite element technique [21] and a special element for the structural, inertia and aerodynamic properties of the swept tip was developed. Typical results showing blade equilibrium and stability for hover are presented in Fig. 7 and 8, for a stiff-in-plane rotor blade at a thrust coefficient of $C_T = 0.005$ (corresponding to a collective pitch setting of $\theta_0 = 0.1432$). Figure 7 shows the static blade equilibrium in flap, lag and torsion for zero precone and $\beta_p = 3$ degrees. It is evident that presence of precone significantly changes the equlibrium position in torsion as the 0.10R portion of the blade is gradually swept back. The curve of the torsional equilibrium ($\phi$) has a characteristic concave shape. Precone interacts with sweep to change the nose-down torsional moment due to lift and the nose up moment due to centrifugal force. The influence of sweep and precone on the root-locus of first torsion and first lag mode are presented in Fig. 8. For zero precone frequency coalesce occurs between the first lag and first torsion mode. As frequencies coalesce, the torsional damping increases considerably, while the lag mode becomes unstable. This lag instability is not eliminated by small amounts of structural damping [4] indicating that this is a strong instability. For $\beta_p$ = 3 degrees, increasing sweep <u>increases</u> the imaginary part of the first torsion eigenvalue, instead of decreasing it, as was the case for zero precone. Thus frequency coalesce does not occur, and the lag mode remains stable. Many additional results are presented in Ref. 28, which represents an important contribution to the literature since it contains the first detailed and systematic study of the effect of sweep on blade stability in hover and in forward flight.

Another new study by Kosmatka [30] combines a capability of modeling highly curved and swept blades undergoing moderate deflections, with the ability to deal with blades having a general, anisotropic, composite construction. This model was developed for the structural dynamic modeling of advanced composite propellers (prop-fans) however it is equally applicable to the analysis of composite, pretwisted, rotor blades. A curved pretwisted blade, is modeled by straight beam elements which are aligned with the curved line of shear centers of the blade. Each straight pretwisted beam finite element is derived, using Hamilton's principle, assuming that the beam undergoes moderate deflection, is composed of anisotropic materials, has an arbitrary shaped cross section, and rotates about a vector in space. Combined with this beam model a companion isoparametric eight node quadrilateral finite element model has been developed which is capable of calculating the shear center and the structural constants of an arbitrary shaped cross section, built up from anisotropic materials. The finite element model can also predict shear stress distribution over the composite cross section and it provides insight on the effects of ply orientation and material selection on the stress distribution within the cross section. A representative example used in this study was a composite blade cross section which consists of uni-directional Kevlar, laminated Kevlar and aluminum strip shown in Fig. 9. The location of the shear, area and mass center for different ply orientations are shown in Fig. 10. The shear center location can be easily moved within the cross section by varying the ply orientation. The structural properties of the blade can be also greatly modified by varying the ply orientation as indicated in Fig. 11. The axial stiffness and bending stiffness decrease as ply orientation is increased. On the other hand the torsional stiffness of the blade increases significantly with ply orientation.

Another interesting study associated with the structural dynamics of rotating pretwisted beams was the recognition that the use of twisted principal coordinates can lead to increased effectiveness in frequency and mode shape calculations [31].

From the studies reviewed in this section it is evident that very substantial advances in structural modeling capabilities have taken place during the time period considered.

### 3. Aerodynamic Modeling

Accurate modeling of the unsteady aerodynamic loads required for aeroelastic analyses continues to be one of the major challenges facing the analyst. An excellent review of some of major issues in blade unsteady aerodynamics have been presented in a paper written by Dat [32]. The accuracy with which the unsteady aerodynamic loading phenomena environment needs to be determined depends to a large extent on the dynamic problems which are investigated. Thus for example the coupled flap-lag-torsional aeroelastic instability in hover is frequently investigated using quasi-steady aerodynamics [5] whereas the periodic loads in forward flight must be evaluated using more precise aerodynamic models. The combination of the blade advancing and rotational speeds is a formidable source of complexity for the flow. At large values of advance ratio, the aerodynamic field around the blades undergoes such variations that there are problems of transonic flow, with shock waves, at the advancing blade tip, problems of speed reversal and low speed unsteady stall on the retreating blade, and problems due to the high blade sweep angle in the fore and aft positions. Furthermore the geometry of the wake, which is an important source of vibration and noise, is much more complicated than the fixed wing wake geometry.

The empirical and semi-empirical treatment of the unsteady, two dimensional, dynamic stall problem has played an important role in rotary wing aeroelastcity during the last twenty years. The review of three relatively recent dynamic stall models can be found in Ref. [5]. Continued research on these dynamic stall models has led to improved predictive capabilities which are described below.

Beddoes [33] has continued his work on indicial formulation of unsteady lift which was a basic ingredient in his dynamical stall model during the attached flow regime. Subsequently this work was incorporated by Leishman and Beddoes [34] in an improved generalized model for airfoil unsteady aerodynamic behavior and dynamic stall using the indicial method. This improved model provides the methodology for the computation of two dimensional unsteady airfoil lift, pitching moment and drag for an airfoil undergoing arbitrary forcing in the time domain, using an indicial response formulation. The linearized unsteady aerodynamic response on the attached flow regime is separated into two components, namely circulatory and impulsive loading, which are computed independently using indicial aerodynamic transfer functions. In the separated flow regime the nonlinear lift characteristics of the trailing edge separation was evaluated using the concept of Kirckhoff flow. The Kirchoff model was also used to evaluate the nonlinear effects on chordwise force and pressure drag response. The onset of of vortex shedding during dynamic stall was captured using a generalized criterion for the onset of leading edge or shock induced separation. Furthermore this leading edge separation was also coupled with the trailing edge separation calculation. These feature were incorporated in a general numerical algorithm for predicting airfoil unsteady aerodynamic behavior and dynamic stall, for arbitrary forcing or motion in the time domain. Extensive validation of the model was conducted by comparing it with other available analytical results and two dimensional unsteady test data. This model represents a major improvement on the previous model developed by Beddoes. One of the important attributes of the new new model is the capability for simulating, in a fairly accurate manner, the unsteady drag hysteresis loop of an airfoil undergoing either light or deep stall.

Gangwani [36] continued developing his original dynamic stall model which was initially reviewed in Ref. 5. The most important contribution of this new study was synthesized unsteady drag data which provides a basis for the computation of unsteady pressure drag of airfoils and rotor blades, in the time domain. A typical unsteady drag coefficient loop data for the SC1095 airfoil, at M = 0.30, a mean angle of attack of $\alpha_o$ =12° and oscillation amplitude of $\bar{\alpha}$ = 8.0° at a reduced frequency of k = 0.10 is shown in Fig. 12. The method for generating the unsteady aerodynamic coefficients for such an airfoil depends on the predictions of three major events associated with dynamic stall, namely: (1) stall onset; (2) vortex at trailing edge; and (3) reattachment. These events are usually closely related with the unsteady pitching moment coefficient, and they are also shown in Fig. 12. The unsteady drag coefficient $C_{DU}$ is represented by an algebraic relation which depends on eight coefficients which are computed by using linear least square curve fitting with experimentally determined unsteady drag data. Numerous results presented in Ref. 35 indicate good agreement with experimental data. This method is considerably less sophisticated than that described in Ref. 34, however one of its attractive features is its relative simplicity compared to other dynamic stall models.

Among the three dynamic stall models reviewed in Ref. 5 the model developed at ONERA by Dat, Tran and Petot had a number of features which caused it to be suitable for inclusion in rotary wing aeroelastic response and stability calculations, because it is a time domain theory for an airfoil performing completely arbitrary motions. Furthermore the model utilizes the properties of differential equations to simulate the different effects which can be identified on an oscillating airfoil such as pseudo elastic, viscous and inertial effects, and the effect of the flow time history [32]. The theory also recognizes that in the linear range of airfoil motions the Theodorsen lift deficiency function represents the aerodynamic transfer function for the airfoil relating the three quarter chord downwash velocity to circulatory lift. Furthermore the theory is based on approximating the aerodynamic transfer function by rational fractions. For convenience, the input variables for the system of equations which are plunge (h) and pitch ($\theta$), are combined in a single almost equivalent variable defined as $\alpha = \mathring{h} + \theta$, where $\mathring{h}$ is nondimensionalized in a suitable manner which is, range the model consists of a system of differential equations containing the angle of attack or downwash at the forward quarter chord. In the nonlinear range the model consists of a system of differential euqaitons containing unsteady linear terms whose coefficients are functions of the angle of attack and steady flow nonlinear terms. A lucid description of this model together with an outline for imbedding it in unsteady aerodynamic calculation including the effects of three dimensional flow can be found in a paper by Dat [32].

Further work aimed at an improved physical understanding of this model was carried out by Rogers [36] and Peters [37]. Rogers considered primarily the equation for unsteady lift and verified the validity of the model by reproducing previously published lift hysteresis data [39]. He also considered simplifications to the model and concluded that a term representing apparent mass in the model could be neglected without influencing the results. To determine some of the practical aspects of the model he incorporated it into a very simple dynamic model representing single section flapping dynamics of a rotor blade in forward flight. He concluded that Floquet theory, based upon linearized perturbation equations, could provide useful information on stability behavior when the ONERA model is used to represent the unsteady aerodynamics.

More recently Peters [37] continued to study the unsteady lift equation associated with the ONERA model. He noted that the original model lumped the pitch and plunge into one variable which can cause difficulty when attempting to compare the model to classical Theodorsen or Greenberg theory [5]. He introduced certain modifications in the theory so that it reduces to Greenberg's theory for small angles of attack, and it reduces to Theodorsen's theory for steady free stream. He also introduced modifications which remove certain ambiguities in the model at large angles of attack.

The ONERA stall model was also used in a recent test [38] to compare the measured and calculated stall flutter behavior of a one bladed model rotor. The agreement between the calculated and measured results showed good agreement, further illustrating the utility of the model for aeroelastic calculations.

It appears that the ONERA dynamic stall model is gaining acceptance in rotary wing aeroelasticity as other researchers introduce refinements in the model. Recent research by Leiss [40,41] has emphasized the role of unsteady sweep in the semi-empirical simulation of rotor blade aerodynamic loading. Thus the introduction of sweep effects into the ONERA model could produce another potential improvement.

Another significant portion of recent research in unsteady aerodynamics has been aimed at developing two dimensional unsteady airfoil theories in the time domain. Two dimensional aerodynamic theories, which provide analytic expressions for unsteady loads on a moving airfoil are usually based on the assumption of simple harmonic motion. Representative theories of this category are Theodorsen's and Greenberg's theories for fixed wings and Loewy and Shipman and Wood's theory for rotary wings [5]. These theories which deal with the linear, attached flow regime, have a significant deficiency when applying them to aeroelastic stability calculations, since the assumption of simple harmonic motion, upon which they are based, implies that they are strictly valid only at the stability boundary, and thus they provide no information on system damping before or after the flutter condition is reached. Thus standard stability analyses, such as the root locus method cannot be used in conjunction with these theories. Another important limitation of these theories is evident when one tries to apply them to the rotary-wing aeroelastic problem in forward flight, which is governed by equations with periodic coefficients. In this case the complex lift deficiency function associated with frequency domain unsteady aerodynamics is not consistent with the numerical methods employed in the treatment of periodic systems [5]. Thus many rotary-wing analyses in forward flight are based upon quasi-steady aerodynamics [5]. To remedy this situation a number of recent studies [42-48] were aimed at developing arbitrary motion unsteady aerodynamic theories in the time domain. In these studies the term arbitrary motion was used to denote motion with growing or decaying oscillation with a certain frequency.

In Ref. 42 the basic procedures for generalizing Greenberg's and Loewy's theories to the time domain, for airfoils undergoing arbitrary motion were presented. In Ref. 43 the generalized Greenberg theory was incorporated in the simple flap-lag analysis of a hingeless rotor blade in forward flight and the influence of unsteady aerodynamics on blade response and stability was obtained. When using a second order rational approximant for the generalized Theodorsen lift deficiency function, the finite state time domain representation of the circulatory aerodynamic lift and moment can be written in the following form [43]

$$L_c(t) = 2\pi \rho_A b R U_T(t) \left[ 0.00685 (U_{T0}/bR)^2 X_1(t) + 0.10805 (U_{T0}/bR) X_2(t) \right]$$
$$+ \pi \rho_A b R U_T(t) Q(t) \qquad (1)$$

$$M_c = bR(0.5 + \bar{a}) bR L_c(t) \qquad (2)$$

These expressions are written in terms of the downwash velocity $Q(t)$ at the ¾ chord, which depends on the airfoil degrees of freedom and their time derivatives, and two additional augmented state variables $X_1$ and $X_2$. The augmented state variables are governed by a system of first order differential equations which depend on $Q(t)$, as shown below

$$\begin{Bmatrix} \dot{X}_1(t) \\ \dot{X}_2(t) \end{Bmatrix} = \begin{bmatrix} 0 & 1 \\ -0.01365(U_{TO}/bR)^2 & -0.3455(U_{TO}/bR) \end{bmatrix} \begin{Bmatrix} X_1(t) \\ X_2(t) \end{Bmatrix} + \begin{Bmatrix} Q \\ Q(t) \end{Bmatrix} \quad (3)$$

The additional augmented state variables $X_1$ and $X_2$ convey information regarding the unsteady wake history. Such an aerodynamic theory provides a good approximation to both low frequency and high frequency regimes of blade motion. Furthermore it should be noted that this time domain unsteady aerodynamic model bears a close resemblance to the unsteady aerodynamic loads, used for the attached flow case, in the dynamic stall model for rotor blades developed by Dat, Tran and Petot [5,32,37].

To assess the influence of arbitrary motion unsteady aerodynamics on blade aeroelastic stability and response in forward flight a simple problem consisting of an offset hinged, spring restrained model of an isolated hingeless blade was selected [43]. Typical results for blade response and stability are shown in Figs. 13 and 14. Figure 13 illustrates the steady state flap response of the blade over one revolution using both time domain unsteady aerodynamics and quasisteady aerodynamics. There is a pronounced unsteady aerodynamic effect on the flap response. The effect of phase lag and amplitude modulation associated with unsteady aerodynamics are both evident in Fig. 13. In Ref. 43 the same effect, on the lag degree of freedom, was also examined and found to be small. The influence of unsteady aerodynamics on blade stability is shown in Fig. 14, where the real part of the characteristic exponents (which is a measure of damping in a periodic system) is plotted as a function of the advance ratio $\mu$. The interesting result in this plot is the instability in the flap degree of freedom which occurs at an advance ratio of $\mu = 0.45$. When quasisteady aerodynamics are used this instability does not occur. It was shown [43,45] that this instability can be associated with an unsteady lift deficiency function which represents the ratio between unsteady lift and quasisteady lift. The important conclusion from these plots is that unsteady aerodynamics influences primarily the flap response of the blade [48].

It was also shown in Refs. 42 and 44 that generalizing a rotary-wing theory such as Loewy's to the time domain is more complicated that the extension of Greenberg's theory. To overcome this problem a novel technique for formulating high quality finite state unsteady aerodynamic models, based upon the Bode plot, was developed [46,47]. This technique is based on recognizing that the circulatory portion of the lift, per unit span, of the airfoil in the Laplace domain can be written as

$$L_c(\bar{s}) = \alpha \rho_A bRUC(\bar{s})Q(\bar{s}) \quad (4)$$

where $Q(\bar{s})$ represents the Laplace transform of the ¾ chord downwash velocity. The Bode plot method used in control systems engineering is a useful tool for constructing approximations to complicated transfer functions. It can also be used to construct approximations to lift deficiency function, which has the role of an aerodynamic transfer function, according to Eq. (4). Using this technique good approximations to Loewy's lift deficiency function were obtained [46,47] and used to obtain a rotary-wing indicial response function. These indicial response functions are oscillatory and thus are different from fixed wing indicial response functions which are non oscillatory [46,47].

Another contribution made in Refs. 44 and 45 was the development of an arbitrary motion unsteady cascade airfoil theory for helicopters rotors in hover.

It was pointed out in Ref. 48 that dynamic inflow also represents an arbitrary motion type of approximate unsteady aerodynamic theory, which captures low frequency aerodynamic effects associated with the wake. A comprehensive review of the dynamic inflow models available in hover and forward flight together with their correlation with experimental data was presented by Gaonkar and Peters [49].

In addition to the theories considered in this section, more complicated theories such as unsteady prescribed wake models, lifting surface models and more sophisticated models based on computational fluid mechanics are also needed for more accurate aeroelastic stability and response calculations. A newly developed unsteady prescribed wake model for helicopter rotor blades in hover and forward flight was presented by Rand and Rosen [50]. Unsteady lifting surface theories were also considered in Ref. [32]. A detailed survey on the role of computational fluid mechanics for rotorcraft was given by Davis [51].

## 4. Aeroelastic Problem Formulation

The derivation of equations of motion for aeroelastic stability and response calculations, for an isolated rotor blade in forward flight including geometrical nonlinearities, is a relatively complicated task from an algebraic point of view. When the fuselage degrees of freedom are added to the problem this task tends to become very arduous, even when an ordering scheme is used to simplify the equations. Good representative examples showing the complexity of the equations which model a coupled rotor/fuselage system in forward flight can be found in Refs. 52-54. The solution process of such equations leads to additional complications since use of a global Rayleigh-Ritz or Galerkin method, combined with the multiblade coordinate transformation, frequently used in coupled rotor/fuselage analyses, requires considerable algebraic effort [5]. Substantial increases in raw computing power, as represented by high computational speeds and availability of large core memory at low cost, which have taken place during the last five years imply that the time has come to delegate these algebraic tasks to the computer. Only a few papers were published on the automatic generation of helicopter equations of motion using computers, however, the number of such papers is increasing. From these papers it is evident that two different approaches are being used to achieve the same goal.

One approach is based on generating equations of motion in <u>explicit form</u>. This can be accomplished by developing special purpose symbolic manipulators written in FORTRAN to automatically generate equations of motion for rotary-wing aeroelastic applications. One of the first studies based on this approach was done by Nagabhushanam, Gaonkar and Reddy [55]. Using this approach the complete equations of motion are obtained in fully <u>explicit</u> nonlinear form, directly from the computer.

The approach developed originally in Ref. 55 has been extended to the problem of coupled flap-lag-torsional dynamics of hingeless rotor blades in forward flight [56,57]. A detailed description of the symbolic processor program principles was presented by Reddy [57]. The program generates the steady state and linearized perturbation equations in symbolic form and then codes them into FORTRAN subroutines. These equations are obtained in explicit form. The coefficients for each equation and for each mode are identified through a numerical program. A Lagrangian formulation is used to obtain equations in generalized coordinates. The coupled flap-lag-torsion equations with dynamic inflow are converted to equations in a multiblade coordinate system by deriving explicit multiblade equations in symbolic form. The whole process, from derivation to numerical calculation, is automated with minimum user interface. The equations have been carefully validated in Ref. 57 by comparing results obtained for hover with other results available in the literature. Many useful results for forward flight were generated with the program [56,57]. These results will be discussed in the next section of this paper.

Another *explicit* approach is discussed by Crespo Da Silva and Hodges [58]. This approach is based on utilizing a commercially available symbolic manipulation program called MACSYMA, running on a dedicated LISP workstation. The general methodology of deriving flexible blade equations using MACSYMA are discussed and the process is illustrated by a simple example associated with the flap motion of a rotor blade in forward flight.

The second approach to generating rotary-wing aeroelastic equations of motion is based on the *implicit approach*. In this approach one generates automatically the coefficient matrices for equations of motion linearized in perturbation coordinates about an equilibrium approach. This approach, which was used by Done and his associates in two recent papers [59,60], does not require that the equations be explicitly written out at any stage of the analysis. The first paper by Gibbons and Done [59] presented the theoretical background for the method. The procedure consists of writing down the appropriate transformations governing the dynamics of a mass point and combining it with Lagrange's equations to obtain the mass, aerodynamic and stiffness terms needed to calculate the equilibrium position. Subsequently perturbation equations about this equilibrium position are generated. The differentiations and integrations required in this process are performed numerically. The equations generated are in numerical form, and their solution is obtained by iterative algorithms. Only a few simple results in hover were used to validate the program. In a second paper [60] three practical examples were treated by the computer program which was developed and results were compared to results generated by Westland Helicopters. Among these the most complicated example was a Lynx ground resonance calculation and comparison between the two sets of results was satisfactory.

Finally it is important to mention that implicit formulations have been used in recent finite element analyses of rotary-wing aeroelastic problems [14,28]. The implicit nature of Ref. 28 is principally associated with two features:

1. The algebraic expressions for the aerodynamic loads are not expanded explicitly . They are coded separately in the computer program and combined numerically with the inertia and structural terms during the solution of the response problem.
2. The approximate set of generalized coordinates obtained during one iteration of the solution procedure is used to generate the aerodynamic loads for the next iteration.

The implicit nature of the GRASP program [14] is primarily due to its hybrid finite element/multibody nature which allows the theatment of complicated configuration, without explicitly writing out the governing equations.

## 5. Aeroelastic Analyses in Forward Flight

The general methodology for the aeroelastic analysis of rotor blades in forward flight was reviewed in detail in Ref. 5. This aeroelastic problem is governed by nonlinear equations with periodic coefficients and furthermore the aeroelastic problem is coupled to the trim state of the helicopter. A considerable number of recent studies were aimed at an improved understanding of the coupled flap-lap-torsional problem of an isolated blade in forward flight [56,57,61-63]. A common element among these studies is a solution procedure which is similar to that first presented in Ref. 64. The solution consists of the following steps: (a) calculation of trim, (b) calculation of the nonlinear time dependent equilbirium posisiton, (c) linearization of the perturbation equations about this time dependent equilibrium position, and (d) calculation of blade stability using Floquet theory.

A comprehensive study of the coupled flap-lag-torsional aeroelastic behavior of hingless rotor blades in forward flight was done by Reddy and Warmbrodt [56,57]. They used symbolically generated equations and studied the influence of: dynamic inflow, trim, as well as various approximations to the complete coupled-flap-lag-torsional equations. Figure 15 shows the effect of number of degrees of freedom used in trim analysis on lead-lag damping plotted as a function of the advance ratio. It can be seen that a flap-lag-torsion stability analysis based upon a flap trim [64] tends to underpredict the lead-lap damping. Another feature of this plot is the instability, in the lag degree of freedom observed in a stiff-in-plane blade configuration when $\mu > 0.40$. This behavior was also observed in Ref. 64. The effect of torsion and dynamic inflow on lead lag regressing mode damping is shown in Fig. 16. Dynamic inflow seems to have a relatively small effect in this case. Furthermore the damping predicted by a flap-lag model is much lower than that predicted by a coupled flap-lag-torsional model, this was also noted in Ref. 64.

The feasibility of simplifying coupled lag-flap-torsional models for blade stability analyses in forward flight was studied in Ref. 61 and it was concluded that the only reliable model under various conditions is the fully coupled model.

Panda and Chopra [62] have also studied flap-lag-torsion stability in forward flight using an offset hinged spring restrained model of a hingeless blade. The effects of pitch-flap and pitch-lag coupling, torsional stiffness and dynamic inflow were considered. The results also confirmed those obtained in Ref. 56, 57 and 64. Subsequently the same authors [63] studied the behavior of hingeless and bearingless rotors in forward flight, including dynamic inflow and using a previously derived finite element method [22]. The results indicated that stiff-in-plane configurations were destabilized by forward flight, while soft-in-plane blade configurations were stabilized by forward flight, which was also found in Refs. 56 and 64.

An experimental and analytical study of the flap-lag stability in forward flight of an isolated, three bladed hingeless model rotor, having a diameter of 1.62 meters was performed in Ref. 65. The rotor was not trimmed and many data points were obtained for high advance ratios and high shaft angles. The purpose of this paper was to determine the adequacy of the linear quasisteady aerodynamic model with dynamic inflow. The results of this correlation study were somewhat inconclusive. Because in some cases the use of dynamic inflow improved the correlation between theory and experiment while in other cases it did not. Fig. 17 shows the lag regressing mode damping at an advance ratio of $\mu = 0.30$ and increasing shaft angles. The lack of correlation for this case was not explained in a convincing fashion and was attributed to stall.

In addition to aeroelastic stability studies in forward flight a smaller number of studies dealt with the aeroelastic response problem due to gusts. Gust response of a coupled rotor/fuselage system in hover and forward flight was studied by Bir and Chopra [66,67]. The blades were represented by a fully coupled flap-lag-torsional model including, moderate deflections. The fuselage had three translational and two rotational (pitch and roll) degrees of freedom. Gusts were represented by a deterministic three dimensional gust field. Some of the more important conclusions of this study were that a complete coupled flap-lag-torsional model of the blade is needed for an accurate response analysis because when the vehicle encounters a gust the blades respond quickly, absorbing the initial impact of the gust. Another important conclusion was that using dynamic inflow can be important for gust response calculations, otherwise the blade response can be overestimated.

The effect of random air turbulence on flap-lag stability in forward flight was considered in Ref. 68, using a random process analysis. It was found that in absence of elastic coupling, turbulence is stabilizing. As indicated by Refs. 56,57 and 61-64 the damping in the flap-lag model is lower (sometimes 300% lower) than in the coupled flap-lag-torsional model. Thus results based on the flap-lag model tend to exhibit excessive sensitivity to gusts. Therefore it appears that the influence of turbulence on blade stability is small.

The rotary-wing aeroelastic problem in forward-flight (after spatial discretization) is governed by nonlinear ordinary differntial equations with periodic coefficients. The numerical treatment of stability and response of such periodic systems is a key ingredient in the solution of these problems. During the last five years a number of reliable efficient numerical schemes for dealing wth such problems have become available and these are described in Ref. 69. Recently the finite element method in the time domain [70] was applied to the solution of periodic systems by Borri [71]. This method is based upon Hamilton's weak principle and consists of the time discretization of the linearized version of this principle. The time discretization utilizes appropriate interpolation functions in time, such as cubic polynomials for example. Application of this method to a periodic system yields a system of linear alegebraic equaitons which have to be solved in an iterative manner to obtain the response of the system. This method can be also used to obtain the transition matrix at the end of one period.

From the discussion presented above it is evident that our analytical understanding of blade behavior in forward flight is improving. However there is considerable need for high quality experimental data on isolated, <u>trimmed</u> hingeless and bearingless rotor blades having simple configurations, i.e., uniform mass stiffness, with zero or linearly varying pretwist and without sweep or droop. Availability of such data, for an advance ratio range of $0 < \mu < 0.45$, could provide a sound basis for verifying and improving forward flight analyses in a systematic manner.

## 6. Coupled Rotor/Fuselage Aeromechanical Analyses

The aeromechanical instability of a helicopter, on the ground or in flight, is caused by coupling between the rotor and body degrees of freedom. This instability is commonly denoted air resonance when the helicopter is in flight and ground resonance when the helicopter is on the ground. The phyiscal phenomenon associated with this instability is quite complex. The rotor lead-lag regressing mode usually couples with the body pitch or roll to cause an instability. The nature of the coupling which is both aerodynamic and inertial is introduced in the rotor by body or support motion. The importance of developing a mathematically consistent model capable of representing the coupled rotor/fuselage dynamic system has already been discussed in previous reviews [5,6]. A considerable number of such coupled rotor/fuselage analyses which were developed are described below. A number of these models yield good correlation with experimental data.

A relatively comprehensive study by Nagabhushanam and Gaonkar [72] was aimed at determining the influence of various dynamic inflow models and aeroelastic coupling effects on the air resonance problem in forward flight. The model consisted of a number of centrally hinged spring restrained blades having flap and lag degrees of freedom for each blade combined with a fuselage having pitch and roll degrees of freedom. Some of the results obtained were consistent with other results available in the literature. One of the conclusions, namely the deterioration of regressing lag mode damping of soft-in-plane rotors, with increases in advance ratio appears to be somewhat contradictory to other results available.

A much more general coupled rotor/fuselage analysis is one of the many options available in a computer program developed by Johnson [53,73], which had acquired the name CAMRAD (for Comprehensive Analytical Model for Rotorcraft Aerodynamics and Dynamics). This model was used by NASA Langley Research Center to calculate hingeless rotor aeromechanical stability [74]. The model was tested in the Transonic Dynamics Tunnel. The model was a soft-in-plane, four bladed, hingeless rotor with flexures to accommodate flap and lead-lag motion combined with a mechanical feathering hinge to allow blade pitch motion. The support had body pitch and roll motions. The analysis included these degrees of freedom and the dynamic inflow model. The correlation covered the influence of pitch-flap coupling, blade sweep, blade droop, and blade precone as a function of $\mu$, rotor speed and collective pitch. Figure 18 shows the correlation obtained, which was quite good. This code was also used by NASA Ames for hover stability tests of a full scale hingeless rotor [75], and good correlation was obtained.

Johnson also used this code to model the influence of unsteady aerodynamics on hingeless rotor ground resonance [76]. He compared his results with the high quality experimental data obtained by Bousman [77] and obtained the remarkable result that inflow dynamics introduces an additional "inflow mode", which explained previously unresolved questions about the correlation between the theory and the test.

Venkatesan and Friedmann [78,79] developed a mathematical model capable of modeling aeromechanical problems associated with multirotor vehicles, where the two rotors were connected by a flexible supporting structure which also had rigid body degrees of freedom. The blades were modeled as offset hinged spring restrained blades, including geometric nonlinearities. Each blade had flap, lag and torsional degrees of freedom.

A subset of this model, consisting of a three bladed hingeless rotor with flap and lag degrees of freedom for each blade mounted on a gimbal which could pitch and roll, was used in Ref. 80 to simulate the experimental data obtained by Bousman [77]. The results obtained [80], using quasisteady aerodynamics, were in good agreement with the experimental data obtained in Ref. 77, except that the quasisteady model was incapable of predicting the "dynamic inflow mode" found by Johnson [76]. Subsequently both perturbation inflow and dynamic inflow aerodyanmics were incorporated in the coupled rotor/fuselage model [81] and the result obtained with dynamic inflow produced good agreement with the experimental data. Furthermore the "inflow mode" obtained by Johnson was also reproduced. Results illustrating this unsteady aerodynamic effect are shown in Figs. 19 and 20 [81]. Figure 19 shows the variation, of modal frequencies as a function of rotor speed, at zero collective pitch setting, using quasisteady aerodynamics. All frequencies except the one corresponding to 0.7 Hz. are predicted well. When perturbation inflow and dynamic inflow are included the results shown in Fig. 20 indicate, that with dynamic inflow all frequencies are predicted well. Furthermore the "inflow mode", associated with the augmented states introduced but the dynamic inflow model, is also predicted. It is shown in Refs. [48,81] that the identification of this mode is relatively complicated.

Another new program capable of predicting rotorcraft aeromechanical problems, as well as other dynamic problems, is the RDYNE program developed by Sopher and Hallock [82]. This program uses a time-history analysis for rotorcraft dynamics based on dynamical substructures and nonstructural mathematical and aerodynamic components. The program contains both geometrical and aerodynamic nonlinearities and used component mode synthesis to combine various structural elements. The program was applied to ground resonance problems and performed very well.

A modern and modular program, named GRASP, was completed recently [14]. GRASP combines the finite element and multibody approaches and incorporates multiple levels of substructures to provide a powerful tool for the analysis of bearingless rotor aeromechanical problems. GRASP has been designed around the concept of a collection of flexible and rigid bodies connected in an arbitrary manner. The element library of the program contains three elements: (1) an aeroelastic beam element which contains no small angle approximations; (2) an air mass element; and, (3) rigid body mass element. Results for a coupled rotor/body model were obtained, and the eigenvalues of the regressing lag mode damping were compared with results obtained by Ormiston [83]. The correlation between the two sets of results was good. This program was written using modern programming methods, emphasizing clarity and modularity. Despite its many attractive features the program is somewhat limited since it cannot treat blades made of composites, nor can it deal with a variety of problems which lead to equations with periodic coefficients, such as fuselage mass offset from the axis of rotation and blade dissimilarities.

The majority of the studies cited above dealt with a rotor/body system where the blades were identical. The interesting effect of blade-to-blade dissimilarities on rotor/body lead-lag dynamics was studied by MuNulty [84]. The most noticeable effect of these dissimilarities was the appearance of additional peaks in the frequency spectrum.

The influence on nonlinear damping on helicopter ground resonance was studied by Tang and Dowell [85]. The analytical model included a three bladed articulated rotor, with each blade having only lead-lag motion, combined with a fuselage which could pitch and roll. The formulation contains both a nonlinear blade damper and a nonlinear landing gear damping. The analytical results were compared with experiments conducted on a model and good agreement was obtained.

## 7. Active Controls and Their Application to Vibration Alleviation and Blade Stability Augmentation

The use of active controls whereby the pitch of the rotor blade is modified by an automatic control system so as to alleviate dynamic effects represents a typical aeroservoelastic problem. The level and scope of the activity in this area was very substantial. The use of active controls to provide reduction of vibratory loads at the hub, reduction of vibratory loads in the fuselage, gust load alleviation, alleviation of effects due to dynamic stall, stability augmentation in the lead-lag degree of freedom and suppression of coupled rotor/fuselage instabilities were only some of the potential applications considered. A complete review of this subject would require a separate review article. The main objective of this section is to present a concise review of some of the more interesting recent developments.

Two basic approaches for the active control of rotor dynamic problems were considered and implemented. In the first approach the time dependent pitch control is introduced in the fixed system through a conventional swash plate. The majority of studies, which are described in the first part of this section, are based upon this approach. In the second approach the time dependent pitch control, of a particular blade, is introduced in the rotating reference frame and is denoted individual blade control (IBC).

The most important topic, from a practical point of view, is vibration reduction in forward flight using higher harmonic controls (HHC). This approach produces reduced vibration levels in the fuselage, or at the hub, by tailoring the vibratory aerodynamic loads on the blades. Thus vibratory forces and loads are modified, at their source, before they reach the airframe. This is in contrast with conventional means of vibration control [7,8] which deal with vibratory loads after they have been generated. A particularly successful approach to this problem was an adaptive control system which combines recursive parameter estimation with linear optimal control theory. A class of such algorithms has been discussed in detail by Johnson [86] and was also analytically investigated by Molusis, Hammond and Cline [87]. Subsequently wind tunnel tests [88], flight tests [89] and digital simulations [90] have shown that an algorithm, denoted as the "cautious controller" provided good performance. A brief discussion of such an adaptive control system is given below.

The need for an adaptive control system, in which parameters describing the helicopter model are identified on line, follows from the inability of current analytical tools to predict vibration characteristics with sufficient accuracy. Furthermore the sensitivity of vibration characteristics to changes in aircraft configuration and flight condition implies that a constant gain control system might be ineffective. The HHC input in its most general form consists of a harmonic variation of collective and cyclic pitch components

$$\theta_{HHn} = \theta_{HHO} \sin n\psi + \left[\theta_{HHC} \sin(n\psi + \phi_c)\right] \cos\psi +$$
$$\left[\theta_{HHS}(\sin n\psi + \phi_s)\right] \sin\psi \qquad (5)$$

For a four bladed rotor n = 4; this input in the non-rotating system results in 3,4 and 5/rev oscillations in the rotating system.

It is assumed that the helicopter can be represented by a linear, quasistatic frequency domain model relating the output vector Z, consisting of harmonics of vibration, to the input vector θ, consisting of harmonics of blade pitch control at time $\psi_i = i\Delta\psi$, where $\psi_i$ is the sampling time, thus

$$Z_i = Z_o + T\theta_i \qquad (6)$$

where T is the nxm transfer matrix relating output vibration response to input higher harmonic control angles. The sampling interval $\Delta\psi$, should be sufficiently large for the transient to die out and the harmonics to be measured, usually it is taken as once per revolution. The uncontrolled vibration level $Z_o$ and the transfer matrix T are not known, because analytical methods for their prediction are not sufficiently accurate. These quantities are therefore estimated using a Kalman filter, the details of this estimation are presented in Refs. 86, 87 and 90.

The objective function, to be minimized, is the expected value of the performance index

$$J = E(Z_i^T W_z Z_i) + \theta_i^T W_\theta \theta_i + \Delta\theta_i^T W_{\Delta\theta} \Delta\theta_i \qquad (7)$$

usually the weighting matrices $W_z$, $W_\theta$ and $W_{\Delta\theta}$ are diagonal and the control law is found by setting $\partial J/\partial\theta_i = 0$. Solution of this relation produces a cautious controller. The control law for such a formulation can be found in Refs. 86, 87 and 90.

When Eq. (6) is based on the uncontrolled vibration level $Z_o$, sometimes denoted as the open-loop or global model, caution introduces an effective limit on control amplitudes. An alternative approach where Eq. (6) is replaced by

$$Z_i = Z_{i-1} + T(\theta_i - \theta_{i-1}) \qquad (8)$$

known as a closed-loop or local model produces an alternative control law where caution introduces limits on the rate. In either case the cautious controller introduces control limits which compensate for the uncertainty in the parameter estimates. A schematic diagram showing the implementation of such a control system for a digital simulation of control laws [90] is shown in Fig. 21.

Hammond [88] conducted extensive wind tunnel tests, on an aeroelastically scaled, four bladed, articulated helicopter rotor model. A number of alternative algorithms were tested, and it was found that the cautious controller gave very good performance. A typical result [88] showing the variation of the vibratory vertical force with advance ratio is presented in Fig. 22. Reduction between 70-90%, for this vibratory component, were obtained over the range of advance ratios tested. The results also indicated that HHC inputs produce increased edgewise bending moments, torsional moments, and control loads. The increased loads experienced during the tests, were within the design loads. This wind tunnel test was intended to support a subsequent full scale flight demonstration test of OH-6A aircraft equipped with a higher harmonic control system. The results of the full scale tests, which took place in the summer of 1983, were presented in a landmark paper by Wood, Powers, Cline and Hammond [89]. The aircraft was flown from zero airspeed to 100 knots, with the HHC system operated both open loop (manually) and closed loop (computer controlled). Flight test results exhibited significant reduction in helicopter vibrations without undue penalties in blade loads and aircraft performance. Six months later, in 1984, the flights resumed with an improved Kalman filter implementation combined with some hardware improvements. These modifications resulted in improved system performance [89]. Figure 23 shows the closed loop HHC-4P vertical acceleration, at the pilot seat, as a function of airspeed. Comparing Figs. 22 and 23, and recognizing that vertical vibrations at the pilot seat are different from vertical forces measured in the wind tunnel, shows that the full scale tests produced vibration reductions similar to those obtained in the wind tunnel tests.

A comprehensive digital simulation of such an HHC system was conducted by Davis [90]. This study, was a continuation of an earlier study [91], and it was aimed at a comparative study of three basic control algorithms: (1) deterministic, (2) cautious, and (3) dual. A diagram of this system is shown in Fig. 21. Reduction of vibration levels between 75-95% were achieved with HHC angles of less than one degree. The effect of nonlinearity and interharmonic coupling were also considered. This is an important problem because Eqs. (6) or (8) imply a linear or linearized model. A detailed study of the role of nonlinearities (both geometric and aerodynamic) in HHC systems for helicopter vibration was first presented by Molusis [92]. Both Refs. 90 and 92 utilized the G400 [93] helicopter aeroelastic simulation program, to generate their results. Molusis concluded that under certain conditions, nonlinearities could be sufficiently important so as to require modifications in the control algorithms. The simulations conducted at UTRC provided good guidelines for the implementation of a full scale flight demonstration of a HHC-system on the S-76 helicopter [94]. These tests, conducted in the open loop mode only, were the first demonstration of HHC on a 10,000 lb helicopter at speeds up to 150 knots. These successful tests, conducted in early 1985, will eventually lead to flight demonstration of the closed loop system.

Flight tests of an experimental HHC system on a SA349 Gazelle were also conducted in France in 1985 [95]. A detailed description of both the simulations and the flight tests are presented in Ref. 95. The control algorithms were similar to those used in Ref. 87 and 89 and a reduction of 80% in cabin vibrations at an airspeed of 250 km/h was demonstrated.

The higher harmonic control model and the wind tunnel and flight tests discussed above do not imply that these are the only viable approaches for dealing with vibration reduction, gust alleviation and potential performance enhancement. Many other studies have considered alternative approaches and fundamental problems associated with HHC. Shaw et al. [96] have demonstrated a closed-loop HHC system on a dynamically scaled model of a three bladed CH-47D rotor in the Boeing Vertol wind tunnel. Very effective multicomponent vibration suppression was demonstrated up to flight speeds of 188 knots. This vibration suppression was demonstrated with a fixed-gain feedback control which was much simpler and faster than the adaptive control laws used in the studies cited previously. A different approach, based on mathematical programming techniques to determine the HHC angles was presented by Jacob and Lehmann [97]. Wind tunnel test results performed on a model of a BO-105 four bladed hingeless rotor, using a relatively simple cost function, were presented by Lehmann [98]. The special HHC testing facility, discussed in Ref. 98, has a number of unique capabilities. Another different approach to vibration reduction in the fuselage using state-feedback vibration control was proposed and evaluated by DuVal, Gregory and Gupta [99]. While this approach appears to be promising and it needs further study.

All the studies cited above were aimed at vibration reduction in forward flight. Active control systems also offer the potential for cost effective solutions to other dynamics problems. A natural extension of adaptive control system approach is to apply it to gust alleviation. The feasibility of such a system was studied analytically by Saito [100], using a four bladed aritculated blade model. Response to step and sinusoidal gusts was considered. Gust response alleviation, between 50-100% was obtained.

Two fundamental studies on the use of active controls to augment rotor/fuselage stability were recently completed. Straub and Warmbrodt [102] performed an analytical study of ground resonance with airloads. The control system was modeled using state variable feedback with appropriate gain and phase. The analytical model of the coupled rotor/fuselage system was represented by an offset hinged spring restrained, three bladed hingeless rotor with flap and lag degrees of freedom. The rigid fuselage had pitch, roll, lateral and longitudinal translations. The model was based on quasisteady aerodynamics and contained geometrical nonlinearities. The configuration analyzed is shown in Fig. 24. The feedback was applied through a conventional swash plate, active pitch input to the $k^{th}$ blade was

$$\theta_{Ak} = \theta_{AC(\psi)}\cos\psi_k + \theta_{AS(\psi)}\sin\psi_k \tag{9}$$

To control the linearized constant coefficient system written multiblade coordinates, three "active" actuators in the fixed system were used. Using state variable feedback this control problem was explored in detail to obtain good physical understanding of the problem. The study assumed that all states are known. It was found that a 1% augmentation in critical damping of the regressing lag mode could be obtained with a 0.3 degree of blade cyclic lead-lag feedback.

In a second paper, Straub [102] used the same mathematical model to study the linear optimal control problem (LQG) of a four bladed articulated rotor in ground resonance. The solution for the control law was the deterministic optimal controller with linear feedback of all the state variables. The optimal gain was obtained from the solution of the algebraic Ricatti equation. Analytical results were generated in order to simulate the behavior of an articulated four bladed, H-34 rotor, with 4% hinge offset mounted on the Rotor Test Apparatus (RTA) in the 40 x 80 wind tunnel. Figure 25 illustrates, that a simple reduced state controller, using only control involving position and velocity of the hub $u = u(x,y,\dot{x},\dot{y})$, and without gain scheduling, yields a stable system at all speeds. This result should be compared with u = o (3200 Nms) which corresponds to nominal lag dampers, whereas all other result in this plot are based on lag damping reduced by a factor of ten ($C_\zeta$ = 320 Nms). The influence of this suboptimal controller, with four feedback loops u = u(x,y,ẋ,ẏ), on the modal frequencies is shown in Fig. 26. It is seen that only small changes occur at the coalesce rotor speeds and that frequencies are not changed at other rotor speeds. This clearly indicates that improvements in system stability, as a result of active control, are strictly due to increased regressing lead-lag mode damping.

An alternative to control through the conventional swash plate is the individual blade control (IBC) approach in which each blade is individually controlled in the rotating frame over a wide range of frequencies. This control concept was pioneered by Kretz [103] however a considerable amount of the more recent work in this area was done by Ham and his associates [104]. Reference 104 contains a detailed review of this work. Using a simple wind tunnel model combined with the concept of modal control, a number of important applications of this method were considered. These applications were: (1) gust alleviation, attitude stabilization and vibration alleviation [105]; (2) lag damping augmentation [106]; (3) stall flutter suppression [107]; and (4) flapping stabilization in forward flight [108].

An important contribution in this area was the recognition that by multiblade coordinate transformation, individual blade control laws could be implemented through a conventional swashplate [105]. The practical applications of this control concept are currently being evaluated by industry, Ref. 109 is representative of such a feasibility study conducted at Bell Helicopters.

## 8. Application of Structural Optimization to Vibration Reduction

The higher harmonic control, for vibration reduction, discussed above, modifies the unsteady aerodynamic loads acting on the blade and thus reduces the aeroelastic response of the blade. A somewhat similar goal can be accomplished by "aeroelastic tunning" of the blade, using changes in blade twist, sweep and mass or stiffness distribution [110]. Both methods are similar because they reduce the vibrations at the source, namely the rotor. Instead of the conventional design approach, used in Ref. 110, one can use modern structural optimization to reduce vibrations in rotorcraft [111]. In Ref. 111 the various techniques available for vibration reduction in rotorcraft using structural optimization were explored and reviewed with considerable detail. The only prudent approach for reduction of blade vibrations in forward flight, when changing mass and stiffness distribution, combined with changes in blade tip sweep is one in which aeroelastic constraints on blade stability are enforced. This requirement complicates the problem because a fully coupled aeroelastic stability and response analysis in forward flight has to be coupled with a structural optimization program. Therefore only a few studies having this capability are available [111-114]. In Ref. 112-114 modern structural optimization was used to reduce vibration levels in forward flight. The objective function minimized was the oscillatory vertical hub shear at $\mu$ = 0.30. Behavior constraints are the frequency placements of the blade in flap, lag and torsion, combined with the requirement that aeroelastic stability margins in hover remain unaffected by the optimization process. Stiff-in-plane, hingeless, blade optimization is discussed in Refs. 112 and 113, a detailed treatment of the soft-in-plane configuration is given in Ref. 112 and 114. Figure 27 shows the influence of structural optimization on the vertical hub shears after two stages of optimization, $D_o$ refers to the intial design and $D_{II}$ indicates the final design. Other results obtained indicated hub shear reductions between 15-40%, and blade weight reductions between 9-20%.

Structural optimization was also used in a systematic study by Davis and Weller [115] in which a hierarchy of dynamic problems were considered. These were: (1) maximizing a bearingless rotor inplane structural damping, due to shearing of the elastomer; (2) frequency placement of blade natural frequencies; (3) minimizing hub vibratory shear using a simplified model for rotor aerodynamics (40% reduction in hub shears was obtained); and (4) minimizing rotor vibration indices. Various optimization algorithms, problem formulations and solution strategies were also considered in this comprehensive study. This excellent study could be extended to include aeroelastic constraints, without too much difficulty.

## 9. Aeroelastic Analysis and Testing of Special Configurations

The purpose of this section is twofold. First the aeroelastic analysis of some special configurations is described. Often those configurations are unique and thus it is inconvenient to include them in any of the previous topics described in this paper. The second part of this section describes recent dynamic tests which have been performed to validate various analyses and designs.

The aeroelastic stability of two somewhat unusual rotor configurations, a constant lift rotor (CLR) and a free tip rotor (FTR), were analyzed by Chopra [116] for the case of hover. The CLR-configuration employs a pitch control input to rotate several independent airfoil sections which are free to pivot around a continuous spar, allowing them to change their pitch so as to obtain the desired lift. For this blade rigid body flap and lag motion was assumed at the root hinge, and each strip was assumed to undergo independent torsional motion. Stability boundaries are obtained in a conventional manner using a linearized stability analysis in hover [5]. The influence of several parameters on blade dynamics was examined. The free-tip rotor blade consists of two sections: an inboard section similar to that of a conventional blade with a pitch control system and a small outboard section (about 5-10% of radius) freely pitching on its spar with control input of pitch motion. Thus a free tip rotor is similar to a constant lift rotor with two sections. Under appropriate conditions both configurations were found to be free of aeroelastic instabilities.

The analysis of circulation control rotors was also considered by Chopra [117,118]. In Ref. 117 the aeroelastic stability of a hingeless circulation control rotor blade, in hover, was analyzed using finite elements [22]. The airfoil characteristics associated by the blowing, which produces circulation control, was taken from experimental data. With the exception of the effects of blowing this analysis was similar to Ref. [22]. The importance of including the second lag mode in the analysis was noted and it was found that blowing can have significant effects on blade stability. In Ref. 118 a much more comprehensive study of bearingless circulation control rotors was conducted, using tabular aerodynamics and dynamic inflow. The influence of dynamic inflow was found to be very small on this type of rotor. It was concluded that the expected levels of internal structural damping appear adequate to stabilize the mildly damped fundamental and second lag mode.

The aeroelastic stability of a two bladed rotor on flexible unsymmetrical support was analyzed by Chen [119]. Due to the complicated interaction between the aeroelastic and parametric excitation problems, present in this configuration, it was difficult to obtain conclusive results.

Using the mathematical model developed in Ref. 54 Venkatesan and Friedmann [120] have analyzed the aeromechanical stability of a hybrid heavy lift multirotor vehicle in hover. This model was intended to simulate the dynamic behavior of the hybrid heavy lift helicopter built by the Piasecki Aircraft Co., which crashed during flight tests, in the early summer of 1986. This dynamic model consisted of two rotors, connected by a flexible supporting structure and combined an envelope providing buoyant lift. Each four bladed articulated rotor was modeled using fully coupled, nonlinear, flap-lag-torsional dynamics, the vehicle had six rigid body degrees of freedom and the supporting structure had two bending and one torsional degree of freedom. The aeromechanical stability model had a total of 31 degrees of freedom, and thus it represents one of the more complicated aeromechanical problems considered. The results obtained indicated a potential for air and ground resonance type instabilities.

An interesting case study of a coupled pitch-flap-lag instability involving the coupling of higher chord and flap bending modes combining with a reactionless torsion mode, observed in the tests of an experimental rotor in hover, was presented by Neff [121].

A thorough design oriented treatment of the aeromechanical aspects of hingeless/bearingless rotor system was presented in Ref. 122. The importance of blade parameters such as droop, control system flexibility and blade stiffnesses are discussed. It is noted that lead-lag damping levels in bearingless main rotors is lower than that for hingeless rotors. This property of bearingless rotors is illustrated by Fig. 28 where comparisons of measured and predicted lead-lag damping (in the rotating system) are shown. The data was obtained in whirl tower tests of actual full scale rotors. A somewhat similar study on scaled bearingless main rotors was also done by Weller and Peterson [123].

A very comprehensive experimental investigation of bearingless model rotor stability was undertaken by Dawson [124]. The emphasis was on isolated blade stability. Five different configurations were tested and a significant body of high quality data was obtained. The experimental results were compared with analytical results obtained by the FLAIR analysis [125], in general the agreement between experiment and theory was acceptable except in cases when blade flexibility played a strong role. Most of the differences between predicted and experimental damping data occurred at high pitch angles. Figure 29 compares the lead-lag damping of Configurations 4 and 5 [124] as pitch angle is varied. For Configuration 4 the blade is preconed 2.5 deg. with respect to the flexbeam, while for Configuration 5 the flexbeam is preconed by 2.5 deg. with respect to the hub. Measurements of lead-lag damping at low pitch could not be obtained due to flutter. The theoretical model underpredicts damping at the higher pitch angles.

Additional test results obtained on similar two and three bladed bearingless rotors are described by Bousman and Dawson [125]. These results are particularly interesting because a pitch-flap type of flutter, attributed to unsteady aerodynamic effects, was observed.

A design oriented parameteric study of the aeromechanical stability, in air resonance, of bearingless rotors in hover was conducted by Hooper [127] using the FLAIR program [125]. In this study it was found that precone angle of the blade relative to the flexbeam and vertical offset of the snubber attachment point can produce beneficial blade dynamic behavior. This theoretical, design oreinted study, also complements the previously mentioned combined experimental/analytical studies [122-124].

The behavior of bearingless rotors, for tail rotor applications, was also studied. Thus, Ref. 122 also contains useful information on the dynamic behavior of bearingless tail rotors. The aeroelastic characteristic of the AH-64 bearingless tail rotor were presented by Banerjee [128]. The elastomeric shear attachment of the flexbeam to the hub introduces beneficial damping and modal characteristics which yields an aeroelastically stable rotor. This rotor was extensively tested in wind tunnel covering most operating conditions.

Finally, it should be mentioned that another potential tool for aeroelastic research on rotors are dynamically scaled wind tunnel models. The development and testing of a 27% dynamically scaled model of AH-64 main rotor is described in Ref. 129. This scaled down version, of an existing full scale rotor offers the potential for studying in detail the aeromechanical behavior of the rotor and can be also used to simulate other aeroelastic problems.

## 10. Concluding Remarks

It is evident from this survey that the level of activity in rotary-wing aeroelasticity during the last few years was very substantial. Comparing this activity to its fixed-wing counterpart gives one the impression that the center of gravity in the field of aeroelasticity is shifting from fixed-wing configurations to rotary-wing configurations. This is not surprising in view of the historical fact that helicopters are thirty years behind fixed-wing aircraft in development. Judging by the recent research, discussed in this paper, it appears that this thirty year gap might be narrowing. However, the level of complexity present in rotary-wing systems, which imposes stringent demands on the level of sophistication required in the analysis, implies that rotary-wing aeroelasticity is far from being a mature field of research. Much additional research is needed before rotary-wing aeroelasticity will achieve the level of maturity which currently exists in the fixed-wing field.

A number of topics where additional research has the potential for significant payoffs in terms of improved helicopter designs are:

1. Generation of an experimental data base for the validation of both <u>isolated</u> blade and <u>coupled rotor/fuselage</u> in forward flight. This test data should be obtained for hingeless, bearingless and articulated rotor configurations with simple geometries and properties (i.e., no sweep and droop, constant mass and stiffness distribution, and zero or linear pretwist).

2. Correlation studies based on this data to validate forward flight analyses.

3. Improved unsteady aerodynamics, in the time domain, for compressible and transonic regimes which are suitable for incorporation in aeroelastic analyses in hover and forward flight.

4. Development of improved methods for dynamic load predictions, on blades, which could lead to multidisciplinary optimization of rotor systems with simultaneous aeroelastic performance and acoustic constraints.

Acknowledgement

This research was partially funded by NASA Grants NAG 2-209 and NAG 2-226, the support of the grant monitors Dr. W. Warmbrodt and Dr. H. Miura, from NASA Ames Research Center, is gratefully acknowledged.

References

1. Loewy, R.G., "Review of Rotary-Wing V/STOL Dynamic and Aeroelastic Problems," Journal of the American Helicopter Society, Vol. 14, No. 3, pp. 323, 1969.
2. Dat, R., "Aeroelasticity of Rotary Wing Aircraft," Agard Lecture Series, No. 63 on Helicopter Aerodynamics and Dynamics, Chapter 4, 1973.
3. Hohenemser, K.H., "Hingeless Rotorcraft Flgith Dyanamics," Agardograph, No. 197, 1974.
4. Friedmann, P.P., "Recent Developments in Rotary-Wing Aeroelasticity," Journal of Aircraft, Vol. 14, No. 11, November 1977, pp. 1027-1041.
5. Friedmann, P.P., "Formulation and Solutions of Rotary-wing Aeroelastic Stability and Response Problems," Vertica, Vol. 7, No. 2, pp 101-104, 1983.
6. Ormiston, R.A., "Investigation of Hingeless Rotor Stability," Vertica, Vol. 7, No. 2, pp. 143-181, 1983.
7. Reichert, G., "Helicopter Vibration Control - A Survey," Vertica, Vol. 5, No. 1, pp. 1-20, 1981.
8. Loewy, R.G., "Helicopter Vibrations: A Technological Perspective," Journal of American Helicopter Society, Vol. 29, No. 4, pp 4-30, October 1984.
9. Johnson, W., "Recent Developments in the Dynamics of Advanced Rotor Systems - Part I," Vertica, Vol. 10, No. 1, pp. 73-107, 1986.
10. Johnson, W., "Recent Developments in the Dynamics of Advanced Rotor Systems _ Part II," Vertica, Vol. 10, No. 2, 1986.
11. Alkire, K., "An Analysis of Rotor Blade Twist Variables Associated with Different Euler Sequences and Pretwist Treatments," NASA TM-84394, 1984.
12. Hodges, D.H., Ormiston, R.A. and Peters, D.A., "On the Nonlinear Deformation Geometry of Euler Bernoulli Beams," NASA TP-1566, April 1980.
13. Hodges, D.H., "Nonlinear Equations for the Dyanamics of Pretwisted Beams Undergoing Small Strains and Large Rotations," NASA TP-2470, May 1985.
14. Hodges, D.H., Hopkins, A.K., Kunz, D.L. and Hinnant, H.E., "Introduction to GRASP - General Rotorcraft Aeromechanical Stability Program - A Modern Approach to Rotorcraft Modeling," Proceedings, $42^{nd}$ Annual Forum of the American Helicopter Society, June 2-4, 1986, Washington, D.C., pp. 739-756.
15. Crespo DaSilva, M.R.M. and Hodges, D.H., "Nonlinear Flexure and Torsion of Rotating Beams, with Application to Helicopter Rotor Blades - I. Formulation," Vertica, Vol. 10, No. 2, 1986.
16. Crespo DaSilva, M.R.M. and Hodges, D.H., "Nonlinear Flexure and Torsion of Rotating Beams, with Application to Helicopter Rotor Blades - II. Results for Hover," Vertica, Vol. 10, No. 2, 1986.
17. Rosen, A. and Rand, O., "Formulation of a Nonlinear Model for Curved Rods," Proceedings of $26^{th}$ Israel Conference on Aeronautics and Astronautics," February 1984, pp. 244-256.
18. Rosen, A. and Rand, O., "A General Model of the Dyanmics of Moving and Rotating Rods," Computers and Structures, Vol. 21, No. 3, 1985, pp. 543-561.
19. Rosen, A. and Rand, O., "Static Aeroelastic Behavior of Curved Helicopter Blades in Hovering and Axial Flight," Vertica, Vol. 7, No. 3, 1983, pp. 241-257.

20. Sivaneri, N.T. and Chopra, I., "Dynamic Stability of a Rotor Blade Using Finite Element Analysis," AIAA Journal, Vol. 20, No. 5, May 1982, pp. 716-723.
21. Friedmann, P.P. and Straub, F., "Application of the Finite Element Method to Rotary-Wing Aeroelasticity," Journal of the American Helicopter Society, Vol. 25, January 1980, pp. 36-44.
22. Sivaneri, N.T. and Chopra, I., "Finite Element Analysis of Bearingless Rotor Aeroelasticity," Journal of the American Helicopter Society, Vol. 29, April 1984, pp. 42-51.
23. Fu, C. and Wang, S., "Aeroelastic Stability of a Rotor by Lifting Surface Theory and Finite Element Method," Paper No. 68, Proceedings Eleventh European Rotorcraft Forum, London, September 10-13, 1986.
24. Mansfield, E.H. and Sobey, A.J., "The Fibre Composite Helicopter Blade, Part I: Stiffness Properties by E.H. Mansfield, Part 2: Prospects for Aeroelastic Tailoring by A.J. Sobey, Aeronautical Quarterly, May 1979, pp. 413-449.
25. Hong, C.H. and Chopra, I., "Aeroelastic Stability Analysis of a Composite Rotor Blade," Journal of the American Helicopter Society, Vol. 30, April 1985, pp. 57-67.
26. Hodges, D.H. and Dowell, E.H., "Nonlinear Equations of Motion for Elastic Bending and Torsion of Twisted Non-uniform Blades, NASA TN D-7818, December 1974.
27. Tarzanin, F.J. and Vlaminck, R.R., "Investigation of Blade Sweep on Rotor Vibratory Loads," NASA CR166526, October 1983.
28. Celi, R., "Aeroelasticity and Structural Optimization of Helicopter Rotor Blades with Swept Tips," Ph.D. Dissertation, Mechanical, Aerospace and Nuclear Engineering Department, University of California, Los Angeles, October 1986.
29. Shamie, J. and Friedmann, P., "Effect of Moderate Deflections on the Aeroelastic Stability of a Rotor Blade in Forward Flight," Paper No. 24, Porceedings of Third European Rotorcraft and Powered Lift Aircraft Forum, Aix-en-Provence, France, September 1977.
30. Kosmatka, J., "Structural Dynamic Modeling of Nonisotropic Blades by the Finite Element Method," Ph.D. Dissertation, Mechanical, Aerospace and Nuclear Engineering Department, University of California, Los Angeles, October 1986.
31. Rosen, A., Loewy, R.G. and Mathew, M.B., "Use of Twisted Principal Coordinates in Blade Analysis," Paper presented at the International Conference on Rotorcraft Basic Research, Research Triangle Park, NC, February 19-21, 1985.
32. Dat, R., "Development of Basic Methods Needed to Predict Helicopter Aeroelastic Behavior," Vertica, Vol. 8, No. 3, 19084, pp. 209-228.
33. Beddoes, T.S., "Practical Computation of Unsteady Lift," Vertica, Vol. 8, No1. 1, 1984, pp. 55-71.
34. Leishman, J.G. and Beddoes, T.S., "A Generalized Model for Airfoil Unsteady Aerodyanmic Behaviour and Dynamic Stall Using the Indicial Method," Proceedings of the 42$^{nd}$ Annual Forum of the American Helicopter Society, June 2-4, 1986, Washington, D.C., pp, 243-265.
35. Gangwani, S.T., "Synthesized Airfoil Data Method for Prediction of Dynamic Stall and Unsteady Airloads," Vertica, Vol. 8, No. 2, 1984, pp. 93-118.
36. Rogers, J.P., "Application of an Analytic Stall Model to time History and Eigenvalue Analysis of Rotor Blades," Journal of the American Helicopter Society, Vol. 29, January 1984, pp. 25-33.

37. Peters, D.A., "Toward a Unified Model for Use in Rotor Blade Stability Analyses," Journal of the American Helicopter Society," Vol. 30, July 1985, pp.32-42.
38. Bergh, H. and Van DerWekken, A.J.P., "Comparison Between Measured and Calculated Stall Flutter Behaviour of a One Bladed Model Rotor," Paper No. 67, Proceedings Eleventh European Rotorcraft Forum, London, England, September 10-13, 1985.
39. Tran, C.T. and Petot, D., "Semi-Empirical Model for the Dynamic Stall of Airfoils in View of their Application to the Calculation of Responses of a Helicopter Rotor Blade in Forward Flight," Vertica, Vol. 5, No. 1, 1981, pp. 35-53.
40. Leiss, U., "A Consistent Mathematical Model to Simulate Steady and Unsteady Rotor-Blade Aerodynamics," Paper No. 7, Proceedings of the Tenth European Rotorcraft Forum, The Haag, The Netherlands, September 1984.
41. Leiss, U., "Unsteady Sweep a Key to Simulation of Three-Dimensional Rotor Blade Airloads," Paper No. 25, Proceedings of Eleventh European Rotorcraft Forum, London, England, September 1985.
42. Dinyavari, M.A.H. and Friedmann, P.P., Unsteady Aerodynamics in Time and Frequency Domains for Finite Time Arbitrary Motion of Rotary Wings in Hover and Forward Flight," AIAA Paper 84-0988, Proceedings AIAA/ASME/ASCE/AHS $25^{th}$ Structures, Structural Dynamics and Materials Conference, Palm Springs, California, May 1984, pp. 266-282.
43. Dinyavari, M.A.H. and Friedmann, P.P., "Application of the Finite State Arbitrary Motion Aerodynamics to Rotor Blade Aeroelastic Response and Stability in Hover and Forward Flight," AIAA Paper 85-0763, Proceedings of AIAA/ASME/ASCE/AHS, $26^{th}$ Structures, Structural Dynamics and Materials Conference, Orlando, Florida, April 15-17, pp. 522-535.
44. Dinyavari, M.A.H. and Friedmann, P.P., "Finite-Time Arbitrary-Motion Unsteady Cascade Airfoil Theory for Helicopter Rotors in Hover," Paper No. 26, Proceedings of Eleventh European Rotorcraft Forum, London, England, September 1985.
45. Asghar-Hessari-Dinyavari, M., "Unsteady Aerodyanmics in Time and Frequency Domains for Finite-Time Arbitrary Motion of Helicopter Rotor Blades in Hover and in Forward Flight," Ph.D., Dissertation, Mechanical, Aerospace and Nuclear Engineering Department, University of California, Los Angeles, California, March 1985.
46. Friedmann, P.P. and Venkatesan, C., "Finite State Modelling of Unsteady Aerodynamics and Its Application to a Rotor Dynamic Problem," Paper No. 72, Proceedings of Eleventh European Rotorcraft Forum, London, September 1985.
47. Venkatesan, C. and Friedmann, P.P., "A New Approach to Finite State Modeling of Unsteady Aerodynamics," AIAA Paper 86-0865CP, Proceedings of AIAA/ASME/ASCE/AHS $27^{th}$ Structures, Structural Dynamics and Materials Conference, May 1986, San Antonio, Texas, pp. 178-191.
48. Friedmann, P.P., "Arbitrary Motion Unsteady Aerodynamics and Its Application to Rotary-wing Aeroelasticity," Proceedings $42^{nd}$ Annual Forum of the American Helicopter Society, Washington, D.C., June 1986, pp. 757-776.
49. Gaonkar, G. and Peters, D., "Effectiveness of Current Dynamic-Inflow Models in Hover and in Forward Flight," Journal of the American Helicopter Society, Vol. 31, April 1986, pp. 47-57.
50. Rand, O. and Rosen, A., "An Unsteady Prescribed Wake Model for a Helicopter Rotor in Forward Flight," Journal of the American Helicopter Society, Vol. 30, October 1985, pp. 11-21.
51. Davis, S.S and Chang, I.C., "The Critical Role of Computational Fluid Dynamics on Rotary-Wing Aerodynamics," AIAA Paper 86-0336, AIAA $24^{th}$ Aerospace Sciences Meeting, Reno, Nevada, January 1986.

52. Warmbrodt, W. and Friedmann, P., "Formulation of Coupled Rotor/Fuselage Equations of Motion," *Vertica*, Vol. 3, 1979, pp. 254-271.
53. Johnson, W, "A Comprehensive Analytical Model of Rotorcraft Aerodynamics and Dynamics, Part I: Analysis Development," NASA TM-81182, June 1980.
54. Vankatesan, C. and Friedmann, P., "Aeroelastic Effects in Multirotor Vehicles with Aplication to Hybrid Heavy Lift System, Part I: Formulation of Equations of Motion," NASA CR-3822, August 1984.
55. Nagabhushanam, J., Gaonkar, G.H. and Reddy, T.S.R., "Automatic Generation of Equations for Rotor-Body Systems with Dynamic Inflow for a Priori Ordering Schemes," Paper No. 37, Proceedings of Seventh European Rotorcraft Forum, Garmisch-Partenkirchen, September 1981.
56. Reddy, T.S.R. and Warmbrodt, W., "The Influence of Dynamic Inflow and Torsional Flexibility on Rotor Damping in Forward Flight from Symbolically Generated Equations," *Rotorcraft Dynamics 1984*, Proceedings of the 2nd Decennial Specialists Meeting on Rotorcraft Dynamics, NASA Ames, Nov. 7-9, 1984, NASA CP-2400, November 1985, pp. 221-239.
57. Reddy, T.S.R., "Symbolic Generation of Elastic Rotor Blade Equations Using a FORTRAN Processor and Numerical Study on Dynamic Inflow Effects on the Stability of Helicopter Rotors," NASA TM-86750, June 1986.
58. Crespo Da Silva, M.R.M. and Hodges, D.H., "The Role of Computerized Symbolic Manipulation in Rotorcraft Dyanmic Analysis," *Computers and Mathematics with Applications*, Vol. 12A, No. 1, 1986, pp. 161-172.
59. Gibbons, M.P. and Done, G.T.S., "Automatic Generation of Helicopter Rotor Equations of Motion," *Vertica*, Vol. 8, No. 3, pp. 229-241, 1984.
60. Patel, M.H. and Done, G.T.S., "Experience With a New Approach to Rotor Aeroelasticity," *Vertica*, Vol. 9, No. 3, pp. 285-294, 1985.
61. Nilakantan, G.R. and Gaonkar, G.H., "Feasibility of Simplifying Coupled Flap-Lag-Torsion Models for Rotor Blade Stability in Forward Flight," *Vertica*, Vol. 9, No. 3, 1985, pp. 241-256.
62. Panda, B. and Chopra, I., "Flap-Lag-Torsion Stability in Forward Flight," *Journal of the American Helicopter Society*, Vol. 30, October 1985, pp. 30-39.
63. Panda, B. and Chopra, I., "Dynamic Stability of Hingeless and Bearingless Rotors in Forward Flight," *Computers and Mathematics with Applications*, Vol. 12A, No. 1, 1986, pp. 111-130.
64. Friedmann, P.P. and Kottapalli, S.B.R., "Coupled Flap-Lag-torsional Dynamics of Hingeless Rotor Blades in Forward Flight," *Journal of the American Helicopter Society*, Vol. 27, October 1982, pp. 28-36.
65. Gaonkar, G., McNulty, M.J. and Nagabhushanam, J., "An Experimental and Analytical Investigation of Isolated Rotor Flap-Lag Stability in Forward Flight," Paper No. 66, Proceedings of Eleventh European Rotorcraft Forum, London, England, September 1986.
66. Bir, G.S. and Chopra, I., "Gust Response of Hingeless Rotors," *Journal of the American Helicopter Society*," Vol. 31, April 1986, pp. 33-46.
67. Bir, G.S. and Chopra, I., "Prediction of Blade Stresses Due to Gust Loading," Paper No. 73, Proceedings of Eleventh European Rotorcraft Forum," London, England, September 1985.
68. Prussing, J.E., Lin, Y.K., and Shiau, T.N., "Rotor Blade Flap-Lag Stability and Response in Forward Flight in Turbulent Flows," *Journal of the American Helicopter Society*, Vo. 29, No. 24, October 1984, pp. 81-87.
69. Friedmann, P.P., "Numerical Methods for Determining the Stability and Response of Periodic Systems with Application to Helicopter Rotor Dynamics and Aeroelasticity," *Computers and Mathematics with Applications*, Vol. 12A, No. 1, 1986, pp. 131-148.
70. Zienkiewicz, O.C., *The Finite Element Method*, Chapter 21, Third Edition, McGraw-Hill, 1977

71. Borri, M., "Helicopter Rotor Dynamics by Finite Element Time Approximation," *Computers and Mathematics with Applications*, Vol. 12A, No. 1, pp. 149-160.
72. Nagabhushanam, J. and Gaonkar, G.H., "Rotorcraft Air Resonance in Forward Flight with Various Dynamic Inflow Models and Aeroelastic Couplings," *Vertica*, Vol. 8, No. 4, 1984, pp. 373-394.
73. Johnson, W., "Assessment of Aerodynamic and Dynamic Models in Comprehensive Analysis of Rotorcraft," *Computers and Mathematics with Applications*, Vol. 12A, No. 1, 1986, pp. 11-28.
74. Yeager, W.T., Hamouda, M.H. and Mantay, W.R., "Aeromechanical Stability of a Hingeless Rotor in Hover and Forward Flight: Analysis and Wind Tunnel Tests," NASA TM-85653, August 1983, also available as Paper No. 54, Proceedings of Ninth European Rotorcraft Forum, Stresa, Italy, September 1983.
75. Warmbrodt, W. and Peterson, R.L., "Hover Test of a Full-Scale Hingeless Rotor," NASA TM-85990, August 1984.
76. Johnson, W., "Influence of Unsteady Aerodynamics on Hingeless Rotor Ground Resonance," *Journal of Aircraft*, Vol. 19, August 1982, pp. 668-673.
77. Bousman, W.G., "An Experimental Investigation of the Effects of Aeroelastic Couplings on Aeromechanical Stability of a Hingeless Rotor Helicopter," *Journal of the American Helicopter Society*, Vol. 26, January 1981, pp. 46-54.
78. Venkatesan, C. and Friedmann, P.P., "Aeroelastic Effects in Multirotor Vehicles with Application to a Hybrid Heavy Lift System, Part I: Formulation of Equations of Motion," NASA CR-3822, August 1984.
79. Venkatesan, C. and Friedmann, P.P., "Aeroelastic Effects in Multirotor Vehicles, Part II: Method of Solution and Results Illustrating Coupled Rotor/Body Aeromechanical Stability," Low Number NASA CR-, September 1986, (in press).
80. Friedmann, P.P. and Venkatesan, C., "Coupled Rotor/Body Aeromechanical Stability Comparison of Theoretical and Experimental Results," *Journal of Aircraft*, Vol. 22, February 1985, pp. 148-155.
81. Friedmann, P.P. and Venkatesan, C., "Influence of Unsteady Aerodynamic Models on Aeromechanical Stability in Ground Resonance," *Journal of the American Helicopter Society*, Vol. 31, January 1986, pp. 65-74.
82. Sopher, R. and Hallock, D., "Time-History Analysis of Rotorcraft Dynamics Based on a Component Approach," *Journal of the American Helicopter Society*, Vol. 31, January 1986, pp. 43-51.
83. Ormiston, R.A., "Rotor-Fuselage dynamic Coupling Characteristics of Helicopter Air and Ground Resonance," Proceedings of the Theoretical Basis of Helicopter Technology, Nanjing Aeronautical Institute, Nanjing, China, Nov. 6-8, 1985.
84. McNulty, Michael, J., "Effect of Blade-to-Blade Dissimilarities on Rotor-Body Lead-Lag Dyanmics," Paper No. 64, Proceedings of the Eleventh European Rotorcraft Forum, London, England, September 1985.
85. Tang, D.M. and Dowell, E.H., "Influence of Nonlinear Blade Damping on Helicopter Ground Resonance," *Journal of Aircraft*, Vol. 23, No. 2, 1986, pp. 104-110.
86. Johnson, W., "Self-Tunning Regulators for Multicyclic Control of Helicopter Vibrations," NASA Technical Paper 1996, 1982.
87. Molusis, J.A., Hammond, C.E. and Cline, J.H., "A Unified Approach to Optimal Design of Adaptive and Gain Scheduled Controllers to Achieve Minimum Helicopter Rotor Vibration," *Journal of the American Helicopter Society*," Vol. 28, April 1983, pp. 9-18.

88. Hammond, C.E., "Wind Tunnel Results Showing Rotor Vibratory Loads Reduction Using Higher Harmonic Blade Pitch," *Journal of the American Helicopter Society*, Vol. 28, January 1983, pp. 1983.
89. Wood, E.R., Powers, R.W., Cline, J.H. and Hammond, C.E., "On Developing and Testing a Higher Harmonic Control System," *Journal of the American Helicopter Society*, Vol. 30, January 1985, pp. 3-20.
90. Davis, M.W., "Refinement and Evaluation of Helicopter Real-Time Self-Adaptive Active Vibration Controller Algorithms," NASA CR3821, November 1983.
91. Taylor, R.B., Zwicke, P.E., Gold, P. and Miao, W., "Analytical Design and Evaluation of an Active Control System for Helicopter Vibration Reduction and Gust Response Alleviation," NASA CR 152377, July 1980.
92. Molusis, J.A., "The Importance of Nonlinearity on the Higher Harmonic Control of Helicopter Vibration," Proceedings 39th Annual Forum of the American Helicopter Society, May 1983, pp. 624-647.
93. Bielawa, R.L., "Aeroelastic Analysis for Helicopter Rotor Blades with Time-Variable, Nonlinear Structural Twist and Multiple Structural Redundancy-Mathematical Derivation and User's Manual," NASA CR-2638, October 1976.
94. Miao, W., Kottapalli, S.B.R. and Frye, H.M., "Flight Demonstration of Higher Harmonic Control (HHC) on the S-76," Proceedings of 42nd Annual Forum of the American Helicopter Society," June 1986, Washington, D.C., pp. 777-791.
95. Polychroniadis, M. and Achache, M., "Higher Harmonic Contrl: Flight Tests on an SA349 Research Gazelle," Proceedings of 42nd Annual Forum of the American Helicopter Society, June 1986, Washington, D.C., pp. 811-820.
96. Shaw, J., Albion, N., Hanker, E.J. and Teal, R., "Higher Harmonic Control: Wind Tunnel Demonstration of Fully Effective Vibratory Hub Force Suppression," Proceedings 41st Annual Forum of the American Helicopter Society, Fort Worth, Texas, May 1985, pp. 1-15.
97. Jacob, H.G. and Lehmann, G., "Optimization of Blade Pitch Angle for Higher Harmonic Control," *Vertica*, Vol. 7, 1983, pp. 271-286.
98. Lehmann, G., "The Effect of Higher Harmonic Control (HHC) on a Four-Bladed Hingeless Model Rotor," *Vertica*, Vol. 9, 1985, pp. 273-284.
99. DuVal, R.W., Gregory, C.Z. and Gupta, N.K., "Design and Evaluation of a State-Feedback Vibration Controller," *Journal of the American Helicopter Society*, Vol. 29, July 1984, pp.30-37.
100. Saito, S., "Application of an Adaptive Blade Control Algorithm to a Gust Alleviation System," *Vertica*, Vol. 8, No. 3, 1984, pp. 289-307.
101. Straub, F.K. and Warmbrodt, W., "The Use of Active Controls to Augment Rotor/Fuselage Stability," *Journal of the American Helicopter Society*, Vol. 30, July 1985, pp. 13-22.
102. Straub, F.K., "Optimal Control of Helicopter Aeromechanical Stability," Paper 77, Proceedings of the Eleventh European Rotorcraft Forum, London, England, September 1985.
103. Kretz, M., "Research in Multicyclic and Active Control of Rotary Wings," *Vertica*, Vol. 1, No. 2, 1976, pp. 95-105.
104. Ham, N.D., "Helicopter Individual-Blade-Control Research at MIT 1977-1985," *Vertica*, Vol. 10, No. 4, 1986 (to be published).
105. Ham, N.D., "Helicopter Gust Alleviation, Attitude Stabilization and Vibration Alleviation Using Individual-Blade-Control Through a Conventional Swashplate," Paper 75, Proceedings of Eleventh Europenan Rotorcraft Forum, London, England, September 1985.

106. Ham, N.D., Behal, B.L. and McKillip, R.M., "Helicopter Rotor Lag Damping Augmentation Through Individual-Blade-Control," Vertica, Vol. 7, No. 4., 1983, pp. 361-371.
107. Quackenbush, T.R., "Testing of a Stall Flutter Suppression System for Helicopter Rotors Using Individual-Blade-Control," Journal of the American Helicopter Society, Vol. 29, 1984, pp. 38-44.
108. McKillip, R.M., "Periodic Control of the Individual-Blade-Control Helicopter Rotor," Vertica, Vol. 9, No. 2, 1985., pp. 199-224.
109. Guin, K.F., "Individual Blade Control Independent of a Swashplate," Journal of the American Helicopter Society, Vol. 27, July 1982, pp. 25-31.
110. Blackwell, R.H., "Blade Design for Reduced Helicopter Vibration," Journal of the American Helicopter Society, Vol. 28, July 1983, pp. 33-41.
111. Friedmann, P.P., "Application of Modern Structural Optimization to Vibration Reduction in Rotorcraft," Vertica, Vol. 9, No. 4, 1985, pp. 363-373.
112. Shanthakumaran, P., "Optimum Design of Rotor Blades for Vibration Reduction in Forward Flight," Ph.D., Dissertation, University of California, Los Angeles, California, 1982.
113. Friedmann, P.P. and Shanthakumaran, P., "Aeroelastic Tailoring of Rotor Blades for Vibration Reduction in Forward Flight," AIAA Paper No. 83-0914, Proceedings AIAA/ASME/ASCE/AHS 24th Structures, Structural Dynamics and Materials Conference, Lake Tahoe, Nevada, Vol. II, May 1983, pp. 344-359.
114. Friedmann, P.P. and Shanthakumaran, P., "Optimum Design of Rotor Blades for Vibration Reduction in Forward Flight," Journal of the American Helicopter Society, Vol. 29, October 1984, pp. 70-80.
115. Davis, M.W. and Weller, W.H., "Application of Design Optimization Techniques to Rotor Dynanmics Problems," Proceedings 42nd Annual Forum of the American Helicotper Society, June 1986, Washington, D.C., pp. 27-44.
116. Chopra, I., "Dynamic Analysis of Constant-Lift and Free Tip Rotor," Journal of the American Helicopter Society, Vol. 28, January 1983, pp. 24-33.
117. Chopra, I., "Aeroelastic Stability of an Elastic Circulation Control Rotor Blade in Hover," Vertica, Vol. 8, No. 4, 1984, pp. 353-371.
118. Chopra, I., "Dynamic Stability of a Bearingless Ciruclation Control Rotor Blade in Hover," Journal of the American Helicopter Society, Ocotber 1985, pp. 40-47.
119. Chen, S.Y., "Stability of Two Bladed Aeroelastic Rotors on Flexible Supports," Journal of the American Helicopter Society, Vol. 28, January 1983, pp. 34-41.
120. Venkatesan, C. and Friedmann, P., "Aeromechanical Stability Analysis of a Hybrid Heavy Lift Multirotor Vehicle in Hover," Journal of Aircraft, Vol. 22, November 1985, pp. 965-972.
121. Neff, J.R., "Pitch-Flap-Lag Instability of Elastic Modes of an Articulated Rotor Blade," Proceedings 40th Annual Forum of the American Helicopter Society, Arlington, VA., May 1984, pp. 573-579.
122. Kloppel, V., Kampa, K. and Isselhorst, B. "Aeromechanical Aspects in the Design of Hingeless/Bearingless Rotor Systems," Paper No. 57, Proceedings of Nith European Rotorcraft Forum, Stresa, Italy, September 1983.
123. Weller, W.H, and Peterson, R.L., "Inplane Stability Characteristics for an Advanced Bearingless Main Rotor Model," Journal of the American Helicopter Society, Vol. 29, July 1984, pp. 45-53.
124. Dawson, S., "An Experimental Investigation of the Stability of a Bearingless Model Rotor in Hover," Journal of the American Helicopter Society, Vol. 28, October 1983, pp. 29-34.

125. Hodges, D.H., "Aeromechanical Stability of Helicopters with Bearingless Main Rotor - Part I: Equations of Motion," NASA TM 78459," - "Part II: Computer Program," NASA TM-78460, February 1978.
126. Bousman, W.G. and Dawson, S., "Experimentally Determined Flutter from Two and Three-Bladed Model Bearingless Rotors in Hover," Journal of the American Helicopter Society, Vol. 31, July 1986, pp. 45-53.
127. Hooper, W.E., "A Parametric Study of the Aeromechanical Stability of a Bearingless Rotor," Journal of the American Helicopter Society, Vol. 31, January 1986, pp. 52-64.
128. Banerjee, D., "Aeroelastic Characteristics of the AH-64 Bearingless Tail Rotor," Vertica, Vol. 8, No. 3, 1984, pp. 263-287.
129. Straub, F.K., Johnston, R.A. and Head, R.E., "Design and Development of a Dynamically Scaled Model AH-64 Main Rotor," Vertica, Vol. 9, No. 2, 1985, pp. 165-180.

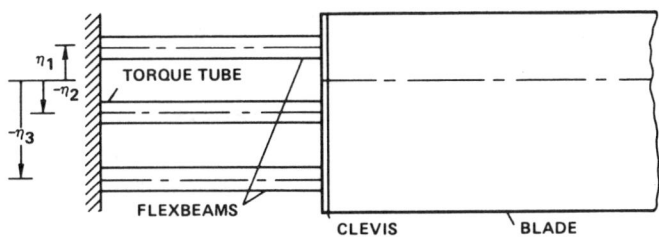

**Figure 1.** Analytical model of a bearingless blade Ref. (22)

**Figure 2.** Finite element model showing nodal degrees of freedom (Ref. 22)

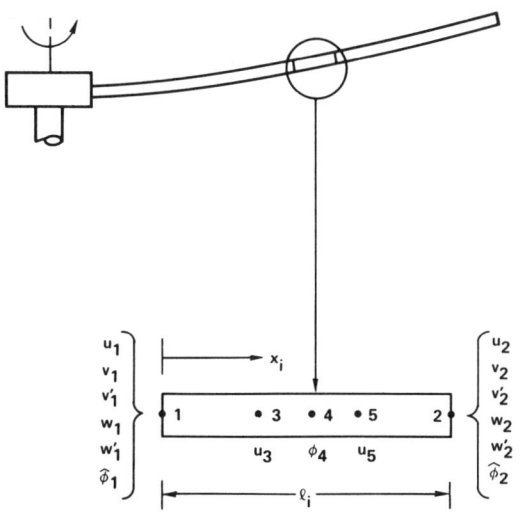

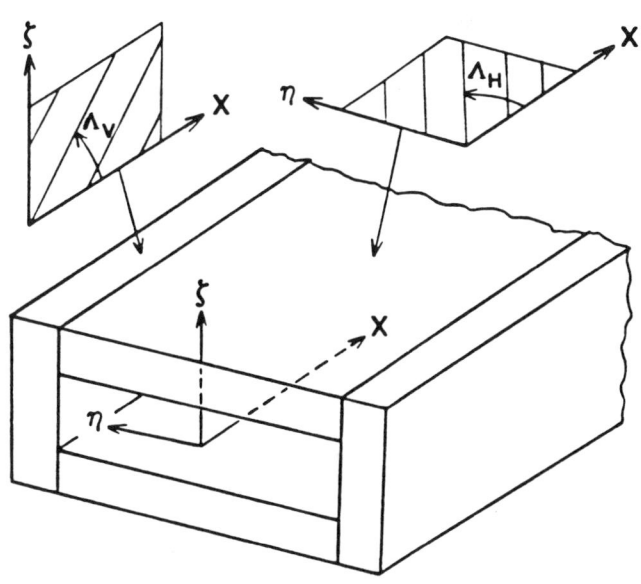

**Figure 3.** Composite box beam representing blade spar. Ply angle orientation with respect to reference coordinate: $\Lambda_V$ - vertical laminates, $\Lambda_H$ for horizontal laminates (Ref. 25)

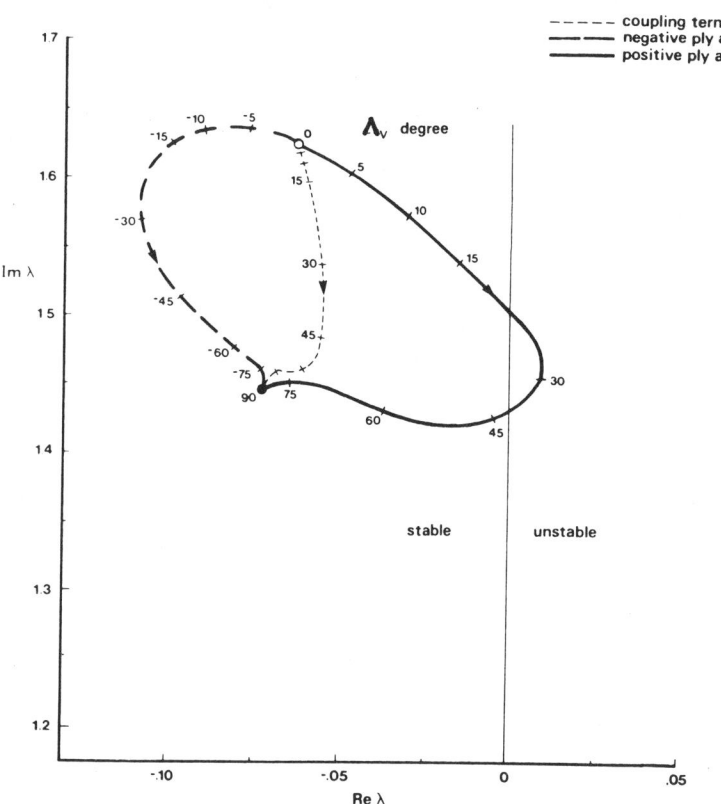

**Figure 4.** Root locus (eigenvalue) plot for lag mode of a composite blade with symmetric laminates, $C_T/\sigma = 0.10$ (Ref. 25)

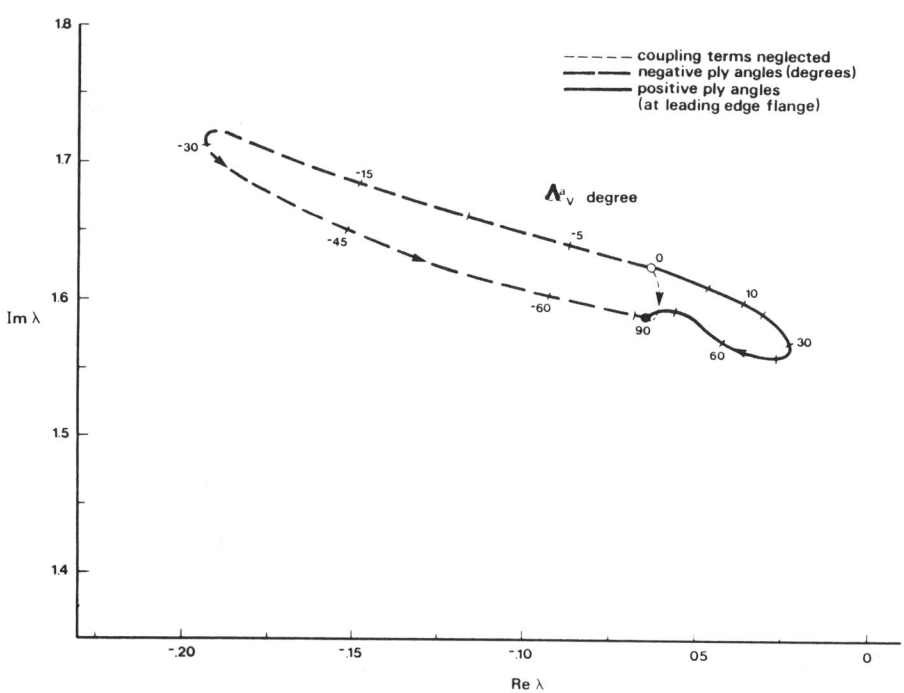

**Figure 5.** Root locus (eigenvalue) plot for lag mode of a composite blade with antisymmetric laminates, $C_T/\sigma = 0.10$ (Ref. 25)

**Figure 6.** Swept-tip hingelesss rotor blade model (Ref. 28)

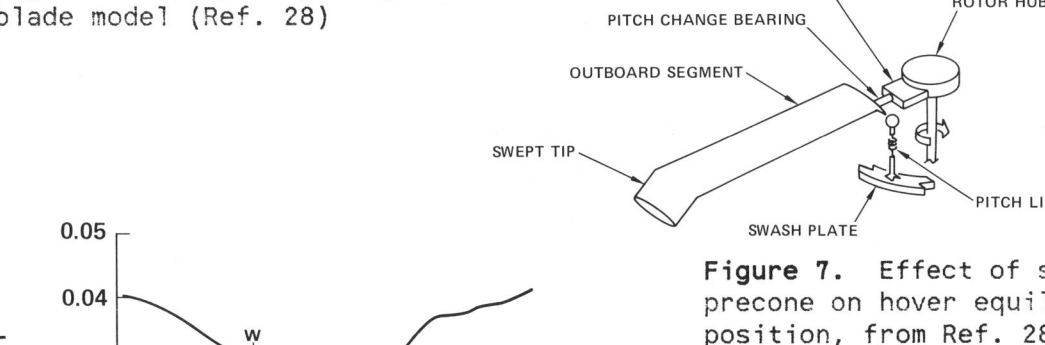

**Figure 7.** Effect of sweep and precone on hover equilibrium position, from Ref. 28 ($\omega_{L1}$ = 1.147; $\omega_{F1}$ = 1.125; $\omega_{T1}$ = 3.176 $\gamma$ = 5.5, $\ell_T$ = 0.1R; $\Theta_o$ = 0.1432; $\sigma$ = 0.07)

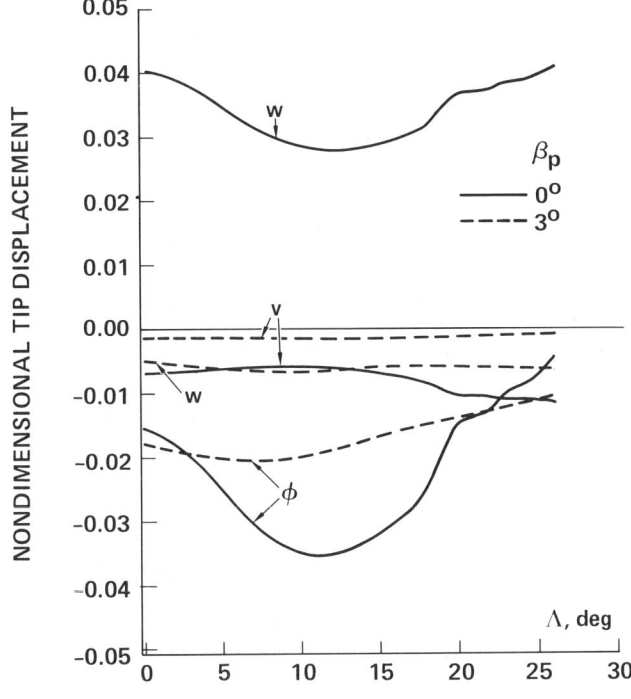

**Figure 8.** Effect of sweep and precone on aeroelastic stability in hover, (or real and imaginary part of eigenvalue) from Ref. 28, with same data as Fig. 7

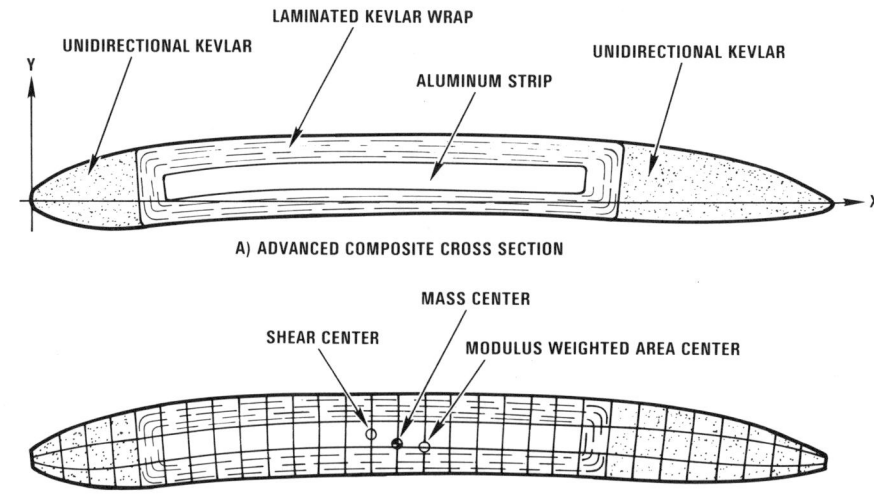

Figure 9. Analysis of an advanced composite airfoil cross section (Ref. 30)

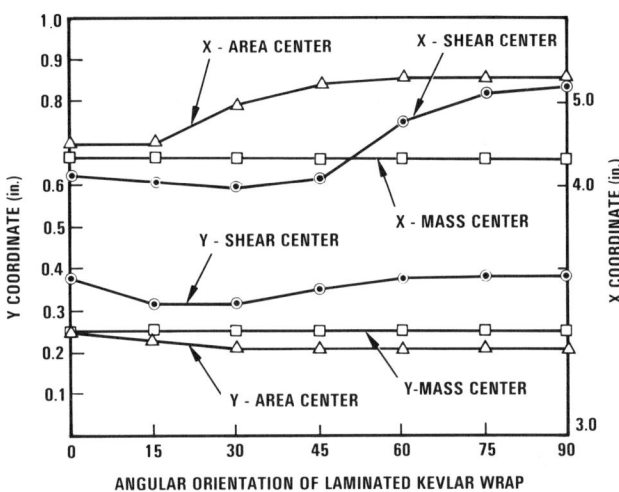

Figure 10. Location of mass, area and shear centers on an advanced composite cross section with different angles of wrap (Ref. 30)

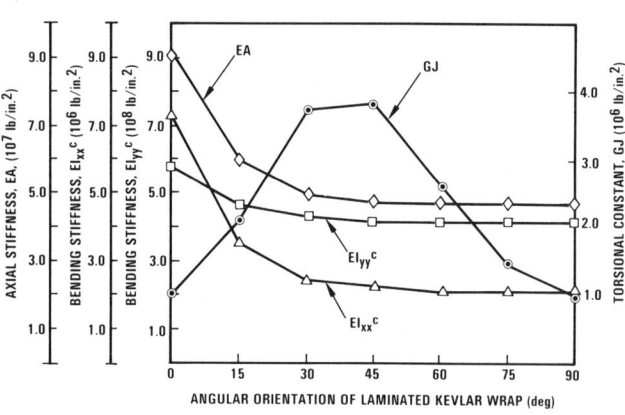

Figure 11. Structural constants of an advanced composite cross section with different angles of wrap (Ref. 30)

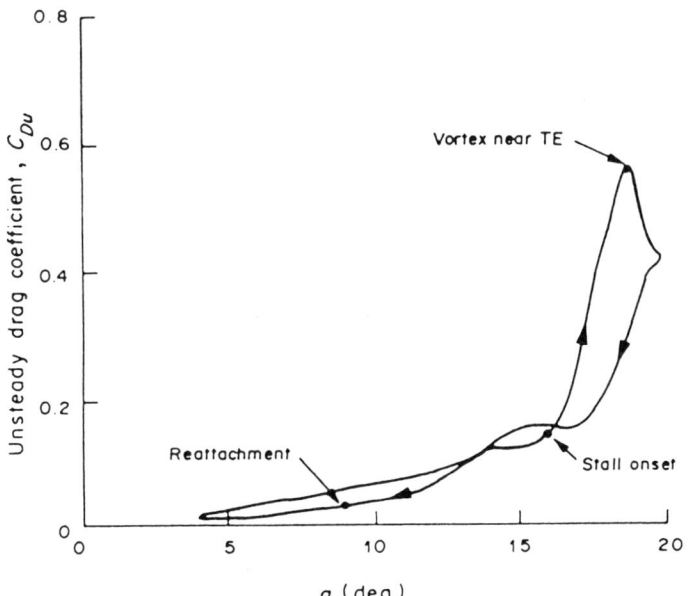

**Figure 12.** Typical unsteady drag coefficient loop data, SC1095 airfoil, $\alpha_o = 12$ deg; $\bar{\alpha} = 8$ deg; k = 0.10 deg (Ref. 35)

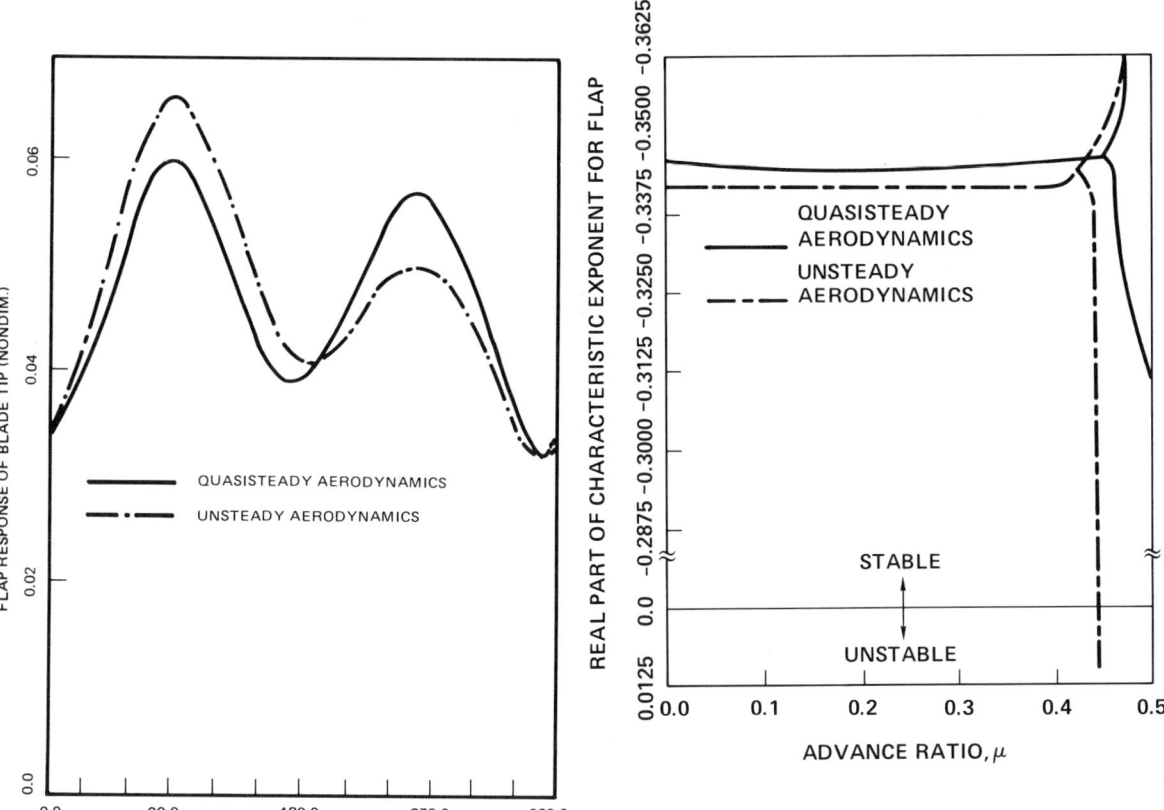

**Figure 13.** Comparison of flap response calculated using quasisteady and unsteady aerodynamics for $\mu = 0.40$, from Ref. 43 ($C_W = 0.005$; $\gamma = 5.5$; $\sigma = 0.07$; $\bar{\omega}_{F1} = 1.125$; $\bar{\omega}_{L1} = 0.732$; $R_C = 0.0$)

**Figure 14.** Real part of characteristic exponent for flap calculated using quasisteady and unsteady aerodynamics, same data at Fig. 13, (Ref. 43)

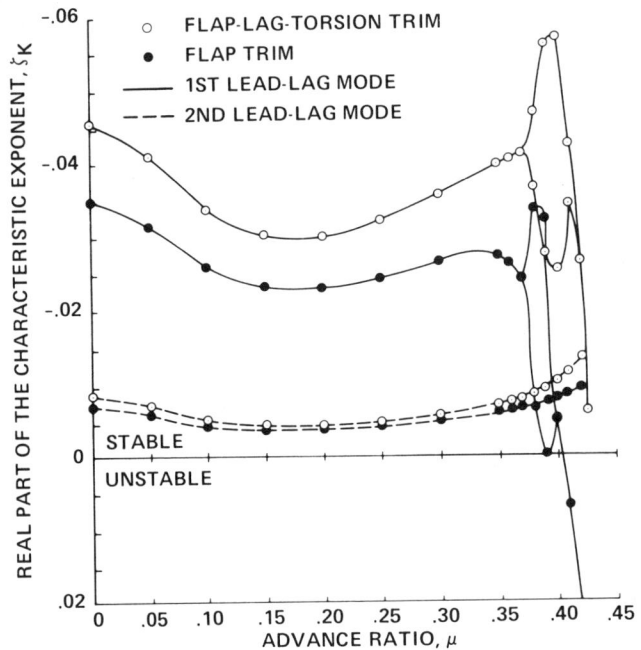

Figure 15. The effect of number of degrees of freedom used in trim analysis on lead-lag damping versus advance ratio, stiff-in-plane blade, from Ref. 56 ($\bar{\omega}_{L1} = 1.40$; $\bar{\omega}_{F1} = 1.15$; $\bar{\omega}_{T1} = 3.0$; $\sigma = 0.10$; $R_C = 1.0$; propulsive trim)

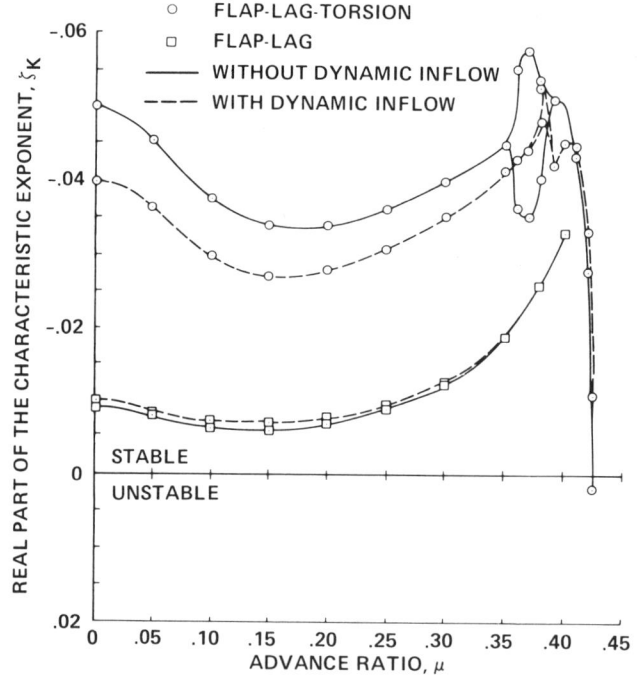

Figure 16. The effect of torsion and dynamic inflow on lead-lag regressing mode damping versus advance ratio, stiff-in-plane blade, from Ref. 56 (same data as Fig. 15)

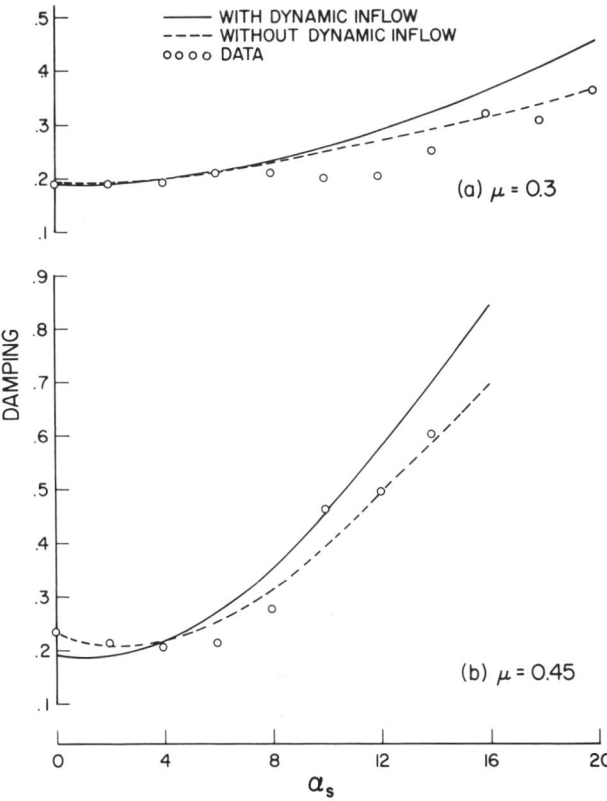

Figure 17. Lag regressing mode damping correlations in substall and stall, from Ref. 65 ($\Omega$ = 1000 RPM; $R_C$ = 0; $\theta_O$ = 0 deg.)

Figure 18. Blade regressing lag mode damping ratio as a function of collective pitch at an advance ratio of $\mu$ = 0.30; from Ref. 74, ($\delta_3$ = 42.5 degrees; R = 1.38 m; $\Omega$R = 90 m/s in Freon 12; $\sigma$ = 0.10)

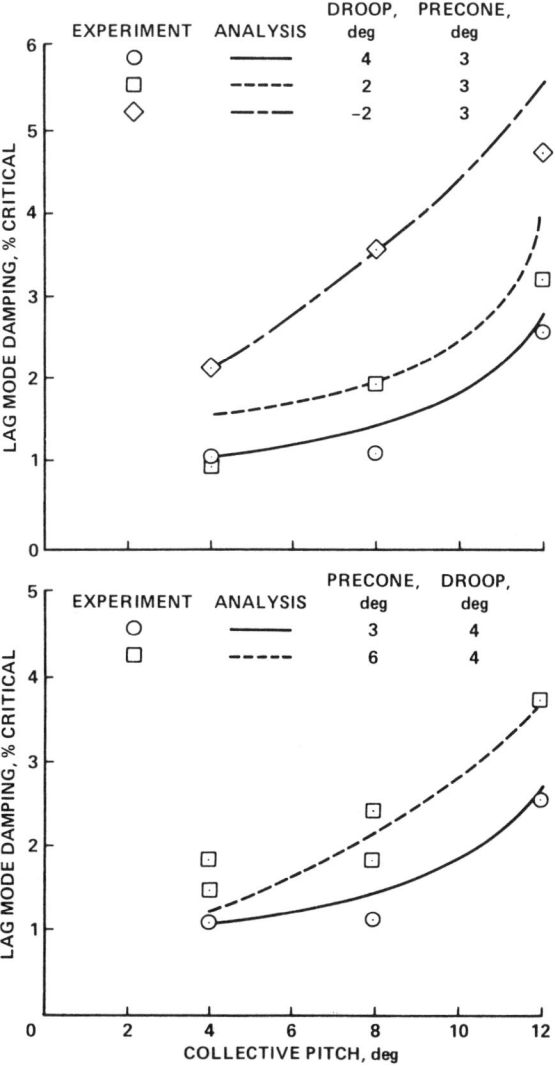

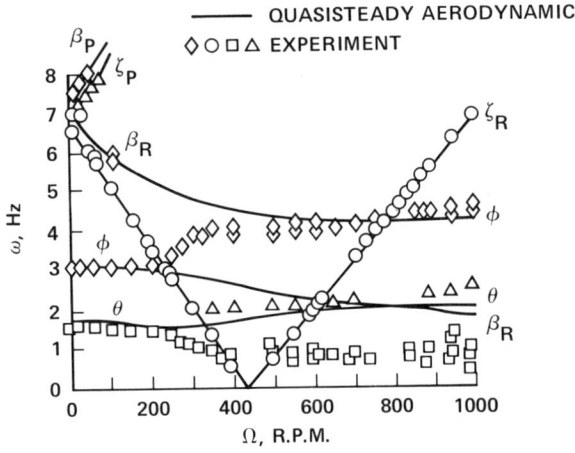

**Figure 19.** Variation of modal frequencies with $\Omega$; $\theta_c = 0$, Configuration 4, (Ref. 81)

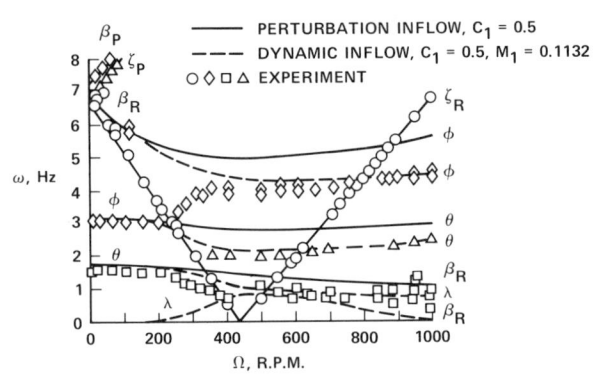

**Figure 20.** Variation of modal frequencies with $\Omega$; $\theta_c = 0$, Configuration 4, (Ref. 81)

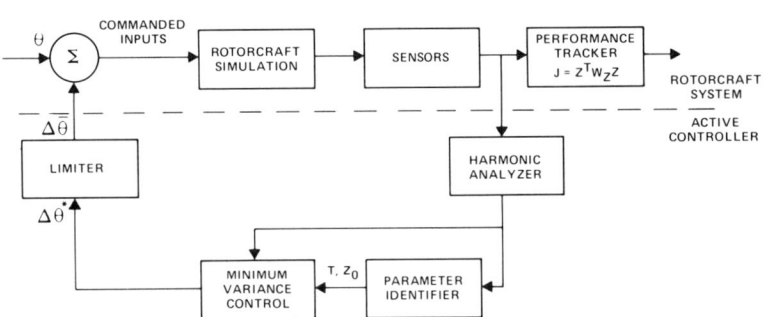

**Figure 21.** Schematic block diagram of closed-loop adaptive HHC system, used in simulations (Ref. 90)

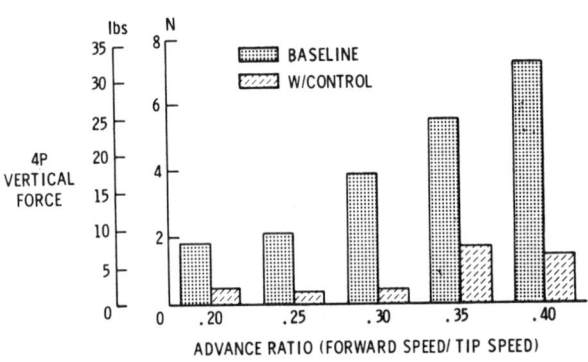

**Figure 22.** Variation of vibratory vertical force with advance ratio, using adaptive HHC in wind tunnel test (Ref. 88)

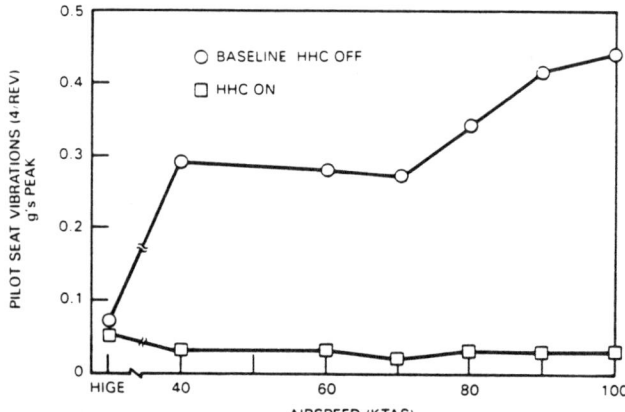

**Figure 23.** Vertical Vibration reduction at pilot set, 1984 Software (Ref. 89)

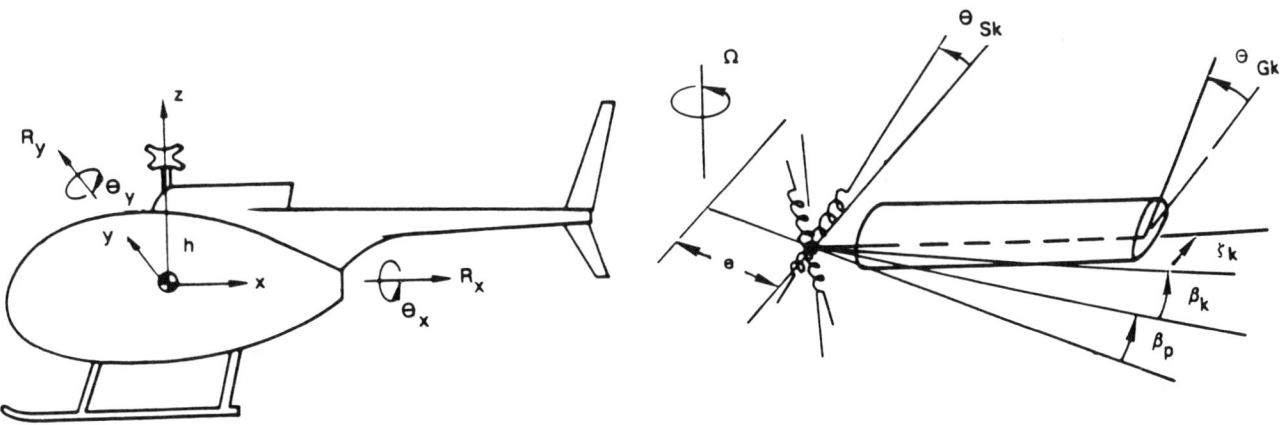

**Figure 24.** Fuselage and rotor model used in coupled rotor/body simulation for active controls (Ref. 101)

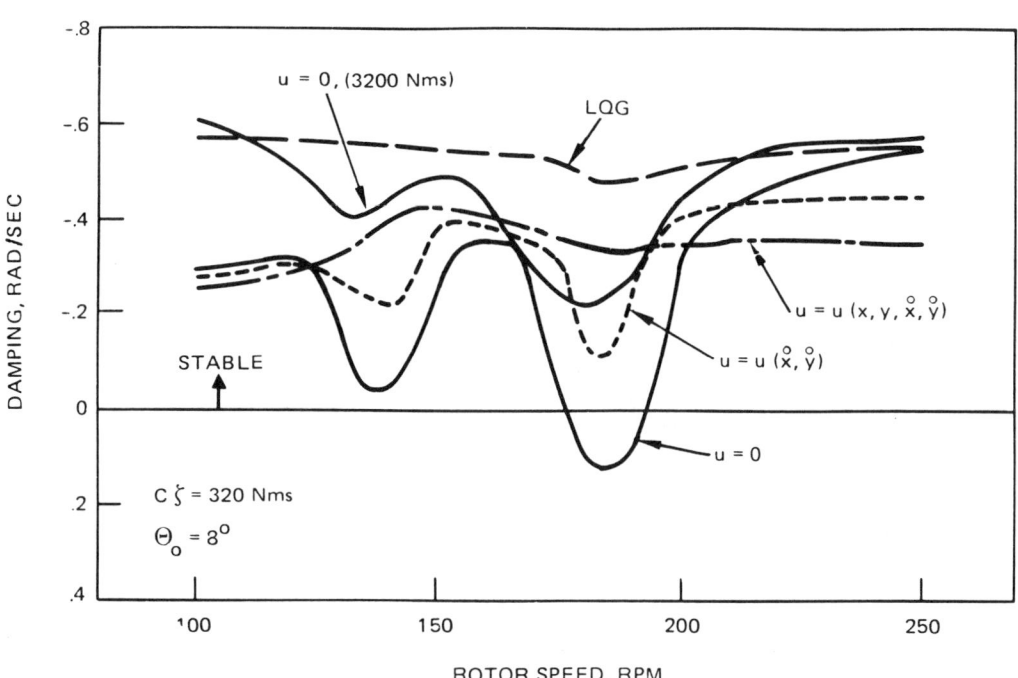

**Figure 25.** Aeromechanical stability with optimal controller and three reduced state, constant gain feedback systems for H-34/RTA (Ref. 102)

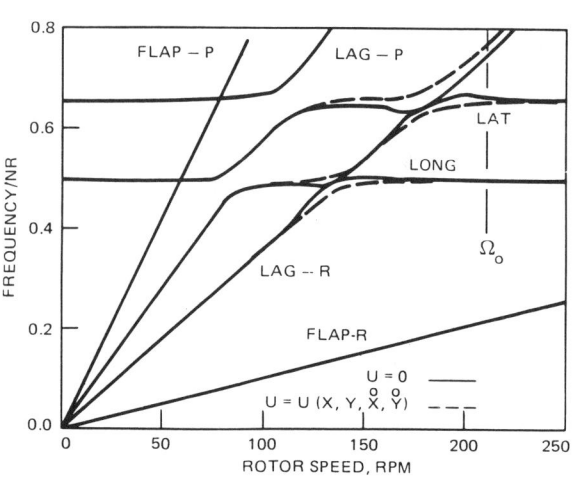

**Figure 26.** Effect of feedback control on H-34/RTA modal frequencies (Ref. 102)

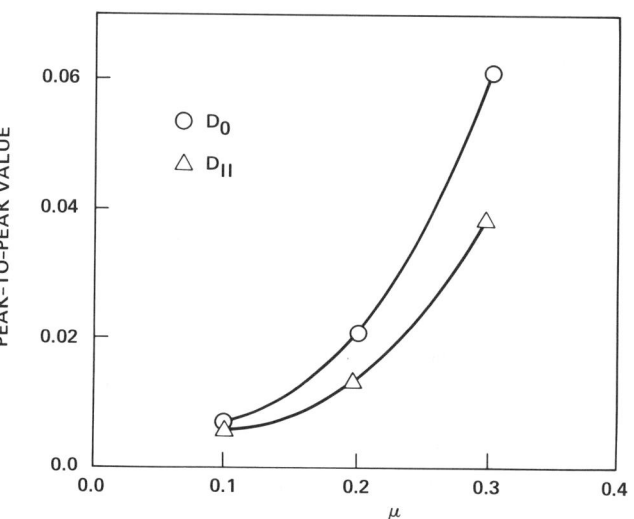

**Figure 27.** Vertical hub shears, nonlinear, peak-to-peak values, (nondimensionalized), comparison of initial and final designs after two stages of optimization, soft-in-plane blade (Ref. 114)

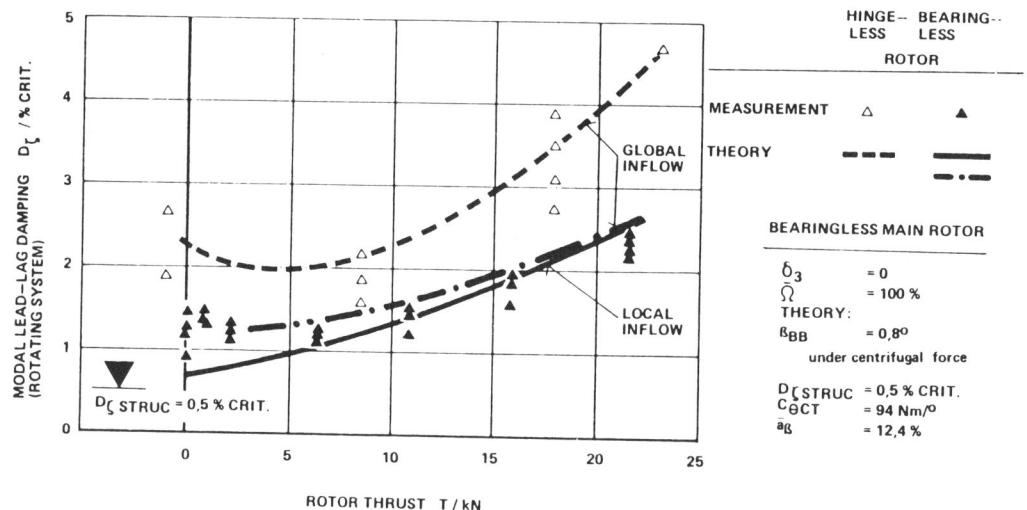

**Figure 28.** Comparison of lead-lag damping of experimental bearingless main rotor and BO105 hingeless rotor, obtained on whirl tower (Ref. 122)

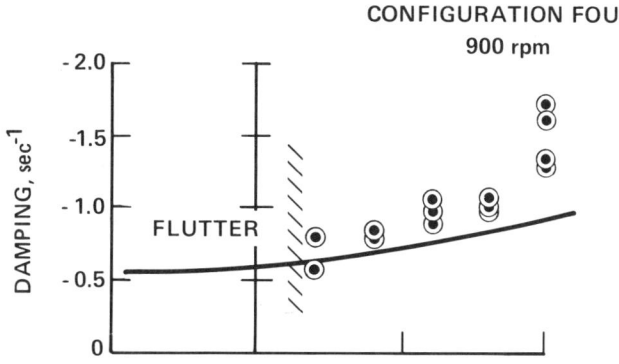

**Figure 29.** Lead-lag damping versus blade pitch angle at $\Omega$ = 900 RPM for configurations 4 and 5 (Ref. 124)

# A SURVEY OF CURRENT PROBLEMS IN TURBOMACHINE AEROELASTICITY

F. Sisto
Stevens Institute of Technology
Hoboken, New Jersey 07030

INTRODUCTION

Aeroelasticity as a field of enquiry extends back in time with accounts of flutter occurences in the early "iron" bridges in England, ca. 1818. Aircraft empennage and wing flutter became a recognized problem at the time of World War I, and development of the field as applied to external aerodynamics of lifting and control surfaces of aircraft and missiles continued into the early 1960s. In this time period several excellent English-language textbooks appeared, the most notable being Bisplinghoff, Ashley and Halfman [1] and Bisplinghoff and Ashley [2]. Subsequently the state of the art has continued to mature with emphases on supersonic flow, transonic flow, missile skin panels, control surfaces and helicopter rotors. The substance of the present Symposium may be expected to bring this up to date. It is interesting to note that no English language textbook dealt at any length with the special field of aeroelasticity as applied to turbomachinery until the relatively recent appearance of Dowell et. al. [3].

The emergence of turbojet powerplants during World War II coincided with the first documentation of axial compressor blade flutter [4]. Isolated instances of flutter probably had occurred earlier in compressors and in the condenser stages of steam turbines. Blade vibrations attributable to partial admission of steam in impulse turbines [5] were forced oscillations of aeroelastic nature. Many of the fundamental concepts related to the vibration mode of bladed disc assemblies currently being intensively researched, have their roots in that earlier steam turbine experience of the 1920s.

The development of turbomachine aeroelasticity has continued since World War II as set out selectively in the remainder of this survey. It is important to summarize the basic difference that gives rise to new phenomena in turbomachines. Clearly the major feature differentiating the aircraft wing and the axial flow turbomachine stage is the large multiplicity of identical, evenly spaced and mutually interfering airfoils in the latter. The aerodynamic coupling amongst the airfoils in an annular row of vibrating blades is extremely complex and depends on many governing parameters, particularly the interblade phase angle, σ (sigma). The structural coupling is also uniquely different in the turbomachine, being strongly affected by the twist of the airfoils, their attachment to the disk and to each other. Rotor blades are inertially coupled by operation in a strong rotational body force field. The importance of modeling the rotor structure as a bladed-disk assembly has characterized the development of aeroelasticity in axial turbomachines as contrasted with that in aeronautics.

In Shannon's pioneering report [4] it was noted that axial flow compressors were prone to stall flutter vibrations. These cantilever rotor blade oscillations occurred at part speed when the operating line traversed a region of the compressor map where high incidence on the first several stages was combined with moderately high relative airspeed. See Region I in the schematic compressor map, or characteristic, comprising Figure 1.

This graphical representation of compressor performance, with its aerodynamic parameters of aeroelastic significance, is explained in Chapter 8 of Dowell et al [3]. (For example contours of the first rotor tip incidence may be plotted in Figure 1). However, the compressor map identification of flutter regions is necessary but not sufficient. The map is purely an aerothermodynamic descriptor and gives no information about structural dynamics.

Forced vibrations were also observed where there was a clear coincidence of the natural frequency of the rotor blade with the frequency of encountering disturbances in the airstream, or the rotor passing frequency in the case of a stator vane vibration. One development has been in the application of unsteady aerodynamics to relate the magnitude, frequency, orientation and phase of the periodic gust to the resulting unsteady blade lift and moment and hence, unsteady stress.

Static aeroelasticity has only a minor exposure in the turbomachinery literature, the chief applications being to blade "untwist" and "lean" which stem from the combined effects of steady aerodynamic and centrifugal loadings.

The early attempts to predict and thus avoid stall flutter in practice resulted in empirical "design rules" which limited the reduced velocity to certain critical values which should not be exceeded. A rather fanciful depiction circa 1954 of an empirical flutter boundary appears in the artistic rendering of Figure 2; the disadvantages and advantages related to crossing or non-crossing of the boundary are graphically portrayed. Stall flutter explanation rested upon the previous single wing experience documented by Studer [6], Mary Victory [7] and others. Although the design rules ignored the importance of aerodynamic coupling noted previously, they were moderately successful, primarily due to the fact that the rules were empirical to begin with and also, that at the inception of flutter, stresses are low and all blades vibrate at their individual "as manufactured" frequencies with disparate amplitudes. Before entrainment of frequencies occurs at higher flutter stress levels a unique interblade phase angle cannot be defined and the aerodynamic coupling, when averaged over time, is not a strong influence. The stress level for entrainment is subject to the degree of mistuning present.

Beginning in the early 1950s the research on blade flutter was advanced by Billington [8], Lilley [9], Pearson [10], Carter [11], and others in England, and Mendelson [12], Sisto [13], Carta [14] and others in the United States. This work had several thrusts, not the least useful of which was the extension of the earlier design rules to include correlations for the effect of blade geometry such as aspect ratio, thickness and aerothermodynamic quantities such as stage pressure ratio, number of stages and most particularly incidence. Much of this work was proprietary.

An important thrust in the 1950s was hardware development with the objective of avoiding stall flutter. One such item was the part-span shroud, or snubber, introduced to control the vibrations of higher Mach number compressors and fan components of the then nascent turbofan engines. Some controversy exists as to the reason for the effectiveness of these devices. There is an unquestioned stiffening effect on the structure thus providing an increase in reduced frequency and modification of the vibration mode, both aeroelastically important. During vibration it is also true that the interfacial surfaces between butting shroud segments may introduce mechanical damping. Both benefits accrue and one effect or the other may be optimized by the particular philosophy employed in the shroud design. The turbine component has made use of tip shrouds as a viable option from the earliest gas turbine engine development, probably as a carry-over from prior steam turbine practice.

In this same time period variable guide vanes were introduced in compressor components. One objective was to control the incidence and thereby inhibit stalling and the consequent stalling flutter. The research on the aeroelastic implications of these devices was largely experimental and improved stall flutter avoidance was one of the major practical results.

Another major thrust was the analytical formulation of the unsteady incompressible aerodynamics of two dimensional cascades restricted to conditions of small thickness, camber and incidence. It proved to be possible to include the important interblade phase angle as a parameter in the analysis. Although flutter was not being encountered in practice under those restrictive conditions (low incidence and subsonic flow) it was possible to study the effect of interblade phase angle analytically and observe the strong, even dominant effect of this parameter on aerodynamic work and the discrimination of aeroelastic stability. The central role of sigma ($\sigma$) in cascade aeroelasticity, firmly established at that time, continues to the present day. Thus, for the first time, the interblade phase angle was conceived of as a variable, or a parameter, with a strong influence on the aerodynamic reactions. Before this development at MIT, $\sigma$ had been considered to be roughly

180°, more or less specified in advance of any predictive calculation and denoted widely as "antiphase" behavior.

The entire question of the influence of interblade phase angle was put on a firm analytical basis with the publication a "phase theorem" by Lane [15] at New York University. The following proposition was proved, subject to the assumptions of linearity and identical blades equally spaced about a common rotor: i) permissible values of the interblade angle are $\sigma = 2\pi n/N$, where N is the number of blades in the annular row and n is an integer, $0 \leq n < N$; and ii) the flutter inception point may be determined by minimizing the flutter speed of a simple equivalent blade with respect to $\sigma$, a discrete variable. These formulations put the flutter prediction of turbomachinery stages on a rational foundation.

At the end of the second decade of turbomachine aeroelasticity the analytical/theoretical body of knowledge and the practical/experimental developments were quite unrelated. It was not possible to use the incompressible small incidence theory to predict flutter and elimination of stall flutter relied almost exclusively on empiricism. The transonic compressor made its debut in this time period and, as supersonic tip relative Mach numbers appeared, so also did supersonic flutter. With the introduction of compressibility in the analysis [14], [16] theory and practice began to intersect. Subsequently the treatment of subsonic, supersonic and finally transonic (M = 1) flows have characterized the more practically useful analyses that have appeared.

## ROTATING STALL AND STALL FLUTTER

In continuing to survey problems of aeroelasticity as applied to axial turbomachines, the phenomenon known as rotating stall, or propagating stall, must be introduced. Rotating stall was observed experimentally by the Whittle Jet Engine Group in 1938 and proceeded to be involved in an interesting and important controversy.

A row of blades does not stall uniformly as the incidence is increased. Instead, zones of low velocity or even reversed flow appear evenly spaced about the annulus; the patches rotate in the annulus in the same direction as the rotor but at lower speed. It is clear that the loading on rotor blades and stator vanes will change with a frequency that depends on the relative angular velocity of stall propagation and the number of patches. This frequency is in general non-synchronous, i.e., it does not coincide with an engine order line on the Campbell diagram.

This diagram, a typical example of which appears as Figure 3, is a classical device for diagnosing rotor blade vibrations. With the abscissa of rotational speed and ordinate of vibration frequency, blade natural frequencies may be plotted as a function of rotor speed. Straight lines radiating from the origin correspond to engine order lines. Forced vibration will usually occur where the natural frequency line of the resonating rotor blade crosses an <u>integral</u> engine order line. High vibratory stress at the natural frequency and lying between integral order lines may be due to rotating stall or else a self-excited instability such as flutter.

Propagating stall, identified in this manner, was researched initially by groups at NACA [17], CalTech [18] and Harvard/MIT [19]. A strong controversy arose concerning the relative significance for blade vibration of propagating stall vis a vis stalling flutter. At one time there was a serious question whether separate mechanisms were involved and whether, in fact, there was such a distinct phenomenon as stall flutter. Tentative agreement now distinguishes between the two types of stall-provoked vibrations by noting that propagating stall may occur without any significant blade vibration, or may result in forced rotor blade vibration much as from any other non-axisymmetric flow. The vibration of the annularly cascaded blades usually exerts no influence on the propagating speed of the stall zones, and hence on the forcing frequency. Propagating stall is an oscillatory instability of the fluid. Stalling flutter, on the other hand, is a true self-excited vibration of the structure, as is conventionally implied by the term "flutter", and may occur with steady axisymmetric mean flow.

This resolution of the controversy is quite provisional and not without its paradoxes. Since the nonstationary stalling process is nonlinear there is a small interval of frequencies bracketing blade resonance within which entrainment of frequeny takes place; the stall frequency departs from its continuous dependence on rotor speed and "locks in" to the blade natural frequency. Furthermore, a stall flutter condition in a blade row, with time invariant interblade phase angles, will generate a periodic pattern of flow separations that may be viewed as a travelling wave propagating peripherally relative to the blade row.

Recently so-called vortex methods are beginning to be used to model two-dimensional unsteady cascade flows with the effect of viscosity confined to the boundaries and to the diffusion of discrete vortices. As yet these methods have not been applied to the flutter situation where the boundaries are executing small harmonic displacements.

Lewis & Porthouse [20] describe their vortex method as one of calculating the potential flow past the cascaded profiles under the influence of the mean flow and the previously shed and discretized vorticity distributed throughout the domain. This is followed by calculation of the drift velocity of each free vortex element in the field, and hence its displacement over a time increment $\Delta t$. New elements of free vorticity are shed (over the same time increment) from the separation points. Viscous diffusion calculations result in a random redistribution of all shed vortices and the entire sequence of computations is repeated for each subsequent time step.

Spalart [21] using a similar vortex method, couples it to an integral boundary layer solver to determine the instantaneous separation points, and with several other innovations is able to attain a condition of propagation of the stall zones in an infinite two-dimensional linear cascade with periodicity enforced over each subset of either 3, 4 or 5 blades. The solution for the periodic flow displays moving separation points and a definite interblade phase angle. Thus the velocity of propagation is predicted. Steps have to be taken to limit the number of discrete vortex "blobs" that can be tracked on the finite computational domain. Other details and cautions may be found in the references. However the method is viable for predicting periodically stalled flow about cascades subject, naturally, to some limitations. A time sequence of instantaneous streamlines in propagating stall is reproduced using Spalart's method in Figure 4. The stall cell definition and propagation are dramatically portrayed.

Propagation rates as well as the critical incidence at which the instability appears may be computed for an imposed wavelength. Time dependent components of the aerodynamic drag, moment and normal force can be calculated. The lift and drag are shown in the same Figure. By considering multiple wavelengths in successive calculations a stability criterion could be developed relating both propagation rate (also in Figure 4) and wavelength of the stall zones (and hence frequency of encounter) to cascade geometry and incidence.

The most important limitation from the vantage point of the aeroelastician is the complete fixity of the airfoils comprising the boundary. Attention is now being given to allowing for small harmonic motion of the airfoils in plunge, surge and pitch, and how such a computational algorithm could be incorporated into a stall flutter prediction system.

## SYSTEM MODES AND MISTUNING

The fan component of turbofan engines and the front stages of transonic compressors have rotors of low hub/tip ratio. The rigidity of these disk structures is much lower than the drums of high radius ratio stages. Decrease of the reduced velocity is found to be aeroelastically necessary and one or two rings of part-span shrouds are routinely applied to these front end rotor rows. The resulting structural system is one in which the vibration modes of the entire blade-disk-shroud system must be considered.

Although steam turbine designers have analyzed this type of structure, in which the modes are identified by the number of diametral and circumferential nodal lines, Carta [22] first reported the significant application to aeroengine components. It was shown that if the intrablade phase angle between pitching and plunging is $-\pi/2$ (i.e., the torsional motion lags the bending motion by $90°$), then instability is observed to occur typically with a small number, say 4, of diametral nodes. The particular choice of intrablade phase angle corresponds to displacement waves relative to rotor coordinates travelling in the direction of rotor rotation, i.e., "forward" travelling waves. This initial study of system mode instabilities has led subsequently to a great proliferation of important investigations. In particular the current and very significant aeroelastic developments associated with so-called "mistuned" bladed disk assemblies rests upon this prior work on system modes.

Recently the mistuning concept of Whitehead [23] has been studied and exploited most intensively. The small geometric and structural variations from blade to blade naturally give rise to an aeroelastic system in which these multiple elements are not quite identical nor periodically disposed, and to which Lane's Theorem in its primitive form cannot be applied. The nonuniformities of spacing and setting angle in the blade flow annulus imply that mistuning also may be aerodynamic in nature. The general conclusion seems to be that controlled mistuning is generally a favorable effect in that it raises the flutter speed [23], [24], [25], [26]. Furthermore, a mistuning strategy [27] has been developed to optimize the distribution and degree of mistuning of a basic set of blades under certain simplifying assumptions. The analyses have usually employed the travelling wave modal description although the standing wave approach common to propeller and helicopter rotor work has been introduced to turbomachinery by Dugundji [28]. Figure 5 shows the complex eigenvalues at flutter for a specific case of an unshrouded supersonic fan [27] and compares the tuned and optimally mistuned contours. The parameter is interblade phase angle. The suppression of points in the right half plane by mistuning represents the avoidance of flutter. A simpler alternate mistuning strategy would result in two separated contours, both of which would have to lie entirely in the left half plane to avoid flutter.

The SST was a strong stimulus to the modern field of aeroacoustics. The acoustic approximation can be effectively employed in the small perturbation analysis of unsteady compressible flow through cascades. Virtually the same formulation that describes acoustic radiation, also predicts the unsteady flow field properties of cascades at small incidence [29]. Thus the development of unsteady compressible flow in the 1960s provided a synergism with the growing field of aeroacoustics. The phenomenon of "cutoff" to describe frequencies above which disturbances do not propagate upstream at a particular subsonic axial Mach number is an acoustical concept. So also is the phenomenon of aerodynamic resonance, where disturbances from one vibrating blade arrive in phase with the similar vibration of the neighboring blade. Both types of behavior have important consequences in the formulation of unsteady pressure distribution and hence unsteady lift and moment.

The entire lift engine development in the 1960s led to many interesting composite blade materials and types of construction. Aeroelastic benefits in the area of flutter prevention were achieved but were insufficient to overcome the problem of foreign object damage (FOD) and fatigue behavior in forced vibration. This forcing was due to the severe flow distortion presented to the wing-mounted lift engine, and also in some measure, to the increased use of buried engine installations requiring bifurcated and/or tortuous inlet ducting, supersonic inlets operating off-design and a trend to lower cantilever blade frequencies.

## COMPRESSIBILITY EFFECTS

From about 1970 onward, several supersonic flutter regimes have been encountered in practice; see Regions III, IV and V in Figure 1. Only Region III flutter, in either pitching or plunging, will usually be encountered along a normal operating line, and then only at corrected overspeed conditions. Supersonic aerodynamic theories were developed that were adequate to explain and confirm Region III flutter. Low incidence formulations were reported by a number of investigators [30], [31], [32], [33], [34] and others, with greatest interest being attached to onset flows having a subsonic axial component. The survey papers by Platzer [35], [36], [37] provide a recent historical perspective of this modern aspect of turbomachine aeroelasticity.

Regions IV and V in Figure 1 are at higher compressor pressure ratio, above the normal equilibrium operating line, and in Region V, may involve stalling at supersonic blade relative Mach number. Unsteady aerodynamic analyses appropriate to this regime have been presented in [38] and [39]. For the first time account was taken of the effect of shock waves which may appear when the surface Mach number exceeds unity. Flutter observed in these regions has been mostly flexural although not exclusively. In Region V stalling of the flow has been implicated since the region is in the neighborhood of the surge or stall limit line. Hence Region V is provisionally termed "supersonic bending stall flutter" and it is assumed that there is a detached bow shock at each blade passage entrance; i.e. the passage is unstarted. By contrast the flutter mechanism in Region IV is thought to involve an in-passage shock wave whose oscillatory movement is essential for the instability mechanism.

A formulation including finite shock strength and movement is presently the subject of intense investigation. The attempt is to show that the oscillatory movement of the passage shock, or, alternatively, the periodic starting and unstarting of a bow shock configuration due to changing blade passage area results in a fluctuating pressure distribution. The integrated pressures can feed aerodynamic work into the initially assumed vibration of the blades forming the passage and thus account for flutter. Tanida and Saito [40], considering a single thick airfoil oscillating in a windtunnel of small height, were able to slow both analytically and experimentally that flutter is a distinct possibility, critically involving shocks which fluctuate in position and strength. The analysis is based on one-dimensional flow with oscillation limited to torsion.

A number of additional contributions have appeared since then culminating most recently in a paper by Bendiksen [41] in which the "shock-doublet" method of Ashley [42] is used to simulate the effects of the passage shock. Figure 6 taken from [41] demonstrates the combination of parameters that can account for flutter associated with finite shock strength and motion. A linearization of this problem, in the sense of suppressing shock motion and strength variations would have eliminated the possibility of flutter in this instance.

A counterclockwise continuation around Figure 1 returns one to Region I which now appears should be divided into more than one subregion. The so-called system mode instability seems to be associated with the upper end of this region and, although the blade loading is high, flutter may not involve flow separation as an essential part of the mechanism. Instead it has been hypothesized [43] that even with a subsonic onset flow the surface Mach number can exceed unity locally and oscillating shocks may help explain the appearance of negative aerodynamic damping. It seems that these instability mechanisms (separation, oscillating shocks) may both appear in this general region of the fan of compressor map, although not both at the same time in a particular machine. Thus the non-aerodynamic factors, which are not revealed by the map parameters may determine which if any of these flutter types will manifest itself in any particular instance. The clarification of this matter is still required so that Region I is now provisionally labelled Subsonic/Transonic Stall Flutter and System Mode Instability.

Region II, of relatively lesser importance, is associated with choking of the passage and is labelled Choke Flutter. As such the role of oscillatory shock waves is again indicated to be important. Hence for relatively low negative incidence and high enough relative subsonic Mach numbers, appropriate to a middle stage of a multistage compressor, the mechanism of choke flutter has many similarities to the transonic stall flutter of Region I. The choke flutter mechanism is still controversial. It may involve the type of machine (fan, compressor or turbine), type of stage (front, middle or rear) and structural details (shrouded vs. unshrouded, disk vs. drum, etc.).

CONCLUDING COMMENTS

Three dimensional unsteady cascade flow was first formulated by Namba [44], and Salaun [45] and this important area continues to receive significant attention. In order to apply two-dimensional theory to the aeroelastic problems of real blade systems one must either use a representative section analysis or else apply the strip hypothesis; the aerodynamics at one radius is uncoupled from the aerodynamics at any other radius. In particular it is known that at "aerodynamic resonance" the strip theory breaks down and the acoustic modes are strongly coupled radially.

Along with aerodynamic advances the structural description of the bladed-disk assembly [46], [47] has received a great impetus, and the important of forward and backward travelling waves has been firmly established. Within a particular number of modal diameters, coupling between modes has been shown to be significant [48] and the role of the "twin modes" (i.e., $\sin n\phi$ and $\cos n\phi$) in determining propagation has been clarified. Ford & Foord [49] have used the twin mode concept in both analysis and flutter measurement. Furthermore, the number of nodal diameters affects the associated natural frequencies slightly so that they cluster together. Coupling of modes with closely spaced frequencies by aerodynamic means therefore becomes appreciable and the resulting flutter mode may contain significant content from two or three modes with consecutive numbers of diametral nodes.

Engine, rotating rig and cascade experiments have continued to provide important information for direct use in design and also for guiding analytical work. Especially to be noted is the pioneering work of Stargardter [43], Kurkov [50] and in the use of sophisticated optical and electrical methods for gathering rig data from fluttering rotor blades. Important advances in the structural dynamics of axial-turbomachine blading have occurred over the past 40 years. Stationary laboraory experiments have usually simplified the structural features of the nonrotating blade models so that these apparatuses tend to emphasize the aerodynamic information related to aeroelasticity rather than structure. However, the modal description (eigenfrequencies and eigenfunctions) of tapered twisted blades of thin but arbitrary cross-sections in centrifugal and gyroscopic fields [51] has kept pace with the need for increasingly accurate information of this nature. The finite element method has been of great utility in most of this structural description of blades. vanes and disks, particularly when numerical methods must be resorted to early in the analysis.

Studies of material damping and slip damping have led to increased understanding of these effects which are usually beneficial in high vibratory environments. This new knowledge has resulted in the application to turbines of mechanical dampers and special high damping materials such as chromium-based stainless steels, and to compressor/fan blades built up of laminates and composites. Although of importance when operating in regions of potential flutter, the influence of damping is most highly critical in the presence of forced vibration. In the latter case the accumulation of fatigue damage due to lightly damped resonant or near-resonant operation can occur in a very short interval of time.

Other developments not introduced in this survey and subjects receiving attention very recently that have not been treated fully, if at all, include such topics as thick and highly cambered blades in a compressible flow and the effects of curvilinear wakes and voriticity transport. These and other large amplitude and therefore nonlinear perturbations which prevent the linear superposition implicit in classical modal analysis have certain implications relative to the traditional solutions of the aeroelastic eigenvalue problem. Although a linearized treatment of three dimensional unsteady flow is available some of these expected refinements, presumably first to be developed for two dimensional flow, need subsequently to be applied for the full three-dimensional annular geometry.

## REFERENCES

1. Bisplinghoff, R.L., Ashley, H. & Halfman, R.L. 1955 <u>Aeroelasticity,</u> Addison-Wesley Pub. Co., Reading,MA.
2. Bisplinghoff, R.L. & Ashley, H., 1962 <u>Principles of Aeroelasticity,</u> John Wiley & Sons, New York.
3. Dowell, E.H., Curtiss, H.C., Jr., Scanlan, R.H., & Sisto, F., 1978 <u>A Modern Course in Aeroelasticity.</u> Sitjthoff & Noordhoff, The Netherlands.
4. Shannon, J.F. 1945 Vibration Prob. in Gas Turbine, Centrifugal and Axial Flow Comp., Brit. A.R.C., R&M 2226.
5. Campbell, W. 1924 Protection of Steam Turbine Disc Wheels from Axial Vibration, ASME Paper No. 1920.
6. Studer, H.L. 1936 Experimentelle untersuchungen uber flugelschwingungen. Mitteilungen aus dem Institut fur Aerodynamik, Eidgenossiche Technische Hochschule , 4/5, Zurich.
7. Victory, M. 1943 Flutter at High Incidence. Brit. ARC R&M 2048.
8. Billington, A.E. 1949 Aerodynamic Lift and Moment for Oscillating Airfoils in Cascade. Rept. E.63, Aero. Council of Scientific & Industrial Research, Australia.
9. Lilley, G.M. 1952 An Investigation of the Flexure-Torsion Flutter Characteristics of Airfoils in Cascade. Rept. No. 60, The College of Aeronautics, Cranfield.
10. Pearson, H. 1953 The Aerodynamics of Compressor Blade Vibration, Fourth Anglo-Amer. Aeronautical Conf., London, pp. 127-162.
11. Carter, A.D.S. 1955 A Theoretical Investigation of the Factors Affecting Stalling flutter of Compressor Blades. N.G.T.E., A.R.C. Tech. Rept. No. R. 172, C.P. No. 265.
12. Mendelson, A. 1949 Aerodynamic Hysteresis as a Factor in Critical Flutter Speed of Compressor Blades at Stalling Conditions. Journal of the Aeronautical Sciences, 16, 11, pp. 646-652.
13. Sisto, F. 1952 Flutter of Airfoils in Cascade, Doctoral Dissertation, Mass. Inst. of Tech.
14. Carta, F.O. 1957 Unsteady Aerodynamic Theory of a Staggered Cascade of Oscillating Airfoils in Compressible Flow. United Aircrat Corp. Res. Dept. Rept. R-0582-19, East Hartford.
15. Lane, F. 1956 System Mode Shapes in the Flutter of Compressor Blade Rows. Jour. of the Aeronautical Sci., 23, 1, pp. 54-66.
16. Lane, F., and Friedman, M. 1958 Theoretical Investigation of Subsonic Oscillatory Blade-Row Aerodynamics. NACA TN 436.
17. Huppert, M.C. & Benser, W.A. 1953 Some Stall and Surge Phenomena in Axial-Flow Compressors. J. Aero. Sci., 20, 12, 835-145.
18. Iura, T. & Rannie, W.D. 1953 Observations of Propagtating Stall in Axial-Flow Compressors. Calif. Inst. of Tech Mech. Eng. Dept. Lab Rept. No. 4, Pasadena, CA.
19. Emmons, H. et al 1954 Compressor Surge and Stall Propagation. ASME Paper No. 53-A-65.
20. Lewis, R.I. and Porthouse, D.T.C. 1982 A Generalized Numerical Method for Bluff Body and Stalling Aerofoil Flow. ASME paper 82-GT-70. London.
21. Spalart, P.R. 1984 Two Recent Extensions of the Vortex Method. AIAA Paper No. AIAA-84-0343. Reno.
22. Carta, F.O. 1967 Coupled Blade-Disk-Shroud Flutter Instabilities in Turbojet Engine Rotors. Trans. ASME Journal of Engineering for Power, 89, 3, 419-426.
23. Whitehead, D.S. 1966 Effect of Mistuning on the Vibration of Turbomachine Blades Induced by Wakes. J. of Mech. Eng. Science, 8, 1, 15-21.

24. Srinavasan, A.V. 1980 Influence of Mistuning on Blade Torsional Flutter. NASA CR-165137, August.
25. Kaza, K.R.V. & Kielb, R.E. 1982 Flutter and Response of a Mistuned Cascade in Incompressible Flow. AIAA Journal 20, 8, 1120-1127.
26. Bendiksen, O.O. 1983 Flutter of Mistuned Turbomachinery Rotors, ASME Paper 83-GT-153, Phoenix, AZ.
27. Crawley, E.F. & Hall, K.C. 1984 Optimization and Mechanisms of Mistuning in Cascades, ASME Paper 84-GT-196.
28. Dugundji, J. & Bundas, D.J. 1983 Flutter & Forced Response of Mistuned Rotors Using Standing Wave Analysis. AIAA Paper 83-084.
29. Goldstein, M.E. 1974 <u>Aeroacoustics</u> NASA Special Pub. SP-346.
30. Verdon, J.M. 1973 The Unsteady Aerodynamics of a Finite Supersonic Cascade with Subsonic Axial Flow, J. of Applied Mechanics, Vol. 95, Trans. of the ASME, Series E, Vol. 40, No. 3, pp. 667-671.
31. Brix, C.W. & Platzer, M.F. 1974 Theoretical Investigation of Supersonic Flow Past Oscillating Cascades with Subsonic Leading Edge Locus. AIAA Paper 74-14, Washington, DC.
32. Kurosaka, M. 1974 On the Unsteady Supersonic Cascade with Subsonic Leading Edge--An Exact First Order Theory; Parts I and II. Proc. ASME J. Engrg for Power, 96, 1, 13-31.
33. Nagashima, T. & Whitehead, D.S. 1976 Linearized Supersonic Unsteady Flows in Cascades. Aeronautical Research Council, R&M 3811.
34. Adamczyk, J.J. & Goldstein, M.E. 1978 Unsteady Flow in a Supersonic Cascade with Subsonic Leading-Edge Locus. AIAA Journal 16, 12, 1248-1254,.0.
35. Platzer, M.F. 1975 Transonic Blade Flutter-A Survey. Shock & Vib. Digest, 7,7, 97-106.
36. Platzer, M.F. 1977 Unsteady Flows in Turbomachines-A Review of Current Developments. In AGARD-CP-227 Unsteady Aerodynamics, Ottawa.
37. Platzer, M.F. 1978 Transonic Blade Flutter-A Survey of New Developments, Shock & Vib. Digest, 10,9, 11-20.
38. Adamczyk, J.J. 1978 Analysis of Supersonic Stall Bending Flutter in Axial-Flow Compressors by Actuator Disc Theory. NASA Tech. paper 1345.
39. Adamczyk, J.J., Stevans, W. & Jutras, R. 1982 Supersonic Stall Flutter of High-Speed Fans. Trans. ASME J. of Eng. for Power, 104, 3, 675-682.
40. Tanida, Y. & Saito, Y. 1976 A Study on Choking Flutter. IUTAM Symposium on Aeroelasticity in Turbomachines, Paris.
41. Bendiksen, O.O. 1985 Role of Shocks in Transonic/Supersonic Compressor Rotor Flutter. Proc. 7th Int'l Symposium on Air Breathing Engines. Beijing, pp. 691-701.
42. Ashley, H. 1980 Role of Shocks in Sub-Transonic Flutter Phenomena J. of Aircraft, v. 17, n. 3, p.187-197.
43. Stargardter, H. 1979 Subsonic/Transonic Stall Flutter Study Final Report, NASA CR-165256, PWA 5517-31.
44. Namba, M. 1972 Lifting Surface Theory for a Rotating Subsonic or Transonic Blade Row. Aero. Res. Council, R&M 3740.
45. Salaun, P. 1974 Pressions aerodynamiques instationairs sur une grille annulair en ecoulement subsonique. Publication ONERA N$^o$158.
46. Ewins, D.J. 1973 Vib. Characteristics of Bladed Disc Assemblies, J. of Mech. Eng. Sci, 15,3,p. 165-186.
47. Srinavasan, A.V., ed., 1976 Structural Dynamic Aspects of Bladed Disk Assemblies, Proc. ASME Winter Annual Meeting, New York.
48. Chi, R.M. & Srinavasan, A.V. 1984 Some Recent Advances in the Understanding and Prediction of Turbomachine Subsonic Stall Flutter. ASME Paper 84-GT-151.

49. Ford, R.A.J. & Foord, C.A. 1979 An Analy. of Aeroengine Fan Flutter Using Twin Orthogonal Vib. Modes, ASME paper 79-GT-126.
50. Kurkov, A.P. 1983 Measurement of Aerodynamic Work During Fan Flutter, Trans. ASME J. Eng. for Power, 105, 1, 204-211.
51. Sisto, F. & Chang, A.T. 1984 A Finite Element for Vibration Analysis of Twisted Blades Based on Beam Theory. AIAA Journal, v. 22, n. 11, 1646-1651.

FIGURES

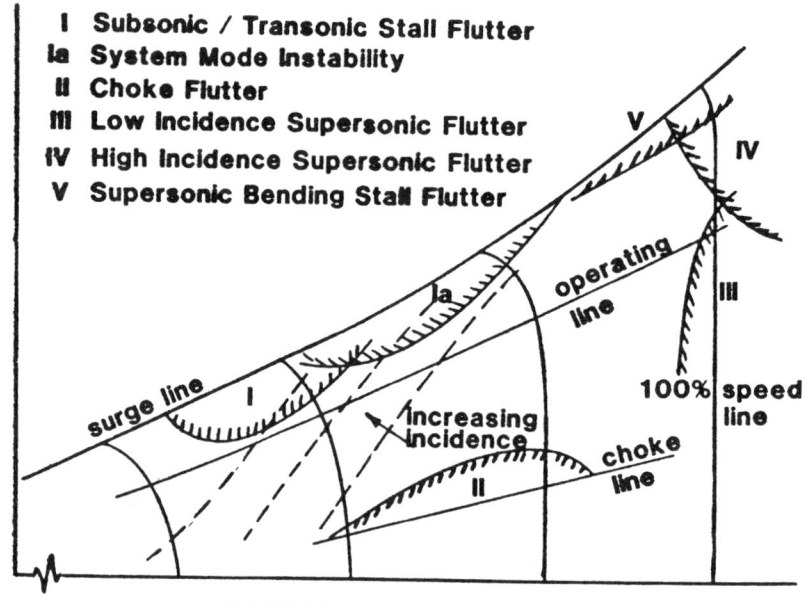

Fig. 1  Compressor Map Showing Principal Flutter Regions by Type

Fig. 2  Allegorical Portrayal (ca 1954) of Perils Near a Stall Flutter Boundary

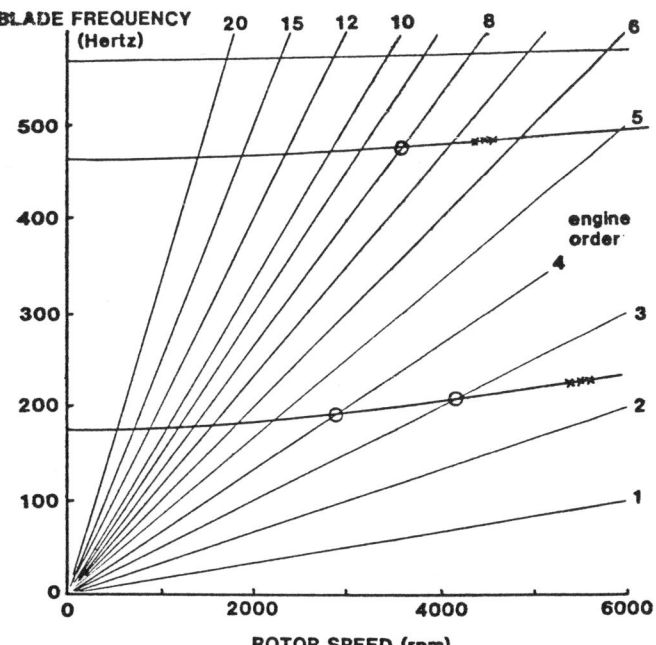

Fig. 3 Campbell Diagram for Rotor Blades (xxxstresses imply either flutter of propagating stall, ooo stresses indicated forced resonance).

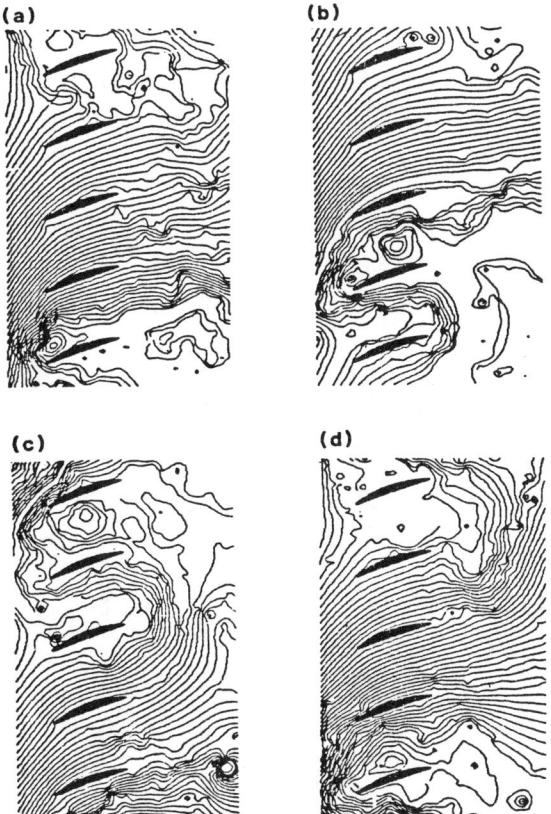

Fig. 4a Streamlines in a Linear Cascade of Cambered Airfoils Undergoing Stall Propagation: (a)t = 40, (b)t = 44, (c)t = 50, (d)t = 54

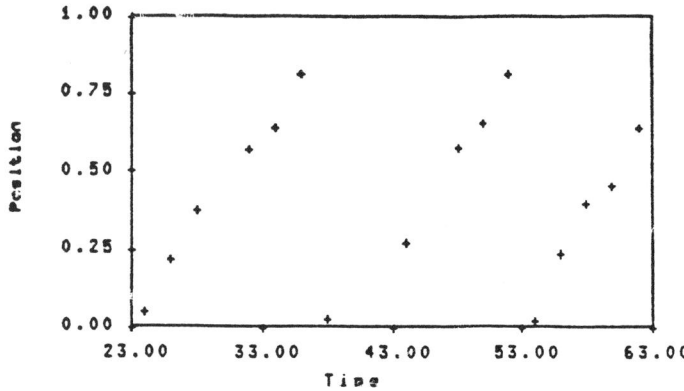

Fig. 4b  Stall Cell Position as a Function of Time for a Cambered Airfoil Cascade

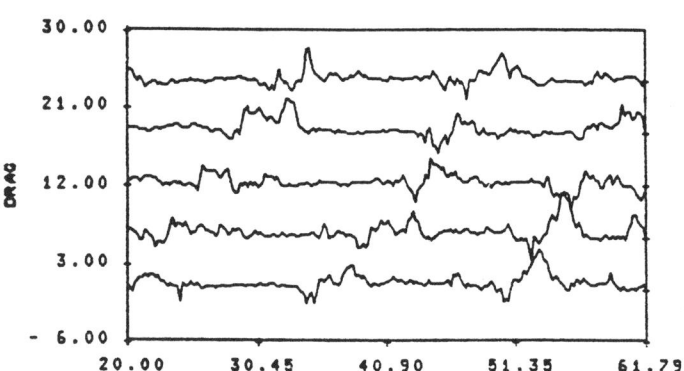

Fig. 4c  Drag as a Function of Time for the Same Cascade

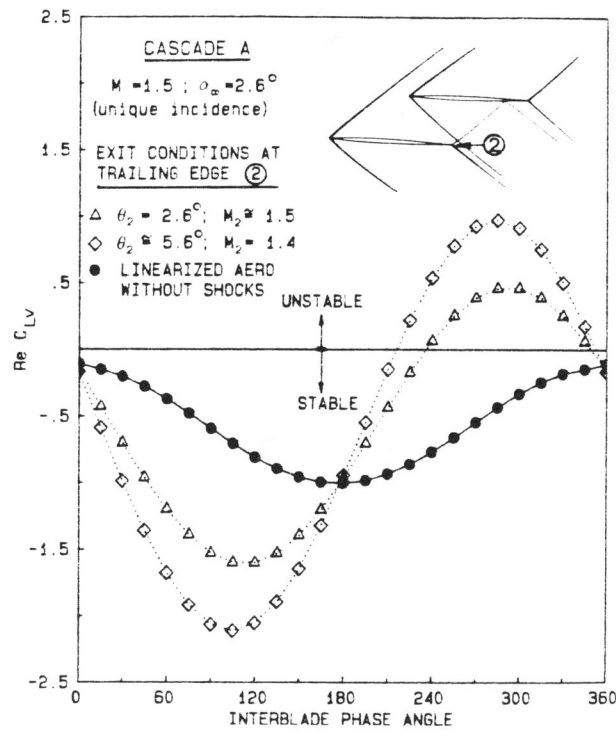

Fig. 6  Typical Effect of Shock Motion on Real Part of $C_{L_V}$, Representative of Damping in Bending

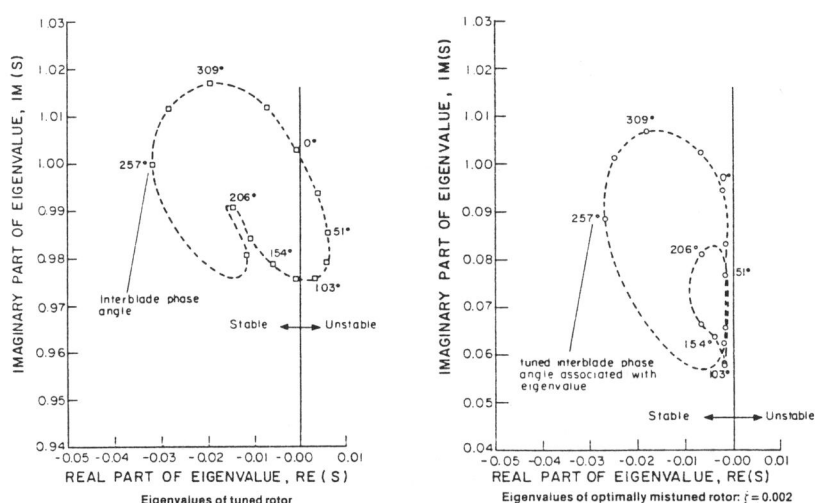

Fig. 5  Eigenvalue Contours for Tuned and Optimally Mistuned Fan

EXPERIMENTAL CLASSICAL FLUTTER RESULTS OF A

COMPOSITE ADVANCED TURBOPROP MODEL


O. Mehmed and K.R.V. Kaza
National Aeronautics and Space Administration
Lewis Research Center
Cleveland, Ohio 44135


ABSTRACT

Experimental results are presented that show the effects of blade pitch angle and number of blades on classical flutter of a composite advanced turboprop (propfan) model. An increase in the number of blades on the rotor or the blade pitch angle is destablizing which shows an aerodynamic coupling or cascade effect between blades. The flutter came in suddenly and all blades vibrated at the same frequency but at different amplitudes and with a common predominant phase angle between consecutive blades. This further indicates aerodynamic coupling between blades. The flutter frequency was between the first two blade normal modes, signifying an aerodynamic coupling between the normal modes. Flutter was observed at all blade pitch angles from small to large angles-of-attack of the blades. A strong blade response occurred, for four blades at the two-per-revolution (2P) frequency, when the rotor speed was near the crossing of the flutter mode frequency and the 2P order line. This is because the damping is low near the flutter condition and the interblade phase angle of the flutter mode and the 2P response are the same.


INTRODUCTION

The unconventional features of the propfan add complexity to its aeroelastic analysis. The blades are thin and flexible, thus deflections due to centrifugal and aerodynamic loads are large. Hence, analyses require geometric nonlinear theory of elasticity. Also, the blades are of low aspect ratio and large sweep, and operate in subsonic, transonic, and possibly supersonic flows. Therefore, three-dimensional, unsteady aerodynamic theory is required for accurate analysis. The blades have large sweep and twist, which couples blade bending and torsion motions, and are plate-like structures because of their low aspect ratio. These factors require a finite element structural model for accurate analysis. Then, there are six or more blades on the rotor which means aerodynamic coupling between blades or cascade effects should be considered. These features require the use of both experimental and analytical work to understand propfan aeroelastic phenomena and to develop aeroelastic analysis methods for the design of propfans.

Classical flutter of a propfan occurred unexpectedly during an earlier wind tunnel experiment on a model with ten highly swept titanium blades. For that model, the blade tip relative flow at flutter was transonic, between Mach 0.95 and 1.05, and the flutter frequency was almost equal to the blade first normal mode frequency. Also, based on the three-quarter radius and the semichord, the reduced flutter frequency was between 0.17 and 0.19 and the blade-to-air mass ratio was 115. The measured flutter speed was much lower than predicted by two-dimensional, subsonic, unsteady, isolated airfoil aerodynamic theory with a beam structural model for the blade. The difference between theory and experiment prompted an experimental and analytical effort. To evaluate the effect of aerodynamic coupling on flutter, two-, five-, and ten-blade configurations of the highly swept model were tested. These experiments revealed that aerodynamic coupling significantly lowered the flutter boundaries, and that the flutter was classical with coupling between the normal modes of the blade. The test results provided guidance to refine analytical models. The first phase of refinements included incorporation of two-dimensional, subsonic, unsteady, cascade aerodynamics with a sweep correction, based on similarity laws, and an idealized beam structural model of each blade. Analytical and experimental results from that study were correlated in reference [1]. The second phase of analytical refinements included the use of mode shapes and frequencies from a finite-element plate structural model in a modal flutter analysis with the same aerodynamics as in reference [1]. References [2] and [3] document some of this analytical work and correlate analytical and experimental results. The correlation between theory and experiment in references [1,2 and 3] varied from poor to good. Hence, there was a need to further understand the phenomena and to get data in the subsonic flow regime in order to develop and validate analytical models.

The present study was then planned to obtain flutter data in subsonic relative flows at the blade tip and to further investigate the phenomena of propfan unstalled bending-torsion flutter. A blade design was tailored by the use of composite material. This paper describes the model blades, the experiment and the results.


FLUTTER MODEL BLADES

A flutter model, designated SR3C-X2, made from graphite-ply/epoxy-matrix material was designed with the flutter analysis described in reference [2]. The blade was designed for research purposes only and was not dynamically scaled from a large blade. Under a cooperative effort, NASA Lewis Research Center developed the finite element model and provided the modal data for the flutter analysis, Hamilton Standard performed the flutter analysis and NASA Ames fabricated the blades.

The design was tailored by the orientation of the unidirectional tape material used for blade construction. Figure 1 shows the ply directions. The blades had 80 percent of their plys oriented along the 0° axis shown. The remaining plys were oriented along the ±22.5° directions.

Figure 2 shows a photo of the blade. It has a geometric midchord sweep of 45° at its tip and a nominal tip diameter of 0.62 m (2 ft). The blade-to-air mass ratio is 33.

Figure 3 shows the variation with rotational speed of natural frequencies, in vacuum, for the SR3C-X2 blade from an MSC NASTRAN finite element plate model. The measured average bench natural frequency of eight blades is also indicated for comparison. Figure 4 shows hologram photos of the measured bench mode shapes and corresponding natural frequencies of one of the blades. Here, the black fringes represent constant displacement contours and the whitest fringes are nodes or areas of near zero displacement. It can be seen from the displacement contours that the first mode is primarily flexural but also has a large degree of torsion and the second mode is primarily torsional near the tip. It will be shown later that flutter occurred between these two modes.

## FLUTTER EXPERIMENT

The experiment was conducted in the Lewis 8-by-6-ft (2.44-by-1.83 m) wind tunnel at tunnel Mach numbers from 0.36 to 0.75 and rotor speeds up to 8000 rpm. Eight- and four-blade rotors were tested to investigate the degree of aerodynamic coupling between blades. The blades were mounted in a hub which can be considered rigid. Figure 5 shows the rotors in the wind tunnel. The rotor was driven by an air turbine and its axis was aligned with the tunnel flow.

Blade mounted strain gages provided the vibration data. Each blade had at least one gage, since the blade amplitudes at flutter were expected to differ. Figure 6 shows the instrumentation installed for the two rotors. The number of strain gages was limited since only ten signals could be taken from the rotor. The gages were located at the points of maximum strain for the first four normal blade modes, as determined by finite element analysis. Only dynamic strain signals were recorded and monitored during the test. Gage 2 usually had the maximum response at flutter for both rotors but Gage 3 gave a greater response when the flutter mode changed with the eight blade rotor. The flutter modes are discussed later.

The test variables were tunnel Mach number, rotor speed and propeller blade pitch. The blade pitch angle, $\beta_{0.75R}$ - the acute angle that the blade chord makes with the plane of rotation at the 0.75 blade radius, was locked manually. The wind tunnel was then started and a Mach number was set. The propeller was left in the unpowered or windmilling condition during tunnel speed changes. The rotor was then powered and its speed slowly increased until flutter occurred or an operating limit was reached.

Flutter was not always reached because of three operating limits. At some conditions the rpm was limited by the power available from the rotor air drive system. Then, rotor speed was limited to 8000 rpm because of blade strength. Also, because of blade strength there was an intended limit of 700 microstrain (0-peak) on blade dynamic strain it was exceeded at times during flutter. To get out of high stress conditions during flutter the following was done: in the powered rotor conditions, the rotor speed was dropped; in the unpowered rotor condition, the tunnel Mach number was dropped.

EXPERIMENTAL RESULTS

Figure 7 summarizes the operating boundaries of the experiment. The range of test variables are: tunnel Mach number, 0.36 to 0.75; rotor speed, windmill rpm to 8000 rpm; and blade pitch angle, 56.6° to 68.4°. The operating procedure described above traces a vertical path on the figure from the windmilling rpm to the boundary. Two types of blade activity are identified. Points labeled "F" designated blade flutter. Points labeled "2P" designate a blade forced response at a two-per-revolution frequency. The rig power limit existed at points labeled "P" and the blade rpm limit at points labeled "R". Open symbols represent the four-blade rotor and closed symbols the eight-blade rotor.

Comparing the flutter boundaries (points labeled "F") for the two rotors at the same blade angle and Mach number, it is seen that the critical rpm is higher for four than for eight blades at all conditions. This indicates that aerodynamic coupling (the cascade effect) reduces the flutter speed. The flutter occurred at small angles-of-attack of the blades (windmilling points), as well as large angles-of-attack at all blade pitch angles.

The flutter came in with explosive suddenness. The amplitude would grow from very low to very high levels with an increase in rpm of about 1 percent. Figure 8(a) and (b) illustrate the rapid increase in stress amplitude, at two different conditions. Figure 8(a) is for a tip torsion gage and figure 8(b) is for a flexure sensitive gage. In one case, the strain amplitude reached about three times its intended limit, and in the other over four times. Most of the flutter points in figure 7 are shown at the rotational speed where the sudden stress rise occurred. In some cases the authors became cautious and used the unsteadiness of the flutter frequency amplitude peak, from an on-line spectral analyzer, to infer the proximity to the explosive growth point. Nevertheless, it is estimated that all flutter points in figure 7 are within 80 rpm of the explosive condition. The arrows of figure 8 indicate the direction of rpm change.

Time history records at the flutter condition show a limit amplitude was reached at the large displacements experienced. A limit amplitude is possible due to aerodynamic and/or structural nonlinearities which occur at large displacement amplitudes. Nonlinear response was also evidenced by the presence of harmonics of the flutter frequency in the spectra at these conditions. Although high stresses were reached no blades were damaged during the test.

Figure 9 shows the stress amplitudes during flutter at typical conditions. The blade amplitudes varied because the individual blade properties differed, that is, the rotor was mistuned.

Figure 10 shows the flutter data of figure 7 replotted to display the effect of blade pitch angle on the flutter boundaries. The flutter boundaries for eight and four blades are shown on separate plots. Note, a blade tip rotational Mach number ordinate scale is included, based on an airflow static temperature of 529° R. It can be seen that an increase an blade pitch angle decreases the critical rotational speed at a constant tunnel Mach number for all cases, except for four blades at windmill conditions (marked "W"). This decrease is due to a change in aerodynamic coupling between blades, a change in the blade loading and/or a change of the blades normal mode shapes and frequencies under centrifugal loading.

For a rotor of N identical blades at flutter, the phase angle between adjacent blades is the same. Thus, in a rotating frame of reference, a traveling wave can describe the flutter. A forward traveling wave is defined as one traveling in the direction of rotation. The phase angle between blades (interblade phase angle), in degrees, for each of the possible flutter modes is given by

$$\sigma_k = \frac{360 \, k}{N} \; ; \quad k = 0,1,2,\ldots,N-1 \tag{1}$$

where k is the phase angle index.

Figure 11(a) is a plot of the measured phase angle at flutter of each blade relative to blade 1 for the four-blade rotor. The possible interblade phase angle modes, given by equation 1, are represented by the lines thru the origin. The x-axis represents the 0° phase angle. The figure shows that the predominant phase angle is 180°. This corresponds to a 2 nodal diameter pattern (the number of diametral node lines around the rotor). Some data points fall off the 180° mode line, hence, the actual phase angle between blades varied around the rotor. This is because the rotor was mistuned, and more than one interblade phase angle mode participated simultaneously.

Similarly, figure 11(b) is a plot of the measured phase angle of each blade relative to blade 1 at flutter for the eight-blade rotor. Again, the interblade phase angle mode lines are shown. The eight blade data also shows evidence of mistuning. In addition, it displays a change of the predominant flutter mode with blade pitch. At a blade pitch of 61.6° the 225° mode is predominant, which is the same as -135°, and corresponds to a backward traveling wave of 3 nodal diameters. Then, at a blade pitch of 68.4° the 180° mode, 4 nodal diameter pattern, is predominant. Furthermore, at 56.6° blade pitch both the 180° and the 225° modes are evident.

Figures 12(a) and (b) give the measured flutter frequencies for the four-blade and the eight-blade rotors respectively. The harmonic order excitation lines of rotational speed and the analytically predicted first and second natural mode lines are also shown. For both rotor configurations the flutter mode frequency falls between 254 and 284 Hz and is between the first two normal modes indicating aerodynamic coupling of the normal modes. In contrast, the flutter frequency for the rotor of titanium blades described in reference [1] was very close to the first blade normal mode frequency. This indicates a weaker aerodynamic coupling between normal modes for the titanium blades. The difference is caused by the lower blade-to-air mass ratio for the composite blades. Note, the flutter frequencies are nearly the same for four and eight blades, showing only a small effect of cascade aerodynamics on the flutter frequencies.

A significant blade response occurred with the four-blade rotor at the 2P frequency. Referring to figure 12(a), this occurred when the flutter mode was near the crossing with the 2P order excitation line. The points labeled "2P" on figure 7 designate these respondse points. This response occurs because the damping is low near the flutter condition and the interblade phase angle of the flutter mode and the 2P response are the same for four blades. Of course, a source of excitation at the 2P frequency is implied. This excitation source is not understood at this time. No corresponding blade response occurred with the eight-blade rotor. This is because the flutter frequency did not approach as near to the 2P excitation line (see figure 12(b)) and the interblade phase angle of the flutter mode and a 2P response is different for eight blades.

Figure 13(a) shows a spectrum of the strain amplitude near the 2P/flutter mode speed crossing of the four-blade rotor. The 1P (127 Hz) and 2P (254 Hz) frequencies are labeled, as well as frequencies at 264 and 274 Hz respectively ("F"). The strongest response is at the (2P) 254 Hz frequency but there is also a weaker response at 274 Hz which is the flutter mode frequency. In addition, there is a lower amplitude peak labeled 264 Hz. It is possible to have more than one interblade phase angle mode close to instability at the same time. The 264 Hz frequency is inferred to be such a dual flutter mode. Typically, a second frequency peak, of lower amplitude, was observed in the data near the flutter frequency.

Figure 13(b) shows a strain amplitude spectrum of the four-blade rotor, at 5160 rpm, away from the 2P crossing. Here, a large amplitude exists only at the 264 Hz flutter frequency. There was no evidence of a 3P response with either four or eight blades, although the flutter mode crosses the 3P order line. This indicates that a source of 3P excitation was not present.

Figure 14 shows the reduced flutter frequency, $k_f$, plotted against relative Mach number for both rotors. The reduced flutter frequency is based on semichord and the relative Mach number are calculated at the blade 0.75 radius. The data shows a decrease of reduced flutter frequency with an increase of relative Mach number. The only exception is with four blades at the windmill points, labeled "W". The reduced flutter frequency falls between 0.34 and 0.41 for eight blades, and 0.31 and 0.38 for four blades.

Figure 15 shows the blade tip relative Mach number at flutter plotted against rotational speed for both rotors. The data indicate that the flow is in the high subsonic regime. The trend of the data for eight blades at each of the three blade angles shows a rise and then a fall of blade tip relative Mach number as rotor speed is increased. Whereas, the data for four blades is more random and shows no such trend. The blade tip relative Mach number at flutter falls between 0.77 and 0.87 for eight blades and 0.80 and 0.90 for four blades. In general, flutter occurred at lower relative Mach numbers for eight than four blades at the same blade angle.

OBSERVATIONS AND CONCLUSION

An experimental study was performed with a propfan model to obtain flutter data in subsonic relative flows at the blade tip and to investigate the phenomena of propfan unstalled bending-torsion flutter. Composite material was used to tailor the blade structural properties to obtain flutter in subsonic relative flows at the blade tip. Based on the results of this study the following observations and conclusions are made:

1. Classical bending-torsion unstalled flutter was observed. The flutter frequency was between the first two normal modes, indicating an aerodynamic coupling effect between the normal modes, and the flutter occurred from small to large angles-of-attack of the blades.

2. With eight blades flutter occurred predominantely in either the 180° (four nodal diameter) or 225° (three nodal diameter, backward) rotor flutter mode, whereas, with four blades flutter occurred predominantely in the 180° (two nodal diameter) rotor flutter mode. This indicates an aerodynamic coupling (cascade effect) between the blades and that the interblade phase angle mode at flutter is affected by the blade pitch angle.

3. The flutter frequencies were identical on all the blades but the strain amplitudes were not. The strain amplitude variation is attributed to blade frequency mistuning. This also shows an aerodynamic coupling effect between the blades.

4. Increasing the number of blades on the rotor is destablizing. This is inferred to be due to a difference in the aerodynamic coupling between blades.

5. Increasing the blade pitch angle is destablizing. This may be due to a change in aerodynamic coupling between blades, a change in the blade loading and/or a change of the blade normal mode shapes and frequencies under centrifugal loading.

6. The reduced flutter frequency was between 0.34 and 0.41 for eight blades and between 0.31 and 0.38 for four blades.

7. The blade tip relative Mach number at flutter was between 0.77 and 0.87 for eight blades and between 0.80 and 0.90 for four blades.

8. A strong blade response occurred with four blades at the two-per-revolution frequency when the rotational speed was near the crossing of the flutter mode frequency and the 2P order line. This is because the damping is low near the flutter condition and the interblade phase angle of the flutter mode and the 2P response are the same.

REFERENCES

1. Mehmed, O., et al.: Bending-Torsion Flutter of a Highly Swept Advanced Turboprop. NASA TM-82975, 1982.

2. Elchuri, V., and Smith, G.C.C.: Flutter Analysis of Advanced Turbopropellers. 24th Structures, Structural Dynamics and Materials Conference; Part 2, AIAA, 1983, pp. 160-165.

3. Turnberg, J.E.: Classical Flutter Stability of Swept Propellers. AIAA Paper 83-0847-CP; May 1983.

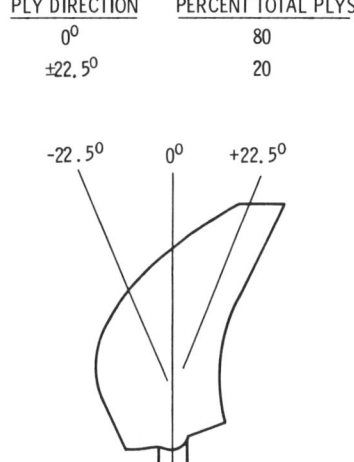

Figure 1. - Ply directions for flutter blade.

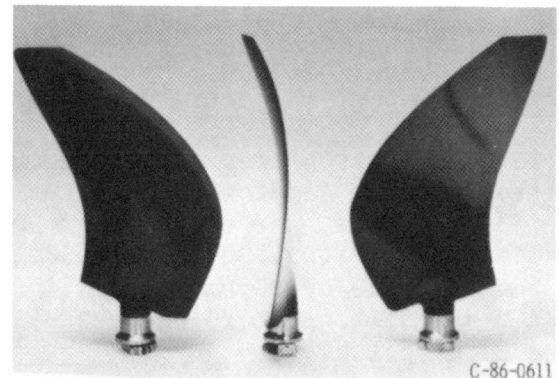

Figure 2. - SR3C-X2 flutter model blade.

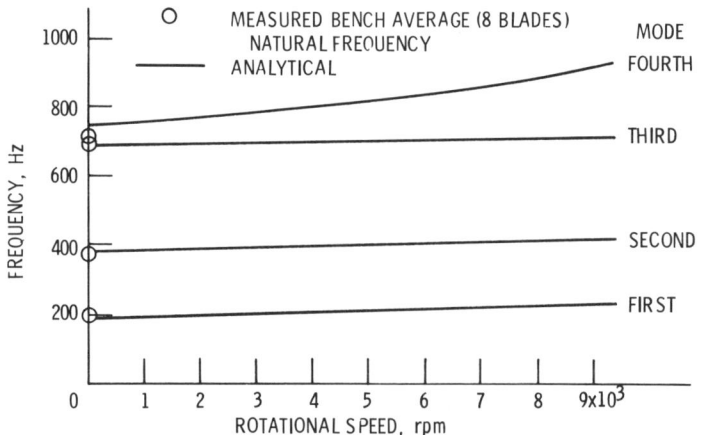

Figure 3. - Variation of natural frequency with rotational speed, $\beta_{0.75R} = 57°$, SR3C-X2 model blades.

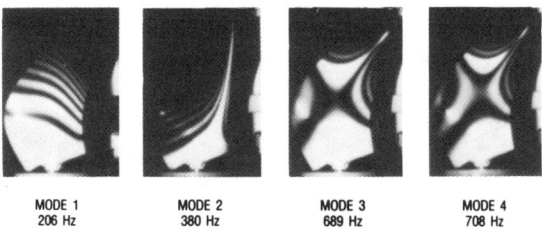

Figure 4. – Bench measured natural frequencies and mode shapes, SR3C-X2 model, blade number 6.

(a) 8-blade rotor.

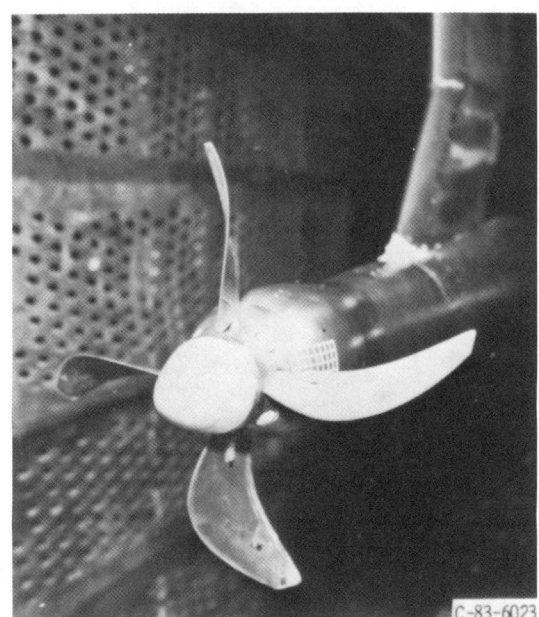

(b) 4-blade rotor.

Figure 5. – SR3C-X2 model installation in the Lewis 8x6 ft wind tunnel.

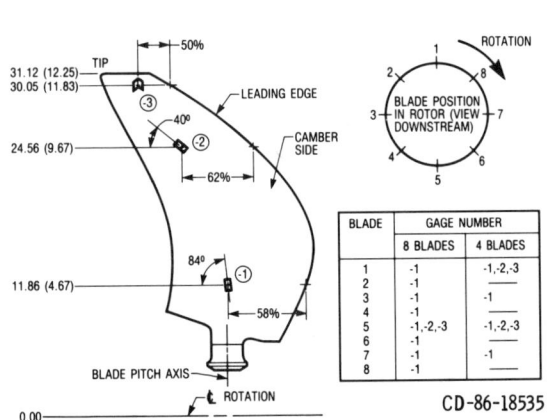

Figure 6. – Blade strain gage installation, SR3C-X2 model.

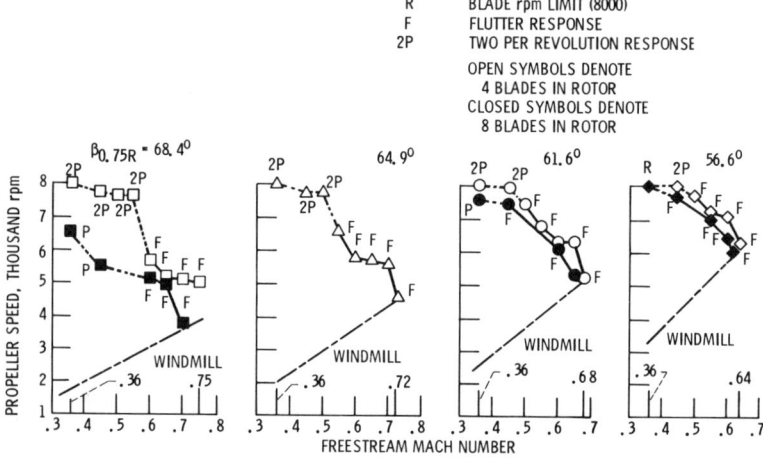

Figure 7. – Operating boundaries of the SR3C-X2 flutter experiment, 8x6-ft wind tunnel.

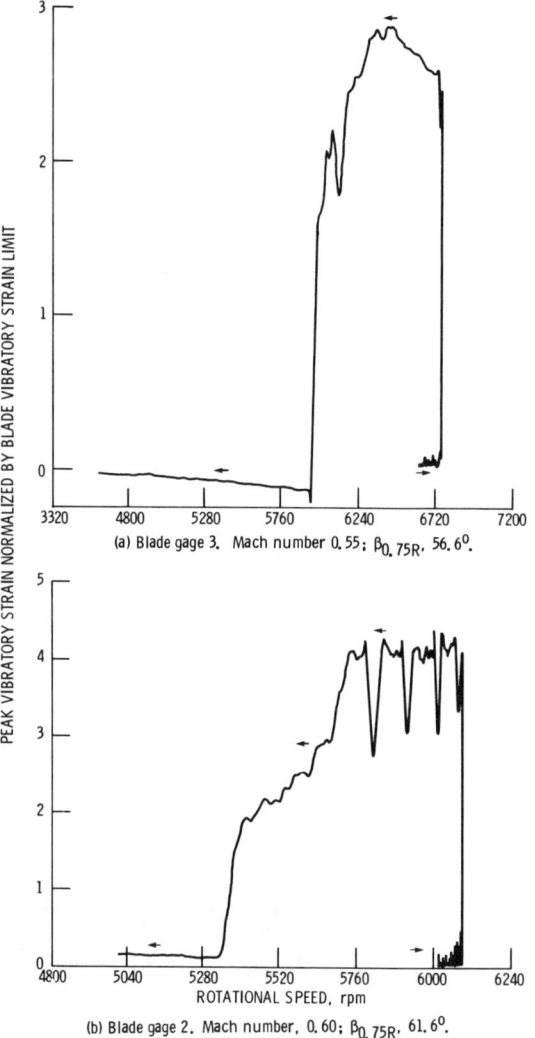

(a) Blade gage 3. Mach number 0.55; $\beta_{0.75R}$, 56.6°.

(b) Blade gage 2. Mach number, 0.60; $\beta_{0.75R}$, 61.6°.

Figure 8. – Peak strain amplitude variation with rpm at flutter, 8-blade rotor, SR3C-X2 model.

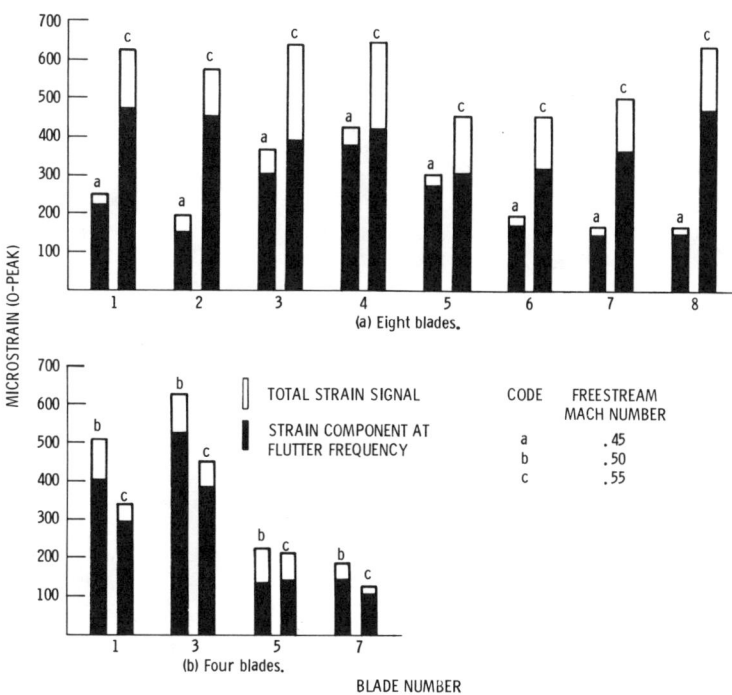

Figure 9. – Typical blade-to-blade variation in strain amplitudes at flutter, $\beta_{0.75R} = 56.6°$, blade gage 1.

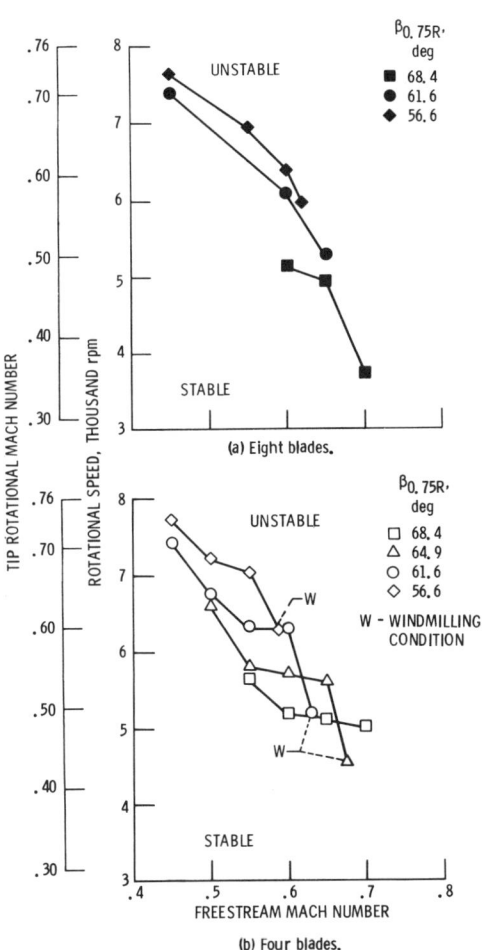

Figure 10. – Flutter conditions, SR3C-X2 model.

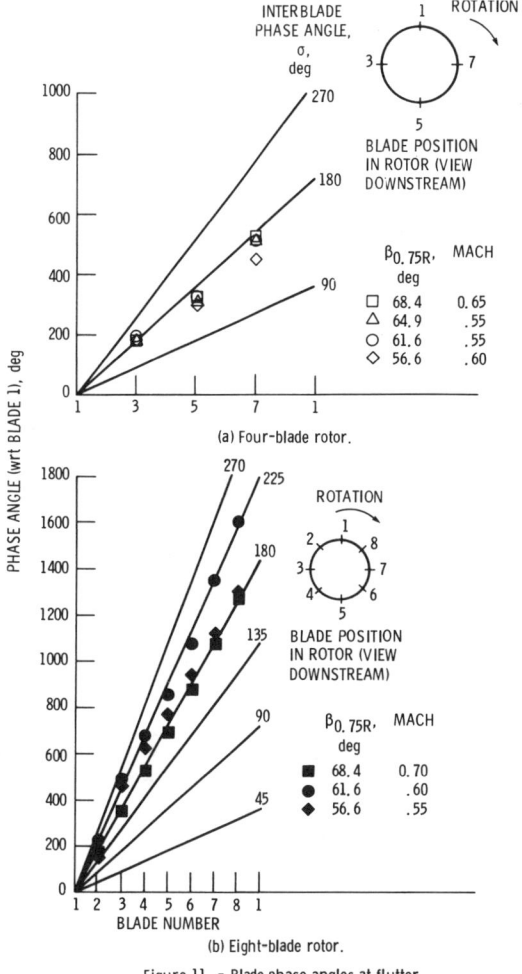

Figure 11. – Blade phase angles at flutter.

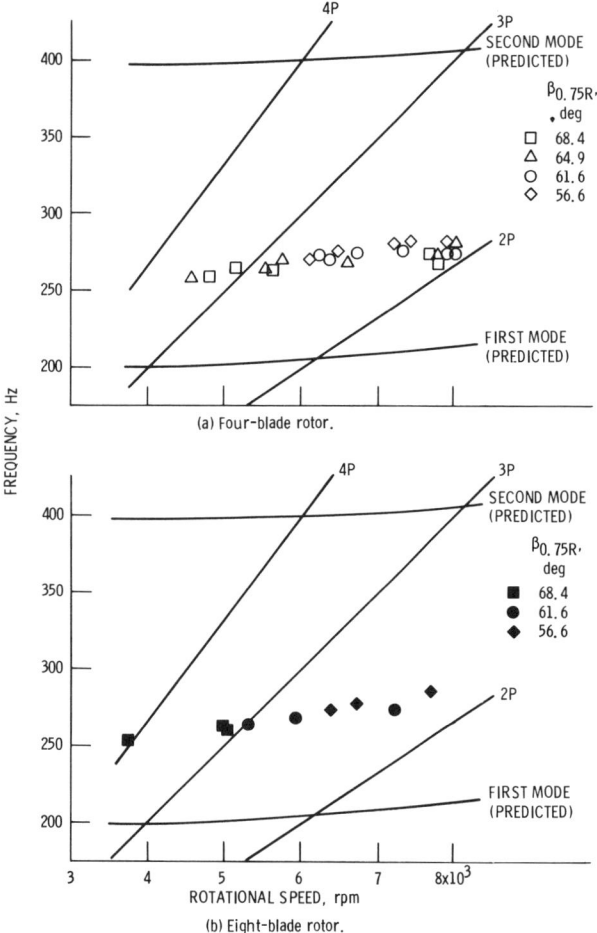

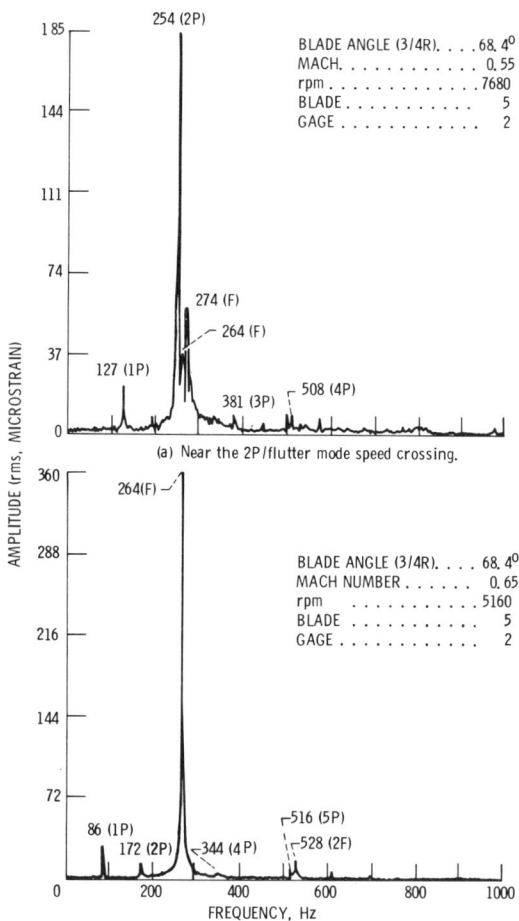

Figure 12. – Measured flutter frequencies, SR3C-X2 model.

Figure 13. – Strain amplitude spectrum, 4-blade rotor, SR3C-X model.

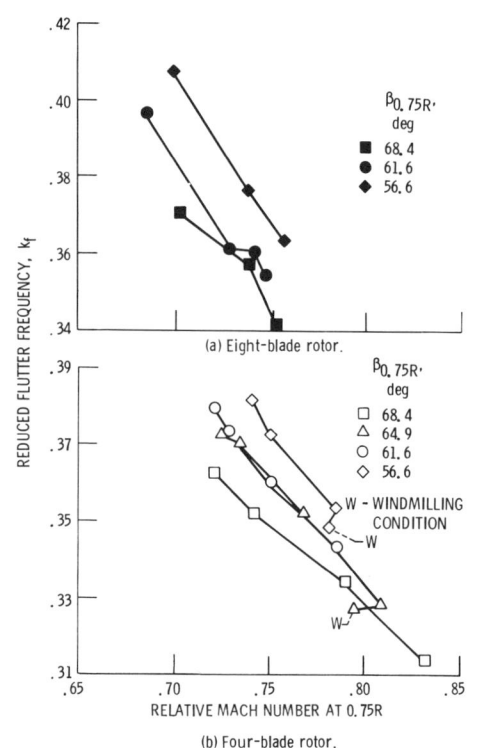

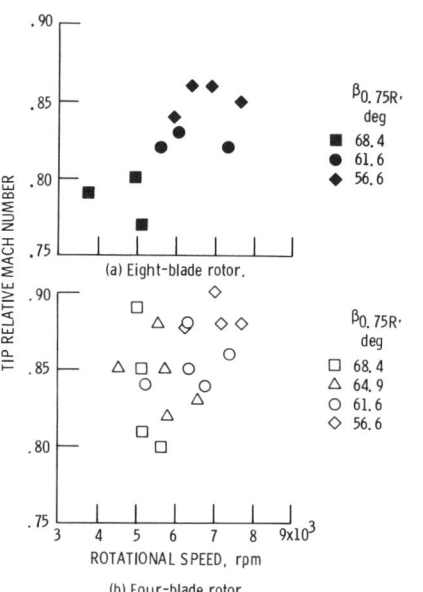

Figure 14. – Flutter reduced frequency, SR3C-X2 model.

Figure 15. – Tip relative Mach number at flutter, SR3C-X2 model.

WHIRL FLUTTER OF SWEPT TIP PROPFANS

Professor M. I. Young and C. A. Harper
Department of Mechanical Engineering
The University of Delaware
Newark, DE 19716

ABSTRACT

The catastrophic aeroelastic whirl flutter instabilities encountered by the Lockheed Electra aircraft and the dynamics of similar propeller-nacelle systems are briefly reviewed [1,2]. Advanced propeller and turbofan engine propeller systems projected for widespread commercial aircraft applications as soon as 1991 [3] are discussed, and their aeroelastic whirl flutter dynamics are presented. The principal innovative aspects of such propfans as they are called are (1) aerodynamic sweep of the outer portions of the blades to provide enhanced efficiency at high subsonic flight speeds well beyond those of previous propeller flight applications, (2) trailing ("pusher") as well as the more conventional forward ("tractor") configurations and (3) counter-rotating, dual propeller systems to permit maximum solidity/activity factors for needed take-off, climbing and cruise performance, and to obviate the destabilizing effects of gyroscopic coupling of nacelle pitching and yawing type motions. This paper concentrates on the influences of sweep, tip loss effects, density altitude and their interactions with the structural design of the nacelle mounting as typified by the uncoupled pitching and yawing frequencies of oscillation.

INTRODUCTION

This paper examines the dynamic instability known as propeller whirl flutter. First experienced in flexibly mounted propeller/nacelle systems, whirl flutter entails either a self-sustained oscillation or, in the extreme case, a divergent whirling precessional motion in the propeller plane.

Propeller whirl flutter was first cited as a possible instability by Taylor and Browne [4] in 1938. In 1950 two papers from Rensselaer Polytechnic Institute were published [5,6] that investigated the gyroscopic effect of rotating propellers on vibration modes. It was not until 1960, after the loss of two Lockheed Electra turboprop aircraft, that a series of experimental and analytical studies into the mechanism of whirl flutter was begun. Throughout the early sixties several papers on the subject were written [refs. 7-10].

In addition to whirl flutter of conventional propeller designs, flutter of unconventional propellers was also being investigated throughout the sixties and seventies. The concept of tilt-rotor aircraft has been in existence since the early fifties and several scaled-down models have been built and tested. In 1962, during testing at the NASA Ames 40 x 80 ft wind tunnel, a proprotor/pylon instability (similar to propeller whirl flutter) was encountered. Bell Helicopter Co. initiated programs to investigate tilt-rotor instabilities in 1962. [11-13] With the advent of the Composite Aircraft Program in 1965 came the emergence of interest in V/STOL aircraft. Two approaches were pursued: the tilt-proprotor design of Bell Helicopter and designs incorporating propellers with hinged blades. [14-16]

With the recent interest in propfans, attributed in part to the high potential for projected fuel savings, propfan type propulsion systems are being studied with renewed vigor. Of primary interest are the new swept-tip composite blade types. The configurations being considered include both tractor and pusher type propulsion systems with either single stage or counter-rotating propellers. This paper limits its discussion to the single stage tractor type.

This paper reports on ongoing research [17] on the aeroelastic whirl flutter dynamics of propfans. The primary innovative features of this work are (1) the aerodynamic sweep of the blade tips which provides greater efficiency at high subsonic speeds, (2) the investigation of pusher as well as tractor configurations, and (3) the examination of the counter-rotating, dual propeller design on system stability. It is not the intent of this paper to precisely model an actual production type wing/nacelle propfan propeller system, but to model the essence of the propeller innovations and their effects on system dynamic stability. It is an updating and report of progress in developing and expanding the insights of [18].

ANALYSIS

A linearized perturbation analysis of the dynamic system leads to two second order, linear differential equations. In matrix form the equations can be written as

$$\begin{bmatrix} m_{11} & m_{12} \\ m_{21} & m_{22} \end{bmatrix} \begin{Bmatrix} \ddot{q}_1 \\ \ddot{q}_2 \end{Bmatrix} + \begin{bmatrix} c_{11} & c_{12} \\ c_{21} & c_{22} \end{bmatrix} \begin{Bmatrix} \dot{q}_1 \\ \dot{q}_2 \end{Bmatrix} + \begin{bmatrix} k_{11} & k_{12} \\ k_{21} & k_{22} \end{bmatrix} \begin{Bmatrix} q_1 \\ q_2 \end{Bmatrix} = \begin{Bmatrix} 0 \\ 0 \end{Bmatrix} ,$$

where $q_1 = \theta$, the pitch angle and

$q_2 = \psi$, the yaw angle

The details of the derivation can be found in Appendix A and [17]. Expanding the system characteristic polynomial yields,

$$\beta^4 + \eta_3 \beta^3 + \eta_2 \beta^2 + \eta_1 \beta + \eta_0 = 0 ,$$

where the $\eta_i$ coefficients, in terms of matrix components, are as follows:

$$\eta_3 = \left(\frac{m_{11} c_{22} + m_{22} c_{11} - m_{21} c_{12} - m_{12} c_{21}}{m_{11} m_{22} - m_{21} m_{12}}\right)$$

$$\eta_2 = \left(\frac{m_{11} k_{22} + c_{11} c_{22} + m_{22} k_{11} - m_{21} k_{12} - c_{12} c_{21} - m_{12} k_{21}}{m_{11} m_{22} - m_{21} m_{12}}\right)$$

$$\eta_1 = \left(\frac{c_{11} k_{22} + c_{22} k_{11} - c_{21} k_{12} - c_{12} k_{21}}{m_{11} m_{22} - m_{21} m_{12}}\right)$$

$$\eta_0 = \left(\frac{k_{11} k_{22} - k_{21} k_{12}}{m_{11} m_{22} - m_{21} m_{12}}\right) ,$$

where the matrix components are defined in Appendix B.

Of the many possible numerical approaches, a digital computer solution to show the parametric influence was chosen. A computer program was established whereby a number of the twenty-three system parameters could be varied independently throughout a particular range. A computerized version of Lin's Method [19] was employed to solve for the roots of the resulting quartic equations

DISCUSSION

The computations for predicting the influence of key system parameters have been carried out employing the preliminary design concept illustrated in Figure 1. The propfan is assumed to be of the size illustrated with an application to commercial transport aircraft. It is taken as a single stage tractor device having strong gyroscopic coupling between its pitching and yawing degrees of freedom. To gain insight into the influence of sweep, the cases of 0°, 20°, and 40°, of aftward sweep are presented here. Also, the influence of an assumed tip loss factor varying from nothing (i.e. 1.00) to as much as the outer ten percent of span is also included (i.e. 0.90). The aircraft is assumed to be flying at maximum speed at 35,000 feet density altitude, on a standard day. Accordingly the helical tip Mach number is almost unity at a flight path Mach number of 0.80. Reference to Figure 2 shows a summary of the system dynamic stability boundaries in terms of the neutral stability frequency ratios in nacelle pitch and yaw at 1100 RPM of the propfan. It is seen that the three dimensional flow aspects typified by the increasing tip loss reduce the frequency-mounting stiffness requirements at a given angle of sweep. In turn at a given tip loss factor, the increasing angle of sweep is seen to reduce the frequency-stiffness requirements.

Since the data from a normal cruising altitude of 35,000 feet may not represent the worst case scenario in terms of whirl flutter instability, flight at lower altitudes is also considered. The data is presented in Figure 3 for the nominal case of 40° of sweep and five percent tip loss (i.e. 0.95). It is assumed that the aircraft lift to drag ratio and power available characteristics permit flight at the same flight path speed as at the 35,000 foot cruising altitude. At this speed of 778 feet per second the helical tip Mach number is less than that at 35,000 feet, but as to be expected the flight dynamic pressure and perturbation airloads are significantly greater. Accordingly, the stability boundaries shown in Figure 3 imply that flight at maximum available power at sea level is likely to impose the most severe structural requirement.

CONCLUSIONS

It is seen that swept tip propfans can postpone the dynamic instability known as whirl flutter to greater flight speeds for a given sweep angle and nacelle mounting stiffness, or pitch and yaw frequency ratios. On the other hand the complexity of the true unsteady, three dimensional perturbation aerodynamics as represented in part by an effective tip loss factor may lead to a significant overestimate of the appropriate minimal nacelle mounting stiffness in the pitching and yawing freedoms. Other factors such as virtual pivot distance ratio (to the ultimate extreme of "pusher" rather than "tractor" configurations), size/scale effects and staging with counter-rotating propfans are considered at length in reference [18]. It is also seen that a considerable analytical and experimental effort is required in the future to capitalize fully on the performance-economic potential of advanced turbofan engines employing swept tip propfans, while successfully avoiding the dreaded flight dynamic instability known as whirl flutter.

REFERENCES

1. Houbolt, John C. and Wilmer Reed, III: "Propeller-Nacelle Whirl Flutter", Journal of the Aerospace Sciences, March 1962, pp. 333-346.

2. Reed, Wilmer H., III: "Review of Propeller-Rotor Whirl Flutter", NASA TR R-264, February 1967.

3. Banks, Howard: "The Next Step", Forbes, May 7, 1984, pp. 31-33.

4. Taylor, E. S. and K. A. Browne: "Vibration Isolation of Aircraft Power Plants", Journal of the Aeronautical Sciences, vol. 6, no. 2, December 1938, pp. 43-49.

5. Scanlan, R. H. and J. C. Truman: "The Gyroscopic Effect of a Rigid Rotating Propeller on Engine and Wing Vibration Modes", Journal of the Aeronautical Sciences, vol. 17, October 1950, pp. 653-659, 666.

6. Brower, W. B. and R. H. Lassen: "The Effects of Gyroscopic Coupling of Propulsion Units on the Vibration Modes of a Dynamically Similar Model of the Lockheed Constitution Airplane", Masters Thesis, Renssalaer Polytechnic Institute, Troy, New York, June 1950.

7. Houbolt, J. C. and W. H. Reed, III: "Propeller-Nacelle Whirl Flutter", Journal of Aerospace Sciences, vol. 29, March 1962, pp. 333-346.

8. Sewall, J. L.: "An Analytical Trend Study of Propeller Whirl Instability", NASA TN D-996, 1962.

9. Reed, W. H. III and S. R. Bland: "An Analytical Treatment of Aircraft Propeller Precession Instability", NASA TN D-659, 1961.

10. Bennett, R. M. and S. R. Bland: "Experimental and Analytical Investigation of Propeller Whirl Flutter of a Power Plant on a Flexible Wing", NASA TN D-2399, 1964.

11. Young, M. I. and R. T. Lytwyn: "The Influence of Blade Flapping Restraint on the Dynamic Stability of Low Disk Loading Propeller-Rotor Systems", Journal of the American Helicopter Society, October 1967.

12. Hall, W. E.: "Prop-Rotor Stability at High Advance Ratios", Journal of the American Helicopter Society, June 1966.

13. Edenborough, H. K.: "Investigation of Tilt-Rotor VTOL Aircraft Rotor-Pylon Stability", AIAA Paper 67-17, January 1967.

14. Brandt, D. E.: "Aeroelastic Problems of Flexible V/STOL Rotors", Presented at the AGARD 34th Flight Mechanics Panel Meeting on "Aeroelastic Effects from a Flight Mechanics Standpoint", Marseilles, France, April 1964.

15. Tiller, F. E. and R. Nicholson: "Stability and Control Consideration for a Tilt-Fold-Proprotor Aircraft", presented at the 26th Annual National Forum of the American Helicopter Society, June 1970.

16. Kvaternik, R. G.: "Studies in Tilt-Rotor VTOL Aircraft Aeroelasticity", Doctoral Dissertation, Case Western Reserve University, June 1973.

17. Harper, Cheryl A.: "Whirl Flutter Dynamic Stability of Advanced Propeller Systems", Master of Mechanical and Aerospace Engineering Thesis, University of Delaware, Newark, Delaware, May 1986.

18. Harper, C. A. and M. I. Young: "Whirl Flutter Dynamics", Proceedings of the 22nd Annual Meeting of the Society of Engineering Science, Preprint ESP22/85016, October 1985.

19. Chestnut, Harold and Robert W. Mayer: Servomechanisms and Regulating System Design, vol. 1, John Wiley & Sons, Inc., New York, 1951, pp. 131-133.

20. Glauert, H.: The Elements of Aerofoil and Airscrew Theory, Macmillan Co., New York, 1926, pp. 208-221.

21. Nikolsky, Alexander A.: Helicopter Analysis, John Wiley & Sons, Inc., New York, 1951, pp. 1-26.

22. Gessow, Alfred and Garry C. Myers, Jr.: Aerodynamics of the Helicopter, Macmillan Co., New York, 1952, pp. 55-65.

23. McCormick, Barnes W., Jr.: *Aerodynamics of V/STOL Flight*, Academic Press, New York, 1967, pp. 79-93.

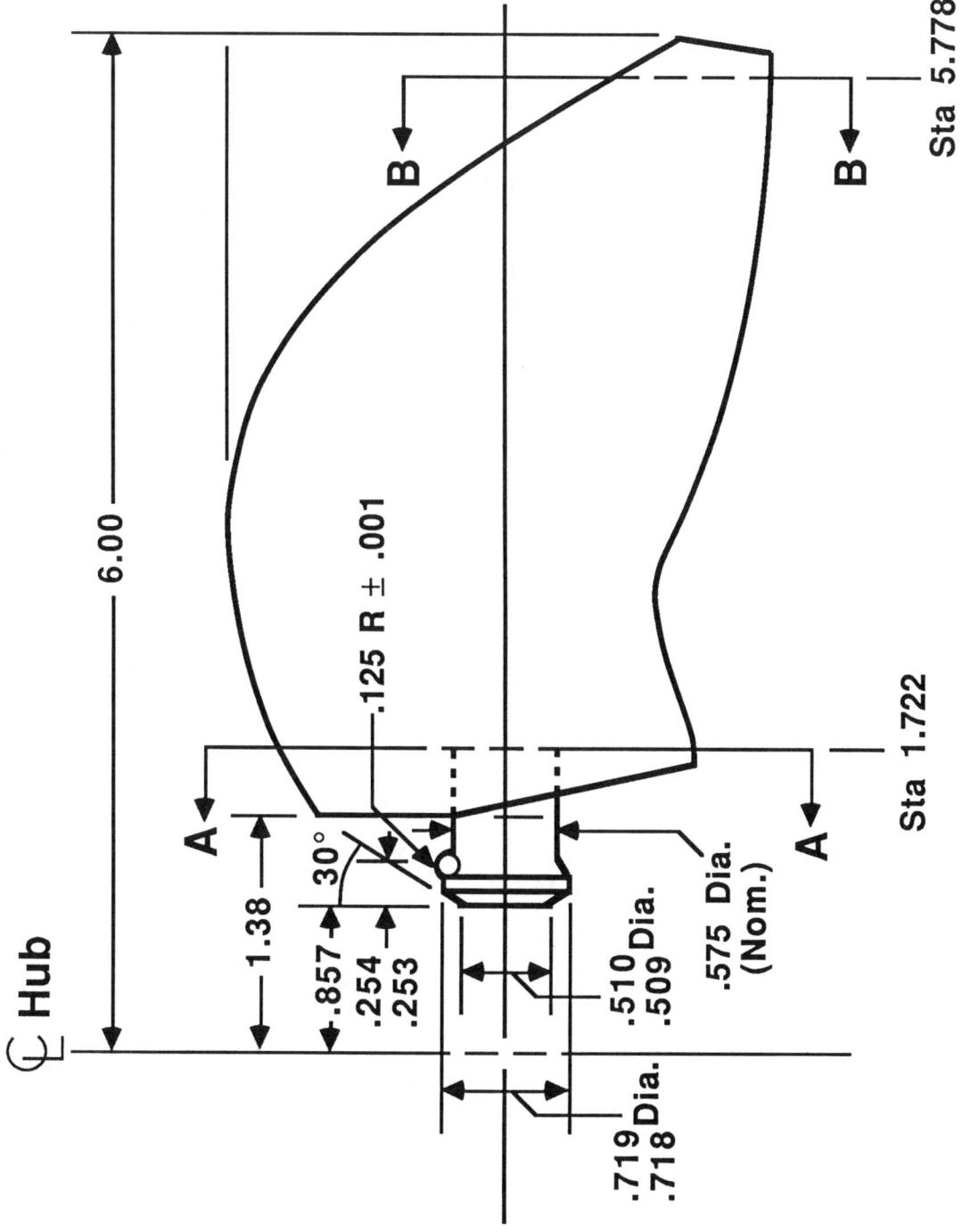

Propfan Blade - Planform

Figure 1

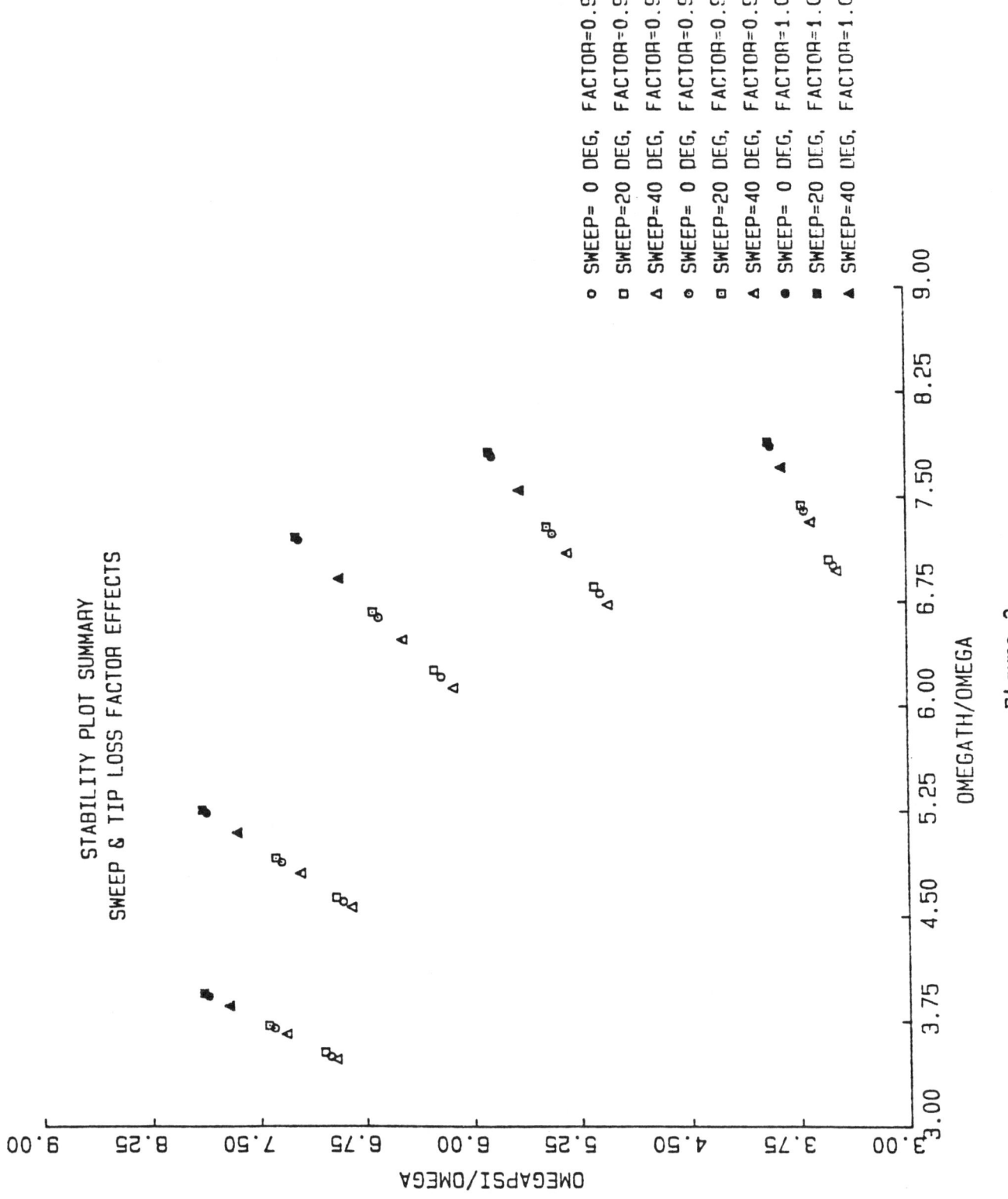

Figure 2

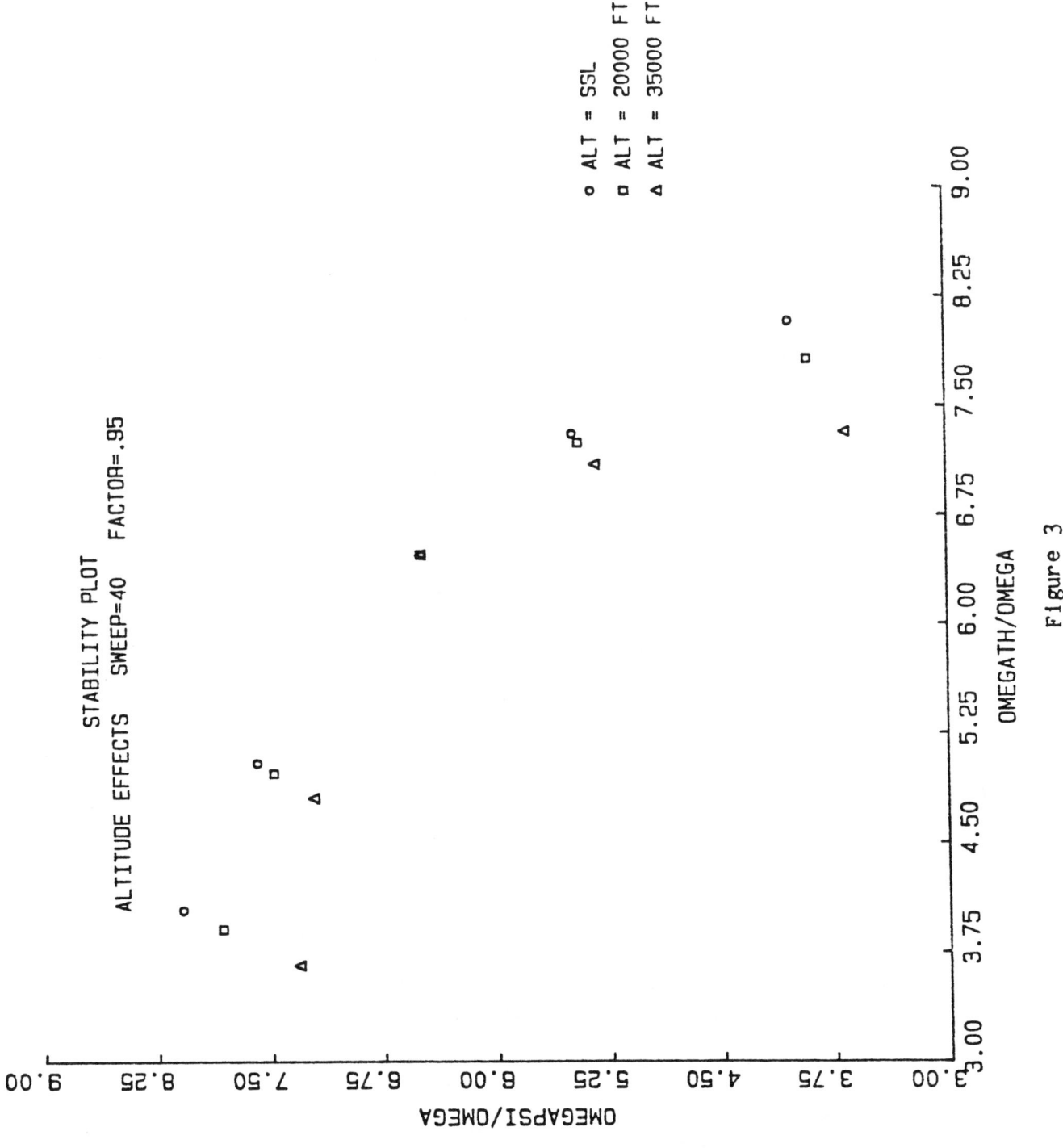

Figure 3

## APPENDIX A

The components of the mass, damping and stiffness matrices can be found in terms of system parameters by using the Principle of Angular Momentum.

$$\underline{M}(O) = \underline{\dot{H}}(O)$$

(The resultant moment about point "O" equals the time rate of change of the angular momentum about "O").

In this paper it is assumed that the propeller-nacelle rigid body pivots at some point "O" along the engine axis (a distance 'e' behind the propeller plane). Motion is restrained in the pitch and yaw directions by springs with respective stiffnesses $K_\theta$ and $K_\psi$ (see Fig. 4).

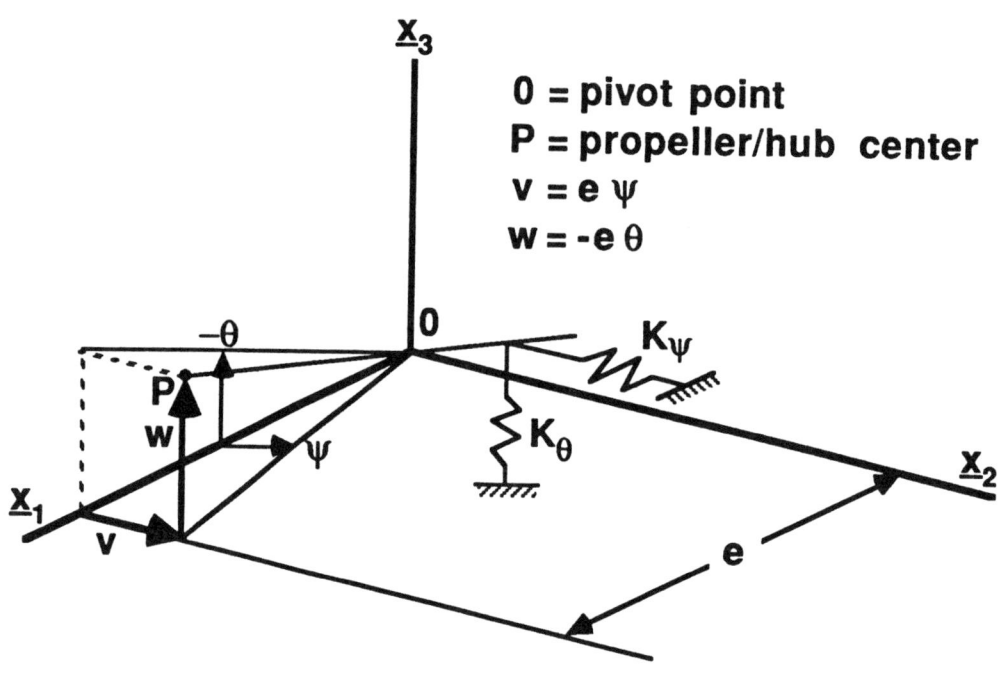

Figure 4

Neglecting gravity, the only external moments and forces acting at "P" are those due to aerodynamics. The resultant moment about "O" can be broken into three parts: aerodynamic, elastic and apparent.

$$\underline{M}(O) = \underline{M}(O)_{aero} + \underline{M}(O)_{elas} + \underline{M}(O)_{app}$$

The apparent moment is equivalent to the gyroscopic moment due to the rotation of the propeller. The elastic moment is the moment induced as a result of the restraining springs. The aerodynamic moment about "O" consists of the actual aerodynamic moments as well as the moments that result from the aerodynamic forces.

The angular momentum about "O" can be separated into two components: that of a rigid body (i.e. a non-rotating propeller system) and a relative angular momentum due to the rotation of the propeller.

$$\underline{H}_{(O)} = \underline{H}_{(O)\text{rigid body}} + \underline{H}_{(O)\text{relative}}$$

With a constant angular propeller velocity, the relative time rate of change of the angular momentum is zero.

$$\underline{\dot{H}}_{(O)\text{rel}} = 0$$

The rigid body angular momentum consists of the propeller/hub and nacelle/engine angular momenta about "O".

The difficulty in obtaining the equations of motion lies in developing expressions for the aerodynamic forces and moments. The aerodynamic analysis is based on blade element-strip theory [20-23] with the underlying assumption that each local strip is independent of all the others. The primary justification for this assumption is that the secondary perturbations which are induced in the propeller velocity field are minor compared to the kinematic effects; moreover, the mean induced velocity itself is insignificant in stability analysis for a propeller in high speed forward flight.

APPENDIX B

In non-dimensional terms, the components of the mass, damping and stiffness matrices are:

$$m_{11} = \frac{1}{\lambda^2}\left(I_{1\,\text{Prop Diam}} + I_{1\,\text{Prop Pivot}} + I_{\text{Nacelle/Engine Pivot}}\right)$$

$$m_{12} = 0$$

$$m_{21} = 0$$

$$m_{22} = \frac{1}{\lambda^2}\left(I_{1\,\text{Prop Diam}} + I_{1\,\text{Prop Pivot}} + I_{1\,\text{Nacelle/Engine Pivot}}\right)$$

$$c_{11} = -e\,\kappa_{13} - \kappa_{34}$$

$$c_{12} = \frac{1}{\lambda^2} I_{1\,\text{Prop Spin}} + e\,\kappa_{14} - \kappa_{33}$$

$$c_{21} = -\frac{1}{\lambda^2} I_{1\,\text{Prop Spin}} - e\,\kappa_{14} + \kappa_{33}$$

$$c_{22} = -e\,\kappa_{13} - \kappa_{34}$$

$$k_{11} = \frac{1}{\lambda^2} e^2 K_\theta + e L_1 - e \kappa_{11} - \kappa_{32}$$

$$k_{12} = e \kappa_{12} - \kappa_{31}$$

$$k_{21} = - e \kappa_{12} + \kappa_{31}$$

$$k_{22} = \frac{1}{\lambda^2} e^2 K_\psi + e L_1 - e \kappa_{11} - \kappa_{32}$$

# RECENT DEVELOPMENTS IN FLUTTER SUPPRESSION TECHNIQUES FOR TURBOMACHINERY ROTORS

O. O. Bendiksen
Department of Mechanical and Aerospace Engineering
Princeton University
Princeton, NJ 08544

## ABSTRACT

Recent research on passive flutter control techniques for rotors and cascades is reviewed. The proposed methods include the use of periodicity-breaking imperfections, aeroelastic tailoring, dry friction damping, and mode shape control. Numerical results are presented to illustrate the effectiveness of each method in controlling various types of flutter, and to evaluate the feasibility of incorporating the techniques in current and future technology rotors. Mistuning may be the only feasible approach for existing engines, but aeroelastic tailoring offers promising advantages for future engines.

## INTRODUCTION

During the past five years, several researchers have investigated the feasibility of suppressing or controlling flutter in turbomachinery rotors. This paper presents an attempt to review the ideas behind the proposed methods, and to interpret the main results obtained.

The motivation for this research can best be understood by looking at a typical compressor map, Fig. 1, with the common flutter boundaries superimposed. During transients, such as accelerations and decelerations, the dynamic operating line undershoots or overshoots the steady-state operating line shown, and may therefore intersect the stall or choke flutter boundary. The operating line is also sensitive to the matching of the individual compressors and turbines, and often shifts appreciably as the engine deteriorates. Thus ensuring that an engine rotor remains flutter free throughout its flight envelope and operating life is by no means an easy task.

In the early days of the jet engine, stall and choke flutter were the most common aeroelastic problems encountered. This is not surprising, since compressor designs were less sophisticated and the stall margins were often inadequate. Today, the most troublesome type of flutter occurs in large fan rotors operating at supersonic tip Mach numbers, and is labeled "supersonic unstalled flutter" in Fig. 1. This type of flutter is very dangerous because it intersects the operating line at or near take-off power, during the most critical part of the flight envelope.

The review will only cover _passive_ flutter control and suppression techniques. Most of the proposed methods fall into three main categories; according to whether 1) mistuning, 2) aeroelastic tailoring, or 3) dry friction damping and/or mode shape control is used to delay the onset of flutter or suppress its amplitude to an acceptable level. While the use of aeroelastic tailoring is well established in isolated wing aeroelasticity, the two other methods are unique to the turbomachinery field.

AEROELASTIC STABILIZATION BY DISORDER AND IMPERFECTIONS

The theory of elastic stability has provided a number of classical examples of imperfection-sensitive structures. In certain buckling problems, the presence of even small imperfections can result in a dramatic reduction of the critical buckling load. It is therefore quite surprising to find that in certain _dynamic_ stability problems in aeroelasticity, the reverse occurs, i.e., small imperfections actually have a stabilizing effect. This is typically what happens in turbomachinery rotors and cascades, where small variations in structural properties of the individual blades can have a first-order effect on the flutter boundaries.

This anomalous behavior was observed by Whitehead [1] over twenty years ago, but the idea of intentionally introducing imperfections to stabilize a rotor is relatively new. In the turbomachinery field, any imperfection (or disorder) that breaks the blade-to-blade periodicity of the rotor is commonly referred to as "mistuning". It turns out that there is a dark side to the mistuned rotor, revealed through its undesirable forced response characteristics. Even small amounts of mistuning within normal manufacturing tolerances can result in a large scatter -- often exceeding 100 percent -- in the individual blade amplitudes. This is a highly dangerous situation from a fatigue standpoint, since high cycle fatigue is always a concern in highly stressed turbine and compressor airfoils. Because mistuning offers a plausible explanation for socalled "rogue blades", most of the early studies were concerned with the forced response problem and did not consider flutter [2-4].

In the past five years, the possibility of using mistuning as a passive flutter control technique has been investigated by several researchers. Kaza and Kielb [5] studied the mistuned cascade in incompressible flow, using Whitehead's unsteady cascade theory [6] and modeling each blade with two degrees of freedom (bending and torsion). Structural coupling between the blades was neglected. In later work, they considered cascades in subsonic and supersonic flow [7], nonuniform blades with several degrees-of-freedom [8], and the effect of disk flexibility [9]. Typical results from their work are shown in Fig. 2, for a representative cascade in incompressible flow, and illustrate the strongly stabilizing effect of mistuning.

Whitehead concluded in Ref. 1 that mistuning should always have a stabilizing effect on cascade flutter, and presented a proof to this effect for single-degree-of-freedom (SDOF) torsional flutter. Bendiksen [10] later showed that the SDOF case is special in that respect, and that the conclusion does not necessarily hold for multi-degree of freedom blades. Mistuning was found to be stabilizing in systems governed by _normal_ matrices (or operators) and that, in general, only cascades and rotors with a single degree of freedom per blade belong to this class[†].

If the aeroelastic eigenvalue problem is written in the general form,

$$[B]\{q\} = \lambda [K]\{q\} \tag{1}$$

where

$$[B] = [M] + (1/\mu)[A] \tag{2}$$

$$\lambda = (\omega_0/\omega)^2 \tag{3}$$

then flutter occurs when $\text{Im}\lambda > 0$. Here $[M]$, $[K]$, and $[A]$ are the mass, stiffness, and aerodynamic matrices, respectively, corresponding to the generalized coordinates $\{q\}$, and $\mu = m/\pi\rho_0 b^2$ is the blade mass ratio. The imaginary part of the nth eigenvalue corresponding to the eigenvector $\{\phi\}_n$ can be written as

$$\text{Im}\lambda_n = \frac{\{\phi\}_n^*[B]_2\{\phi\}_n}{\{\phi\}_n^*[K]\{\phi\}_n} \tag{4}$$

where

$$[B]_2 = \frac{1}{2i}([B] - [B]^*) \tag{5}$$

and the asterisk denotes conjugate transpose. If $[B]$ is a normal matrix, one can then show [10] that Eq. (4) behaves as a Rayleigh quotient in the vicinity of the unperturbed eigenvectors $\{\phi\}_n$, i.e. it is stationary. Consider now a perturbation of the mass matrix, then $[K]$ and $[B]_2$ remain fixed if $k = \omega b/U$, Mach No. M, and $\mu$ do, and from the extremum properties of Rayleigh's quotient, it follows that

$$\min(\text{Im}\lambda_i) \leq \text{Im}\,\tilde{\lambda}_n \leq \max(\text{Im}\,\lambda_i) \tag{7}$$

where $\tilde{\lambda}_n$ is the nth perturbed eigenvalue, and $\lambda_i$ denotes the collection of unperturbed eigenvalues. If the tuned rotor is stable, then $\max(\text{Im}\lambda_i) \leq 0$, and it follows from Eq. (7) that the mistuned rotor is also stable. If $\min(\text{Im}\lambda_i) > 0$, on the other hand, then mistuning cannot stabilize the rotor. It can be shown that the same conclusions hold for perturbations of the stiffness matrix, although Eq. (7) does not.

---

[†]A normal matrix commutes with its Hermitian adjoint (conjugate transpose), and its left and right eigenvectors form complex conjugate pairs.

The stabilizing effect of mistuning comes from the fact that unstable modes become coupled to stable modes when the periodicity of the cascade is broken. For maximum benefit, one should try to couple the most unstable mode as strongly as possible to the most stable mode. Recently, Bendiksen [11] has shown that the stabilization mechanism is closely connected with the phenomenon of mode localization, which often occurs in periodic structures when the strict periodicity is broken by disorder or imperfections. This phenomenon is of considerable importance in solid state physics, as it explains many of the transport properties of disordered solids. In the framework of localization theory, the stabilization mechanism can be explained by the fact that the original monochromatic flutter wave is scattered by the imperfections into waves of different and more stable wavelengths.

Based on localization theory, the sensitivity of the flutter boundaries to mistuning should decrease as the interblade coupling is made stronger. Thus rotors where the coupling is entirely aerodynamic (unshrouded blades on a stiff disk) would be expected to be most strongly stabilized by imperfections. Results from flutter calculations support this conclusion [11]. In Fig. 3 the strong stabilizing effect of alternating blade stiffness mistuning is illustrated for two supersonic cascade configurations, typical of large fan rotors. The structural model is as shown in Fig. 4, but with no structural coupling between the blades ($K_c = 0$) and the disk degrees of freedom $q_{n+N}$ eliminated ($M_d = 0$).

Figure 5 shows the effect of interblade coupling on the mistuning required to stabilize a rotor with a relatively flexible disk, $K_d/K_b = 1$, if the mistuning is of the "alternating blade" type, i.e., adjacent blades have $\Delta K_b/K_b = \pm \bar{\epsilon}$ and every other blade is identical. Note that the required magnitude $\bar{\epsilon}_{cr}$ increases rapidly as $\bar{K}_c = K_c/K_b$ is increased from zero (as in Fig. 3), and also as the reduced frequency parameter $\bar{k} = kM/(M^2-1)$ is lowered. It is also interesting to note that $\bar{\epsilon}_{cr}$ is not a monotonic function of reduced frequency (or velocity), and that certain "resonance bands" exists within which the rotor cannot be stabilized by mistuning.

Crawley and Hall [12] have cast the mistuning problem as a minimization of a suitable cost function,

$$\phi(\epsilon) = \{\frac{1}{N} \sum_{i=0}^{N-1} \epsilon_i^4\}^{\frac{1}{4}} \tag{8}$$

subject to inequality constraints on the stability margin of the individual modes. They find that, although alternate blade mistuning is not as cost effective as the calculated optimum, it is more robust. This is shown in Fig. 6 for a 14-bladed rotor with a supersonic typical section (M = 1.317). Structural coupling between the blades is neglected.

Recently Hoyniak and Fleeter [13] have considered the novel approach of using "aerodynamic mistuning" as a flutter control. The basic idea is again to break the periodicity of the structure, except that in this case it is the cascade passages between the blades that are made different, as illustrated in Fig. 7(a). Typical results are shown in Fig. 7(b), for SDOF torsional flutter of Verdon's Cascade B. At a constant Mach number, stability is generally enhanced by "disorder" in the blade spacing, with the largest effect apparent at low supersonic Mach numbers. Fleeter and Hoyniak later extended their analysis to consider coupled bending-torsion flutter [14], and Topp and Fleeter considered aerodynamic detuning using splitter blades [15].

AEROELASTIC TAILORING

The introduction of laminated composites has provided the aircraft designer with interesting "tailoring" possibilities. By varying fiber orientation and ply stacking configuration, flutter and divergence speeds can often be increased substantially. The subject of aeroelastic tailoring of isolated wings has received considerable attention [16-20], but this has not been the case in the turbomachinery field. Although the subject was mentioned in a paper sixteen years ago [21], there exists almost no published data on aeroelastic tailoring of composite turbomachinery blades in the open literature. This lack of data is no doubt related to the unsuccessful attempt by Rolls-Royce to use carbon fiber composites in the RB.211 engine fan blades. One must suspect that the failure of the blades to meet bird strike requirements discouraged futher use of composites in aircraft compressors.

Today the technology of composite materials is sufficiently advanced to warrant reconsidering the unique advantages such materials offer. In a 1976 study by Troha and Swain, composite inlays were used to increase the flutter speed of TF43-A100 first stage fan blades [22]. Although the inlays covered only a fraction of the chord, a moderate improvement was demonstrated in both flutter margin and aerodynamic performance of the modified blade over the original titanium blade. However, only one fiber orientation (radial) was tested, and the tailoring question was not pursued.

In a very recent study, Bendiksen and White [23] demonstrated the feasibility of stabilizing critical aeroelastic modes through a judicious choice of fiber sweep and laminate stacking schemes. A global Rayleigh-Ritz formulation was used, with the blades modeled as thin, multilayered composite plates symmetric about the middle surface. Five elastic degrees of freedom were used to represent the lateral deflection of each blade, associated with two bending modes (1B,2B), two torsion modes (1T,2T), and a chordwise (camber bending) mode (1C). Three of these modes are illustrated in Fig. 8.

The linearized two-dimensional cascade theory of Bendiksen [24] was modified to account for chordwise bending, and applied in a strip theory fashion to calculate the unsteady aerodynamic forces. Of the many different types of flutter indicated in Fig. 1, supersonic unstalled flutter is the most challenging from a control standpoint. In addition to being the most dangerous, this type of flutter imposes hard constraints on fan blade thickness and chord for a given rotor diameter. In the results shown in Figs. 9-12 the stability behaviors of various laminated AS/3501 graphite/epoxy blades are compared to a titanium reference or "baseline" blade, and the potential for aeroelastic tailoring is explored. The rotor has 24 blades and a Verdon Cascade A typical section, and the blade thickness (3%) and aspect ratio (1) was chosen to be representative of the tip sections of large fans, outboard of the shrouds.

The baseline titanium blade was intentionally designed to be unstable in supersonic flow, as shown in Fig. 9. In fact, three of the five branches become unstable in the Mach number range of practical interest, and one mode (1T) is unstable for all supersonic Mach numbers. In this and subsequent figures, flutter is indicated whenever the real part $\bar{p}_R = p_R/\omega_{ref}$ of the time dependence, exp(pt), becomes greater than zero for a branch. Also, since $|\bar{p}_R| \ll 1$, $\bar{p}_R$ is a good representation of the relative damping of a mode, albeit not its damping ratio.

As a test of aeroelastic tailoring concepts, attempts were made to stabilize the highly unstable baseline blade section by switching to composite materials and choosing favorable fiber sweeps and ply stacking configurations. Two stacking schemes were studied, one with all fibers swept at an angle (pos. aft) to the spanwise direction, and the second with crossed plies that had alternating forward and aft swept fibers in the stacking sequence. The central one-third of the plies always had their fibers in the spanwise direction to provide longitudinal stiffness to support the centrifugal force acting on the blade.

Figure 10 shows the effect of fiber sweep on the important aeroelastic modes, at a constant Mach number of M = 1.5. Note that fiber sweep is generally destabilizing on the first bending branch, especially forward sweep up to about -60 degrees. Forward fiber sweep is also destabilizing on the first torsion branch, while moderate aft sweep is stabilizing. At zero fiber sweep, the chordwise mode is extremely unstable because of insufficient chordwise bending stiffness. Figure 10 shows that either forward or aft fiber sweep is effective in stabilizing this mode. The chordwise mode is essentially a plate type mode, which has recently been found to play an important role in flutter of thin blades of low to moderate aspect ratio [24,25]. Chordwise bending also has a significant effect on the stability of the first torsion branch, as can be seen from Fig. 11.

The results obtained in the study of Bendiksen and White suggest that cross-ply configurations are more effective than swept fiber configurations in stabilizing the first torsion branch without destabilizing other branches. Figure 12 shows that a blade with ±45° fiber cross-ply angles is stable up to a Mach number of about 1.55, at which point the first torsion branch becomes unstable. The corresponding titanium blade is unstable for all supersonic Mach numbers, as illustrated in Fig. 9. From these results, it would appear that aeroelastic tailoring is well worth pursuing.

FLUTTER "QUENCHING"

If the flutter instability is weak, as is often the case with subsonic stall and choke flutter, it may be feasible to try to quench or limit its amplitude to an acceptable level. The difficult part is to predict what the limit cycle amplitude will be, and determine whether the resultant dynamic stresses can be tolerated.

Dry friction dampers have been used effectively to control forced response in highly stressed turbine airfoils [27], and have found extensive application in current technology civilian and military engines. They have not, to the author's knowledge, been used successfully to suppress flutter in fan or compressor blades. In two recent studies, Sinha and Griffin [28,29] investigated the feasibility of using dry friction dampers for this purpose. Lumped parameter models were used, with one degree-of-freedom per blade, and any coupling between different nodal diameter modes that might arise from the nonlinear friction force was ignored. The aerodynamic force was modeled as a negative viscous damping force, thus ignoring the important dependence on reduced frequency and interblade phase angle. The results were encouraging, but difficult to interpret in view of the aerodynamic approximation used.

In a later study, Sinha, Griffin and Kielb [30] incorporated realistic unsteady cascade theories in their analysis. Figure 13 illustrates a typical stability diagram for one mode of a 3-bladed rotor in incompressible flow. The lower line is a stable limit cycle, whereas the upper line is an unstable limit cycle above which the flutter amplitude becomes unbounded. They also studied a supersonic 28-bladed NASA rotor, and concluded that blade-to-ground dampers would allow the rotor to operate up to a Mach number of 1.6, whereas the undamped stage would flutter at a Mach number of 1.2.

Bendiksen [31] has investigated the feasibility of using dry friction damping at the slipping shroud interfaces to quench supersonic unstalled flutter. The aeroelastic model is a shrouded cascade, Fig. 14, with the blades modeled as typical sections attached to a finite element model of the shroud. Typical results are shown in Fig. 15, for a 24-bladed rotor with Verdon's Cascade A typical section. The lower solid curves are stable limit cycles, whereas the upper dashed curves represent unstable limit cycles above which the motion becomes unbounded. Here $\mu$ = coefficient of friction; and $N_0$ = mean shroud preload. The flutter mode is a two-nodal-diameter coupled bending-torsion mode. With the shrouds welded together ($\mu = \infty$) the rotor flutters at a Mach number of 1.65, while if the shrouds slip freely ($\mu = 0$) it flutters at M = 1.185. Note that flutter quenching is achieved, up to realistic blade amplitudes, for $\mu$ and $N_0$ in the practical range. Also note that basing the calculations on SDOF torsion leads to overly optimistic results (top curve).

The flutter boundary was found to be quite sensitive to the shroud contact angle $\phi$, Fig. 14, making the optimization of this angle an important design consideration. This arises from the fact that important aeroelastic parameters such as bending/torsion amplitude ratio, Fig. 16, modal frequency, and slip amplitude at the shroud interfaces, Fig. 17, depend on $\phi$. Additional results, including the effect on mistuning, can be found in the M.S. Thesis by Valero [32], and in the paper by Valero and Bendiksen [33]. For shrouded blades, the effect of mistuning is much less than for unshrouded blades, especially in the strongly coupled bending-torsion aeroelastic modes.

Finally, the possibility of using mode shape control to suppress flutter should be mentioned. In the study of the shrouded cascade, Fig. 14, several instances were observed of unstable nonslipping modes becoming absolutely stable as soon as the shrouds broke loose and started slipping. The quenching mechanism here is *not* dry friction damping, but a change in the aeroelastic mode shape (bending/torsion ratio). Thus it might be possible, through a clever shroud design, to quench flutter by altering the mode shape of the flutter mode.

## CONCLUDING REMARKS

Of the three main techniques for controlling or suppressing flutter in compressor rotors, the author believes that aeroelastic tailoring holds the most promise for future engines. For existing engines, mistuning has the advantage of not requiring design changes for its implementation. Flutter quenching is perhaps the method least appealing to the designer, because of the inherent risk if the quenching mechanism should fail in service. Experimental verification of the various methods is needed before they can be considered ready for implementation in production engines.

## ACKNOWLEDGMENT

A large part of this research was supported by NASA Lewis Research Center under Grant NAG 3-308. The author gratefully acknowledges this support.

## REFERENCES

1. Whitehead, D. S., "Torsional Flutter of Unstalled Cascade Blades at Zero Deflection," Great Britain, A.R.C. R&M 3429, 1964.
2. Dye, R. C. F., and Henry, T. A., "Vibration Amplitudes of Compressor Blades Resulting from Scatter in Blade Natural Frequencies," ASME Journal of Engineering for Power, Vol. 91, July 1969, pp. 182-188.
3. El-Bayoumy, L. E., and Srinivasan, A. V., "Influence of Mistuning on Rotor-Blade Vibrations," AIAA Journal, Vol. 13, No. 4, Apr. 1975, pp. 460-464.
4. Whitehead, D. S., "Effect of Mistuning on the Vibration of Turbomachine Blades Induced by Wakes," Journal of Mechanical Engineering Science, Vol. 8, No. 1, Mar. 1966, pp. 15-21.
5. Kaza, K. R., and Kielb, R. E., "Flutter and Response of a Mistuned Cascade in Incompressible Flow," AIAA Journal, Vol. 20, No. 8, Aug. 1982, pp. 1120-1127.
6. Whitehead, D. S., "Force and Moment Coefficients for Vibrating Aerofoils in Cascade," Great Britain A.R.C. R&M 3254, 1960.
7. Kielb, R. E., and Kaza, K. R. V., "Aeroelastic Characteristics of a Cascade of Mistuned Blades in Subsonic and Supersonic Flows," ASME Paper No. 81-DET-122, 1981.
8. Kaza, K. R. V., and Kielb, R. E., "Coupled Bending-Torsion Flutter of a Mistuned Cascade With Nonuniform Blades," AIAA Paper No. 82-0726, New Orleans, 1982.

9. Kaza, K. R. V., and Kielb, R. E., "Vibration and Flutter of Mistuned Bladed-Disk Assemblies," AIAA Paper No. 84-0991, Palm Springs, CA, 1984.
10. Bendiksen, O. O., "Flutter of Mistuned Turbomachinery Rotors," ASME Journal of Engineering for Gas Turbines and Power, Vol. 106, No. 1, Jan. 1984, pp. 25-33.
11. Bendiksen, O. O., "Aeroelastic Stabilization by Disorder and Imperfections," Paper No. 583P presented at the XVIth International Congress of Theoretical and Applied Mechanics, Lyngby, Denmark, Aug. 19-25, 1984.
12. Crawley, E. F. and Hall, K. C., "Optimization and Mechanism of Mistuning in Cascades," ASME Paper No. 84-GT-196, June 1984.
13. Hoyniak, D., and Fleeter, S., "Aerodynamic Detuning Analysis of an Unstalled Supersonic Turbofan Cascade," ASME Paper No. 85-GT-192, Mar. 1985.
14. Hoyniak, D. and Fleeter, S., "The Effect of Aerodynamic Detuning on Coupled Bending Torsion Supersonic Flutter," ASME Paper No. 86-GT-100, June 1986.
15. Topp, D. A., and Fleeter, S., "Splitter Blades as an Aeroelastic Detuning Mechanism for Unstalled Supersonic Flutter of Turbomachine Rotors," ASME Paper No. 86-GT-99, June, 1986.
16. Austin, F. et al., "Aeroelastic Tailoring of Advanced Composite Lifting Surfaces in Preliminary Design," Proc. of the AIAA/ASME/SAE 17th Structures, Structural Dynamics and Materials Conference, King of Prussia, PA, May 1976, pp. 69-79.
17. Weisshaar, T. A., "Aeroelastic Tailoring of Forward Swept Composite Wings," Journal of Aircraft, Vol. 18, No. 8, Aug. 1981, pp. 669-676.
18. Sherrer, V. C., Hertz, T. J., and Shirk, M. H., "Wind Tunnel Demonstration of Aeroelastic Tailoring Applied to Forward Swept Wings," Journal of Aircraft, Vol. 18, No. 11, Nov. 1981, pp. 976-983.
19. Weisshaar, T. A., and Foist, B. L., "Vibration and Flutter of Advanced Composite Lifting Surfaces," Proc. of the 24th AIAA/ASME/ASCE/AHS Structures, Structural Dynamics and Materials Conference, Lake Tahoe, NV, May 1963, pp. 498-508.
20. Weisshaar, T. A., and Ryan, R. J., "Control of Aeroelastic Instabilities Through Elastic Cross-Coupling," Proc. of the 25th AIAA/ASME/ASCE/AHS SDM Conference, Palm Springs, CA, May 1984, pp. 226-235.
21. Goatham, J. I., "Design Considerations for Large Fan Blades," SAE Transactions, Vol. 78, Sec. 3, Apr. 1969, pp. 1397-1403.
22. Troha, W., and Swain, K., "Composite Inlays Increase Flutter Resistance of Turbine Engine Fan Blades," ASME Paper No. 76-GT-29, Mar. 1976.
23. Bendiksen, O. O., and White, J. F., III, "Aeroelastic Tailoring of Advanced Composite Compressor Blades," Proc. of the 27th AIAA/ASME/ASCE/AHS SDM Conference, San Antonio, TX, May 1986, pp. 684-692.
24. Bendiksen, O. O., and Friedmann, P., "Coupled Bending-Torsion Flutter in a Supersonic Cascade," AIAA Journal, Vol. 19, No. 6, June 1981, pp. 774-781.

25. White, J. F., and Bendiksen, O. O., "Aeroelastic Behavior of Low Aspect Ratio Metal and Composite Blades," ASME Paper No. 86-GT-243, June, 1986.
26. White, J. F., "Aeroelastic Behavior of Low Aspect Ratio Metal and Composite Blades," MSE Thesis, Department of Mechanical & Aerospace Engineering, Princeton University, Princeton, NJ, July 1985.
27. Griffin, J. H., "Friction Damping of Resonant Stresses in Gas Turbine Engine Airfoils," ASME Journal of Engineering for Power, Vol. 102, Apr. 1980, pp. 329-333.
28. Sinha, A., and Griffin, J. H., "Friction Damping of Flutter in Gas Turbine Engine Airfoils," Journal of Aircraft, Vol. 20, No. 4, Apr. 1982, pp. 372-376.
29. Sinha, A., and Griffin, J. H., "Effects of Friction Dampers on Aerodynamically Unstable Rotor Stages," AIAA Journal, Vol. 23, Feb. 1985, pp. 262-270.
30. Sinha, A., Griffin, J. H., and Kielb, R. E., "Influence of Friction Dampers on Torsional Blade Flutter," ASME Paper No. 85-GT-170, Mar. 1985.
31. Bendiksen, O. O., "Effect of Structural and Damping Nonlinearities on Flutter of Shrouded Fans," AIAA Paper No. 84-0990, presented at the 25th AIAA/ASME/ASCE/AHS Structures, Structural Dynamics, and Material Conference, Palm Springs, CA., May 14-16, 1984.
32. Valero, N. A., "Dynamics of Shrouded Blade Assemblies with Mistuning and Slip," M.S. Thesis, Department of Mechanical and Aerospace Engineering, Princeton University, Sept. 1985.
33. Valero, N. A., and Bendiksen, O. O., "Vibration Characteristics of Mistuned Shrouded Blade Assemblies," ASME Paper No. 85-GT-115, Mar. 1985.

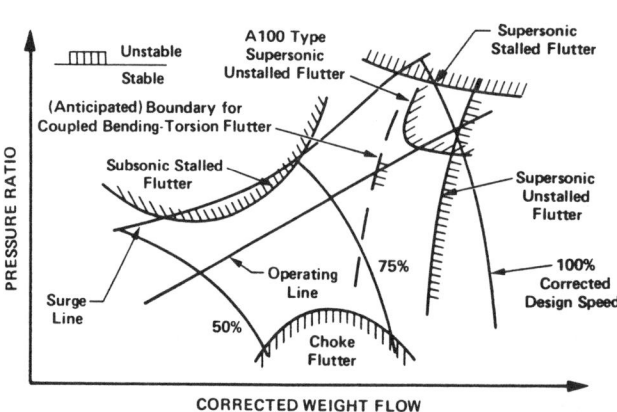

Fig. 1 Compressor map showing typical flutter boundaries.

O. O. Bendiksen / 97

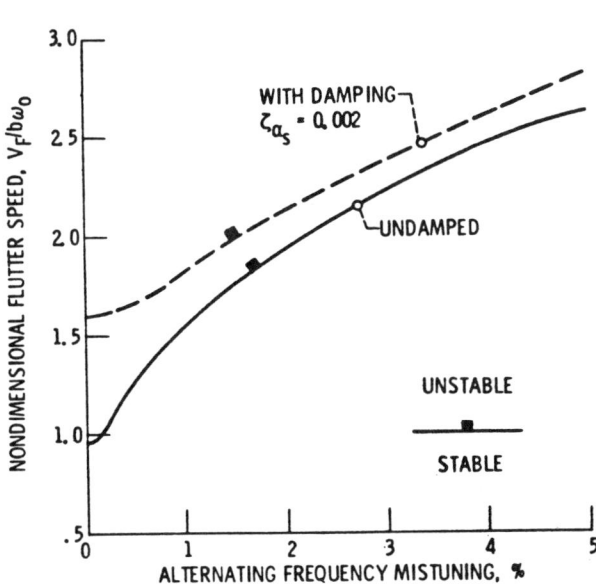

Fig. 2 Torsional flutter speed vs. percent alternating blade mistuning (incompressible flow; results from Ref. 5).

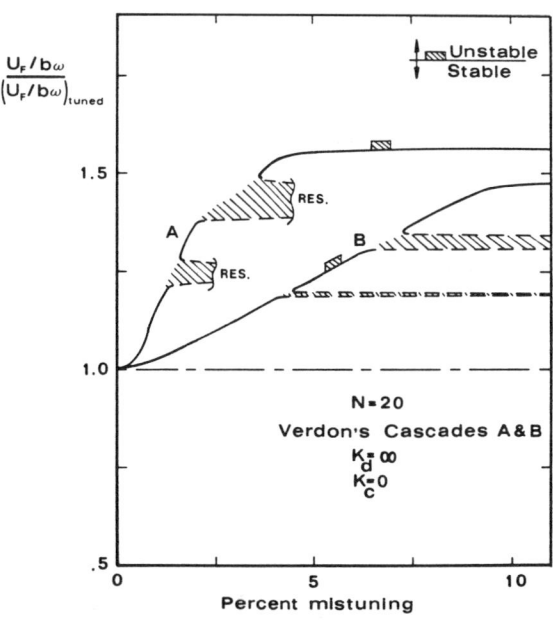

Fig. 3 Stabilization ratio vs. percent mistuning for typical supersonic cascades (torsional oscillations about midchord).

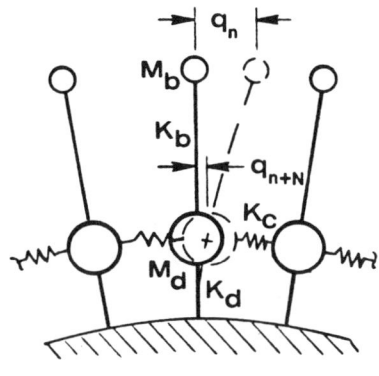

Fig. 4 Simple structural model of bladed disk.

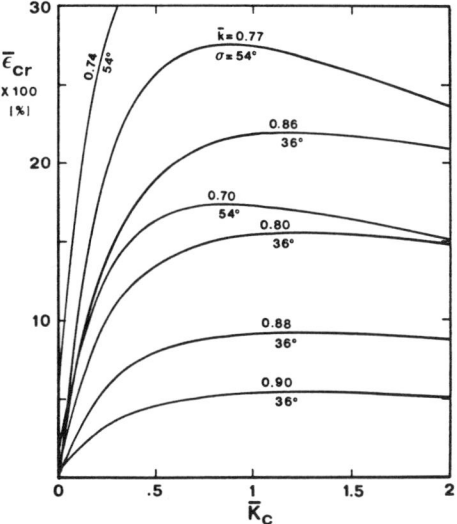

Fig. 5 Effect of interblade coupling strength $\bar{K}_c$ on critical blade stiffness mistuning $\bar{\epsilon}_{cr}$ (N = 20; $K_d/K_b$ = 1; Verdon's Cascade A; SDOF torsion).

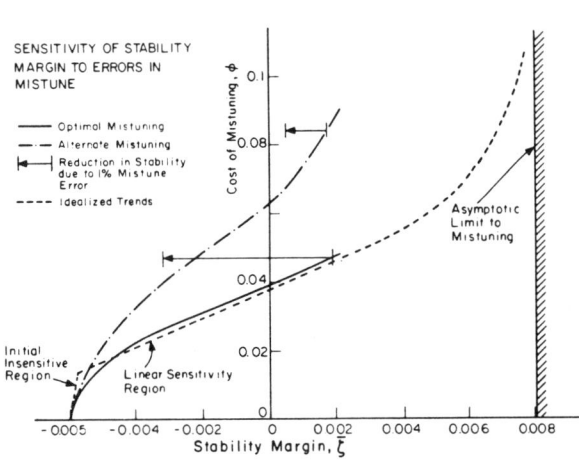

Fig. 6 Sensitivity of stability margin to errors in mistuning (from Ref. 12).

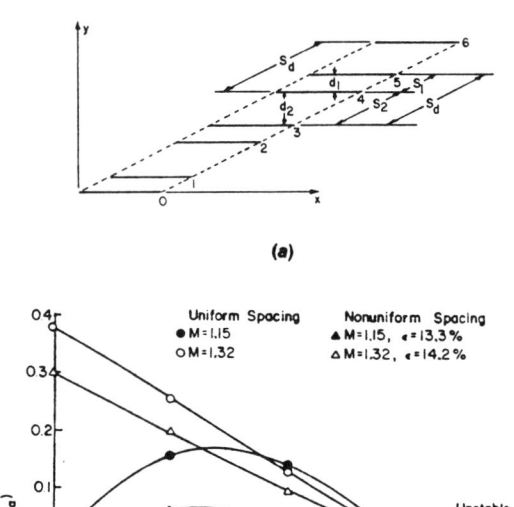

Fig. 7 a) Nonuniform cascade
b) Stability of uniform and nonuniform cascades (from Ref. 13).

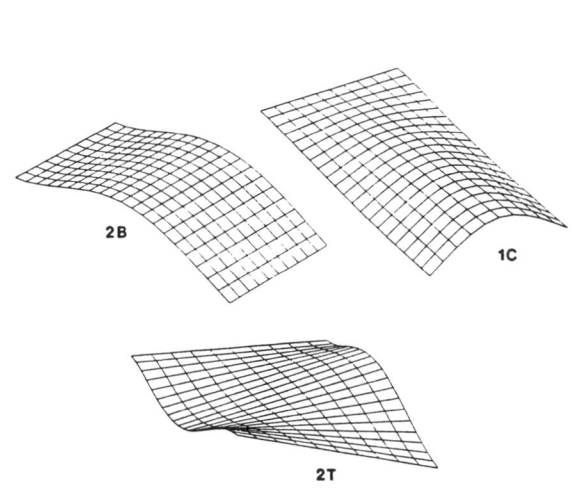

Fig. 8 Three of the five elastic blade modes used in the Rayleigh-Ritz formulation.

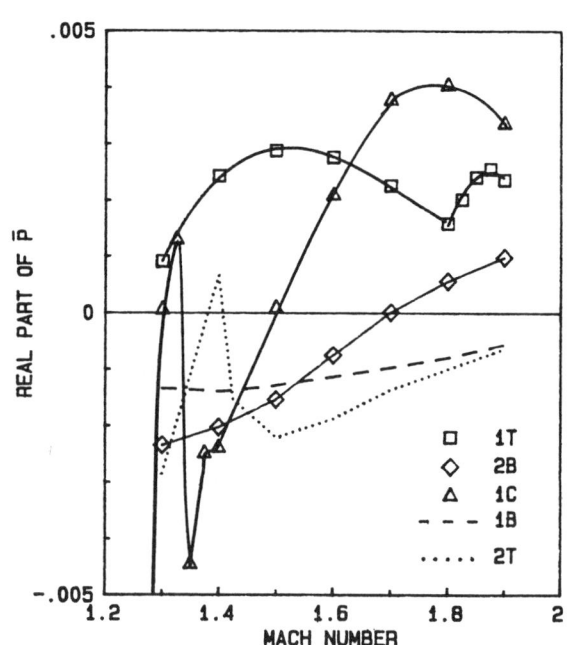

Fig. 9 Stability of baseline titanium blade, plotted as $\bar{p}_R$ vs. M for each aeroelastic mode.

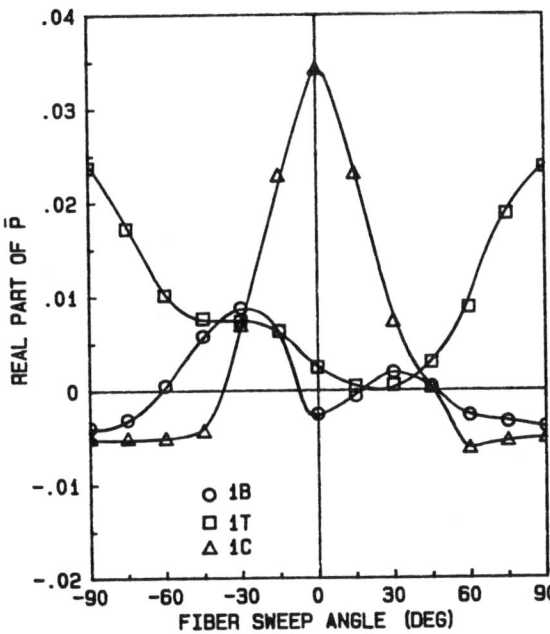

Fig. 10 Effect of fiber sweep on stability of composite blade first bending (1B), first torsion (1T) and first chordwise bending (1C) branches at M = 1.5.

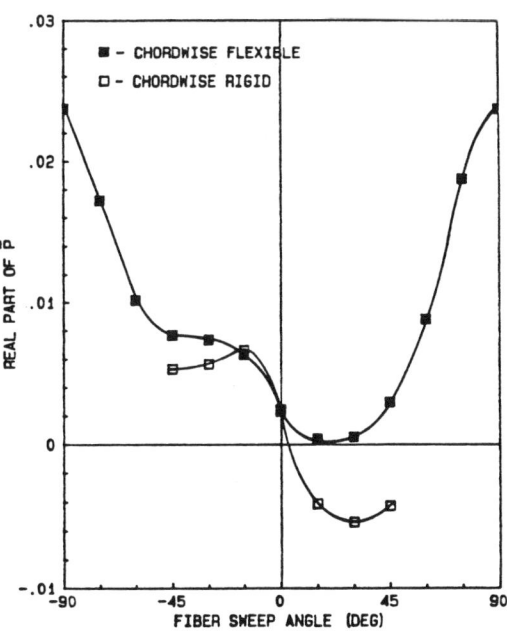

Fig. 11 Effect of chordwise flexibility on stability of first torsion branch of composite blade at M = 1.5.

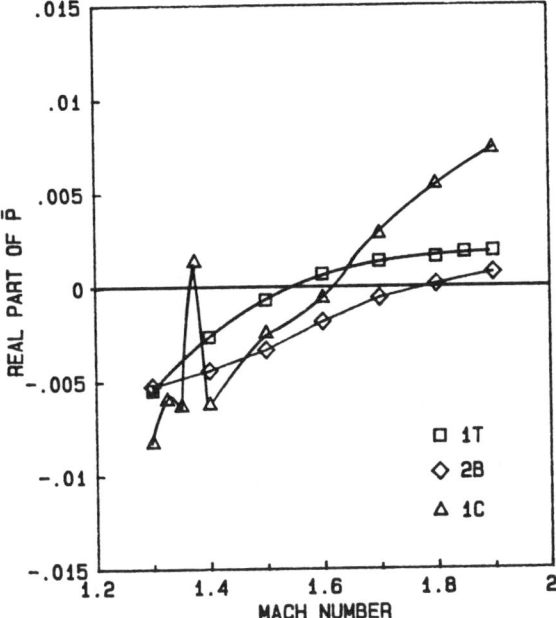

Fig. 12 Stability of ±45° cross-ply composite blade.

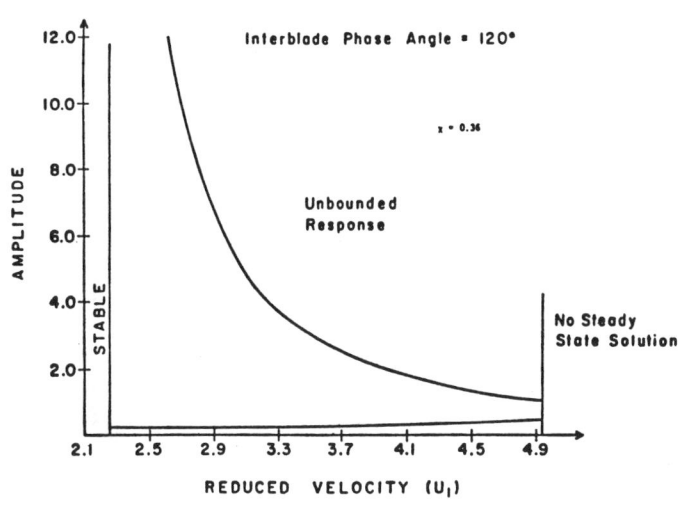

Fig. 13 Amplitude vs. reduced velocity (U/ωc) for typical mode of 3-bladed rotor with blade-to-blade dampers (from Ref. 30; incompressible flow).

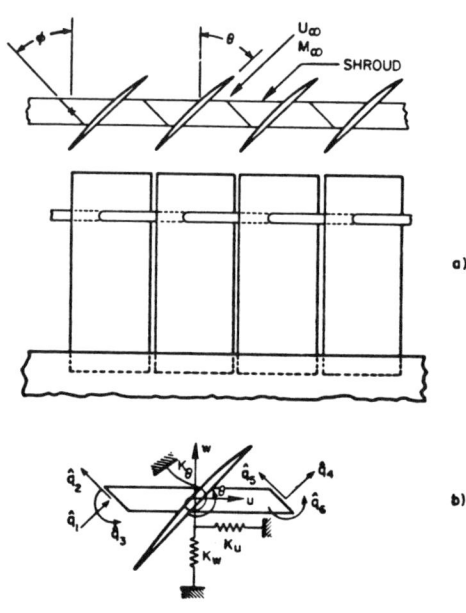

Fig. 14 Structural model of shrouded rotor:
a) infinite shrouded cascade
b) blade-shroud element.

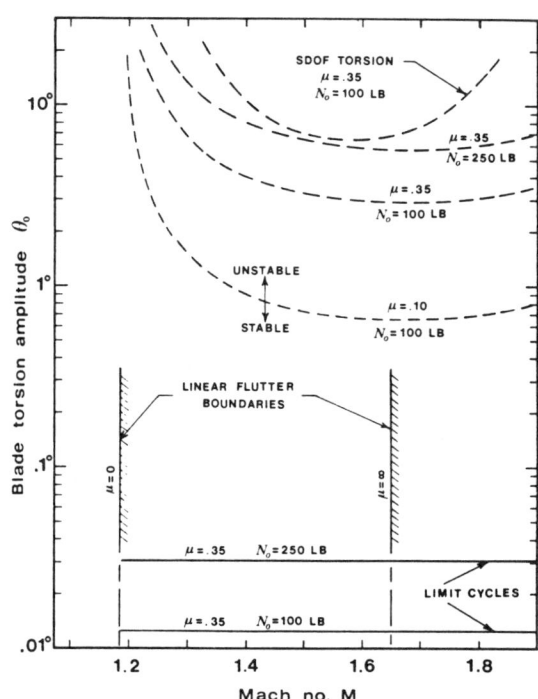

Fig. 15 Linear and nonlinear flutter boundaries for shrouded rotor (N = 24; Verdon's Cascade A typical section).

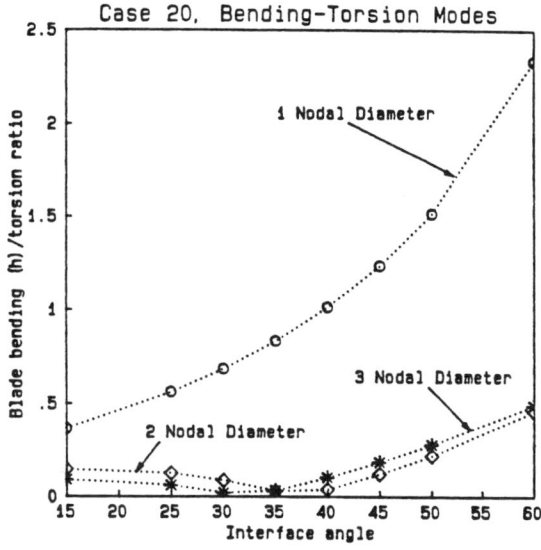

Fig. 16 Typical blade bending to torsion ratio, $h/\ell\theta$, vs. shroud interface angle for first 3 bending-torsion modes. (Slipping shrouds; $\ell$ = shroud element length; from Ref. 32).

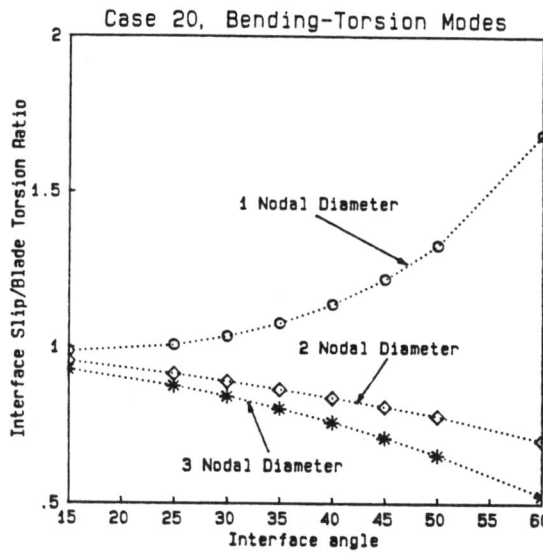

Fig. 17 Typical shroud slip to blade torsion amplitude ratio, $s/\ell\theta$, vs. shroud interface angle (from Ref. 32).

# REVIEW OF FLOQUET THEORY IN STABILITY AND RESPONSE ANALYSES OF DYNAMIC SYSTEMS WITH PERIODIC COEFFICIENTS

G. H. Gaonkar
Florida Atlantic University, Boca Raton, FL 33431

D. A. Peters
Georgia Institute of Technology, Atlanta GA 30332

## ABSTRACT

The applications of Floquet theory to typical engineering problems are briefly reviewed. The emphasis is to identify the fundamental sour.e of barriers which now prevent similar applications to large systems of design interest. Accordingly, toward eliminating those barriers, related computational problems are discussed concerning periodic equilibrium positions, transition matrices, and unsymmetric eigensolutions of transition matrices at the end of one period. New results demonstrate that such problems merit further research and that the feasibility of using Floquet theory in large-scale design analysis offers promise.

## INTRODUCTION

Approximations to a wide range of physical systems lead to "linearized" ordinary differential equations with periodic coefficients. Floquet theory has been extensively applied to such deterministic and stochastic systems for computing stability and response data --- damping levels, frequencies, periodic modal vectors, and forced-response [1-6]. Linearized is in quotes because the applications often refer to nonlinear systems for which nonlinear terms are important but do not dominate the solution (e.g., nonlinearities in lifting rotor blades). That is, the nonlinear effects are adequately manifest in the coefficients of the equations perturbed about a periodic orbit or equilibrium state. That orbit, which insures periodic forced-response, and the state transition matrix (STM) are often computed simultaneously by iterative adaptation of a method developed for the linear case [7-9]. The stability analysis then reduces to the unsymmetric eigenanalysis (eigenvalues and eigenvectors) of the Floquet transition matrix (FTM) which is the STM at the end of one period. Thus, the Floquet analysis basically comprises the periodic equilibrium state, STM over one period, and eigenanalysis of the FTM. However, in some cases, the control settings (or pilot's inputs) for a desired system operation or flight condition as well as the equilibrium state to give a periodic solution of the nonlinear equations (given those controls) have to be computed simultaneously or concomitantly [10]. Here we assume that the equilibrium state vector is augmented with the inclusion of unknown control inputs [10]. We do not address the related approach of reducing a periodic-coefficient system to a constant-coefficient system, what is often referred to as Lyapunov transformation [11]. However, the stability and response data are related to that transformation. Accumulated experience, though limited to relatively small order systems [3], does show that such a transformation is computationally too devious to be viable for large order systems (say, system dimension or order N>50).

To set the stage for the review, we begin with a brief reference to some of the applications of Floquet theory in different fields. Such applications have been the subject of many papers over the past 100 years. The first application seems to have been in the area of satellite stability [12,13]. The periodicity enters due to the periodic variation in gravity that a satellite (natural or artificial) encounters. Probably, most of the more recent research work on periodic-coefficient equations has been the result of the need to analyze and understand lifting rotors and wind-mills [2-10]. Other applications refer to such diverse problems as pendulum stability [14], dynamic stability of columns [15], and wave propagation in spatially periodic structures [1]. The encyclopedic work of Cesari [16] is a case in point. Though not exclusively addressed to Floquet theory, it is one of the earliest monographs which contains a mathematically oriented discussion of Floquet theory with an extensive reference to the applications of Floquet theory before the 1960's. Similarly, among other earlier applications of Floquet theory, the work of Cassedy and Oliver [1] on systems in communications engineering is often quoted [11]. A recent review of Dugundji and Wendell [9] contains a good account of Floquet theory as applied to rotating systems.

The Floquet analysis has the advantage that it is simple to implement, is applicable even to systems with strong periodic coefficients, and does not require explicit expressions for equation coefficients. Further, the periodic coefficient systems have properties that are quite similar to those of constant-coefficient systems. For example, the damping levels refer to exponential stability; the modal vectors, though not constant, are periodic; and the STM over one period determines time history for any length of time. From a computational point of view, these properties are significant, particularly in comparison even with linear systems with non-periodic time-variable coefficients. The importance of Floquet analysis arises from the fact that any nonlinear equation with a periodic orbit will have perturbations about that orbit that are described by linear equations with periodic coefficients [17]. Moreover, recent studies on stochastic second-moment stability of lifting rotor blades [5,6] are based on the assumption that the stochastic equlibrium state is close to the deterministic equilibrium state (an assumption that merits further study). Thus, nonlinear equations with periodic orbits impact on the field of chaos and strange attractors [18] and on the field of stochastic equilibrium and stability [5,6].

Given all these attractive features of Floquet analysis, how is it that it has not been found, as yet, to be viable for large-scale design analysis? This review is directed to that question. The review concerns the fact that the existing barriers are computational with respect to all three components of Floquet analysis --- equilibrium state, STM, and the eigenanalysis of the FTM. Although Floquet analysis is used rather routinely for systems up to 35th order or so for research purposes, design problems of practical interest require applications to systems of at least 50th order. (In fact, 100th order systems are quite common in lifting rotor design). Moreover, if we apply Floquet theory for the stochastic stability of an n-th order system (such as rotor stability in turbulence), then the Floquet analysis must be performed on state equations of order $N(N+3)/2$. In other words, a 50th order system requires a 1,325th order Floquet analysis. This partially explains why the stability of stochastic systems is virtually limited to an isolated rigid flap-lag blade model (N=4, and the order of the Floquet analysis = 14), despite the fact that the theory applies to any order. Even for deterministic systems, computational aspects, particularly machine time constraints and computational reliability, become crucial as the order of the FTM increases beyond 50.

Throughout, our emphasis is to facilitate an appreciation of the computational constraints or barriers which now prevent the use of Floquet analysis in design applications. To avoid ambiguities of definitions and terminologies in the subsequent discussion of computational aspects, we begin with a brief account of the three components of Floquet analysis. To put the existing barriers in proper perspective, we present results on the computational aspects of these components. The results are of an exploratory nature, mostly new, and refer to relatively small order systems. Nevertheless, they show why the feasibility of applying Floquet analysis in design offers promise. They also offer insight to further research as to what must be done to push these barriers even further.

FLOQUET ANALYSIS

We consider an N-dimensional linear system:

$$\{\dot{x}\} = [A(t)]\{x\} + \{C(t)\} \tag{1}$$

where the NxN state matrix A is T-periodic. Since the NxN state transition matrix $\Phi(t)$ is the fundamental matrix with $\Phi(0) = I$, we have:

$$[\dot{\Phi}] = [A(t)][\Phi], \; 0 \leq t \leq T \tag{2}$$

And the Floquet Transition matrix is $\Phi(T)$. It can be shown that, for some positive constant C, and for some norm (we choose the 2-norm throughout), we have:

$$||\Phi(t + nT)|| < C ||(\Phi(T))^n||, \; 0 \leq t \leq T, \; n=1,2,3,\ldots N \tag{3}$$

From the above equation, the basic result is that the system defined by equation (1) is stable if:

$$|\lambda_k| < 1 \text{ for all } k \; (k=1,2,\ldots N) \tag{4}$$

The $\lambda_k$'s, called the characteristic multipliers, are the eigenvalues of $\Phi(T)$; and they are symbolically represented as follows:

$$\lambda_k = \{\alpha [\Phi(T)]\}, \; k=1,2,\ldots N \tag{5}$$

where $\alpha[\;]$ implies the eigenvalues of $[\;]$. We assume throughout that $\Phi(T)$ has distinct eigenvalues. The past experience with small to medium size systems (N> 30) shows that such an assumption is usually reasonable in actual computations, due to rounding errors etc. [4].

For a better physical interpretation, it is good to study the role of $\lambda_k$ vis-a-vis the stability margins and frequencies. For this, invoking Floquet theory, we have:

$$\{x\} = [B(t)] \{e^{pt}\} \tag{6a}$$

where $B(t)$ is an NxN, T-periodic matrix; and $\{e^{pt}\}$ is an Nx1 column vector. For $t=T$,

$$e^{p_k T} = \{\alpha[\Phi(T)]\} = \lambda_k \tag{6b}$$

and,

$$p_k = (\ln|\lambda_k| + i \arg \lambda_k) = n_k + i\omega_k \tag{6c}$$

In the above equation, $n_k$ and $\omega_k$ respectively represent the damping levels or stability margins (growth or decay of the transient solution) and the nonunique frequencies. Thus, the crux of the stability investigation boils down to computing the FTM $[\Phi(T)]$ and its dominant eigenvalue, $(\alpha [\Phi(T)])$ max, as discussed in the sequel following the discussion of forced response, taken up next.

PERIODIC RESPONSE

Referring to the complete or nonhomogeneous equation (1), the initial state to yield periodicity of the steady state, $\{x(0)\} = \{x(T)\}$, is given by [10]:

$$\{x(0)\} = [I-\Phi(T)]^{-1} \{x_E(T)\} \tag{7}$$

where $\{x_E(T)\}$ is the nonperiodic solution of the complete equation (1) at $t=T$ for the zero initial state. Or equivalently we have,

$$[I - \Phi(T)] \{x(0)\} = \{x_E(T)\} \tag{8}$$

Engineering systems are usually nonlinear and can be represented by the equation,

$$\{\dot{x}\} = [A(t)]\{x\} + C(x,\dot{x},t) \tag{9}$$

where $A(t)$ is an NxN, T-periodic matrix. The correspondent initial state to yield periodicity, $x(0)=x(T)$, is obtained by an iterative application of equation (7) in which one of the elements of the estimated initial state $x_E(0)$ is perturbed by a small amount in each iteration as sketched in Table I [10]. The variety of iteration schemes used in the literature are conceptually variants of the classical shooting technique with Newton-Raphson iteration [10]. Computationally, the important ingredient in the forced-response analysis is the role of the $[I-\Phi(T)]$; see equation (8) for the linear case and equation (10) for the nonlinear case. That role is best described by the matrix condition number concept discussed in the section entitled "Computational Reliability". Equation (10), in Table I, is the iterative adaptation of equation (8).

**TABLE 1**

PROCEDURE FOR OBTAINING THE PERIODIC SOLUTION OF THE COMPLETE NONLINEAR EQUATION AS A MODIFIED NEWTON RAPHSON PROCEDURE

1. Assume some N x 1 initial state vector....$x_E(0)$

2. Obtain the N x 1 numerical response vector $\{x_E(t)\}$ of the complete nonlinear equation for the assumed initial state.

3. Sequentially perturb each of the N elements of $\{x_E(0)\}$ by a small amount $\varepsilon_j(j=1,N)$ to obtain a total of N solution vectors $[x(t)]$. Then form:

$$\{\Delta x_E(t)\}_j = \{x(t)_j\} - \{x_E(t)\}$$

$$[\Phi(t)] = \left[ \frac{1}{\varepsilon_1}\{\Delta x_E(t)\}_1 \ldots, \frac{1}{\varepsilon_n}\{\Delta x_E(t)\}_n \right]$$

$$x(t) = x_E(t) + [\Phi]\{\varepsilon\}$$

4. Iterate with a new corrected $x_E(0)$ till convergence is obtained.

That is, after the k-th iteration, we have:

$$[I-\Phi(T)]_k \{x(0)-x_E(0)\}_{k+1} = \{x_E(T)-x_E(0)\}_k \tag{10}$$

Before concluding this section, we offer the following comments concerning equation (8). Theoretically, the FTM is nonsingular, as seen from equation (6b). However, the generated value is FTM + $\delta$(FTM), where $\delta$(FTM) represents computational perturbations. That the generated value will be close to being singular will be well reflected by the condition number of (FTM + $\delta$(FTM)) as pursued in a subsequent section on computational reliability. Further, the matrix $[I-\Phi(T)]$ is theoretically singular only for some pathological cases when the homogeneous system has nontrivial periodic solutions; cases that are not typical of stable or unstable engineering systems of design interest. Further, we also reiterate that the iteration scheme in equation (10) is keyed to the assumption that the nonlinear terms, though important, are not dominant. As mentioned earlier, such an assumption is valid for a wide class of systems and implies that the iteration scheme is convergent.

When pilot control settings are included as unknowns in the problem (in order to trim the helicopter), the iteration scheme in equation (10) includes these unknowns (usually no more than 6) along with the x(0) unknowns. The added errors (up to 6) are errors in force balance rather than in periodicity. Although these additional unknowns do not add significantly to the order of the matrix $I-\Phi(T)$ (which then becomes a partial derivative matrix), they can seriously impact its condition number [10]. Figure 1, taken from [10], shows quality of initial guess ($\rho=1$ implies a perfect guess) versus advance ratio ( a measure of helicopter forward speed). Here, the guess includes 4 initial conditions, x(0), and 4 unknown controls. The shaded areas are initial guesses that result in divergence of equation (10). The irregularity of the curve is a direct result of ill conditioning.

## EIGENANALYSIS

The crux of the Floquet analysis is the eigenanalysis, which becomes more and more demanding with increasing order of the FTM. Due to algorithmic robustness and availability of well-documented computer codes, the generic QR-method (e.g., EISPACK version for a general matrix) is almost exclusively used for the eigenanalysis [19]. Such a usage, when applied to large order systems, raises several issues.

First, for a general matrix, the QR-method is the recommended method for a __complete__ eigenanalysis, as seen from its algorithmic structure (e.g., reduction to Hessenberg form). Therefore, the operation counts, and consequently the machine time requirement, grow cubically with the order of the FTM. However, the Floquet analysis theoretically requires only the dominant characteristic multiplier. (In practice, we need a small subset of the dominant/sub-dominant eigenvalues and the correspondent eigenvectors due to frequency ambivalence and due to the necessity of identifying stability margins of critical modes.) Second, for many cases of time-invariant structural systems, well beyond the 100th order, eigenanalysis has been made with remarkable computational efficiency. That is because algorithmic details are tailored to fit special properties of the state matrix such as symmetry, sparsity, and bandwidth or organized sparsity. The development of the computer codes based on the Lanczos procedure for large symmetric eigenvalue computations is a case in point [20]. By comparison, the FTM is not symmetric. Though unique to a particular system, it is a numerically generated solution to an initial value problem. Numerical experience shows that for N<50 sparsity ceases to be a factor. However, it must be mentioned that sparsity may be a factor for large order systems and that it merits further documentation. Third, compared to the FTM computations of high accuracy, an approximate eigenanalysis is sufficient for design applications. Fourth, higher order systems usually emerge as a result of correlations with test data or comparisons with earlier models or designs. With the benefit of such prior information, a better-than-ad-hoc estimate of the critically damped modes is possible, and it can be well exploited in iterative-type procedures.

Given the above issues, the generic simultaneous iteration (SI) technique for unsymmetric matrices offers considerable promise [21-23]. Its algorithmic structure naturally lends itself to approximately computing a small subset of dominant/subdominant p(p<<N) eigenvalues and the corresponding eigenvectors, as required in the Floquet analysis. Descriptively stated, the eigenanalysis per se centers around an interaction or coupling matrix whose order, say m, is far less than N(p<<N, p<m). The p-m eigenvectors can be used to virtually eliminate the risk of poor convergence. It is true that the SI-method is not as foolproof as the QR-method (defective interaction matrix for some contrived examples) and that the information-base on the SI-method for unsymmetrical matrices is highly limited. However, these limitations should be viewed in the context that the eigenanalysis of large unsymmetric matrices is still an evolving process. Further, the basic issues of the Floquet analysis seem to outweigh the limitations of the SI-method and motivate the search for a viable alternative to the QR-method.

## STATE TRANSITION MATRICES

There are basically two methods which are primary candidates for the FTM computation for large systems. One is that of conventional, initial value codes; and the other is Hamilton's Finite Elements. We wish to discuss both methods briefly, since each has certain advantages for different aspects of the work. Conventional, initial-value codes (time marching) are presently used almost exclusively in the generation of the FTM matrix. These generally take the form of Runge Kutta or fourth-order predictor-correctors. These are well-proven and easy to implement. On the other hand, the new generation of finite elements in time offer much promise as possible aids in extending the applicability of Floquet theory.

In brief, the finite-element (or bilinear) approach expands the solution between two points in time, as

$$q(t) = \sum_{i=1}^{\infty} c_i \, \Theta_i(t) \tag{10a}$$

$$\delta q(t) = \sum_{i=1}^{\infty} d_i \psi_i(t) \tag{10b}$$

where the $\Theta_i(t)$ are appropriately chosen functions and where the $c_i$ are unknown constants. Substitution of (10a) and (10b) into the virtual action principle,

variation (action) + virtual action = 0

for the period in question ($t_0 \leq t \leq t_1$) yields algebraic equations for the unknowns, $c_i$. Several characteristics of this method make it ideal for FTM applications.

First, when we use the correct formulation for $\psi_j$, $\psi_j(T) = 0$, the method can be proven to converge uniformly; and we have verified this numerically. Second, the minimum error on $q(t)$ is at $t=T$ (the exact point we need) and this error converges at a rate much faster than it would with time marching, Third, a Lagrange multiplier associated with the problem can give an estimate of $\dot{q}(T)$ that also converges at a spectacular rate. Since the FTM needs $q(T)$ and $\dot{q}(T)$, this method provides highly accurate values of the transition matrix. Furthermore, the method is adaptable (in terms of number of elements and number of polynomials per element) to give an optimum accuracy. Finally, preliminary computations show that the bilinear formulation may become more efficient (vis-a-vis marching) as the order of the system increases. Thus, the finite elements in time could be one key to application of larger systems.

Figure 2, from [24], provides typical results from initial-value codes of the computing time required for a given accuracy of Floquet analysis. The solid curves represent numerical experiments, and the dashed curves represent results from an a priori count of floating-point multiplications. The N-pass results are for conventional computation of the FTM. The single-pass results refer to a case for which periodic coefficients are calculated only once (not repeated for each set of initial conditions). The horizontal scale is the number of blades. The order of the system is $4b + 3$ where $b$ is the number of blades. The results show how CPU increases as the order of the system increases.

Figure 3 provides a typical plot of error versus computing time for a 2nd order system. Results from the bilinear formulation (one element) are compared with those of Hamming's Predictor-Corrector. Here we see the superiority of finite elements (B2) at strict error levels. The same superiority occurs for large number of degrees of freedom.

COMPUTATIONAL RELIABILITY

The results we presented earlier refer to the computational reliability of the FTM, the emphasis being on high accuracy. By comparison, a similar exercise for periodic initial state and eigenanalysis is much more demanding. We now approach that exercise in three phases. First, we discuss a set of numerical coordinates as a viable means of quantifying reliability. We also touch upon why the computational requirements are somewhat different from those of the FTM. Second, we give a brief account of computing those coordinates. Third, we present numerical results to demonstrate the feasibility of using such coordinates.

The computations of the periodic initial state, as well as the eigenvalues and eigenvectors of the FTM, are subject to numerical perturbations which can become magnified due to already existing errors in the FTM. Therefore, it is necessary to have an objective measure of the goodness of the computations for spot checking purposes. To this end, we introduce 3 numerical coordinates: (1) condition number of [I-FTM], (2) condition number of a characteristic multiplier, (eigenvalue of the FTM) and (3) residual error of an eigenpair (eigenvalue and the corresponding eigenvector).

While the first coordinate is an a priori evaluation, the latter two are a posteriori. Further, the first two condition numbers respectively provide upper bound estimates of how bad the error can be in the computations of the initial state and the characteristic multiplier. For a real symmetric matrix (in general,

Hermitian), the eigenvalue problem is well conditioned, since the right eigenvector is equal to the left eigenvector. This fact is well reflected in the successful development of several computer codes for large symmetric eigenproblems. By comparison, the eigenanalysis of large unsymmetric matrices is not completely understood, and eigenanalysis of large FTMs is even less understood. Further, the data on the reliability of large unsymmetric eigenproblems is limited. Moreover, compared to the overall computational effort of the Floquet analysis, the additional machine time of occasionally computing a left eigenvector for spot checks is not appreciable (the right eigenvector and the eigenvalue are routine outputs in the Floquet analysis). We also mention that the left eigenvector is very valuable if one wishes to use the Floquet solution in order to decompose the system for forced response. These are appealing aspects of the eigenvalue condition number, which is an estimate of how much the perturbations in the elements of the FTM may be magnified and transformed to contamination of the corresponding eigenvalue. In other words, it shows the sensitivity of the eigenvalues to small changes in the elements of the FTM. Though the condition-number concep. has a rigorous analytical basis, the corresponding sensitivity analysis for eigenvectors is too complex. However, a judicious combination of the condition number of an eigenvalue and the relative residual error of the corresponding eigenpair should provide a reasonable estimate of the reliability of the eigenpair computations.

Coming to the condition number of [I-FTM] in particular, we realize that it entails additional computational effort and that the actual error in some cases could be magnified appreciably. However, this condition number (as an a priori estimate) provides a rational means of quantifying the goodness of a simultaneous computation of the FTM and the equilibrium position, which are fundamental blocks of the Floquet analysis.

Given the complexities of the Floquet analysis, the condition numbers that are related to error bounds may overly magnify the actual error, and one coordinate may fail to provide enough information. However, the established trends of the goodness of the Floquet analysis (vis-a-vis the order of the system) should remain valid. Moreover, when such coordinates are tested on the FTMs of increasing order, the results provide direction as to the judicious choice and combination of the coordinates that provide improved bounds and error measures. These results are also required to determine objectively how far the Floquet analysis is practical with increasing number of degrees of freedom and how far computational reliability is an issue.

For simplicity of presentation, we refer to a generic matrix A which can be appropriately identified with the FTM in the eigenanalysis, as in equation (5), or with the matrix [I-FTM] in the analysis for periodic initial state, as in equations (8) and (10). The matrix A and its transpose $A^T$ have the same eigenvalues. We assume that $\lambda$ is one such simple eigenvalue (of multiplicity one) and that x and y are the corresponding right-and left-hand eigenvectors. Therefore we have:

$$Ax = \lambda x \text{ and } A^T y = \lambda y \qquad (11)$$

Then, the condition number of $\lambda$ is given by:

$$\text{Cond.}(\lambda) = | y^T x |^{-1} \tag{12}$$

The above equation merits some comments. For engineering systems, $\lambda$ is usually complex, as are the eigenvectors. It is $y^T$ (and not Hermitian transpose $y^H$) that is used [25]. Further, the vectors are normalized. That is, $||x||=||x^H x||=1=||y||=||y^H y||$. For Hermitian matrices $y^H = y^T$, therefore, cond.$(\lambda)=1$ and the eigenproblem is well-conditioned. The significance of the condition number Cond.$(\lambda)$ is demonstrated by the following relation:

$$|\delta\lambda| < ||\delta A|| \; \text{Cond.}(\lambda) \tag{13}$$

where $||\delta A||$ represents the spectral norm of the matrix of perturbations of A. Further, $\delta\lambda$ represents the perturbation of $\lambda$ due to errors in the computed matrix, A + $\delta$A, instead of A. Thus, Cond.$(\lambda)$ provides a measure of the sensitivity of an eigenvalue, typified by $\delta\lambda$, to small changes in the elements of the generated A, typified by $\delta$A. If small changes in the elements of A can lead to arbitrarily large changes in an eigenvalue, then the eigenvalue problem for that eigenvalue is said to be ill-conditioned. That a matrix is well-conditioned for eigenvalue computations is no guarantee that it is well-conditioned for eigenvector computations. Though the condition number approach is fundamental, a similar condition analysis for eigenvectors is too involved. Therefore, it is expedient to study the reliability of the computed eigenpair $(\lambda,x)$ by the residual error approach, which gives the error measure $\varepsilon$, that is,

$$\varepsilon = \frac{||Ax - \lambda x||}{||\lambda x||} = \frac{||r||}{|\lambda|} \tag{14}$$

where $||r||$ is the Euclidean norm of the residual vector r.

It appears that Cond.$(\lambda)$ and error measure $\varepsilon$ should provide adequate information on the reliability of a computed eigenpair.

The concept of matrix condition number, say, Cond.(A), plays a fundamental role in the solution of equations. Referring to equations (8), we identify the matrix [I-FTM] with A. Since A is unsymmetrical, we have:

Cond.(A) = \{max. eigenvalue of $A^T A\}^{\frac{1}{2}}$ / \{min. eigenvalue of $A^T A\}^{1/2}$

$$> |\lambda_1| / |\lambda_n| \tag{15}$$

where $\lambda_1$ and $\lambda_n$ are the largest and smallest eigenvalues of A in absolute value, ($|\lambda_1|$ being the spectral radius of A.). For symmetric matrices, the above inequality reduces to equality, as widely used in finite element analyses. Further:

$$1 < \text{Cond.}(A) < \infty \tag{16}$$

The role of Cond.(A) merits further elaboration due to computational peculiarities associated with the periodic initial state. We introduce two vectors x and b which are respectively identified with $\{x(0)\}$ and $\{x_E(T)\}$ in equation (8), or with $\{x(0) - x_E(0)\}_{k+1}$ and $\{x_E(T) - x_E(0)\}_k$ in equation (10). Therefore, the <u>actual</u> computed periodic initial state $x+\delta x$ can be typified by the following equation:

$$[A+\delta A] \{x+\delta x\} = \{b+\delta b\} \tag{17}$$

where $\delta A$ represents the computational perturbations from two sources. The first source is in the solution for the FTM. The second source, peculiar to the diagonal elements, is in the simple execution (I-FTM), when the diagonal elements of the FTM are close to one (also see the 15x15 FTM in Table II). Similarly, $\delta b$ in equation (17) represents computational perturbations in computing the response at t=T for assumed initial states. It can be shown that [24]:

$$\frac{||\delta x||}{||x||} < \frac{\text{Cond}(A)}{1-\text{Cond}(A)(||\delta A||/||A||)} \left\{ \frac{||\delta A||}{||A||} + \frac{||\delta b||}{||b||} \right\} \tag{18}$$

With the reasonable assumption that $\text{Cond.}(A)(||\delta A||/||A||) < 1$, as a first order approximation, we have:

$$\frac{||\delta x||}{||x||} \quad \text{Cond}(A) \left\{ \frac{||\delta A||}{||A||} + \frac{||\delta b||}{||b||} \right\} \tag{19}$$

The above equation shows that Cond. (A) represents the maximum possible magnification of the relative errors in A and b in the computed initial state. Thus, the higher is the value of Cond.(A), the more is the sensitivity of the equation (17) to computational perturbations; and consequently the less well-conditioned is the computational problem of finding periodic initial states. It must be appreciated that the above equation on relative error refers to a priori upper-bound estimates.

What is of particular significance, is the expected number of significant figures. To this end, it is reasonable to assume that, for a t-digit precision computer:

$$\frac{||\delta A||}{||A||} + \frac{||\delta b||}{||b||} = 10^{-t}, \tag{20}$$

For s-digit precision in the solution, we set,

$$\frac{||\delta x||}{||x||} = 10^{-s} \tag{21}$$

Then, from equation (19), we have:

$$s > t - \log_{10}(\text{Cond.}(A)) \tag{22}$$

Some aspects of the above equation are touched upon next.

We introduced above a set of numerical coordinates on computational reliability such as matrix condition number, N condition numbers for N eigenvalues, and the residual errors for the N eigenpairs. In Table II, we include only the condition number of the FTM (T) and not of $[I-\Phi(T)]$, although the latter matrix is involved in computing for the periodic initial state. Similarly, only the maximum values of Cond.$(\lambda_i)$ and of residual error are presented for brevity.

It is instructive to study these numerical coordinates on the basis of results related to a physical system. Accordingly, we select the experimental model of a helicopter rotor with three "rigid" blades. The model has six degrees of freedom, since each blade executes flapping (our of plane) and lagging (inplane) motions. In addition to a steady airfoil aerodynamics, the unsteady aerodynamic effects are modeled on a global basis by three first-order aerodynamic feedback systems (the so-called dynamic-inflow feedback). Therefore, we have a 15 x 15 state matrix, a very low-order system when compared to typical aeroelastic models of design interest. Nevertheless, the results in Table II facilitate an appreciation of the role of the numerical coordinates vis-a-vis computational efficiency in stability and response analyses of typical models. The preliminary results also include relatively high-speed forward flight conditions (dimensionless speed parameter $\mu > 0.25$). For $\mu \neq 0$, unsymmetric aerodynamic disturbances introduce dominant periodic coefficients, (the higher the $\mu$, the more is the dominance). For $\mu = 0$, the system has constant coefficients, which is referred to as hovering flight.

Table II shows the 15 x 15 FTM for $\mu = 0.4$, along with the numerical coordinates for $\mu=0$, 0.3, and 0.5. These three cases represent a broad-spectrum of systems, starting from systems with constant coefficients ($\mu=0$) to systems with excessively dominant periodic coefficients ($\mu=0.5$). The third case ($\mu=0.5$) is included to simulate severe test cases and is not typical of forward flight conditions. It is seen that the FTM, which is almost always unsymmetrical, has zero sparsity. This lack of sparsity is due to two reasons. First, sparsity is usually evident for large order systems, say beyond 50-100th order. Second, helicopters represent formidable nonconservative systems with some sort of coupling among most of the modes. Table II also shows that the Cond.(FTM), given in the second column, increases with increasing $\mu$. This means that, for high values of $\mu$, the solution of periodic initial state warrants considerable refinement in the computation of both the FTM and the non-periodic responses, as typified in equation (19). We mention in passing that the Cond.(FTM), even for $\mu=0.5$, is far less than that of the 15 x 15 Hilbert matrix which is "notoriously ill-conditioned".

An appealing aspect of the condition number concept is typified by equations (19) and (22), that is to study the role of Cond.(A) vis-a-vis the estimated number of accurate digits. Assuming that we are using seven-digit arithmetic, we have from equation (22): $s > 7 - \log_{10}(\text{Cond.A})$. Thus, for $\mu=0.3$, we can expect at least two-to-three digit accuracy. According to equation (22), for $\mu=0.5$, the seven-digit arithmetic is not adequate. Though these upper-bound error estimates may overly magnify the actual error, they do provide an objective measure of how bad errors can be in solving for the periodic initial states.

The third column is a lower-bound approximation based on the ratio of $|\lambda_{max}|$ and $|\lambda_{min}|$ of the FTM. For symmetric matrices, the lower-bound inequality reduces to equality. This ratio is a poor approximation to Cond.(FTM), as is typical of unsymmetrical matrices. The coordinates in columns 4 and 5 refer to eigenvalues and eigenpair computations. For Hermitian matrices, Cond.($\lambda$)=1, and the eigenvalue problem is well conditioned. By comparison, it is seen that, with increasing $\mu$, the eigenvalue problem becomes less and less well conditioned. This feature is consistent with data in column 5 which show similar trends with respect to maximum residual errors (see equation (14) for eigenpair computations). It should be appreciated that the state-of-the-art of assessing the computational reliability of the eigenproblems of unsymmetric matrices merits considerable refinements. This problem is particularly acute for large systems concerning which very little is documented.

## CONCLUDING REMARKS

We have seen that Floquet theory is necessary to a wide variety of dynamics problems of periodic-coefficient systems including deterministic stability, stochastic stability, and forced response. However, the n numerical application of Floquet theory requires a transient time solution (over on period) and an eigenvalue analysis. For smaller systems <35th order these have been done; but for large systems (typical of realistic design problem, > 100th order), applications have not yielded reasonable results. It appears, however, that use of very accurate time histories (by finite elements in time) combined with new methods of eigenvalue analysis could alleviate the situation. We recommend the use of condition numbers to determine the best means of moving this analysis tool into the realm of large systems.

## REFERENCES

1. Cassedy, E. S. and Oliver, A.A. "Dispersion Relations in Time-Space Periodic Media: Part I - Stable Interactions", Proceedings of the IEEE, Volume 51, July-Dec. 1963, pp. 1342-1359.

2. Peters, D.A. and Hohenemser, K.H., "Application of the Floquet Transition Matrix to Problems of Lifting Rotor Stability", Journal of the American Helicopter Society, Vol. 16, No. 2, April 1971, pp. 25-33.

3. Schrage, D.P., and Peters, D.A, "Effect of Structural Coupling Parameters on the Flap-lag Forced-Response of a Rotor Blade in Forward Flight Using Floquet Theory," Vertica, Vol. 3, No. 3, 1979, pp. 177-185.

4. Nagabhushanam, J., and Gaonkar, C.H., "Rotorcraft Air Resonance in Forward Flight with Various Dynamic Inflow Models and Aeroelastic Couplings", Vertica, Vol. 8, No. 4, 1984, pp. 373-394.

5. Shiau, T.N., Rotor Blade Flap-lag Stability and Response in Forward Flight in Turbulent Flows, Ph.D. Thesis, University of Illinois at Urbana-Champaign, 1984, Chapter 6.

6. Prussing, J.E., Lin. Y.K., and Shiau, T.N., "Rotor Blade Flap-Lag Stability and Response in Forward Flight in Turbulen Flows," Journal of the American Helicopter Society, Vol. 29, No. 4, October, 1984, pp. 81-87.

7. Panda, Brahmananda, Dynamic Stability of Hingeless and Bearingless Rotor Blades in Forward Flight, Ph.D. Dissertation, University of Maryland, 1985, Chapter 3.

8. Nilakantan, C.R. and Gaonkar, C.H., "Feasibility of Simplifying Coupled Lag-Flap-Torsional Models for Rotor Blade Stability in Forward Flight", Vertica, Vol. 9, 1985, pp. 241-256.

9. Dugundji, J. and Wendell, J.H., "Some Analysis Methods for Rotating Systems with Periodic Coefficients," AIAA Journal, Vol. 21, No. 6, June, 1983, pp. 890-897.

10. Peters, D.A. and Izadpanah, A.P., "Helicopter Trim by Periodic Shooting with Newton-Raphson Iteration", Proceedings of the 37th American Helicopter Society Annual Forum, 1981, Preprint No. 81-23.

11. Pease, M.C., Methods of Matrix Algebra, Academic Press, 1965, Chapter 6.

12. Mathieu, E., "Mémoire sur le Mouvement Vibratoire de une Membrance de Forme Elliptique", J. Math. Pures et Appliquee, 1868.

13. Hill, G.W., "Mean Motion of the Lunar Perigee, "Acta Mathematica, Vol. 8, January 1886.

14. Kane, T.R., and Kahn, M.E., "On a Class of Two-Degrees-Of Freedom Oscillations", ASME Paper No. 68-APM-34, December 1967.

15. Bolotin, V.V., *The Dynamic Stability of Elastic Systems*, Holden-Day, San Francisco, 1964

16. Cesari, L. *Asymptotic Behaviour and Stability Problems in Ordinary Differential Equations*, Third Edition, Springer-Verlag, New York, 1971.

17. Brockett, R.W., *Finite Dimensional Linear Systems*, John Wiley & Sons, New York, 1970, Chapters 1 and 4.

18. Holmes, P.J. and Moon, F.C., "Strange Attractors and Chaos in Nonlinear Mechanics", *Journal of Applied Mechanics*, Vol. 50, 1983, p. 1021.

19. Garbow, B.S., Boyle, J.M., Dongarra, J.J., and Moler, C.B., Matrix Eigensystem Routines -- EISPACK Guide Extension", *Lecture Notes in Computer Science,* Springer-Verlag, New York, 1977.

20. Cullum, J.K. and Willoughby, R.A., *Lanczos Algorithms for Large Symmetric Eigenvalue Computations.* Vol. 1 (Theory) and Vol 2 (Programs), Birkhauser, Boston, 1985.

21. Jennings, A. and Steward, W.J., "Simultaneous Iteration Method for the Unsymmetric Eigenvalue Problem", *Journal of the Institute of Mathematics and its Applications (UK)*, Vol. 8, 1971, pp. 111-121.

22. Jennings, A. and Steward, W.J., "Simultaneous Iteration for Partial Eigensolution of Real Matrices", *Journal of the Institute for Mathematics Applications (UK)*, Vol. 15, 1975, pp. 351-361.

23. Stewart, C.S., "Simultaneous Iteration for Computing Invariant Subspaces of Non-Hermitian Matrices", *Nurmerische Mathematik*, Vol. 25, 1976, pp. 123-136.

24. Gaonkar, G.H., Prasad, D.S. Simha, and Sastry, D., "On Computing Floquet Transition Matrices of Rotorcraft", *Journal of the American Helicopter Society,* Vol. 26, No. 3, July 1981, pp. 55-61.

25. Ortega, J.M., *Numerical Analysis, A Second Course*, Academic Press, New York, 1972, Chapters 2 and 3.

116 / PART I

# TABLE II
## FLOQUET TRANSITION MATRIX ($\mu = 0.4$)

FTM (15 X 15) AND EIGENANALYSIS

| $\mu$ | Cond (FTM) | $\geq |\lambda_{max}|/|\lambda_{min}|$ | max.cond ($\lambda_i$) | Residual Error $\epsilon$ |
|---|---|---|---|---|
| 0 | $1.9 \times 10^2$ | $6.8 \times 10$ | 1.7 | $0.24 \times 10^{-14}$ |
| 0.3 | $2.9 \times 10^4$ | $6.8 \times 10^2$ | 2.5 | $0.80 \times 10^{-11}$ |
| 0.5 | $9.2 \times 10^6$ | $1. \times 10^6$ | 3.8 | $0.17 \times 10^{-8}$ |

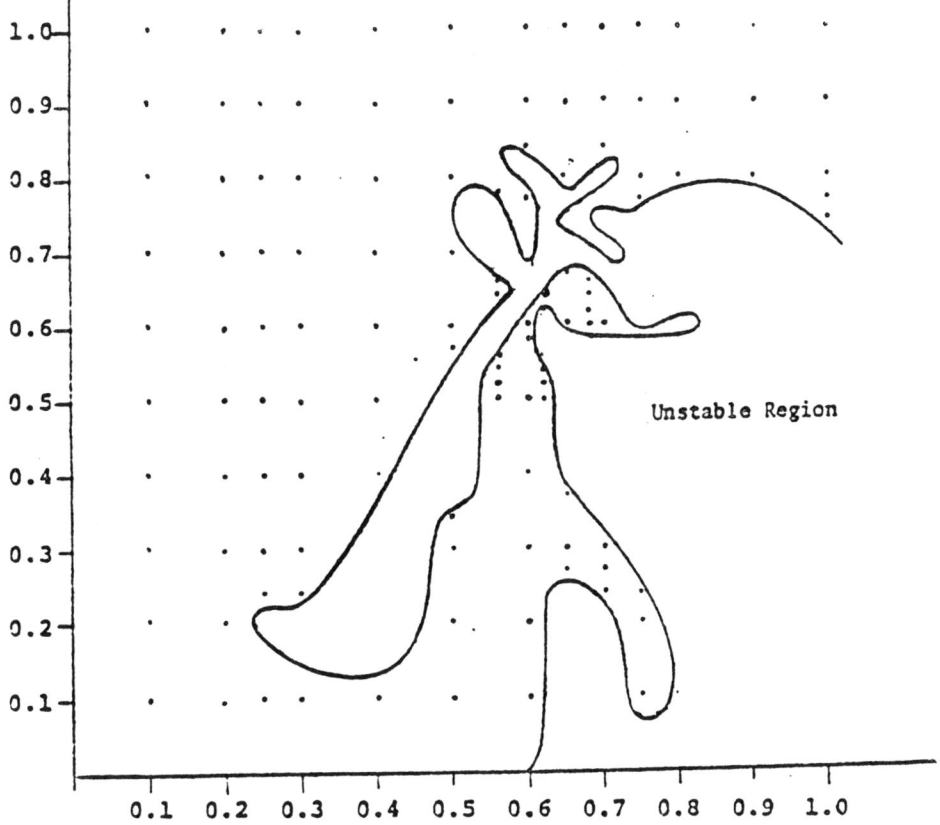

Figure 1
Convergence of Method
versus
Advance Ratio and Accuracy of Initial Guess

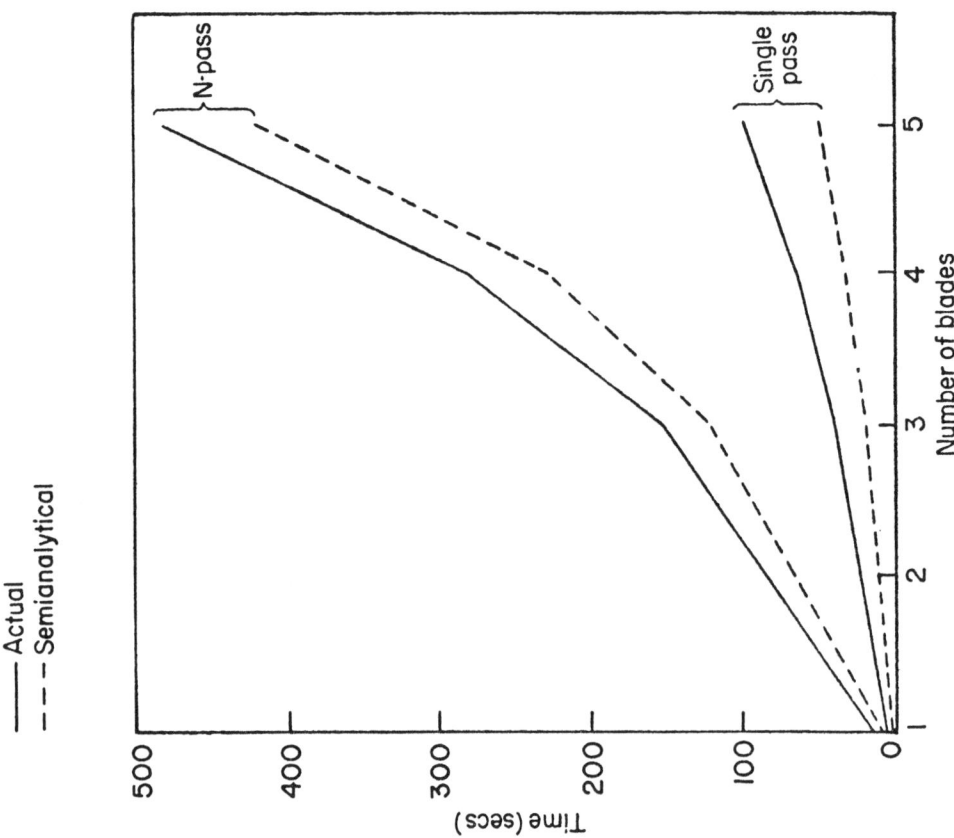

Fig. 2 C.P.U. Time on IBM 360/44 for computing FTM from the Hamming Predictor-Corrector Method (sparseness of state matrix not exploited).

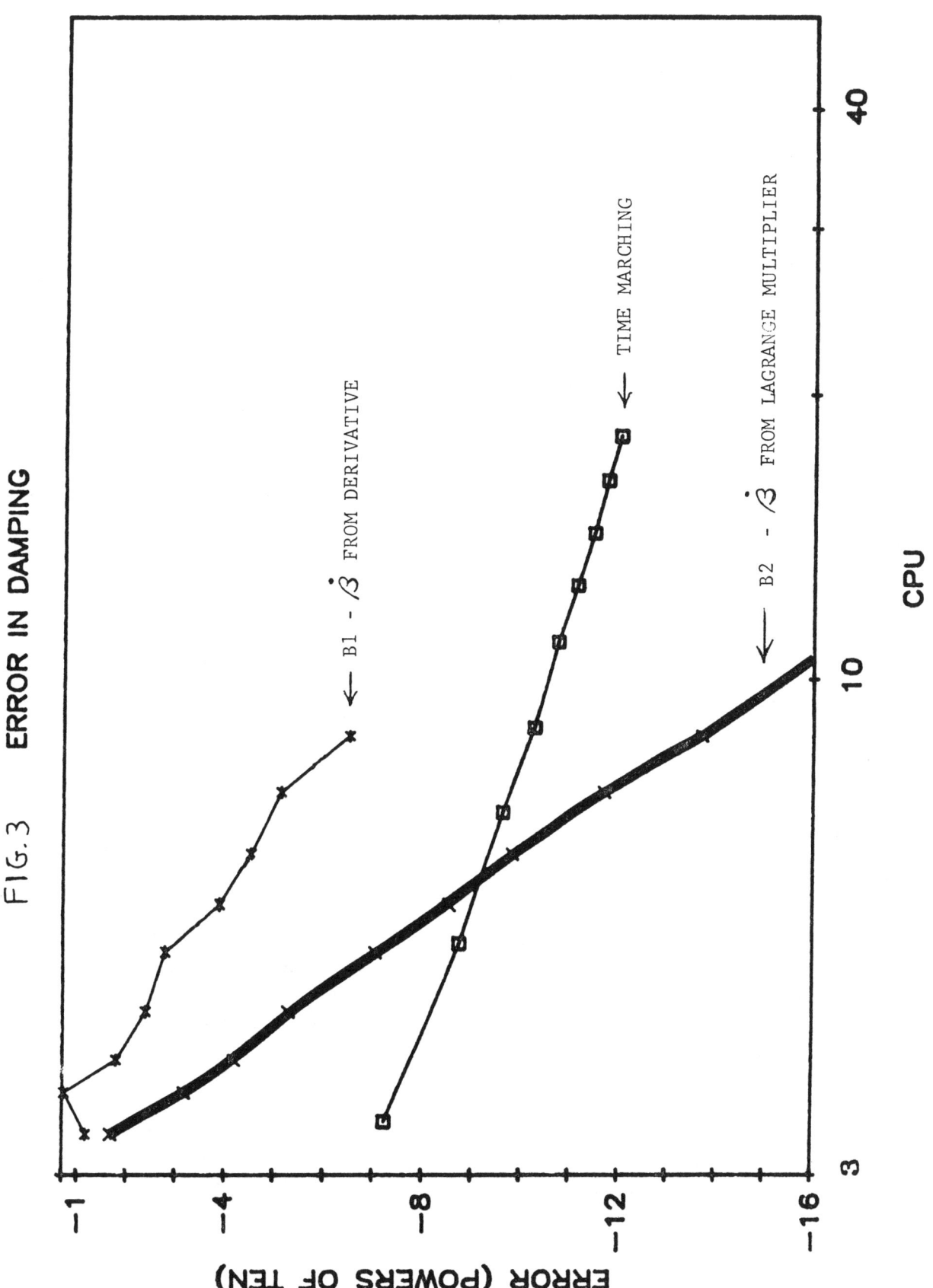

FIG. 3 ERROR IN DAMPING

# IDENTIFICATION OF STRUCTURAL DYNAMIC SYSTEMS

S. Hanagud, M. Meyyappa and J. I. Craig
School of Aerospace Engineering
Georgia Institute of Technology
Atlanta, Georgia

## Abstract

Structural dynamic system identification techniques used to obtain dynamic models are reviewed briefly. Examples are presented to illustrate the basic approaches. Numerical results are given for simple problems to demonstrate and compare the performance of the identification methods. The advantages and disadvantages of using various techniques are discussed along with some applications.

## Introduction

Dynamic response of structural systems is of special concern in the design, construction and operation of aerospace vehicles. Vibrations caused by such response, if excessive, could have a severe effect on crew and passenger comfort, reliability, structural integrity, and the fatigue lives of the components and the structure as a whole. It is not surprising then that a considerable amount of time and effort is spent in trying to minimize the vibration levels. This is usually done in the design stages, where attempts are made to ensure that the design will satisfy certain response requirements. But if a prototype constructed based on the original design fails to fulfill these requirements, it may be necessary to go through several developmental cycles which may prove to be both costly and time consuming. Thus it is essential to improve the modeling and predictive capabilities as much as possible, since better predictive abilities increase the likelihood of the design conforming to expectations.

Techniques for modeling structural dynamic systems are fairly well developed. Simple structures are generally described by partial differential equations, while more complex, built-up structures are modeled using approximate methods. Of the several approximate techniques that are available, the most versatile and the most often used is the finite element method. In both exact and approximate modeling, several assumptions are made in deriving the model, which may directly influence the accuracy of the model predictions. Moreover, in the case of finite element models, additional approximations are involved such as the discretization of a continuous structure for modeling purposes and the choice of particular types of elements. These could cause a further reduction in accuracy, especially if an insufficient number of degrees of freedom is used to represent the structure.

The accuracy of analytical models is usually tested by comparing the response predicted by the model with the response that is observed in tests or during operation. The measured response itself is not exact due to the inherent restrictions imposed by the types of equipment and data processing techniques used. But in general, reasonable bounds can be established from the measured data within which the exact response can be expected to lie. If the analytical model response lies outside these bounds, it follows that the accuracy of the model is inadequate. For complex structures, the difference between the measured and the model response may in fact be so large as to be unacceptable. If this is the case, and if the measured data are considered reasonably accurate, it becomes necessary to somehow obtain an improved model of the structure. Structural dynamic system identification techniques are devised to accomplish this.

For certain systems, it may not be possible to derive a mathematical model for use in analysis. Identification techniques can again be utilized in this case to construct a model from the measured data directly. This is also true of some specific characteristics of systems that are not modeled easily such as damping and nonlinear effects. Identification may provide the only means of quantifying such effects.

System identification thus provides analytical models that can simulate the actual behavior of the system as closely as possible. Identified models are useful in response prediction, loads determination and design modification. Techniques developed for identification can also be employed to predict the effects of, or to detect, changes in structural configuration. In addition, comparing the identified model with the original or a priori analytical model in particular cases may yield some insights into the errors involved in the modeling process. Another possible application of identification consists of obtaining reduced order models for systems that have been originally modeled with a very large number of degrees of freedom. The reduced models can be derived in such a manner that they possess modes with the same characteristics as the lower order modes of the original model or the system. These models are useful for certain applications such as vibration control, where models with fewer degrees of freedom are needed.

Numerous structural dynamic system identification techniques currently employed are reviewed briefly in this article. Attempts have not been made to provide a comprehensive treatment of the various aspects involved. The identification procedures are merely outlined, and simple examples and applications are given to illustrate the use of these procedures in obtaining dynamic models.

## **System Identification**

In studying physical systems for the purpose of investigating their behavior when subjected to external disturbances, identification plays a key role in arriving at an acceptable mathematical description of the system. By facilitating the process of choosing a particular model among all competing models, and by determining the set of parameters for the chosen model which result in the best possible simulation of the physical system, identification leads to a better understanding of the system and its functioning. Model identification, wherein an appropriate form for the mathematical model is deduced from either physical considerations or by observations of the behavior of the system when perturbed by external causes, is usually the first step in system identification. Once the model form is fixed, there remains the task of assigning values to the parameters that may appear in the model equations.

Most of the research activity in identification is confined to the latter task, with the assumption that the form of the model equations is already known. For example, in the case of discrete structural dynamic systems, the model consists of linear second order differential equations. The mass, damping and stiffness matrix elements appearing in these equations constitute the parameters to be identified.

The data utilized in identification may include the input and the output measurements or some system dependent characteristics such as modal parameters, which again are derived from the input-output measurements. All the a priori knowledge about the behavior of the system may also be available in the form of an a priori analytical model. For example, the finite element method may be employed to get the a priori mass and stiffness matrices of a discretized structural system. Identification may in fact yield better results with the use of such an a priori model.

## System Identification Classifications

Most of the identification techniques that have been developed can be grouped into various categories based on any one of several criteria. Some of the criteria used and the categories they give rise to are given below. These, however, are neither exhaustive nor mutually exclusive.

1. Classifications based on the type of data used.

   - time domain data
   - frequency domain data
   - modal data

2. Classifications based on the type of system being identified.

   - distributed(continuous) or discrete systems
   - linear or nonlinear systems
   - stationary or nonstationary systems

3. Classifications based on the identification technique employed

   - iterative or noniterative identification
   - on-line or off-line identification

4. Classifications based on problem formulation

   - equation error approach
   - output error approach
   - minimum deviation approach
   - other approaches

Many other types of classifications are also possible, as discussed, for example, in Refs. 1 and 2.

## Types of Data Used

Measurement data employed in identification may be in time domain(e.g., time histories or impulse response functions) or frequency domain(e.g., Fourier spectra or frequency response functions). For structural dynamic systems, it is also possible to use modal data such as natural frequencies and mode shapes instead of physical time or frequency domain data. When this course is adopted, the modal parameters are first deduced from the physical data. This process itself involves application of parameter identification techniques, where the modal properties are estimated in the time or frequency domain by curve fitting the measurements[3-7].

For a linear time-invariant multidegree-of-freedom structural dynamic system, frequency response functions are often used to obtain the modal parameters. If the system is viscously damped, the frequency response function between the response measured at degree of freedom i for a given excitation at degree of freedom j can be expressed as

$$H_{ij}(\omega) = \sum_{k=1}^{2n} R_{ij}^k / (j\omega - s^k) \tag{1}$$

where $s^k$ and $R_{ij}^k$ are the pole and residue corresponding to mode k given by

$$R_{ij}^k = (\psi_i^k \psi_j^k / a^k)$$
$$s^k = -\zeta^k \omega^k \pm j\omega^k (1-\zeta^{k^2})^{\frac{1}{2}}$$

and

$n$ = number of degrees of freedom

$\psi_i^k$ = complex mode shape coefficient for mode k at degree of freedom i

$a^k$ = complex modal mass for mode k

$\zeta^k$ = viscous damping factor for mode k

$\omega^k$ = natural frequency for mode k

For an underdamped system, the 2n eigenvalues and eigenvectors appear as n pairs of complex conjugate modes. Least squares curve fitting algorithms can be used to fit the expression in Eq. (1) to measured frequency response functions to determine the poles and residues, from which the frequencies, damping factors and mode shape coefficients can be computed[4].

Alternately, the impulse response function given by the inverse Fourier transform of Eq. (1) can be employed and modal parameter estimation can be performed in the time domain. The impulse response is given by

$$h_{ij}(t) = \sum_{k=1}^{n} R_{ij}^k \exp[(-\zeta^k \omega^k + j\omega^k (1-\zeta^{k^2})^{\frac{1}{2}})t] + \overline{R}_{ij}^k \exp[(-\zeta^k \omega^k - j\omega^k (1-\zeta^{k^2})^{\frac{1}{2}})t] \tag{2}$$

where a bar is used to denote complex conjugate. This equation forms the basis of the estimation techniques discussed in Refs. 5-7. Once the modal parameters are determined in either time or frequency domain, they are used in the second stage of identification which eventually leads to the determination of the system parameters appearing in the differential equations.

For a given problem, the type of data to be used depends on various considerations. Response measurements are almost always made in the time domain. No additional processing of the data is needed if the chosen identification method uses the input-output time domain data directly. This is usually the preferred approach in on-line identification problems where the model is updated continuously. Techniques based on time domain data, however, tend to be overly noise sensitive. Furthermore, these are usually restricted to low-order systems in which the number of parameters to be determined is relatively small.

Use of frequency domain or modal data, on the other hand, involves additional processing of measured data, but the effects of measurement noise can be minimized by repeating the tests and averaging the results. This approach is therefore adopted only in off-line identification in general. As in the case of time domain methods, techniques that utilize frequency domain input and response cannot be easily applied to large systems. But when modal data are used, it is possible to use formulations that can handle relatively large systems. This is because of the fact that, for most structures, there is usually a well defined frequency range of interest in which it is required that the analytical model must reproduce the structural response as accurately as possible. Using data for only the modes that lie in this range in identification may suffice to obtain a model with the desired characteristics, and techniques are devised to this end.

### **Types of Identified Systems**

Although most structural systems of practical interest are distributed or continuous in nature, identification techniques developed to date deal almost exclusively with finite-order systems. Structures possesses infinite degrees of freedom, and their behavior is influenced by their properties which are functions of spatial locations. Deriving a finite-size model for these distributed systems requires spatial discretization, which results in a loss of modeling accuracy. Also contributing to this problem are other factors such as using insufficient number of degrees of freedom and choosing inappropriate locations or nodes and/or elements to represent the system. It is theoretically possible to overcome these difficulties by increasing the number of degrees of freedom to a very large value, but this is made impractical by other considerations of implementation. It may thus prove useful to avoid, or atleast delay, discretization as much as possible in any attempt to model the system[8].

On the other hand, not all structures are amenable for treatment as continuous systems. As the level of structural complexity increases, it becomes harder to analyze the problem without discretization or some lumping procedure. In fact, for most large built-up structures, the only feasible modeling approach permitted by the present state-of-the-art remains the finite element method. It follows therefore that most of the research activity in identification is restricted to discrete systems. But for simpler structures which can be described by partial differential equations and appropriate boundary and initial conditions, the advantages of using distributed parameter system identification techniques remain largely unexplored.

If the system under consideration contains nonlinearities, the identification options available become narrower, since far fewer techniques have been developed for nonlinear systems than for linear systems[9-12]. The same is true of nonstationary systems also. In general, only time domain data can be used in such cases. While techniques developed for nonlinear systems can also be employed for linear systems, those that are formulated specifically for linear systems will usually be more efficient.

## Types of Techniques Used

Most of the identification techniques are either direct or iterative in nature. In the case of direct methods, the unknown parameters are obtained in one step, say by solving a set of simultaneous equations. In some instances, the solution may be available in closed form. Iterative methods involve starting with initial guesses for all the parameters to be estimated and successively updating these values based on some convergence criteria. The procedure may or may not converge depending upon the initial values used. As the number of unknowns increases, however, iterative techniques tend to break down. Some direct techniques may be applicable for such systems. But as the number grows even larger these techniques also could become impractical due to restrictions imposed by such factors as the amount of computational resources available.

In on-line identification where estimation is performed in real time, the parameters converge as more and more data are processed. Sequential filtering techniques such as the Kalman filter[12] are examples of on-line identification, but these can also be used off-line, however. Such techniques use mostly time domain data, and are restricted to relatively small systems.

## Types of Formulations Employed

System identification techniques are usually implemented by first defining an objective, performance, cost or criterion function, and then minimizing or maximizing this function as needed. The way in which the objective function is formulated provides another means of classifying the various techniques that have been developed. Each of the possible formulations will now be examined briefly.

### Equation Error Approach

In the equation error approach, the system parameters are estimated so that the error in satisfying the system equations with the given input-output data is minimized. This is accomplished by first defining objective functions which are measures of residual errors in the system equations for given values of the parameters. These functions are then minimized using the least-squares or the weighted least-squares method. In the case where time domain data are utilized, measurement noise in the data is usually not accounted, although confidence factors can be assigned to different equations corresponding to different measurements[13]. It is therefore essential to reduce the noise effects in data as much as possible before using the measurements in identification.

To illustrate the equation error formulation, a single degree-of-freedom system described by the equation

$$m\ddot{x} + c\dot{x} + kx = f \qquad (3)$$

is now considered. The unknown parameters to be estimated are the mass, damping and stiffness values represented by m, c and k respectively. To determine these values in time domain, it is assumed that the excitation f and the displacement, velocity and acceleration given by x and its derivatives are all measured over a time period. The parameters can then be obtained by minimizing, for example,

$$L = \int_0^T ( m \ddot{x}_m + c\dot{x}_m + kx_m - f_m )^2 \, dt \qquad (4)$$

where the subscript m is used to denote the measured or otherwise known quantities, and T denotes the record length of the measurements.

Differentiating Eq. (4) with respect to each of the unknown parameters and equating the resulting integrals to zero leads to a set of simultaneous equations, the solution of which yields the system parameter estimates. This technique therefore belongs to the category of direct methods. In general, the equation error approach leads to noniterative techniques if the model equations are linear in the unknown parameters.

In order to demonstrate the performance of the method, some numerical results are now presented. The system in Eq. (3) was assumed to have the values of 0.2, 0.25 and 150.0 for mass, damping and stiffness respectively. The response quantities were computed from closed form solutions for a step input of 10 units and zero initial conditions. To simulate noisy data that is usually obtained from measurements, the computed input and response records were corrupted with uniformly distributed pseudo-random noise of varying levels. These were then used in identification to determine m, c and k.

Table 1 summarizes the results obtained for a record length of 0.25 second and a sampling interval of 5 milliseconds. It is observed that with increasing levels of noise, the quality of the estimates tends to deteriorate as expected. Other examples of using the equation error approach in time domain can be found in Refs. 9, 10 and 14. Similarly, techniques that use frequency domain data can be derived by starting with the system equations in the frequency domain[15]. When modal data are to be used, equations that define the eigenvalue problem for the system are utilized in the development of the method[16].

Output Error Approach

The output error formulation involves the determination of the model parameters which minimize the difference between some observed system characteristics, such as response data or modal parameters, and the corresponding model characteristics. Since most characteristics used in the formulation are usually nonlinear functions of the model parameters, the identification techniques derived using this approach will be iterative in nature. Analytical model data, needed to compare with the measured data, are computed numerically (or by using closed-form expressions if available) for a given set of model parameters. Appropriate criterion functions are evaluated and the parameters are varied methodically starting from given initial values until the model data matches the measured data to within required tolerance.

The objective functions are generally defined using the least-squares, weighted least-squares or the maximum likelihood criterion. Effects of measurement noise can be taken into account if needed. Weighting matrices which reflect the confidence in the measured data can be incorporated into the objective function for this purpose. One form of weighting often used is the inverse covariance of the measurement noise distribution. If this distribution is Gaussian as is usually assumed, the maximum likelihood approach becomes equivalent to the weighted least-squares approach[17].

In addition to measurement noise in the data, if it is also possible to ascertain the degree of certainty to which the a priori model parameters are known, this information can be incorporated into the identification process by using the mean-square, minimum variance or Bayesian approach to estimation[17-19]. With the Bayesian formulation, for example, the model parameters are constrained by the addition of a quadratic term to the objective function involving the difference between the updated and the a priori parameters. Included in this term is a weighting matrix that describes the confidence levels in the various initial estimates. This weighting matrix may be the inverse covariance of the a priori parameters assumed to be normally distributed. Derivation of objective functions when this formulation is employed is discussed, for example, in Ref. 17. Such an approach ultimately results in the parameters with the highest levels of confidence changing the least during estimation, while the parameters with lesser confidence levels are allowed to vary more. It thus enables taking into account the uncertainty in both the measurements and the analytical model.

In order to discuss the application of the output error approach, the same single degree-of-freedom system example presented in the last section is considered. Assuming that the parameters are to be estimated with time domain response data, the following integral least-squares objective function can be defined.

$$L = \int_0^T (\ddot{x}_m - \ddot{x})^2 \, dt \tag{5}$$

where only acceleration data are employed in identification. For given system parameter values, the analytical model acceleration is obtained by solving Eq. (3) numerically for the measured or specified force history, in this case the step input. The Levenberg-Marquardt nonlinear least-squares method, which is one of the variations of the Gauss-Newton method, is utilized to minimize L with respect to the model parameters. Details of this iterative estimation technique can be found in Refs. 10 and 17 among others.

For computing numerical values of the estimates, acceleration data were generated from closed form solutions, corrupted with pseudo-random noise, and used as measured data as before. Results obtained for the same exact parameter values as used in the previous section are listed in Table 2. Initial estimates assumed to start the iterative estimation scheme are also listed in this table. It is again noted that the error in the estimates increases with increasing noise levels. These values could be improved somewhat by employing the closed form expression for calculating the analytical acceleration rather than solving Eq. (3) numerically. This may also be accomplished by including displacement and velocity data in identification in addition to the acceleration data.

Other applications of the equation error approach in time domain can be found in Refs. 9-11. Frequency domain data are used in Ref. 20. Identification with modal data is discussed in Refs. 21-23. In Ref. 23, a minimum variance formulation is described. Use of extended Kalman filter for nonlinear systems is explored in Ref. 12. If the structure for which identification is to be performed can be divided into several components, it may be possible to simplify the problem by assigning certain structural parameters to the matrices of the component systems and letting only these parameters to vary. The elements of the component matrices are fixed during estimation, which is restricted to the computation of the associated structural parameters[24]. Such an approach is followed in Refs. 25 and 26, where modal data are employed in identification.

## Minimum Deviation Approach

The identification concept used in the minimum deviation approach involves looking for a system model which satisfies the necessary conditions while deviating the least from a given a priori model. In the case of discrete structural dynamic systems described by matrices, weighted matrix norms can be employed as measures of deviation which are to be minimized. The system equations and other conditions to be satisfied are implemented as equality constraints, incorporated into the objective functions by means of suitable Lagrange multipliers. Examples of such constraints, when modal data are used, include the eigenvalue equation and the orthogonality conditions.

Application of this approach for discrete proportionally damped systems is discussed in detail in Refs. 27-31. Several techniques to identify the mass and stiffness matrices for such systems are derived in these references. Identification in this case consists of adding appropriate corrections to the a priori matrices. Required data are the modal parameters, and the identified system is such that the specified modal data are reproduced exactly.

As an example of the minimum deviation approach, one of the several methods described in the cited references will now be outlined. The equation defining the eigenvalue problem can be written as

$$\mathbf{K}\phi_i = \lambda_i \mathbf{M}\phi_i \tag{6}$$

in which $\mathbf{K}$ is the stiffness matrix, $\mathbf{M}$ the mass matrix, and $\lambda_i$ and $\phi_i$ the ith natural frequency and mode shape respectively. Experimental modal data for p modes (p≤n) are assumed to be given, where n is the number of degrees of freedom. The mass matrix is modified first so that it satisfies the orthogonality conditions. This is achieved by minimizing the matrix norm [29]

$$L_M = ||\overline{\mathbf{M}}^{-\frac{1}{2}}(\mathbf{M}-\overline{\mathbf{M}})\overline{\mathbf{M}}^{-\frac{1}{2}}|| \tag{7}$$

with respect to the mass matrix elements subject to

$$\Phi^T \mathbf{M} \Phi = \mathbf{I} \tag{8}$$

where $\overline{\mathbf{M}}$ is the a priori mass matrix, $\Phi$ is the n x p matrix of measured mode shapes normalized with respect to $\overline{\mathbf{M}}$, and $\mathbf{I}$ is the p x p identity matrix. The identified mass matrix is obtained as[29]

$$\mathbf{M} = \overline{\mathbf{M}} - \overline{\mathbf{M}}\Phi\overline{\mathbf{m}}^{-1}(\overline{\mathbf{m}}-\mathbf{I})\overline{\mathbf{m}}^{-1}\Phi^T\overline{\mathbf{M}} \tag{9}$$

with

$$\overline{\mathbf{m}} = \Phi^T \overline{\mathbf{M}} \Phi \tag{10}$$

The modified stiffness matrix is computed next by minimizing[27]

$$L_K = ||\overline{M}^{-\frac{1}{2}}(K-\overline{K})\overline{M}^{-\frac{1}{2}}|| \tag{11}$$

subject to the constraints of Eq. (6) and matrix symmetry expressed as

$$K = K^T \tag{12}$$

where $\overline{K}$ is the a priori stiffness matrix. The corrected matrix is given by[27]

$$K = \overline{K} - \overline{K}\Phi\Phi^T M - M\Phi\Phi^T \overline{K} + M\Phi\Phi^T \overline{K}\Phi\Phi^T M + M\Phi\Lambda\Phi^T M \tag{13}$$

where $\Lambda$ represents the diagonal matrix of specified eigenvalues.

Eqs. (9) and (13) express the identified system matrices in terms of the a priori matrices and the measured modal data. Since these are in closed form, the method can be applied with ease to relatively large systems[30]. A numerical example in which this method is used to identify a simple cantilever beam model is given in a subsequent section.

## Other Approaches to Identification

In some instances, it may be possible to develop identification techniques without using any one of the three approaches discussed above. For example, experimental data may be employed directly to construct a mathematical model without recourse to an a priori model, even though this may not be a desirable procedure. The method of Ref. 32, in which the system matrices are obtained by inverse transformation of a complete set of modal data, belongs to this category. Also included in this are the synthesis procedure of Ref. 33 and the incomplete model discussed in Ref. 34.

The identification technique to be employed in a given problem usually depends on various factors, including the size or the number of degrees of freedom, the type and accuracy of test data, the amount of computational resources available, and the anticipated use for the identified model. Because of all the complexities involved, it is evident that no single method can satisfy the needs for all systems. Consequently, techniques are chosen to meet the requirements of the individual problems.

Several applications of the methods discussed heretofore are considered next. These include the identification of a cantilever beam with test data, design modification of a typical helicopter wing and vibration suppression in helicopters using higher harmonic control.

## **Applications**

### **Cantilever Beam Identification**

The beam used in the analysis was modeled as a 12 node, 24 degree-of-freedom system using beam bending elements(Fig. 1). A 12 degree-of-freedom model, used as the a priori model in identification, was obtained from the original model by condensing out the rotational degrees of freedom. Tests were conducted to determine the frequencies and mode shapes for the first three bending modes. The minimum deviation technique defined by Eqs. (9) and (13) was employed to improve the a priori mass and stiffness matrices.

The efficacy of the identified model was then tested by adding a mass of about 1 kilogram at the tip of the beam and comparing the shifts in the frequencies predicted by the model with the actual values measured from the modified structure. These comparisons are shown in Table 3. The results presented in this table indicate that the identified model performs better than the a priori model in predicting the changes in the frequencies.

## Design Modification With Dynamic Constraints

The original configuration assumed for the helicopter wing, modeled as a cantilever beam, is shown in Fig. 2. Two flexible wing pylons carrying 700 lbs each are attached to the wing at nodes 2 and 4. The properties assigned to the wing model, listed in Table 4, are such that the wing possesses a first vertical bending frequency of 6.2Hz[35].

Modification of the original design is now considered, where the possibility of either increasing the wing store in one of the two pylons or that of adding an extra store at a different location is investigated. It is stipulated that the first bending frequency of the modified wing should not fall below 5 Hz, a constraint imposed by the helicopter excitation frequencies. To determine the maximum load that can be carried at different locations, the technique described in Ref. 25 to identify component structural parameters can be used as follows.

The eigenvalue problem in Eq. (6) is first expressed in its reciprocal form.

$$\bar{\lambda}_i \mathbf{K} \phi_i = \mathbf{M} \phi_i \qquad (14)$$

where $\bar{\lambda}_i$ is the reciprocal of $\lambda_i$. The mass matrix is next expressed as

$$\mathbf{M} = \mathbf{M}_o + v \mathbf{M}_1 \qquad (15)$$

where $\mathbf{M}_o$ is the original mass matrix, $v$ denotes the wing store to be added and $\mathbf{M}_1$ is a matrix containing one on the diagonal element corresponding to the degree of freedom at which the mass is added and zeroes elsewhere. It is assumed that the eigenvectors are normalized with respect to the stiffness matrix ($\phi_i^T \mathbf{K} \phi_i = 1$), so that the reciprocal eigenvalues are given by $\phi_i^T \mathbf{M} \phi_i$.

The output error approach is used to formulate the following objective function in terms of the reciprocals.

$$L = \sum_{i=1}^{p} (\bar{\lambda}_{ic} - \bar{\lambda}_i)^2 \qquad (16)$$

in which $\bar{\lambda}_{ic}$ is the reciprocal eigenvalue corresponding to the ith constraint frequency; p is the number of frequencies constrained, which is equal to one in the present case. When mass is added at any given location, the natural frequencies decrease and approach the constraint frequencies from above. Minimization of Eq. (16) with respect to v yields the maximum mass that can be added so that the modified frequencies are as close to the constraints as possible. This value is given by[25]

$$v = [\sum_{i=1}^{p} (\bar{\lambda}_{ic} - \bar{\lambda}_{io}) d_i] / \sum_{i=1}^{p} d_i^2 \qquad (17)$$

where

$$\bar{\lambda}_{io} = \phi_i^T M_o \phi_i \qquad (18a)$$

$$d_i = \phi_i^T M_1 \phi_i \qquad (18b)$$

An initial value of v is assumed and the reciprocal eigenvalues and eigenvectors are computed. These are used in Eq. (17) to obtain an updated estimate of v, and the procedure is repeated again until sufficient convergence is achieved.

In Table 5, results for a specific case are reproduced from Ref. 35. Before applying the estimation scheme, the a priori finite element mass and stiffness matrices were first modified by the identification procedure of Ref. 16 with pseudo-experimental data and then used as $M_o$ and $K$. From the mass values listed in this table, a decision could be made concerning where an extra store of a given weight could be added so that the frequency constraint is not violated.

## Vibration Suppression Using Higher Harmonic Control

An example in which an identification method is used to minimize vibration levels in rotorcraft with higher harmonic control is now considered. The technique and its performance in reducing vibrations by producing a better control are discussed in more detail in Ref. 36. The basic linear higher harmonic response model for a helicopter is usually approximated by

$$z = z_o + B\theta \qquad (19)$$

where z represents the vector of harmonic amplitudes of the response which maybe hub loads or fuselage accelerations, $\theta$ denotes the input vector of higher harmonic blade pitch amplitudes, $B$ is the transfer matrix that relates the response to the input and $z_o$ is the response when there is no higher harmonic control($\theta = 0$).

If a priori estimates $\bar{z}_o$ and $\bar{B}$ are known through analysis or measurements, an optimum control $\bar{\theta}^*$ can be computed that would minimize the response given by Eq. (19). If this control fails to reduce the response to an acceptable level, however, it may be worthwhile to modify the transfer matrix to make it consistent with the most recent observations so that it may yield a more acceptable control. The observed response of the system is given by

$$z_m = z_{om} + B\bar{\theta}^* \qquad (20)$$

where it is assumed that the uncontrolled response is also measured. The minimum deviation approach can now be used to calculate the modified transfer matrix, which is obtained by minimizing

$$L_B = ||B - \bar{B}|| \qquad (21)$$

subject to Eq. (20), resulting in[36]

$$B = \bar{B} + (z_m - z_{om} - \bar{B}\bar{\theta}^*)(\bar{\theta}^{*T}\bar{\theta}^*)^{-1}\bar{\theta}^{*T} \qquad (22)$$

In Ref. 36, results for several cases are presented using an analytical coupled rotor/airframe model comprising an assumed-mode rotor model and a finite element beam fuselage model. By introducing errors into analytical data, it is shown that the modified transfer matrix is effective in reducing fuselage vibration levels. In addition, the applicability of this technique for the case where the transfer matrix corresponding to one flight condition is treated as the a priori estimate for another and modified subsequently is also demonstrated.

## Summary

Various approaches to structural dynamic system identification were reviewed briefly. Each approach was illustrated with an example. Use of identification techniques in other contexts was demonstrated with applications in the modification of a typical helicopter wing and in higher harmonic control of rotorcraft. Some of the areas in which the field is not yet well developed were discussed. Additional research in these and other areas will be of certain benefit in developing accurate models needed for understanding dynamic structural behavior.

## Acknowledgement

The authors gratefully wish to acknowledge support for this work from the U. S. Army Research Contract DAAG-29-82-K-0094.

## References

1. Pilkey, W. D., ed., "System Identification of Vibrating Structures - Mathematical Models From Test Data," ASME Winter Annual Meeting, 1972.

2. Ibanez, P., "Review of Analytical and Experimental Techniques for Improving Structural Dynamic Models," Welding Research Council Bulletin No. 249, 1979.

3. Kennedy, C. C. and Pancu, C. D. P., "Use of Vectors in Vibration Measurement and Analysis," Journal of the Aeronautical Sciences, Vol. 14, No. 11, 1947, pp. 603-625.

4. Richardson, M. and Potter, R. W., "Identification of the Modal Properties of an Elastic Structure from Measured Transfer Function Data," Proceedings of the 20th ISA Symposium, Albuquerque, New Mexico, 1974, pp. 239-246.

5. Ibrahim, S. R. and Mikulcik, E. C., "A Method for the Direct Identification of Vibration Parameters from the Free Response," Shock and Vibration Bulletin, Vol. 47, Pt. 4, 1977, pp. 183-198.

6. Vold, H., Kundrat, J., Rocklin, G., and Russell, R., "A Multi-input Modal Estimation Algorithm for Mini-computers," SAE Paper No. 820194.

7. Brown, D. L., Allemang, R. J., Zimmerman, R., and Mergeay, M., "Parameter

Estimation Techniques for Modal Analysis," SAE Paper No. 790221.

8. Kubrusly, C. S., "Distributed Parameter System Identification: A Survey," International Journal of Control, Vol. 26, No. 4, 1977, pp. 509-535,

9. Distefano, N. and Rath, A., "System Identification in Nonlinear Structural Seismic Dynamics," Computer Methods in Applied Mechanics and Engineering, Vol. 5, 1975, pp. 353-372.

10. Hanagud, S., Meyyappa, M., and Craig, J. I., "Method of Multiple Scales and Identification of Nonlinear Structural Dynamic Systems," AIAA Journal, Vol. 23, No. 5, 1985, pp. 802-807.

11. McNiven, H. D. and Matzen, V. C., "A Mathematical Model to Predict the Inelastic Response of a Steel Frame," International Journal of Earthquake Engineering and Structural Dynamics, Vol. 6, 1978, pp. 189-202.

12. Yun, C. B. and Shinozuka, M., "Identification of Nonlinear Structural Dynamic Systems," Journal of Structural Mechanics, Vol. 8, 1980, pp. 187-203.

13. Caravani, P., Watson, M. L., and Thomson, W. T., "Recursive Least-Squares Time Domain Identification of Structural Parameters," Journal of Applied Mechanics, Vol. 44, No. 1, 1977, pp. 135-140.

14. Kozin, F. and Kozin, C. H., "Identification of Linear Systems, Final Report on Simulation Studies," NASA CR-98738, 1968.

15. Leuridan, J. M., Brown, D. L., and Allemang, R. J., "Direct System Parameter Identification of Mechanical Structures with Application to Modal Analysis," AIAA Paper No. 82-0767, 1982.

16. Hanagud, S., Meyyappa, M., Cheng, Y. P., and Craig, J. I., "Identification of Structural Dynamic Systems with Nonproportional Damping," AIAA Paper No. 84-0993, 1984.

17. Bard, Y., "Nonlinear Parameter Estimation," Academic Press, 1974.

18. Sorenson, H. W., "Parameter Estimation : Principles and Problems," Marcel Dekker Inc., 1980.

19. Lewis, T. O. and Odell, P. L., "Estimation in Linear Models," Prentice-Hall, 1971.

20. Caravani, P. and Thomson, W. T., "Identification of Damping Coefficients in Multidimensional Linear Systems," Journal of Applied Mechanics, Vol. 41, No. 2, 1974, pp. 379-382.

21. Beliveau, J. G., "Identification of Viscous Damping in Structures from Modal Information," Journal of Applied Mechanics, Vol. 43, No. 2, 1976, pp. 335-339.

22. Chen, J. C. and Garba, J. A., "Analytical Model Improvement Using Perturbation

Techniques," AIAA Journal, Vol. 18, No. 6, pp. 684-690.

23. Collins, J. D., Hart, G. C., Hasselman, T. K., and Kennedy, B., "Statistical Identification of Structures," AIAA Journal, Vol. 12, No. 2, 1974, pp. 185-190.

24. Torkamani, M. A. M. and Hart, G. C., "Earthquake Engineering : Parameter Identification," Preprint No. 2499, ASCE National Structural Engineering Convention, New Orleans, Louisiana, 1975.

25. Natke, H. G. and and Schulze, H., "Parameter Adjustment of a Model of an Offshore Platform from Estimated Eigenfrequencies Data," Journal of Sound and Vibration, Vol. 77, No. 2, 1981, pp. 271-285.

26. Meyyappa, M. and Craig, J. I., "Highrise Building Identification Using Transient Testing," Proceedings of the 8th World Conference on Earthquake Engineering, San Francisco, Vol. 6, 1984, pp. 79-86.

27. Baruch, M., "Optimal Correction of Mass and Stiffness Matrices Using Measured Modes," AIAA Journal, Vol. 20, No. 11, 1982, pp. 1623-1626.

28. Baruch, M., "Methods of Reference Basis for Identification of Linear Dynamic Structures," AIAA Journal, Vol. 22, No. 4, 1984, pp. 561-564.

29. Berman, A., "Mass Matrix Correction Using an Incomplete Set of Measured Modes," AIAA Journal, Vol. 17, No. 10, 1979, pp. 1147-1148.

30. Berman, A. and Nagy, E. J., "Improvement of a Large Analytical Model Using Test Data," AIAA Journal, Vol. 21, No. 8, 1983, pp. 1168-1173.

31. Kabe, A. M., "Stiffness Matrix Adjustment Using Mode Data," AIAA Journal, Vol. 23, No. 9, 1985, pp. 1431-1436.

32. Thoren, A. R., "Derivation of Mass and Stiffness Matrices from Dynamic Test Data," AIAA Paper No. 72-346, 1972.

33. Ross, R. G., "Synthesis of Stiffness and Mass Matrices from Experimental Vibration Modes," SAE Paper No. 710787, 1971.

34. Berman, A. and Flannelly, W. G., "Theory of Incomplete Models of Dynamic Structures," AIAA Journal, Vol. 9, No. 8, 1971, pp. 1481-1487.

35. Hanagud, S., Meyyappa, M., Cheng, Y. P., and Craig, J. I., "Rotorcraft Structural Dynamic Design Modifications," Presented at the 10th European Rotorcraft Forum, The Hague, Netherlands, 1984.

36. Hanagud, S., Meyyappa, M., Sarkar, S., and Craig, J. I., "A Coupled Rotor/Airframe Vibration Model with Higher Harmonic Control Effects," 42nd AHS Forum, Washington, D. C., 1986.

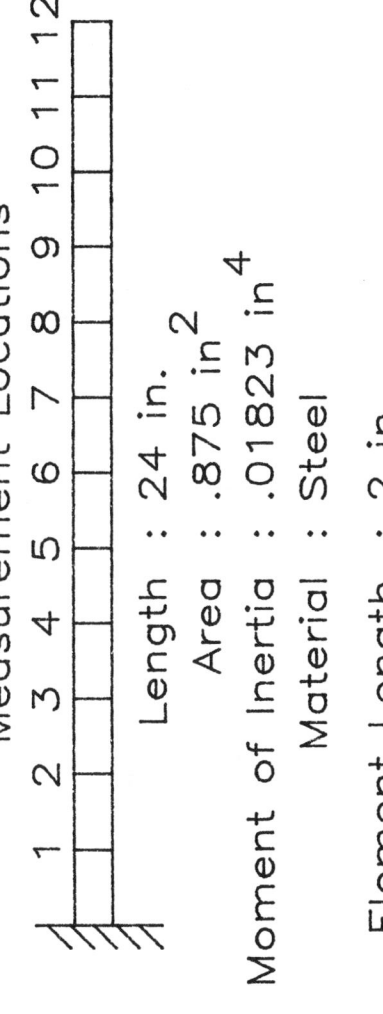

Fig. 1 Cantilever beam

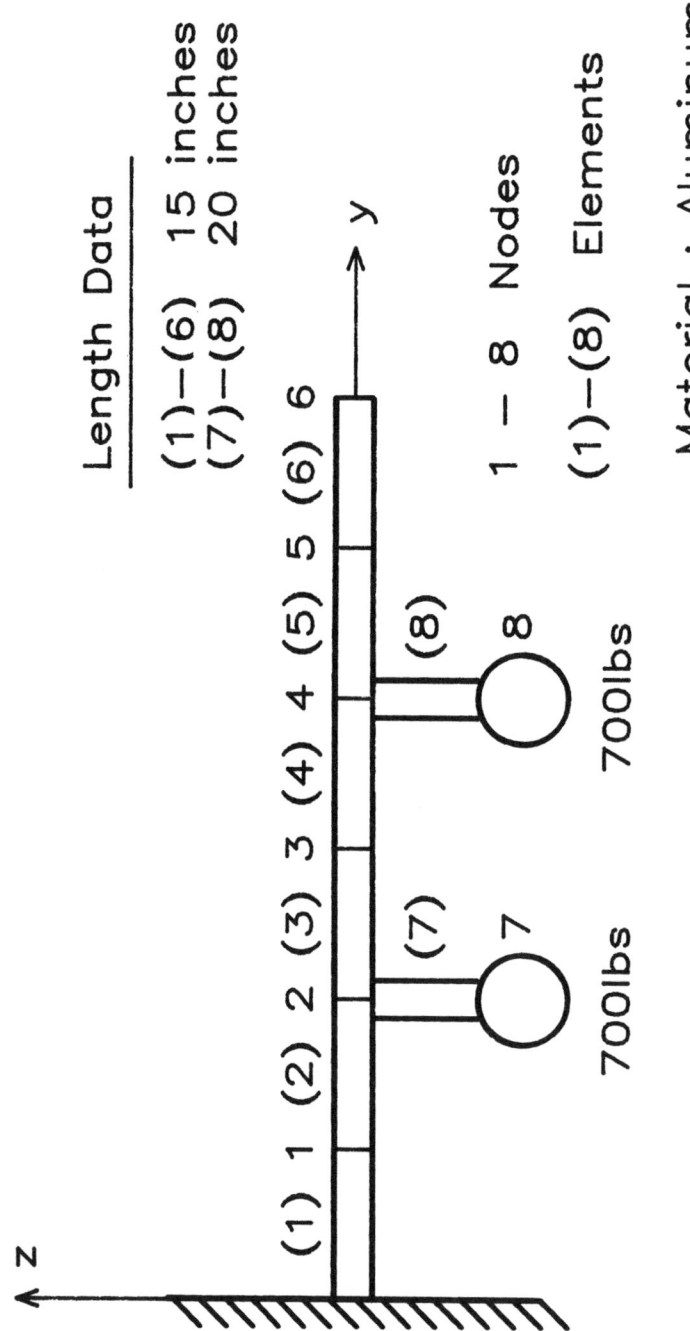

Fig. 2 Helicopter wing model

Table 1  Parameters identified with equation error approach

| NOISE LEVEL | m | c | k |
|---|---|---|---|
| 0% | 0.20 | 0.25 | 150.00 |
| 5% | 0.20 | 0.27 | 148.88 |
| 10% | 0.20 | 0.29 | 147.52 |
| 20% | 0.20 | 0.33 | 144.19 |

Table 2  Parameters identified with output error approach

| NOISE LEVEL | m | c | k |
|---|---|---|---|
| 0% | 0.20 | 0.25 | 150.00 |
| 5% | 0.20 | 0.27 | 151.79 |
| 10% | 0.20 | 0.25 | 155.99 |
| 20% | 0.22 | 0.13 | 172.20 |
| INITIAL ESTIMATES | 0.1 | 0.1 | 50.0 |

Table 3  Cantilever beam frequencies (Hz)

|  | Mode No. | Test | A Priori Model | Identified Model |
|---|---|---|---|---|
| Beam Only | 1 | 25.49 | 27.16 | 25.49 |
|  | 2 | 159.99 | 170.30 | 159.99 |
|  | 3 | 444.94 | 477.00 | 444.94 |
| Beam With Tip Mass | 1 | 16.36 | 17.10 | 16.81 |
|  | 2 | 129.67 | 135.98 | 127.81 |
|  | 3 | 379.32 | 412.26 | 387.32 |

Table 4  Cross-sectional properties of the wing model

| Element No. | Area (in$^2$) A | Moments of Inertia (in$^4$) | | |
|---|---|---|---|---|
| | | $I_{xx}$ | $I_{yy}$ | $I_{zz}$ |
| 1 | 9.29 | 42.9 | 84.6 | 183.1 |
| 2 | 7.93 | 30.1 | 72.6 | 149.1 |
| 3 | 6.71 | 21.8 | 61.8 | 114.7 |
| 4 | 5.62 | 12.1 | 52.4 | 80.0 |
| 5 | 4.66 | 6.9 | 44.4 | 44.8 |
| 6 | 3.84 | 4.5 | 37.7 | 10.0 |
| 7 | 5.62 | 12.1 | 80.0 | 52.4 |
| 8 | 5.62 | 12.1 | 80.0 | 52.4 |

Table 5  Maximum additional wing stores

| Location | Load(lbs) |
|----------|-----------|
| 7 | 3374 |
| 3 | 2397 |
| 4 | 828 |
| 8 | 569 |
| 5 | 373 |
| 6 | 198 |

II. Aeroelasticity and Unsteady Aerodynamics

NONLINEAR AEROELASTICITY: AN OVERVIEW

Earl Dowell
School of Engineering
Duke University
Durham, NC 27706

ABSTRACT

A conceptual framework is provided for the field of Nonlinear Aeroelasticity. Various important physical phenomena are identified and placed in context. Representative methodologies and experimental and theoretical results are discussed.

INTRODUCTION

For many years, linear models, both theoretical and experimental, have served the aeroelastician extraordinarily well. Most of our understanding of aeroelastic phenomena, such as divergence, flutter, control surface reversal and gust response, has been obtained by the study of such models. These models are lucidly presented in the now classic texts by Scanlan and Rosenbaum [1], Fung [2], Bisplinghoff [3,4], Ashley [3,4], and Halfman [3], and more recently by Försching [5] and Dowell, Curtiss, Scanlan and Sisto [6]. Indeed as Ashley [7] has noted, (linear) aeroelasticity is entering the mainstream of aerospace engineering and is now routinely discussed within the framework of broader (or narrower depending on your perspective) fields such as flight mechanics or vehicle dynamics and control. Indeed perehaps the last research frontier of linear aeroelasticity has been the use of the feedback control systems to stabilize or modify the behavior of aeroelastic systems, e.g. increase the flutter speed or reduce the response to gusts [8]. Of course, the ultimate last frontier of any linear system study is optimization. And a good deal of work has been done on this as well, especially within the framework of optimal control of aeroelastic systems and the optimization of structural weight or stiffness distributions [9]. The advent of composite materials has opened up a rich array of possibilities in the latter area.

Here, we consider a different research frontier for aeroelasticity, that is the study of nonlinear aeroelastic systems. Of course, it has been known for many years that nonlinear effects in aeroelasticity may be important. Indeed "nonlinear" effects are often cited as a possible explanation for any difference between theory and experiment. Many times these nonlinear effects are small, hence the major successes of linear models. However sometimes nonlinear effects are more important and occasionally they are crucial.

For example, it has been known for many years that plate or shell flutter [10] leads to limit cycle motions of small amplitude (typically on the order of the plate or shell thickness); thus the flutter motion does not usually lead to immediate catastrophic failure. This is unlike wing flutter where the nonlinear effects normally are small, the flutter motions are large (on the order of a fraction of the wing span), and failure is usually immediate. Control surface flutter may also be of limited amplitude due to actuator nonlinearities [11].

Over the last several decades, the transonic flow regime has been penetrated by aircraft and missile designs and by the aeroelastician with theoretical and experimental models. It was long thought that nonlinear aerodynamic effects are important in the transonic flow regime and, for this reason, theoretical progress was much slower there than in the modeling (by linear theory) of the subsonic and supersonic flow regimes. However finite difference methods for solving nonlinear partial differential equations and the development of a distinct field of study known as "computational fluid dynamics" have provided an effective means for calculating the aerodynamic forces on airfoils and wings which are oscillating in transonic flow [12]. Indeed a pleasant surprise has been the degree to which a transonic (not to say classical subsonic or supersonic) linear model still proves useful to the aeroelastician. However it was only by studying the nonlinear models that this discovery was made [13].

Of course, there are other aeroelastic phenomena which have defied rational modeling, at least until recently, and often this is thought to be due to important nonlinear effects. Among those that are of significant interest are:

- oscillations of bluff bodies in a flowing stream [14]

- large oscillations of streamlined bodies where the flow may separate during at least a portion of the cycle of oscillation [15]

- various structural damping mechanisms including hydraulic and dry friction damping [11,16,17].

Some progress has been made in understanding each of these.

Some of these nonlinear phenomena and associated models will be discussed in the presentation.

CATEGORIZATION OF GENERIC NONLINEAR EFFECTS

It is helpful, before proceeding to the particular discussion of physical phenomena, to distinguish among and define three classes of linear and/or nonlinear models.

- Fully Linear Models

    Here the statically deformed structural shape of the body and the steady flow field about the body may be assumed to have trivial solutions, i.e. the static shape of the body does not significantly influence its dynamic response and the steady flow field deviation from a uniform stream flow due to the presence of the body does not change the unsteady, time-dependent aerodynamic forces which act upon the body when it is set into oscillatory motion. Moreover the dynamic motions are sufficiently small that there are no significant nonlinear effects due to body motion.

In this case, both the static equilibrium problem (usually trivially so) and the dynamic motion problem may be treated by linear models.

- Dynamically Linear Models

Here the statically deformed shape of the body and/or the steady flow field about the body do influence the subsequent dynamic response. For example, if a plate is (statically) buckled from its initial flat configuration this may significantly change its dynamic response from that for the initially flate plate [10]. Another example is the dynamic aeroelastic behavior of a rotor blade which is often significantly modified by its statically deformed shape [18].

An aerodynamic example is that of an airfoil in transonic flow. The (rigid or statically deformed) shape of the airfoil determines the strength and location of steady shock waves. These shock waves which result from the steady flow field have a decisive effect on the unsteady aerodynamic forces which act upon the oscillating airfoil [13,19].

By definition, of course, if the behavior of an aeroelastic system is to be modeled as dynamically linear, the dynamic motion must remain sufficiently small or to say it another way, we must be content to study those phenomena which involve small dynamic motions.

For this class of models, the static problem must be treated by a nonlinear model, but the dynamic motion about this static equilibrium may be treated by a linear dynamic model. The solution, in sequence, of first the nonlinear static model and then the dynamically linear model usually has major conceptual and computational advantages over a fully nonlinear model.

- Fully Nonlinear Models

For this category of models, the dynamic motions themselves are so large that a nonlinear dynamic model must be used. Hence usually there is no advantage to solving the static and dynamic problems separately and sequentially. Examples in this category are

- determination of limit cycle oscillations of either the body or fluid

- study of deliberately large motions to achieve a desired effect, e.g. large control surface oscillations or large wing and body motions (supermaneuvarability)

- deliberate (or inadvertent) use of dynamically nonlinear devices, e.g. dry friction or hydraulic dampers.

In the presentation, both dynamically linear and fully nonlinear models will be discussed. Fully linear models will not be discussed.

## THE EARLIER LITERATURE

There is, of course, a vast array of publications in aeroelasticity on fully linear models. This will not be discussed here; the reader is referred to the cited references. A recent discussion of that literature is contained in Reference 6. A review of much of the linear and nonlinear model literature on plate and shell flutter is contained in Reference 10. The discussion of plate and shell flutter in the presentation will be confined for the most part to the more recent work on chaotic oscillations. The subject of unsteady transonic aerodynamics and aeroelasticity has been reviewed in Reference 13. The present discussion of this subject will be focussed on the most recent findings on flutter analysis techniques for fully nonlinear models and the significance (or lack thereof) of multiple aerodynamic solutions in the transonic flow regime.

## ACKNOWLEDGEMENT

This work was supported in part by the Air Force Office of Scientific Research under Grant 85-0137. Dr. Anthony Amos is the Technical Monitor. Another version of this paper was presented at the Symposium on Dynamic Stability of the Tenth U.S. Congress of Applied Mechanics.

## REFERENCES

1. Scanlan, R. H. and Rosenbaum, R., Introduction to the Study of Aircraft Vibration and Flutter, the Macmillan Company, New York, N.Y., 1951. Also available in Dover Edition.

2. Fung, Y. G., An Introduction to the Theory of Aeroelasticity, John Wiley and Sons, Inc., New York, N.Y., 1955. Also available in Dover Edition.

3. Bisplinghoff, R. L., Ashley, H. and Halfman, R. L., Aeroelasticity, Addison-Wesley Publishing Company, Cambridge, Mass., 1955.

4. Bisplinghoff, R. L. and Ashley, H., Principles of Aeroelasticity, John Wiley and Sons, Inc., New York, N.Y., 1962. Also available in Dover Edition.

5. Försching, H. W., Fundamentals of Aeroelasticity, In German. Springer-Verlag, Berlin, 1974.

6. Dowell, E. H., Curtiss, H. C., Jr., Scanlan, R. H. and Sisto, F., A Modern Course in Aeroelasticity, Sijthoff and Noordhoff, the Netherlands, 1980.

7. Ashley, H., The Constructive Uses of Aeroelasticity, Polish Academy Sciences, Engineering Transactions, Vol. 30, No. 3-4, 1982, pp. 369-396.

8. Many Authors, Proceedings of the Aeroservoelastic Specialists Meeting, AFWAL-TR-84-3105, Vol. I, II, October 1984.

9. McIntosh, S. C. and Ashley, H., On the Optimization of Discrete Structures with Aeroelastic Constraints, Computers and Structures, Vol. 8, 1978, pp. 411-419.

10. Dowell, E. H., Aeroelasticity of Plates and Shells, Noordhoff International Publishing, Leyden, 1975.

11. Breitbach, E., Effects of Structural Nonlinearities on Aircraft Vibration and Flutter, AGARD Report R-665, 1978.

12. Jameson, A., The Evolution of Computational Methods in Aerodynamics, J. Applied Mechanics, Vol. 50, No. 4, Dec. 1983, pp. 1052-1070.

13. Dowell, E. H., Unsteady Transonic Aerodynamics and Aeroelasticity, in Recent Advances in Aerodynamics and Aeroacoustics, Springer-Verlag, New York, 1986.

14. Dowell, E. H., Nonlinear Oscillator Models in Bluff Body Aeroelasticity, J. Sound Vibration, Vol. 75, 1981, pp. 251-164.

15. Chi, R. M. and Srinivasan, A. V., Some Recent Advances in the Understanding and Prediction of Turbomachine Subsonic Stall Flutter, ASME Paper No. 84-GT-151, 1984.

16. Dowell, E. H., Damping in Beams and Plates Due to Slipping at the Support Boundaries, J. Sound Vibration, to appear.

17. Tang, D.-M. and Dowell, E. H., Effect of Nonlinear Damping in Landing Gear on Helicopter Limit Cycle Response in Ground Resonance, J. Am. Helicopter Soc., to appear.

18. Hodges, D. H. and Ormiston, R. A., Stability of Hingeless Rotor Blades in Hover and Pitch-Link Flexibility, AIAA J., Vol. 15, No. 4, April 1977, pp. 476-482.

19. Bland, S. R. and Edwards, J. W., Airfoil Shape and Thickness Effects on Transonic Airloads and Flutter, J. Aircraft, Vol. 21, No. 3, March 1984, pp. 209-217.

# EXPERIMENTAL STUDIES IN AEROELASTICITY OF UNSWEPT AND FORWARD SWEPT GRAPHITE/EPOXY WINGS

John Dugundji
Department of Aeronautics and Astronautics
Massachusetts Institute of Technology
Cambridge, Massachusetts 02139

## ABSTRACT

Three separate experimental studies on the aeroelastic flutter and divergence behavior of stiffness coupled, graphite/epoxy composite wing models are described. These studies were conducted at M.I.T. and dealt with cantilever unswept wings, cantilever forward swept wings, and forward swept wings with rigid body freedoms. For the cantilever wings, bending-torsion flutter and divergence was observed at low angles of attack while torsional stall flutter and bending stall flutter was observed at high angles of attack. For the wings with rigid body freedoms, body freedom flutter, bending-torsion flutter, and a tunnel support related dynamic instability was observed. Good agreement with linear theory was found for all the observed instabilities at low angles of attack. The present studies extend the experimental base for aeroelastic tailoring with composites, and provide further insight into actual aeroelastic behavior of composite wing aircraft in flight.

## INTRODUCTION

At the start of this paper, I would like to say a few words about Raymond L. Bisplinghoff. I was one of those fortunate to have taken courses and to have worked with Professor Bisplinghoff at M.I.T. I will always remember his lecture style in class, where he would get up to the blackboard, write neatly, and present with great clarity the subject matter at hand. As Holt Ashley remarked previously, the notes you took in his class were worth keeping a long time. Also, I will always remember his patience and sympathy listening to students, and helping them with their research and problems.

The use of composite materials in aircraft structures has added another design dimension to the aircraft designer. These composite materials are useful not only for their high strength-to-weight ratio, but also because they give the designer the ability to vary force-deflection and stiffeness-coupling behavior by varying the layup scheme, and thus, they have made some previous impractical design options attractive. In particular, forward swept wings have gained renewed interest because their traditionally low aeroelastic divergence speed can be increased by using appropriate aeroelastically tailored composite material in the wing construction. A good recent comprehensive survey of this subject is given by Shirk, Hertz, and Weisshaar [1].

The present paper describes a series of experimental studies conducted at M.I.T. on the aeroelastic flutter and divergence behavior of stiffness coupled, graphite/epoxy composite wing models. Both low angle-of-attack (linear) and high angle-of-attack (nonlinear) aeroelastic properties were explored. The studies comprised three separate investigations, namely, (a) cantilever unswept wings, by Hollowell and Dugundji [2], (b) cantilever forward swept wings, by Landsberger and Dugundji [3], and (c) forward swept wings with rigid body freedoms, by Chen and Dugundji [4]. Each of these will be described briefly in turn.

## CANTILEVER UNSWEPT WINGS

The wings used in the experiments were cantilevered, flat, 6 ply graphite/epoxy laminate plates. The wings were rectangular in planform with a length of 305 mm and a chord of 76 mm, giving the wing a length-to-chord ratio of 4 (full aspect ratio of 8). Figure 1 shows a layout of the wings with the associated positive directions. Six wings with different ply layups were tested, namely $[0_2/90]_s$, $[30_2/0]_s$, $[-30_2/0]_s$, $[45_2/0]_s$, $[-45_2/0]_s$, and $[\pm 45 /0]_s$. These were selected to give a wide range of stiffness and bending-twist coupling variation, both favorable and unfavorable.

Structural deflections tests were conducted on the wings by placing them in a stiff jig frame, subjecting them to tip forces and moments, and measuring the resulting tip deflections and twist. Also vibration tests were conducted using an electrodynamic shaker to obtain the lowest three natural frequencies and modes of the wings.

All wind tunnel tests were performed in the MIT Acoustic Wind Tunnel. This tunnel has a continuous flow with a 1.5 x 2.3m (5x7 ft) test section and could be varied continuously up to velocities of 30 m/sec (98 ft/sec). The model was mounted vertically on a turntable which could readily be rotated to change the angle of attack. See Fig. 2 for the test set-up in the wind tunnel. The test procedure consisted of setting a given tunnel speed, then slowly increasing the angle of attack of the wing up to 16 degrees until a flutter condition was found. The procedure was then repeated at a higher tunnel speed. Flutter was monitored by bending and torsion strain gages mounted at the wing root and attached to a strip chart recorder, and also by visual observation of the wings.

The flutter and divergence characteristics of these wings are shown in Fig. 3. The $[0_2/90]_s$ wing with no bending-twist coupling exhibited bending-torsion flutter at low angles of attack and torsion stall flutter at high angles of attack. This behavior, showing a large drop in flutter speed with root angle of attack, is similar to that reported earlier by Rainey [5] for thin aluminum wings of 2% and 4% thickness ratios. The wings with large positive bending-twist coupling $[30_2/0]_s$ and $[45_2/0]_s$ exhibited primarily bending-torsion flutter. The flutter velocity here did not drop significantly with increasing root angle of attack because the coupling caused a decrease in tip angle of attack, preventing it from stalling. The $[\pm 45/0]$ wing, which was stiff in torsion and had only a small amount of positive bending-twist coupling, would not flutter at low angles of attack within the 30 m/s maximum speed of the tunnel, but did show torsional stall flutter at the higher angles of attack as did the $[0_2/90]_s$ wings. The wings with large negative bending-twist coupling $[-30_2/0]_s$ and $[-45_2/0]_s$ exhibited divergence with an attendant bending stall flutter near the bending frequency of the wing. The divergence condition was noted when the root angle of attack could not be set small enough to keep the wing from "flipping over" from one side to the other. The bending stall flutter was due to bending motion causing the wing to stall due to the bending-twist stiffness coupling.

The flutter and divergence characteristics at low angles of attack were well predicted by linear theory using a 5 mode Rayleigh-Ritz analysis and 2-dimensional aerodynamic strip theory with a suitable aspect ratio lift slope correction for the divergence. Both the experiments and the analytical results showed the vastly different aeroelastic characteristics possible by changes in the ply layups, for these similar appearing wings. Further details of this investigation can be found in [2].

## CANTILEVER FORWARD SWEPT WINGS

The wings used for this investigation were identical to those used in the previous study, only now, they were mounted to the turntable by means of a large heavy swivel clamp joint which could be rotated to change the angle of sweep. For these tests, only the unswept ($\Lambda = 0°$) and forward swept ($\Lambda = -30°$) positions were investigated. A total of thirteen wings were tested for each sweep position, namely, $[0_2/90]_s$, $[15_2/0]_s$, $[-15_2/0]_s$, $[30_2/0]_s$, $[-30_2/0]_s$, $[45_2/0]_s$, $[-45_2/0]_s$, and the intermediate $[\pm 15/0]_s$, $[\mp 15/0]_s$, $[\pm 30/0]_s$, $[\mp 30/0]_s$, $[\pm 45/0]_s$, $[\mp 45/0]_s$. These wings spanned a wide range of bending stiffness $D_{11}$, torsion stiffness $D_{66}$, and bending-twist coupling ratio $D_{16}/\sqrt{D_{11}D_{66}}$ from -.77 to +.77 approximately.

The wings were checked for their static deflection and vibration characteristics, and again placed in the MIT Acoustic Wind Tunnel as shown in Fig. 2. The same procedures as before were used, only now, more extensive measurements were taken including video movies of the wings, from which the static position and vibrational amplitudes of the wing tips could be determined during the tests. This helped give some information on the nonlinear as well as the linear aeroelastic behavior of these flexible wings.

The flutter and divergence characteristics of these wings are shown in Figs. 4 and 5 for the unswept and forward swept wings, respectively. This data is experimental and was taken in 1 deg. increments in angle of attack. The frequency (Hz) of the resulting oscillations is also indicated on the graphs in order to show the change in flutter modes. The results in Fig. 4 for the unswept wing show roughly the same values and trends as in the previous study, Fig. 3. The results in Fig. 5 for the 30 deg. forward swept wing also exhibit the same four aeroelastic phenomena as for the unswept wing, namely, bending-torsion flutter ($\omega_h < \omega < \omega_\alpha$) and divergence at low tip angles of attack, and torsion stall flutter ($\omega \approx \omega_\alpha$) and bending stall flutter ($\omega \approx \omega_h$) at high tip angles of attack. Also, the results in Fig. 5 indicate that a positive ply layup of $[15_2/0]_s$ seems to have sufficient bending-twist coupling to raise the lowered aeroelastic divergence speed of the untailored $[0_2/90]_s$ forward swept wing.

Figure 6 represents a cross-plot of the results from Figs. 4 and 5 for low root angles of attack. The effects of ply orientation can be seen more clearly here. For negative ply angles $\theta_F$ and forward sweep, divergence dominates the aeroelastic behavior. Going to positive ply angles increases the divergence speed, and a high flutter speed now limits the flight speed. However, too much positive ply angle (+45) results in a low divergence speed again since the bending stiffness becomes too low, and the bending-twist coupling cannot overcome the geometric divergence tendency. A positive ply angle of about 15 deg. seems optimum here. A linear 5 mode Rayleigh-Ritz analysis gave good correlation with these experimental results. It should be mentioned that the trends shown in Fig. 6 are similar to analytical trends given by Hertz et al [6].

Nonlinear steady airload deflections and twist of the wing tips were also obtained. Analytical calculations using a semiempirical nonlinear theory including stalling effects gave reasonable correlation with experiment. Further details of this investigation can be found in [3].

## FORWARD SWEPT WINGS WITH RIGID BODY MODES

Recently, several authors [7]-[9] have pointed out significant effects of rigid body modes in modifying the cantilever aeroelastic behavior of forward swept wings. Looking at Fig. 7, one sees that since the bending frequency of a forward swept wing is lowered due to the approach to wing divergence while the rigid body pitch frequency is increased due to aerodynamic stiffening, a new low frequency "body freedom flutter" which couples wing bending with rigid body aircraft modes becomes possible. This effect was explored on a limited experimental basis by half-plane models in [10] and [11], but generally experimental data on this effect for forward swept wings is sparse.

Accordingly, a generic, full-span, 30-degree forward swept wing aircraft model was constructed and mounted with low friction pitch and plunge support bearings in the MIT Acoustic Wind Tunnel, using nominally the same wings as in the previous cantilever tests. Figure 8 shows a photograph of the actual model constructed. The model was statically stable with a measured static margin of 18%, and the effects of aeroelastic tailoring in "free flight" could be explored by interchanging four different wing layups on the model, namely, $[0_2/90]_s$, $[15_2/0]_s$, $[30_2/0]_s$, and $[-15_2/0]_s$.

Figure 9 shows the MIT Acoustic Wind Tunnel test setup for the model. The model was mounted on a pole by soft springs in plunge and pitch. Snubber cables attached to the nose and tail could be pulled back to hold the model in place if any motions became too violent. Bending and torsion strain gages at the wing root monitored the wing motion, a potentiometer monitored the pitch motion, and a strain gage on a flexible thin steel beam attached to the upper end of the plunge supporting spring monitored the plunge motion. In addition to the strip chart recordings of these signals, video movies were taken of the model during the tests.

Before the flutter tests, static deflection and vibration tests were made to check out the wing and model tunnel support characteristics. Also aerodynamic tests were run at a very low wind speed to obtain the aerodynamic lift, moment, and static margin characteristics of the "rigid" aircraft. The flutter test procedure consisted of setting the canard incidence so that the model would be flying at a slightly positive angle of attack, then slowly increasing the wind speed in small increments until a flutter condition was observed. If the model was stable, a small kick would be given to it by means of a long wooden pole, and the resulting decay motion observed by the recorder traces.

Figure 10 shows an example of body freedom flutter for the $[0_2/90]_s$ wing at a speed V = 20 m/sec. The flutter frequency was 2.8 Hz compared with still air vibration frequencies of .85 Hz for the model pitching frequency and 11 Hz for the wing bending frequency. The wing bending was approximately in phase with the model pitch while the plunge motion was about 180 deg. out of phase, and the flutter mode exhibited a nodal point near the model nose position.

Figure 11 shows the results for the $[15_2/0]_s$ wing. The aircraft model with this wing demonstrated the best tailored aeroelastic behavior of all four wings tested. Instead of body freedom flutter occuring, a bending-torsion wing flutter was observed to set in at the much higher speed of V = 29 m/sec. The flutter frequency here was 32 Hz as compared with still air frequencies of 9 Hz for the wing bending frequency and 45 Hz for the wing torsion frequency.

In addition to these aeroelastic instabilities, there also existed originally a tunnel support related dynamic instability involving the model plunge and pitch motion. This arose because the still air pitch frequency of the model was originally below the still air plunge frequency on the soft spring support. Since the pitch frequency increases with air speed while the plunge frequency remains constant, a coupling instability may occur similar to that shown in Fig. 7 previously. Figure 12 shows an example of the tunnel support dynamic instability for the $[0_2/90]_s$ wing at a speed of V = 10 m/sec, where the model plunged unboundedly until it hit the bottom and top stops.

This dynamic instability occured over a limited velocity range and disappeared at higher tunnel speeds. To eliminate this instability entirely, the still air pitch frequency was increased from .20 Hz to .85 Hz in later tests, so it would be above the still air plunge frequency of .63 Hz.

A summary plot of the results for these forward swept wings with rigid body freedoms is shown in Fig. 13. The open circles show the body freedom flutter points, the open triangles show the bending-torsion flutter, the filled circles show the extent of the tunnel support dynamic instability for the original support system, and the filled triangles show the cantilever wing divergence points, which were obtained by holding the model securely with the snubber cables to eliminate the rigid body freedoms. These results resemble those given in Fig. 6 for the 30 deg. forward swept wing. Good agreement is seen in Fig. 13 between the experimental results and linear analytical calculations for body freedom flutter, bending-torsion flutter, tunnel support dynamic instability, and cantilever wing divergence. Also, it is seen that the body freedom flutter results follow the same aeroelastic tailoring trends as the cantilever divergence results, but with a slightly lower velocity, both experimentally and analytically.

Tests were also conducted on the models restrained in plunge and free in pitch only. These gave somewhat lower body freedom flutter boundaries than did the previous free in pitch and plunge cases. Further details of the rigid body modes investigations mentioned here can be found in [4].

THEORETRICAL ANALYSES

To check the experimental results and to understand the trends, theoretical linear analyses were performed for these various studies. Five mode Rayleigh-Ritz analyses were used to set up the equations of motion for the cantilever wing studies, while seven mode analyses were used for the rigid body mode cases. Figure 14 shows the general procedures and modes used. For the aerodynamic forces, 2-dimensional, incompressible, unsteady, velocity parallel strip theory was used both with and without corrections for 3-dimensional effects (see the specific study reference). The equations were then cast into an appropriate eigenvalue problem as shown in Fig. 15, and the resulting critical speeds determined. In the case of flutter, the earlier cantilever studies used the V-g method of solution, while the rigid body freedom cases used a one pole Padé approximation for the Theodorsen $C(p)$ function in the aerodynamic forces and then cast the equations into standard first order differential equation form to find the eigenvalues. A typical root locus plot of the eigenvalues, p, for the $[0_2/90]_s$ wing is shown in Fig. 16.

CONCLUSIONS

A series of experimental studies have been conducted on the aeroelastic flutter and divergence behavior of unswept and forward swept, stiffness coupled, graphite/epoxy wings. The following general conclusions are drawn.

Large variations in aeroelastic behavior are possible for different ply orientations of the wings.

The tests showed occurences of bending-torsion flutter, divergence, torsional stall flutter, bending stall flutter, body freedom flutter, and tunnel support dynamic instability.

At low angles of attack, bending-torsion flutter, divergence, body freedom flutter, and support dynamic instability were well predicted by linear theory.

At large angles of attack, there was a transition to nonlinear phenomena, i.e., torsional stall flutter and bending stall flutter. These are not well understood.

Forward swept wings develop body freedom flutter rather than divergence when rigid body freedoms are present. This occurs at lower speeds than cantilever divergence, but follows the same aeroelastic tailoring trends.

A tunnel support dynamic instability can develop if the model pitch frequency is lower than the model plunge frequency.

The present studies extend the experimental base for aeroelastic tailoring with composites, and provide further insight into actual aeroelastic behavior of composite wing aircraft in flight.

## ACKNOWLEDGEMENTS

The author wishes to thank the three graduate students, Steven Hollowell, Brian Landsberger, and Gun-shing Chen who conducted this research as part of their theses. Also, he wishes to acknowledge the support of the Materials Laboratory of AFWAL and the U.S. Air Force Office of Scientific Research, Dr. Anthony Amos, technical monitor.

## REFERENCES

1. Shirk, M.H., Hertz, T.J., and Weisshaar, T.J., "Aeroelastic Tailoring - Theory, Practice and Promise," J. of Aircraft, Vol. 23, No. 1 January 1986, pp. 6-18.

2. Hollowell, S.J., and Dugundji, J., "Aeroelastic Flutter and Divergence of Stiffness Coupled, Graphite/Epoxy Cantilevered Plates," J. of Aircraft, Vol. 21, No. 1, January 1984, pp. 69-76.

3. Landsberger, B., and Dugundji, J., "Experimental Aeroelastic Behavior of Unswept and Forward Swept Graphite/Epoxy Wings," J. of Aircraft, Vol. 22, No. 8, August 1985, pp. 679-686.

4. Chen, G-S., and Dugundji, J., "Experimental Aeroelastic Behavior of Forward Swept Graphite/Epoxy Wings with Rigid Body Modes," 27th AIAA/ASME/ASCE/AHS Structures, Structural Dynamics, and Materials Conference, San Antonio, Texas, May 1986, AIAA Paper 86-0971.

5. Rainey, G.A., "Preliminary Study of Some Factors which Affect the Stall-Flutter Characteristics of Thin Wings," NACA TN-3622, March 1956.

6. Hertz, T.J., Shirk, M.H., Ricketts, R.H., and Weisshaar, T.A., "On the Track fo Practical Forward Swept Wings," Aeronautics and Astronautics, Vol. 20, No. 1, January 1982, pp. 40-53.

7. Weisshaar, T.A., Zeiler, T.A., Hertz, T.J., and Shirk, M.J., "Flutter of Forward Swept Wings, Analyses and Tests," 23rd AIAA/ASME/ASCE/AHS Structures, Structural Dynamics, and Materials Conference, New Orleans, Louisiana, May 1982, AIAA Paper 82-0646.

8. Miller, G.D., Wykes, J.H., and Brosnan, M.J., "Rigid Body - Structural Coupling on a Forward Swept Wing Aircraft," J. of Aircraft, Vol. 20, No. 8, August 1983, pp. 696-702.

9. Noll, T.E., Eastep, F.E., and Calico, R.A., "Active Suppression of Aeroelastic Instabilities on a Forward Swept Wing," J. of Aircraft, Vol. 21, No. 3, March 1984, pp. 202-208.

10. Chipman, R., Rauch, R., Rimer, M., Muniz, B., and Ricketts, R.H., "Transonic Tests of a Forward Swept Wing Configuration Exhibiting Body Freedom Flutter," 26th AIAA/ASME/ASCE/AHS Structures, Structural Dynamics, and Materials Conference, Orlando, Florida, April 1985, AIAA Paper 85-0689

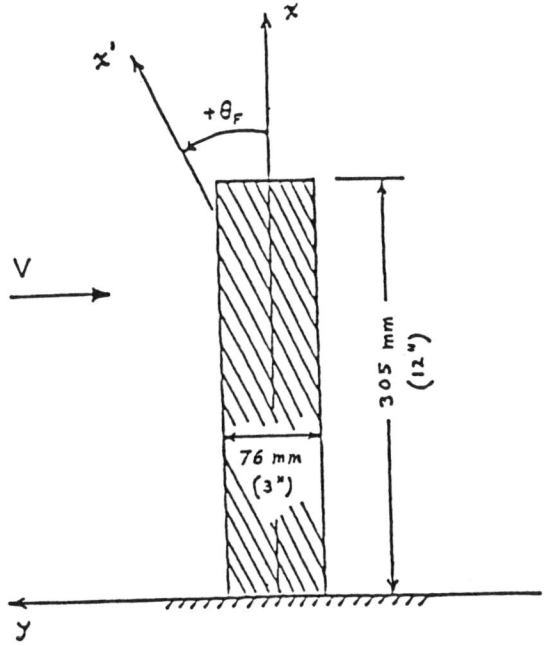

Fig. 1  Cantilever Wing Model

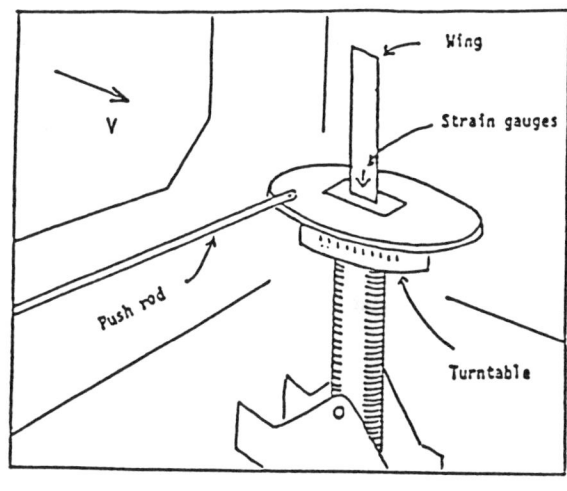

Fig. 2  Test Set-Up in Tunnel (Cantilever Wing)

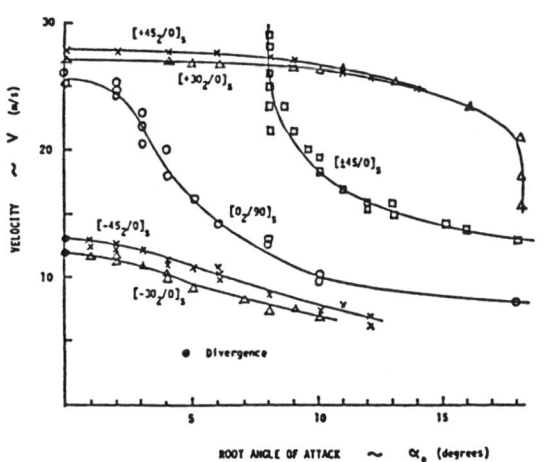

Fig. 3  Flutter and Divergence Results (Unswept, $\Lambda = 0°$)

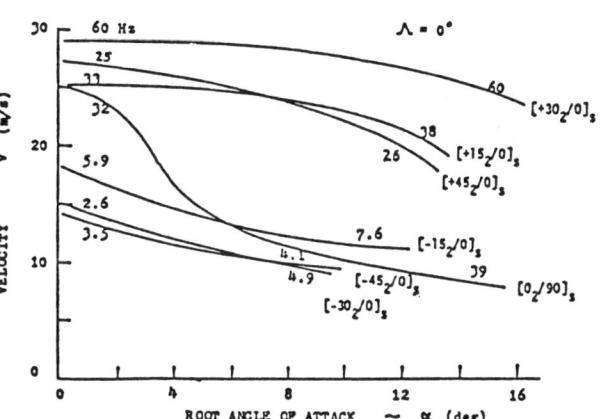

Fig. 4  Flutter and Divergence Results (Unswept, $\Lambda = 0°$)

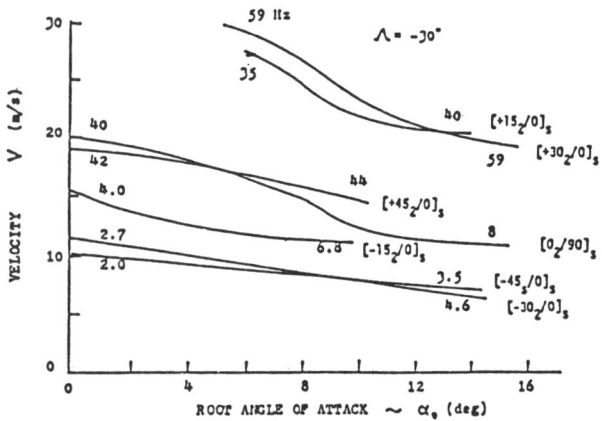

Fig. 5 Flutter and Divergence Results (Swept forward, $\Lambda = -30°$)

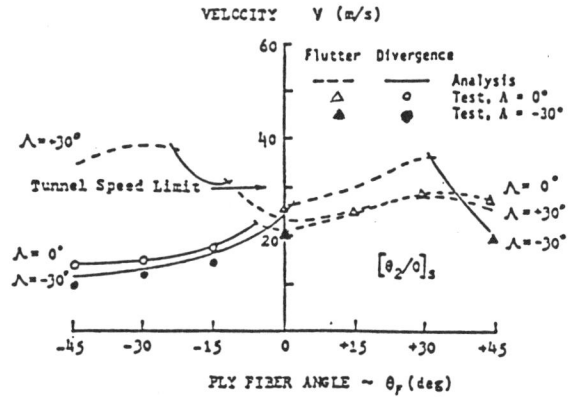

Fig. 6 Flutter and Divergence at Low Angles of Attack

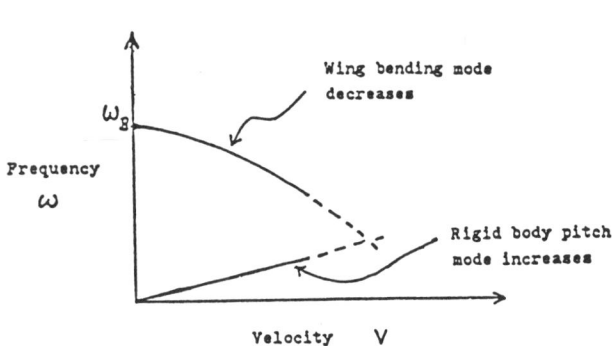

Fig. 7 Rigid Body Model Effects on Forward Swept Wing Flutter

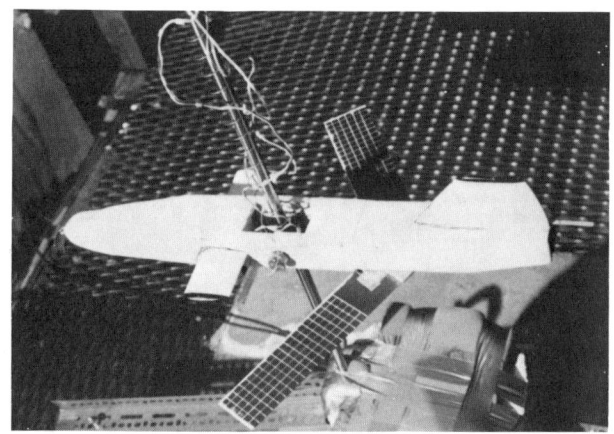

Fig. 8 Forward Swept Wing Aircraft Model

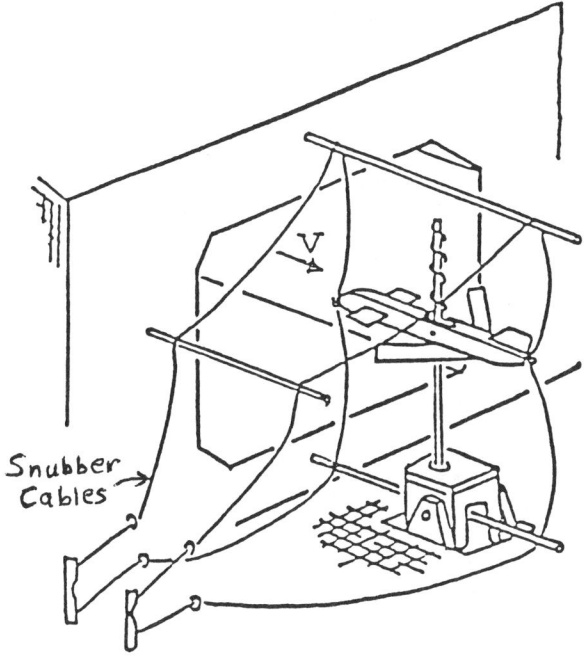

Fig. 9 Test Set-Up in Tunnel (with Rigid Body Modes)

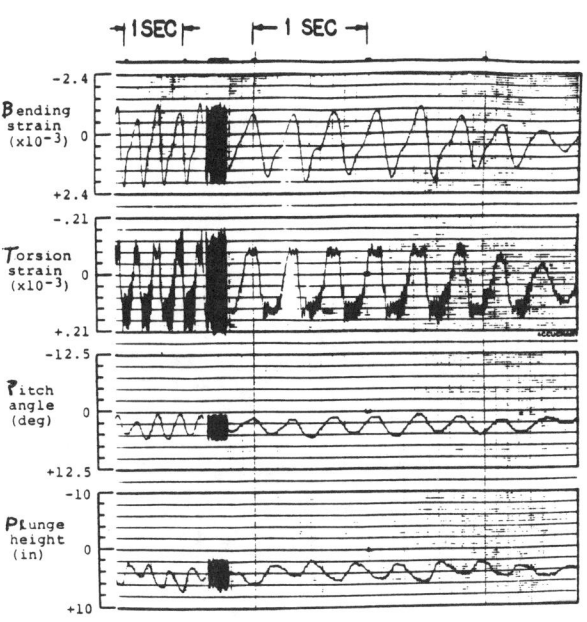

Fig. 10 Body Freedom Flutter, $[0_2/90]_s$ Wing, V = 20 m/sec

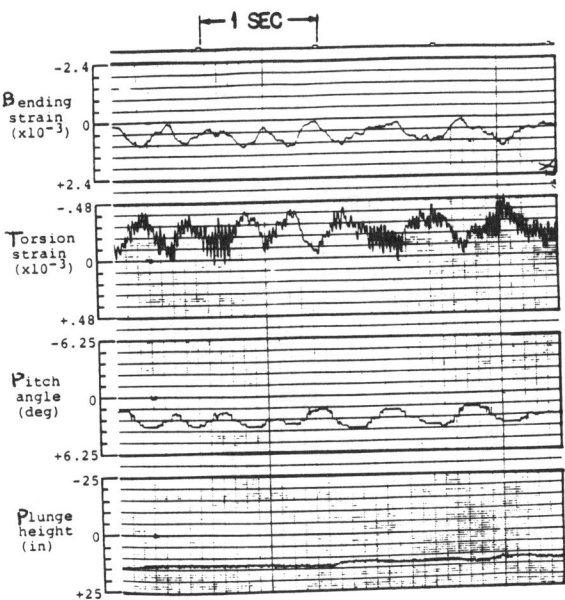

Fig. 11 Bending-Torsion Flutter, $[15_2/0]_s$ Wing, V = 29 m/sec

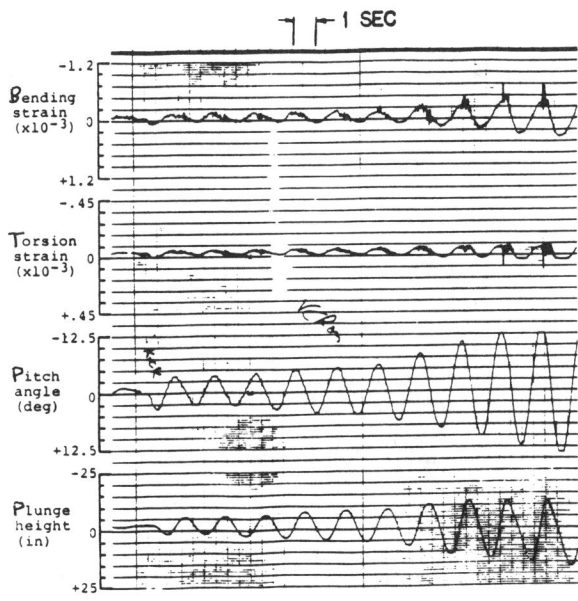

Fig. 12 Support Dynamic Instability, $[0_2/90]_s$ Wing, V = 10 m/sec

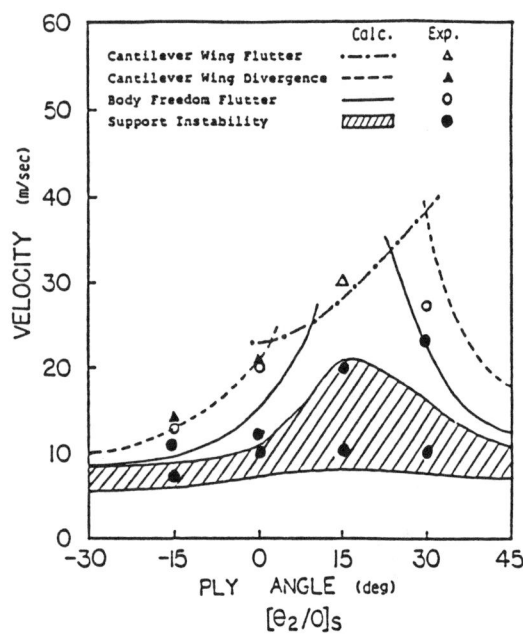

Fig. 13  Flutter and Divergence Results for "Free" Model

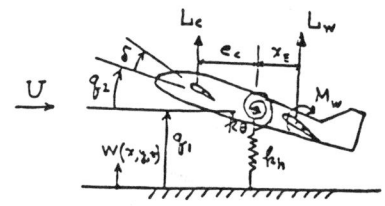

$$W(x,y,t) = \sum_{i=1}^{7} \gamma_i(x,y)\, q_i(t)$$

- $q_1$ = Rigid Transl.
- $q_2$ = Rigid Pitch
- $q_3$ = 1st Cantil. Bend.
- $q_4$ = 2nd Cantil. Bend.
- $q_5$ = 1st Torsion
- $q_6$ = 2nd Torsion
- $q_7$ = 1st Chordwise

$$T = \tfrac{1}{2}\iint m\,\dot{w}^2\,dx\,dy = \tfrac{1}{2}\sum_i\sum_j M_{ij}\,\dot{q}_i\dot{q}_j$$

$$U = \tfrac{1}{2}\iint \{D_{11} w_{xx}^2 + \ldots\}\,dx\,dy + \tfrac{1}{2} k_h q_1^2 + \tfrac{1}{2} k_\theta q_2^2$$
$$= \tfrac{1}{2}\sum_i\sum_j K_{ij}\,q_i q_j$$

$$\delta W = \sum_i Q_i\,\delta q_i$$

Fig. 14  Theoretical Analysis

$$\underline{M}\,\ddot{\underline{q}} + \underline{K}\,\underline{q} = \underline{Q}_{Aero}$$

Vibrations:  $q_i = \bar{q}_i\, e^{i\omega t}$

$$[-\omega^2 \underline{M} + \underline{K}]\,\underline{\bar{q}} = 0 \;\rightarrow\; \text{Solve } \omega,\,\underline{\bar{q}}$$

Flutter:  $q_i = \bar{q}_i\, e^{pt}$

$$\left[p^2\underline{M} + \underline{K} - \tfrac{1}{2}\rho V^2 S\left(\underline{B}_3 p^2 + \underline{B}_2 p + \underline{B}_1 + \underline{B}_0 \tfrac{p}{p+\beta}\right)\right]\underline{\bar{q}} = 0$$
$$\rightarrow \text{Solve } p,\,\underline{\bar{q}}$$

Divergence:  $q_i = \bar{q}_i$

$$[\underline{K} - \tfrac{1}{2}\rho V^2 S\,\underline{B}_0]\,\underline{\bar{q}} = 0 \;\rightarrow\; \text{Solve } V,\,\underline{\bar{q}}$$

Fig. 15  Equations of Motion

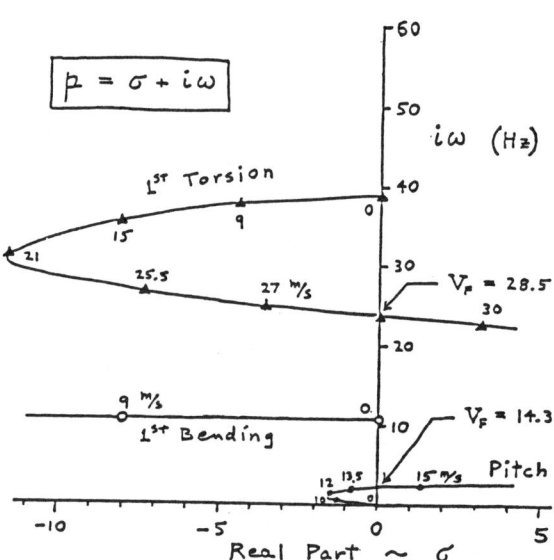

Fig. 16  Root Locus Stability Plot, $[0_2/90]_s$ Wing

# LOW-SPEED AEROELASTICITY OF BRIDGE DECKS

Robert H. Scanlan
The Johns Hopkins University
Baltimore, Maryland, 21218

## ABSTRACT

An introduction and outline of the character of the main aeroelastic problems of bridge decks is given. These problems consist principally of the flutter, vortex excitation, and buffeting of long-span, flexible bridges under the effects of the natural wind. It is pointed out that the field of bridge aeroelasticity, which gained initial impetus and considerable guidance from the older field of aircraft aeroelasticity, has developed a special character of its own due to its principal focus upon the aerodynamic effects of wind in the earth's boundary layer blowing over civil engineering structures which are aerodynamically bluff bodies. The discipline finds world-wide application to long, flexible, suspended-span bridges.

## INTRODUCTION

The Tacoma Narrows incident of 1940 was not the first collapse of a long-span bridge due to wind. Historical review [1] later showed that, as much as a century before, with the Brighton Chain Pier incident, a very analogous self-excited oscillation had resulted in the collapse of a suspended span. Several other bridges have also failed under wind-induced vibration. The Tacoma Narrows failure occurred, however, at a time when aeroelasticity as a recognized discipline (though not yet by that name) was well under way, and the early investigations by Farquharson and by Karman and Dunn, in the United States, and by Scruton in England, clearly categorized the Tacoma Narrows case-- and bridge wind stability problems in general-- as aeroelastic in nature.

The earliest approaches to solution were quite on target: reduced-scale section models of the bridge deck were elastically mounted and directly tested in the wind tunnel. This basic approach-- though with some important nuances of appreciation-- has held to the present time. After a long hiatus it also has now, once again, become fashionable to test reduced-scale full-bridge aeroelastic models in the wind tunnel, although the practice was actually inaugurated in the 1940's by both Farquharson and Scruton.

It may be noted in passing that the "explanation" [2] for the Tacoma Narrows collapse-- that it was "due to vortex shedding"-- has, in the intervening years, been subject to some correction and refinement [3][17]. In fact, the vortex wake shed from the Tacoma Narrows Bridge in its final catastrophic oscillation was a motion-induced, or flutter, wake, and not a natural Karman vortex trail.

Bridge decks, by the nature of their primary design purpose, are usually bluff bodies, from which the cross-wind flow inevitably separates at several points. The aeroelasticity of bodies of this sort is basically nonlinear insofar as the fluid forces are concerned. It is often possible, however, to limit the concerns of the designer to incipient cases of instability and thus invoke an argument of linearity under restricted circumstances. This approach has proven very useful, just as it has been in the case of the classical flutter of thin airfoils.

In fact, the general methods of approach of classical airfoil flutter theory have been a very useful guide-- as to form, though not as to content-- in setting up the theories of bridge instability under wind. Many years ago there were some attempts (see [4] for example) made to bend the application of flat plate flutter theory to bridges, or at least to employ such theory as somehow a background limiting case for bridges. Such attempts were never substantiated, and it has been the subsequent concrete efforts [5][6] to acquire bridge deck flutter derivatives by experimental means, and to compare these results directly with flat plate thin air-foil results, that have definitely shown bridge flutter to differ fundamentally from airfoil flutter [3].

Perhaps the single most dramatic evidence of this lies in the torsional aerodynamic damping which, for bridges, often exhibits a strong sign-reversal characteristic (from stable to unstable) as a function of reduced velocity, while that for airfoils maintains a consistently stable characteristic, unchanged in sign. Viewed alternately in terms of the so-called indicial response function, related as a Fourier transform to the corresponding flutter derivative, the result for airfoils in incompressible flow is the well-known Wagner function, while that for bridge decks-- varying from deck to deck-- takes on a highly different form [3] more related to comparable indicial functions for stalled surfaces.

A further important difference of bridge structures from flight structures is the relative proportions of structure inertial to aerodynamic forces in the two cases. With heavy bridge structures, natural structural frequencies and modal forms are not as greatly affected by aerodynamic forces, and a flutter involving strong frequency coalescence by two or more modes is less likely to be brought about by the relatively low-speed forces of the wind. Civil engineering structures are designed not to fly. What for some time, earlier, was not sufficiently understood by the civil engineer, however, was the intrinsically important role of aerodynamic damping, which, by becoming effectively negative at some reduced velocities, can contribute as the causative factor of bridge instability.

Today such questions have been clarified by the experimental gathering of a large number of flutter derivatives for practical bridges. This is accomplished via the bridge deck section model in a wind tunnel, either spring supported and freely oscillating or mechanically driven. Various methods of system identification-- interesting in their own right-- have been applied in inferring flutter derivatives from wind tunnel tests.

When a bridge deck cross section is first designed with obvious civil engineering objectives in mind, it characteristically may not be at an optimum condition of inherent aerodynamic stability. Subjected then to section model wind tunnel studies under the guidance of an experienced aerodynamicist, the deck form becomes the object of a dialogue with the designer. It was once considered that the deck model in such circumstances needed, in both dynamic and aerodynamic senses, to "duplicate" full bridge behavior, but as the full stability implications of its flutter derivatives are becoming understood and accepted, it is seen that, for flutter testing, strict dynamic similarity need not be required; however, strong emphasis must still be placed upon correct reduced velocity values and on geometric similarity of model to prototype, for obvious reasons.

Bridge decks that undergo evolutionary aeroelastic studies in the design stage-- particularly when their negative aerodynamic damping tendencies are thereby wholly eliminated by judicious choice of form-- become excellent candidates for a service life undisturbed by wind. Unfortunately, a considerable number of bridges lacking such thorough stability examination and assurance were built in the past, or occasionally even now slip through the design stage. Also, many existing bridges escape the natural retribution of a severe wind exposure that tests them to their extremes; but a few do suffer repeatedly for a lifetime. Bridges that have undergone such occasionally tortured existence are exemplified by the Deer Isle in Maine, the Bronx-Whitestone in New York, and even the Golden Gate in San Francisco, all of which have required considerable post-construction attention, because of instability under wind, some of this even extending to the present time.

While a fair number of classical-type catenary suspension bridges are presently under design or in construction stages throughout the world, this older design type is fast giving way to the cable-stayed type for main spans of intermediate size, up to about 500 meters. These newer types have deck forms that entrain new dynamic and aerodynamic considerations. The primary study of their deck contours and of the related flutter derivatives, however, continues to be the wind-stability approach of choice.

While the prevention of catastrophic wind-induced oscillation (flutter) is the primary goal of the aerodynamicist and designer, the bridge may be subject to two other important wind effects: vortex-induced oscillation, both for decks and for free-standing towers in the construction phase, typically at moderate wind speeds; and buffeting by wind turbulence, notably at the higher wind speeds. Both of these require attention to the flutter derivatives of the deck, but special means must also be invoked to account for each of these phenomena beyond the usual framework of flutter dynamics. Both experimental and theoretical approaches to these additional problems have been subjects of more recent researches.

## THEORETICAL FRAMEWORK OF FLUTTER FORCES

To illustrate the general flavor of bridge aeroelastic problems, the flutter problem will be outlined below in detail. Let a symmetric bridge deck cross section be endowed with elastic degrees of freedom (referred to its section c.g.) $h$ (vertical), $\alpha$ (twist), and $p$ (lateral sway).

When turbulent wind approaches the section, the section then may be espressed as

$$L = L_{s.e.} + L_b \tag{1}$$

$$D = D_{s.e.} + D_b \tag{2}$$

$$M = M_{s.e.} + M_b \tag{3}$$

where the subscripts s.e. and b refer, respectively, to "self-excited" and "buffeting". Details of the buffeting forces will not be given in the present article. One form that they may take is suggested in [7].

As a result of several studies [6][7] the linearized self-excited terms have taken on the following conventional form:

$$L_{s.e.} = (1/2)\rho U^2 B[KH_1^*(K)(\dot{h}/U) + KH_2^*(K)(B\dot{\alpha}/U)$$
$$+ K^2 H_3^*(K)\alpha] \tag{4}$$

$$D_{s.e.} = (1/2)\rho U^2 B[KP_1^*(K)(\dot{p}/U) + KP_2^*(K)(B\dot{\alpha}/U)$$
$$+ K^2 P_3^*(K)\alpha] \tag{5}$$

$$M_{s.e.} = (1/2)\rho U^2 B^2[KA_1^*(K)(\dot{h}/U) + KA_2^*(K)(B\dot{\alpha}/U)$$
$$+ K^2 A_3^*(K)\alpha] \tag{6}$$

in which

$\rho$ = air density
$U$ = mean cross-wind velocity
$B$ = deck width
$K = B\omega/U$ = reduced frequency
$\omega$ = oscillation circular frequency
$1/K$ = reduced velocity
$H_i^*$, $P_i^*$, $A_i^*$ are flutter coefficients
($KH_i^*$, etc. play the roles of flutter derivatives)

It is presumed that, for a specific bridge deck, the values of $H_i^*$, $P_i^*$, $A_i^*$, as functions of $K$, are obtained by experiment. The values of $H_i^*$ and $A_i^*$ for deck sections are usually given [5] as functions of $2\pi/K$. Values of $P_i^*$ have not to date been experimentally obtained. It may further be noted that self-excited terms in $\ddot{h}$, $\ddot{p}$, $\ddot{\alpha}$, as well as $h$ and $p$, have been omitted in the context of civil engineering structures as usually being negligible.

## FLUTTER DYNAMICS OF A FULL BRIDGE

Assuming the bridge structure itself to be linearly elastic, the deck will respond in modal components

$$h_i(x), p_i(x), \alpha_i(x),$$

of the dimensionless modal forms of the system, corresponding respectively to the local displacement degrees of freedom, where $x$ designates the spanwise location along the deck.

The total displacement of the local section c.g. will then be given by the components

$$h(x,t) = \sum_i \xi_i(t) h_i(x) B \quad (7)$$

$$p(x,t) = \sum_i \xi_i(t) p_i(x) B \quad (8)$$

$$\alpha(x,t) = \sum_i \xi_i(t) \alpha_i(x) \quad (9)$$

where $\xi_i(t)$ is the generalized coordinate of mode i and the sum is over all modes.

The generalized force $Q_i$ for mode i is then calculated from

$$Q_i(t) = \int_{\text{deck span}} [L(\delta h/\delta \xi_i) + D(\delta p/\delta \xi_i) + M(\delta \alpha/\delta \xi_i)] \, dx \quad (10)$$

and the equation of motion of mode i is

$$I_i [\ddot{\xi}_i + 2\zeta_i \omega_i \dot{\xi}_i + \omega_i^2 \xi_i] = Q_i(t) \quad (11)$$

where $I_i$ is the full-bridge generalized inertia of mode i, $\zeta_i$ and $\omega_i$ are the respective damping and natural circular frequency of mode i. Confining attention to self-excited force components only, $Q_i$ can be expressed by

$$[Q_i(t)]/[(1/2)\rho U^2 B^3] = (K/U) \{ \sum_j \dot{\xi}_j [H_1^* G_{h_j h_i} + H_2^* G_{\alpha_j h_i} + P_1^* G_{p_j p_i} + P_2^* G_{\alpha_j p_i} + A_1^* G_{h_j \alpha_i} + A_2^* G_{\alpha_j \alpha_i}] + (KU/B) \sum_j \xi_j [H_3^* G_{\alpha_j h_i} + P_3^* G_{\alpha_j p_i} + A_3^* G_{\alpha_j \alpha_i}] \} \quad (12)$$

where all of $H_i^*$, $P_i^*$, $A_i^*$ (i = 1,2,3) are functions of the reduced flutter frequency K, and

$$G_{r_m s_n} = \int_{\text{deck span}} r_m(x) s_n(x) \, dx \quad (13)$$

with r,s = h, p, or $\alpha$, and m,n = i or j; ($G_{r_m s_n}$ has the dimension of length).

It is seen that, in principle, the self-excited aerodynamic forces couple all modes of the bridge together. The strength of this coupling remains a matter to be investigated. Note that, in the above formulation, "strip" theory is used and spanwise (aspect ratio) effects are taken as 100% correlated. Modifications to this assumption can evidently be introduced at will, and must be, in fact, in the presence of turbulent flow.

It is a matter of observation, as stated earlier, that the aerodynamic forces on full-span bridge models do not greatly distort the natural vibration modes or their frequencies. This observation is tantamount to asserting that the aerodynamic coupling depicted by equation (12) is weak relative to structural forces. Therefore, one reasonable simplification-- at least for investigative purposes-- may be to neglect coupling, as an approximation, and study the flutter susceptibility of the structure a single mode at a time. This idea brings the generalized self-excited force $Q_i$ to the form:

$$Q_i/[(1/2)\rho U^2 B^2] = (KB/U)\{\dot{\xi}_i[H_1^*(K)G_{h_i h_i} + P_1^*(K)G_{p_i p_i} + A_2^*(K)G_{\alpha_i \alpha_i}] + (KU/B)\xi_i[A_3^*(K)G_{\alpha_i \alpha_i}]\} \quad (14)$$

The equation of motion (11) then becomes

$$I_i[\ddot{\xi}_i + 2\tilde{\zeta}_i \tilde{\omega}_i \dot{\xi}_i + \tilde{\omega}_i^2 \xi_i] = 0 \quad (15)$$

where

$$\tilde{\omega}_i = \{\omega_i^2/[1 + (\rho B^4/2 I_i) A_3^*(\tilde{K}_i) G_{\alpha_i \alpha_i}]\}^{\frac{1}{2}} \quad (16)$$

and

$$\tilde{\zeta}_i = \zeta_i(\omega_i/\tilde{\omega}_i) - (\rho B^4/4 I_i)[H_1^*(\tilde{K}_i)G_{h_i h_i} + P_1^*(\tilde{K}_i)G_{p_i p_i} + A_2^*(\tilde{K}_i)G_{\alpha_i \alpha_i}] \quad (17)$$

with

$$\tilde{K}_i = B\tilde{\omega}_i/U \quad (18)$$

The criterion for flutter at the velocity $U$ will then be given by

$$\tilde{\zeta}_i \leq 0$$

which corresponds to:

$$(\tilde{\omega}_i/\omega_i)(\rho B^4/4 \zeta_i I_i)[H_1^*(\tilde{K}_i)G_{h_i h_i} + P_1^*(\tilde{K}_i)G_{p_i p_i} + A_2^*(\tilde{K}_i)G_{\alpha_i \alpha_i}] \geq 1 \quad (19)$$

This simplified approach may be considered conservative in that it can be applied, mode by mode, to discern those modes that contribute energy to the system via negative damping effects. The approach, however, does not reveal the possible coupling-- and entry into instability-- of two of more individually positively damped modes. This can be investigated by conventional means closely related to those used in aircraft flutter. However, as noted earlier, most bridge flutter, being a bluff-body phenomenon, is not of the classical or coupled kind. Experience with models suggests that, even in cases where two or more modes contribute, one mode, at least, "drives" the system via its negative damping.

## VORTEX-SHEDDING DYNAMICS OF A FULL BRIDGE

Vortex shedding, as a phenomenon, is accompanied by nonlinear aerodynamic forces. Indications of its presence are, however, made known by studies of the linear-model flutter coefficients (notably $H_1^*$) already described above.

Since the proper response to such problems is to create deck section contour changes that eliminate or reduce the problem, it is often sufficient to use the linear indication of vortex shedding merely to corroborate that additional design work is required. The evolution of such a design modification study was exemplified by the work of Wardlaw [8] on the Long's Creek cable-stayed bridge (New Brunswick, Canada).

The question of the spanwise coherence of vortex shedding arises, and the presence of turbulence in the oncoming flow is known to suppress small-amplitude vortex-induced oscillations. However, the well-known "lock-on" of shed vortices to resonant structural motion may dramatically increase spanwise coherence of the effect. Here again, the primary design response should be not merely to develop an analytical model for the expected dynamics, but to suppress the phenomenon itself, which, for large bridges, may be done by either streamlining of the deck cross-section or designing it to "shred" the flow sufficiently to destroy coherence. Open-truss bridge decks often accomplish the latter objective as an incident to their structural form, whereas bluff box and similar sections may require special treatment. Some analytical models for vortex shedding are discussed in [9].

## BRIDGE BUFFETING CONSIDERATIONS

It is first worthy of comment that while turbulence is almost invariably present in winds near the ground, it may on occasion be greatly reduced (to 2 or 3% or less) at the height (typically some 60 to 70 meters) of suspended spans over navigable waterways. This occasional and unusual flow-smoothing phenomenon may possibly be ascribed to certain thermal stratifications that occur in the upstream fetch. In any event, its existence has been corroborated in a few instances [10][11]. This circumstance emphasizes the fact that turbulence should not always be presumed in the flow over long-span bridges. As a result, the stability of bridge decks should be studied under turbulent flow as well as laminar flow.

When turbulence is introduced into the approach flow it has a number of consequences. It modifies the sectional flutter derivatives. It tends to suppress the coherence of vortex shedding. It also causes general loss of spanwise flow coherence. At high wind speeds the combined net effect-- as viewed to date on full-bridge models in the wind tunnel-- appears to be to diminish the tendency (if the latter is present) for flutter to initiate in an abrupt manner at a sharply-defined value of reduced velocity. Instead, buffeting amplitude grows as reduced velocity increases, but in a more steady, less abrupt way. Eventually, true flutter instability may set in, but at a greater reduced velocity than without turbulence; thus turbulence is generally observed to "increase flutter speed", although the last word on the subject has not yet been written.

Under purely laminar flow conditions, the prediction of full-bridge flutter can be done quite accurately from knowledge of the laminar-flow flutter derivatives. This generally proves to be the most conservative case. However, while it is usually unconservative to postulate the presence of turbulence in the oncoming flow when investigating flutter, it is necessary to consider such turbulence when examining how the bridge may be buffeted at any stage of its construction or after completion.

Important characteristics of turbulence are its intensity and its scale (representative wavelength), both for its long-wind and its across-wind components. Taking the important mean, horizontal wind direction to be that which is directed at right angles to the deck roadway, both horizontal and vertical components of turbulence, but especially the latter, play the principal roles in bridge buffeting. Turbulence intensities may vary from zero to as high as 20% in the natural wind. Scale varies widely in the atmosphere, from some 20 to 200 meters. When wind tunnel modeling of turbulence is done it is particularly important to match both the scale and the horizontal and vertical velocity spectra to those of the natural wind.

It is found for wind tunnels that, by typical passive means, such as upstream grids or surface roughness elements, turbulence scale lengths exceeding about 20 cm can rarely be achieved. Since, for correct similarity, the scale of the bridge model itself must match that of the turbulence, this implies that physical models of bridges must range in the scales of 1/100 to 1/1000. This has, in practical terms, implied that either quite small aeroelastic models of bridges are required in ordinary atmospheric wind tunnels, or that the test scale of turbulence ashievable therein is too small for the model.

The result of this situation has been that only a few small-scale, full-bridge aeroelastic models have been studied (perhaps not more than a half dozen at about 1/100 scale in very large boundary layer wind tunnels). Few, if any, section models, of 1/25 to 1/100 scale, have been tested to date under correctly scaled turbulence.

True turbulence being three-dimensional in nature, it is not yet possible to state with certainty the effects of turbulence upon the two-dimensional flutter derivatives. (However, see [15]). The higher frequency turbulence components, with eddies that are small compared to bridge deck width, affect the structure somewhat more like its own "signature" turbulence, tripped off by its windward structural members-- a turbulence effect already present in smooth approach flow. On the other hand, low-frequency, long-wavelength turbulence acts like slowly pulsating quasi-steady laminar flow.

One new approach for section models is an ongoing study, under the direction of the author, to study two-dimensional section models of reasonable scale under a two-dimensional gusting environment with appropriate spectrum created by a flapping airfoil cascade. This device is intended to reveal sensitivities of a local section of a deck to gusting, excluding the effects of spanwise coherence, which is kept effectively at 100% in the device in question. To date [12] at least one model (Golden Gate bridge deck) suggests that turbulence can produce trends in the $A_2^*$ flutter derivative that tend to destabilize the bridge.

Theories of bridge buffeting (for example, [13]), early followed the lead of Liepmann [14] and have been extended to include flutter derivative effects [15]. Recent work [16] suggests that nonlinear stochastic models of buffeting may be appropriate.


CONCLUSION

Theories of low-speed aeroelasticity relative to bluff bodies like bridge decks have benefitted greatly from the pre-existing field of aircraft aeroelasticity. However, while this prior domain has been of inestimable value as guide, stimulus, and definer of investigative routes, the detailed findings in bridge aeroelasticity have proven to have a distinctly different character of their own. Thus, out of this endeavor of the last quarter-century, a separate discipline of low-speed, bluff-body aeroelasticity has emerged. Its range of application is broad, particularly to the numerous long-span flexible bridge structures being built throughout the world.

REFERENCES

1. Farquharson, F.B., ed., Aerodynamic Stability of Suspension Bridges, University of Washington Engineering Experiment Station, Bulletin No. 116, Parts I-V, June 1949-June 1954

2. von Karman, Th. (with Lee Edson), The Wind and Beyond, Little, Brown; Boston, 1967

3. Scanlan, R.H., "On the State of Stability Considerations for Suspended-Span Bridges under Wind", Proc., IUTAM-IAHR Symposium, Karlsruhe, Germany, Sept. 1979, Paper F1, pp.595-618

4. Bleich, F. "Dynamic Instability of Truss-Stiffened Suspension Bridges under Wind Action", Proc., ASCE, Vol. 74, No. 8, Oct.1948, pp.1269-1314; Vol. 75, No.3, March 1949, pp. 413-416; Vol. 75, No.6, June 1949, pp. 855-865

5. Scanlan, R.H. and Tomko, J.J., "Airfoil and Bridge Deck Flutter Derivatives", Jnl.,EMD, ASCE, Vol. 97, No. EM 6, Dec. 1971, pp. 1717-1737

6. Scanlan, R.H., "State-of-the-Art Methods for Calculating Flutter, Vortex-Induced, and Buffeting Response of Bridge Structures", Report RD-80/050, Federal Highway Administration, Wash., D.C., April 1981

7. Scanlan, R.H., "The Action of Flexible Bridges under Wind-- I: Flutter Theory; II: Buffeting Theory", Jnl. of Sound and Vibration, Vol. 60, No.2, 1978, pp. 187-199 and 201-211

8. Scanlan, R.H. and Wardlaw, R.L., "Reduction of Flow-Induced Structural Vibrations", Isolation of Mechanical Vibration, Impact, and Noise, Snowdon and Ungar, Eds., ASME Colloq. Proc., AMD Vol. 1, Cincinnati, Ohio, Sept. 1973, pp. 35-63

9. Simiu, E. and Scanlan, R.H., Wind Effects on Structures, John Wiley and Sons, N.Y.,Second Edition, 1986

10. Teunissen, H., Private Communication, Atmospheric Environmental Service, Downsview, Ontario, Canada, 1979

11. Scanlan, R.H., Unpublished notes: Wind Data taken at the Golden Gate Bridge, June 1982

12. Scanlan, R.H. and Huston, D.R., "Sensitivity of Bridge Decks to Turbulent Wind", *Proc. Asia-Pacific Symposium, Wind Engineering*, Roorkee, India, Dec. 1985

13. Davenport, A.G., "Buffeting of a Suspension Bridge by Storm Winds", *Jnl. Structures Div.*, ASCE, Vol. 88, No. ST 3, June 1962, pp. 233-268

14. Liepmann, H.W., "On Application of Statistical Concepts to the Buffeting Problem", *Jnl. Aeron. Sci.*, Vol. 19, No. 12, Dec. 1952, pp. 793-800 and 822

15. Scanlan, R.H. and Gade, R.H., "Motion of Suspended Bridge Spans under Gusty Wind", *Jnl. Structures Div.*, ASCE, Vol. 103, No. ST 9, 1977, pp. 1867-1883

16. Lin, Y.K., Private Communication, 1986

17. Scanlan, R.H., "Developments in Low-Speed Aeroelasticity in the Civil Engineering Field", *AIAA Jnl.* Vol. 20, No. 6, June 1982, pp. 839-844

# UNSTEADY SUPERSONIC AERODYNAMICS OF PLANAR LIFTING SURFACES ACCOUNTING FOR ARBITRARY TIME-DEPENDENT MOTION

Liviu Librescu[‡]
Virginia Polytechnic Institute and State University
Department of Engineering Science and Mechanics
Blacksburg, VA 24061

## ABSTRACT

This paper deals with the formulation of the unsteady aerodynamic theory of 2D and 3D planar lifting surfaces that undergo arbitrary small motions in a supersonic flow field. In its framework, the analysis is confined to the <u>indirect</u> theory in which framework the determination of the integral equation relating a known downwash distribution to an unknown pressure distribution is of a basic importance. In a former stage, this equation is obtained for the general case of the arbitrary time-dependence of field variables and without any restriction concerning the duration in time of the motion. This allows specialization of the integral equation for the important case of the motion starting from rest at a certain time $t = t_0$, as well as for the well studied case belonging to the harmonic time-dependent motions.

Special attention is payed to the former instance, in which context, for evident reasons, the theory of generalized functions is extensively used.

Its employment allows to derive the integral equation by incorporating in a consistent manner the associated initial conditions and by putting into evidence some properties of the aerodynamic kernel such as e.g. the causality property. The paper includes also the borderline case $M_\infty \to 1$, considered for both 2D and 3D lifting surface theories.

The problem exhibited in this paper could be useful in the evaluation of the structural response of flight vehicles as well as in the active control problem of aeroelastic response.

## INTRODUCTION

During recent years there is an increasing interest for further developoments of the unsteady aerodynamics of lifting surfaces (LS). This renewed interest was prompted by the essential role played by the unsteady aerodynamics in the evaluation of the aeroelastic response of flying vehicles as well as in the design of associated feedback control systems.

---

[‡]On leave from Tel-Aviv University, Faculty of Engineering, Tel-Aviv, Israel.

However, in contrast to the traditional approach of the unsteady aerodynamics, based on the concept of simple harmonic motion (which is fully adequate in flutter analyses), the study of the dynamic structural response requires a more general representation of the fluid-structure motion, which cannot be postulated *a priori*.

In this case, the unsteady aerodynamic theory of LS is to be modelled in the time domain for arbitrary time-dependence of the fluid-structure motion.

Several recent contributions in the field have been reported in [1]-[3] while the available survey-papers underline the importance of this topic in the general context of the design of modern flight-vehicle (see e.g. [4]). The goal of this paper is to generalize the classical indirect 3D and 2D supersonic theories of lifting surfaces, as developed by Watkins and Berman [5] and Ando and Ichikawa [6], i.e., to establish in the time domain the integral equation (IE) relating a known downwash distribution to the unknown pressure distribution.

As it will be shown later, the associated aerodynamic kernel in the time domain is correlated with its counterpart, in the frequency domain by a Fourier Transform (FT). This allows to put into evidence some properties of the time-domain kernel functions, such as e.g., that of the causality property. Owing to this property, the possibility of determining the unsteady pressure field at a certain time t, for the case of the motion starting at a previous time $t=t_o$, is established.

BASIC ASSUMPTIONS. GENERAL EQUATIONS

Let us consider the case of a 3D lifting surface placed in a supersonic flow field (see Fig. 1). We shall postulate that the flowing gas is isentropic, non-viscous, electrically and thermally non-conducting, and chemically non-reacting.

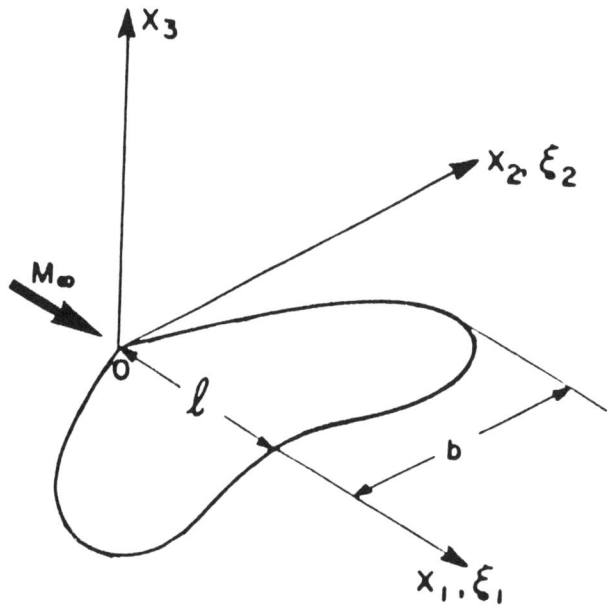

Fig. 1. 3-D Lifting Surface in a Supersonic Flow Field.

Based on these assumptions the field equations governing the small-disturbed motion of a compressible gas flow may be reduced to a wave-propagation type equation. A convenient representation of this equation may be done in terms of the small-disturbance $\hat{\rho} = \hat{\rho}(x_i,t)$ of the free-stream gas density $\rho_\infty$. Under a standard form the equation writes:

$$\frac{D^2\hat{\rho}}{Dt^2} - a_\infty^2 \Delta\hat{\rho} = 0 \qquad (1)$$

where $D/Dt \equiv \partial/\partial t + U_\infty \partial/\partial x_1$; $\Delta$ denotes 3D Laplacean operator; $a_\infty$ is the speed of sound in the undisturbed compressible gas; $U_\infty$ denotes the velocity of the undisturbed gas along the $x_1$-axis; $t$ denotes the time variable. It should be mentioned that variants of Eq. (1), incorporating (i) the electrical conductivity of the gas flow and the presence of a magnetic field and (ii) the chemically-reacting gas properties have been derived in [7,8] respectively. As it was shown in these papers, this representation has some advantages over the standard one, (based on the small-disturbance velocity potential) and in addition its introduction does not imply an explicit statement of irrotationality of the flowfield. The pressure $p = p(x_1,x_2,x_3,t)$ is then defined by:

$$p(x_i,t) = \hat{p}(x_i,t) + p_\infty, \quad (i = \overline{1,3}) \qquad (2)$$

where $\hat{p} = \hat{p}(x_i,t)$ denotes the small-disturbance of the undisturbed pressure $p_\infty$ of the gas flow. In the framework of the small disturbance theory, $\hat{p}$ and $\hat{\rho}$ are connected by (see e.g. the monograph by Bisplinghoff, Ashley and Halfman)

$$\hat{p}(x_i,t) = a_\infty^2 \hat{\rho}(x_i,t) \qquad (3)$$

The solution of Eq. (1) is to be subjected to the impenetrability condition

$$\hat{v}_3\big|_{x_3=0} = -\frac{DZ}{Dt}, \quad (x_1,x_2)\epsilon\Omega_w \qquad (4)$$

and the finiteness condition at infinity. Here $\hat{v}_3 \equiv \hat{v}_3(x_i,t)$ denotes the downwash velocity (positive when downwards); $Z \equiv Z(x_1,x_2,t)$ denotes the vertical displacement at any point $(x_1,x_2)\epsilon\Omega_w$ at the time $t$; $\Omega_w$ denotes the region of the space belonging to the LS.

The pressure differential may be expressed as follows

$$\hat{P} \equiv P\big|_{x_3=0} = \delta p \qquad (5)$$

where

$$\delta(\ ) = (\ )\big|_{x_3=0_+} - (\ )\big|_{x_3=0_-} \qquad (6)$$

denotes the jump between the top $(x_3=0_+)$ and bottom $(x_3=0_-)$ surfaces of LS. It should be underlined that once $\hat{\rho}(x_i,t)$ determined, $\hat{P}(x_1,x_2,t)$ may be found without any difficulty through algebraic operations only.

## PRESSURE EVALUATION

The integral transform technique will be used to determine the equation relating the downwash velocity and the unsteady pressure distribution. Towards this end Eq. (1) will be subjected successively to: (a) a Laplace Transformation, ($L_1T$) with respect to $\bar{x}_1$; (b) a Fourier Transformation ($F_1T$) with respect to $\bar{x}_2$ and (c) a ($F_2T$) with respect to $\bar{t}$, where $\bar{x}_1 \equiv x_1/\ell$; $\bar{x}_2 \equiv x_2/b$ and $\bar{t} \equiv tU_\infty/\ell$ are the dimensionless independent variables, where $\ell$ and $b$ denote a reference chord and the semi-span of the LS, respectively (see Fig. 1).

Let us denote by $\hat{f}^{L_1F_1F_2} \equiv \hat{f}^{L_1F_1F_2}(s,p,\bar{x}_3;r)$ the result of applying the triple integral transformation $L_1F_1F_2$ to the generic disturbance function $f(\bar{x}_1,\bar{x}_2,\bar{x}_3,\bar{t})$.

It is defined as:

$$\hat{f}^{L_1F_1F_2}(s,p,\bar{x}_3;r) \equiv L_1F_1F_2\{\hat{f}(\bar{x}_1,\bar{x}_2,\bar{x}_3;\bar{t})\} \tag{7}$$

$$= \int_0^\infty \int_{-\infty}^{+\infty} \int_{-\infty}^{+\infty} \hat{f}(\bar{x}_1,\bar{x}_2,\bar{x}_3;\bar{t})\exp(-s\bar{x}_1 - jp\bar{x}_2 - jr\bar{t})d\bar{x}_1\,d\bar{x}_2\,d\bar{t} \quad (j \equiv \sqrt{-1})$$

Here s, p and r denote $L_1$, $F_1$ and $F_2$ transform variables associated to their physical counterparts $\bar{x}_1, \bar{x}_2$, and $\bar{t}$, respectively. Applying the triple integral transform (7) to Eq. (1) and considering on physical grounds that $\hat{\rho} = 0$ for $\bar{x}_1 \leq 0$, one obtains:

$$\frac{\partial^2 \hat{\rho}^{L_1F_1F_2}}{\partial \bar{x}_3^2} = \mu^2 \hat{\rho}^{L_1F_1F_2} \tag{8}$$

where

$$\mu^2 \equiv \phi^2 p^2 + B^2 s^2 + 2jM_\infty^2 sr - M_\infty^2 r^2;$$
$$B^2 \equiv M_\infty^2 - 1 \;;\; \phi^2 \equiv \ell^2/b^2, \tag{9}$$

$M_\infty$ denoting the free stream Mach number.

The general solution of Eq. (8) is subjected to the finiteness condition at infinity requiring $\hat{\rho}^{L_1F_1F_2} \rightarrow 0$ as $|\bar{x}_3| \rightarrow \infty$) and to the compatibility condition of the system motion (4). For evident reasons this last one is to be expressed in terms of $\hat{\rho}$.

This will be done by replacing $\hat{v}_3$ as given by (4), into the equation

$$\frac{D\hat{v}_3}{Dt} = -\frac{a_\infty^2}{\rho_\infty}\frac{\partial \hat{\rho}}{\partial x_3} \tag{10}$$

which designates the $x_3$-component of the Eulerian equations of motion (linearized as per the small-disturbance concept). The result is:

$$\left.\frac{\partial \hat{\rho}}{\partial x_3}\right|_{x_3=0} = \frac{\rho_\infty}{a_\infty^2} \frac{D^2 Z}{Dt^2} \qquad (11)$$

This equation represents the impenetrability condition expressed in terms of $\hat{\rho}$. With all these in view, the solution of Eq. (8) writes

$$\left.\hat{\rho}\right|_{x_3=0}^{L_1 F_1 F_2} = -\frac{\rho_\infty M_\infty^2}{u\ell}(s+jr)^2 Z^{L_1 F_1 F_2} \operatorname{sgn} \bar{x}_3 \qquad (12)$$

where $\operatorname{sgn} \bar{x}_3$ denotes signum distribution (being $+1$ on the positive side $\bar{x}_3 > 0$ of $\bar{x}_3 = 0$ plane and $-1$ on the negative side).

It may easily be shown that Eq. (12) may also be expressed in terms of $\hat{v}_3$. Towards this end, use will be made of (7) into (4), thus yielding:

$$Z^{L_1 F_1 F_2} = -\frac{\ell}{U_\infty} \frac{1}{s+jr} \left.\hat{v}_3\right|_{x_3=0}^{L_1 F_1 F_2} \qquad (13)$$

Further employment of (3) and (13) leads to a modified form of Eq. (12):

$$\left.\hat{p}\right|_{x_3=0}^{L_1 F_1 F_2} = \frac{\rho_\infty U_\infty}{B} \frac{s+jr}{\left[(s+j\frac{M_\infty^2}{B^2}r)^2 + \frac{\phi^2 p^2}{B^2} + \frac{M_\infty^2 r^2}{B^4}\right]^{1/2}} \hat{v}_3^{L_1 F_1 F_2} \operatorname{sgn} \bar{x}_3 \qquad (14)$$

This equation expresses (in the $L_1 F_1 F_2$ space) the pressure distribution on 3D lifting surface in terms of the downwash distribution. The presence of sign $\bar{x}_3$, indicates that $\hat{p}$ is an odd function of $\bar{x}_3$. Being an algebraic equation (in the $L_1 F_1 F_2$ space) it may be inverted as to express the downwash distribution in terms of the pressure distribution. Conversion of this last form into the real space will result in the governing integral equation. Its deduction is considered in the next section of the paper.

## THE INTEGRAL EQUATION

The equation, in the transformed space expressing the downwash velocity in terms of the unsteady pressure distribution may be obtained by inverting Eq. (14).

Under a convenient form it will be written as:

$$\bar{v}_3^{L_1 F_1 F_2} = S^{L_1 F_1 F_2} \bar{p}^{L_1 F_1 F_2} \qquad (15)$$

where

$$S^{L_1 F_2 F_2} = \frac{\phi}{2C_1} \frac{\left[p^2 + \frac{B^2}{\phi^2}(s+j\frac{M_\infty^2}{B^2}r)^2 + (\frac{M_\infty r}{\phi B})^2\right]^{1/2}}{s+jr} \qquad (16)$$

while $\bar{v}_3$ and $\bar{p}$ denote the non-dimensional quantities associated to $\hat{v}_3$ and $\hat{p}$, respectively:

$$\bar{v}_3 \equiv \hat{v}_3\big|_{\bar{x}_3=0}/a_\infty M_\infty \quad ; \quad \bar{p} \equiv \hat{P}/(\rho_\infty U_\infty^2/2).$$

Here $C_1$ denotes a tracer taking the values 1 or 2 according to whether the flow is taking place solely on the top face of the LS or simultaneously on both its top and bottom faces, (with the velocity $U_\infty$), respectively.

Let us consider now the inverse Fourier transform ($F_1^{-1}T$) of (16). Making use of its definition, we shall multiply (16) by $(2\pi)^{-1}\exp(jp\bar{x}_2)$ and perform further the integration $\int_{-\infty}^{+\infty}(\ldots)\,dp$ of the resulted quantity. Employment of the identity (see [6,7])

$$\int_{-\infty}^{+\infty} (p^2+N^2)^{1/2} e^{jp\bar{x}_2} dp = \frac{2}{\bar{x}_2} \frac{\partial}{\partial \bar{x}_2} K_0(N|\bar{x}_2|) \tag{17}$$

one obtains the exprssion of $F_1^{-1}\{S^{L_1F_1F_2}\} \equiv S^{L_1F_2}$ given by

$$S^{L_1F_2} = \frac{\phi}{2\pi C_1} \frac{\frac{1}{\bar{x}_2}\frac{\partial}{\partial \bar{x}_2} K_0(N|\bar{x}_2|)}{s+j\frac{M_\infty^2 r}{M_\infty^2-1} - j\frac{r}{M_\infty^2-1}} \tag{18}$$

In Eq. (18) $K_0$ stands for the McDonald function of zero order, $N$ being defined by:

$$N \equiv \left\{ \frac{B}{\phi}\left[ \left(s+j\frac{M_\infty^2 r}{M_\infty^2-1}\right)^2 + \left(\frac{M_\infty r}{M_\infty^2-1}\right)^2 \right]^{1/2} \right\}, \tag{19}$$

In order to obtain $L_1^{-1}T)$ of (18) use will be made of the following relationships:

$$L_1^{-1}\{K_0(b(s^2+\lambda^2)^{1/2}\} = \frac{Y(\bar{x}_1^2-b^2)\cos(\lambda(\bar{x}_1^2-b^2))}{\bar{x}_1^2-b^2} \tag{20}$$

$$L_1^{-1}\left(\frac{1}{s-\alpha}\right) = e^{\alpha \bar{x}_1} Y(\bar{x}_1).$$

In (20) $Y(\bar{x}_1)$ denotes Heaviside distribution whereas the terms denoted in (20) by $\lambda$, $b$, and $\alpha$ may easily be identified by comparing (18) and (19) with (20). Further employment in (18) of the convolution and shifting theorems results in the expression $L_1^{-1}\{S^{L_1F_2}\} \equiv S^{F_2}$, given by:

$$S^{F_2} = \frac{\phi}{2\pi C_1} \frac{1}{\bar{x}_2} \frac{\partial}{\partial \bar{x}_2} \int_0^\infty (e^{-jr\bar{x}_1} \frac{e^{-j\frac{r\xi_1}{M_\infty^2 - 1}}}{(\bar{\xi}_1^2 - \frac{B^2}{\phi^2} \bar{x}_2^2)^{1/2}} Y(\bar{x}_1 - \bar{\xi}_1) Y(\bar{\xi}_1^2 - \frac{B^2}{\phi^2} \bar{x}_2^2) \tag{21}$$

$$\times \cos \frac{M_\infty r}{M_\infty^2 - 1} (\bar{\xi}_1^2 - \frac{B^2}{\phi^2} \bar{x}_2^2)^{1/2} d\bar{\xi}_1$$

where $\bar{\xi}_1$ denotes a dummy variable associated to $\bar{x}_1$.

Further employment of the identity [6]

$$\frac{1}{\bar{x}_2} \frac{\partial}{\partial \bar{x}_2} G(\bar{\xi}_1^2 - \frac{B^2}{\phi^2} \bar{x}_2^2)^{1/2} = \lim_{\bar{x}_3 \to 0} \frac{\partial^2}{\partial \bar{x}_3^2} G[(\bar{\xi}_1^2 - \frac{B^2}{\phi^2}(\bar{x}_2^2 + \bar{x}_3^2))]^{1/2} \tag{22}$$

where G denotes an arbitrary function depending on the indicated argument yields the following form of Eq. (21)

$$S^{F_2}(\bar{x}_1, \bar{x}_2, \bar{x}_3 = 0; r) = \frac{\phi}{2\pi C_1} e^{-jr\bar{x}_1} \lim_{\bar{x}_3 \to 0} \int_{\frac{B}{\phi}(\bar{x}_2^2 + \bar{x}_3^2)^{1/2}}^{\bar{x}_1} \frac{e^{-j\frac{r\bar{\xi}_1}{B^2}}}{} Y(\bar{\xi}_1 - \frac{B}{\phi} \bar{x}_2) \tag{23}$$

$$\times \frac{\partial^2}{\partial \bar{x}_3^2} \{\frac{1}{\bar{\xi}_1^2 - \frac{B^2}{\phi^2}(\bar{x}_2^2 + \bar{x}_3^2)^{1/2}} \cos[\frac{M_\infty r}{M_\infty^2 - 1}(\bar{\xi}_1^2 - \frac{B^2}{\phi^2}(\bar{x}_2^2 + \bar{x}_3^2))^{1/2}\} d\bar{\xi}_1.$$

On the other hand, Eq. (15) converted into ($F_2$T) space reads:

$$\bar{v}_3^{F_2}(\bar{x}_1, \bar{x}_2, \bar{x}_3 = 0; r) = \int_{\Omega_W} S^{F_2}(\bar{x}_1 - \bar{\zeta}_1, \bar{x}_2 - \bar{\zeta}_2; r) \bar{p}^{F_2}(\bar{\zeta}_1, \bar{\zeta}_2; r) d\bar{\zeta}_1 d\bar{\zeta}_2 \tag{24}$$

Multiplication of (24) by $(2\pi)^{-1} \exp(jr\bar{t})$ followed by the integration $\int_{-\infty}^{+\infty}(\ldots)dr$ of its both sides, yields:

$$\bar{v}_3(\bar{x}_1, \bar{x}_2; \bar{t}) = (2\pi)^{-1} \int_{-\infty}^{+\infty} \int_{\Omega_W} S^{F_2}(\bar{x}_1 - \bar{\zeta}_1, \bar{x}_2 - \bar{\zeta}_2; r) \bar{p}^{F_2}(\bar{\zeta}_1, \bar{\zeta}_2; r) e^{jr\bar{t}} dr \, d\bar{\zeta}_1 \, d\bar{\zeta}_2 \tag{25}$$

Making use of the evident relationship

$$F_2\{\bar{p}(\bar{\zeta}_1, \bar{\zeta}_2; \tau)\} \equiv \bar{p}^{F_2}(\bar{\zeta}_1, \bar{\zeta}_2; r) = \int_{-\infty}^{+\infty} \bar{p}(\bar{\zeta}_1, \bar{\zeta}_2; \tau) e^{-jr\tau} d\tau \tag{26}$$

Equation (25) writes as:

$$\bar{v}_3(\bar{x}_1,\bar{x}_2,\bar{t}) = (2\pi)^{-1}\int_{-\infty}^{+\infty}\int\int_{\Omega_W} S^{F_2}(\bar{x}_1-\bar{\zeta}_1;,\bar{x}_2-\bar{\zeta}_2;r)e^{jr(\bar{t}-\bar{\tau})}\bar{p}(\bar{\zeta}_1,\bar{\zeta}_2;\bar{\tau})dr\,d\bar{\tau}\,d\bar{\zeta}_1\,d\bar{\zeta}_2 \quad (27)$$

Under a more convenient form Eq. (27) transcribes as:

$$\bar{v}_3(\bar{x}_1,\bar{x}_2,\bar{t}) = \int_{-\infty}^{+\infty}\int\int_{\Omega_W} K(\bar{x}_1 - \bar{\zeta}_1,\bar{x}_2 - \bar{\zeta}_2;\bar{t} - \bar{\tau})\bar{p}(\bar{\zeta}_1,\bar{\zeta}_2;\bar{\tau})d\bar{\tau}\,d\bar{\zeta}_1\,d\bar{\zeta}_2, \quad (28)$$

where the function K is defined by

$$K(\bar{x}_1 - \bar{\zeta}_1;\bar{x}_2 - \bar{\zeta}_2;\bar{t} - \bar{\tau}) = (2\pi)^{-1}\int_{-\infty}^{+\infty} S^{F_2}(\bar{x}_1 - \bar{\zeta}_1,\bar{x}_2 - \bar{\zeta}_2;r)e^{jr(\bar{t}-\bar{\tau})}dr. \quad (29)$$

The equation (28) relates in the time domain the downwash angle ($\equiv \hat{v}_3/U_\infty$) with the non-dimensional unsteady pressure differential.

The function K entering the Eq. (28) designates the 3D aerodynamic kernel in the time domain.

According to Eq. (29), K may be expressed under a more compact form as

$$K = F_2^{-1}\{S^{F_2}\} \quad (30)$$

## THE RELATIONSHIP BETWEEN TIME AND FREQUENCY DOMAINS KERNEL FUNCTIONS

In the following developments a full interpretation of the Eq. (30) will be given. Towards this end, we shall consider for the moment Eq. (28), which, as is worth remarking, does not imply any restriction on the time-dependence of the involved quantities nor on the extension in time of the motion. As a result, the equation is valid for harmonic time-dependent motions, as well. In this instance, $\bar{v}_3$ and $\bar{p}$ may adequately be expressed in the form:

$$\bar{v}_3(\bar{x}_1,\bar{x}_2,\bar{t}) = \bar{\bar{v}}_3(\bar{x}_1,\bar{x}_2)\exp(j\bar{\omega}\bar{t})$$
$$\bar{p}(\bar{x}_1,\bar{x}_2,t) = \bar{\bar{p}}(\bar{x}_1,\bar{x}_2)\exp(j\bar{\omega}\bar{t}). \quad (31)$$

where $\bar{\omega} \equiv \omega \ell/U_\infty$ denotes the non-dimensional frequency. Insertion of (31) into (27) yields successively:

$$\bar{\bar{v}}_3(\bar{x}_1,\bar{x}_2)e^{j\bar{\omega}\bar{t}} = (2\pi)^{-1}\int_{-\infty}^{+\infty}\int\int_{\Omega_W} S^{F_2}(\bar{x}_1-\bar{\zeta}_1;\bar{x}_2-\bar{\zeta}_2;r)e^{jr(\bar{t}-\bar{\tau})}\bar{\bar{p}}(\bar{\zeta}_1,\bar{\zeta}_2)e^{j\bar{\omega}\bar{\tau}}d\bar{\zeta}_1 d\bar{\zeta}_2 d\bar{\tau}$$

$$(32)$$

$$= (2\pi)^{-1} \int_{-\infty}^{+\infty} \int \int_{\Omega_W} S^{F_2}(\bar{x}_1 - \bar{\zeta}_1, \bar{x}_2 - \bar{\zeta}_2; r) \bar{\bar{p}}(\bar{\zeta}_1, \bar{\zeta}_2) e^{j(\bar{\omega}-r)\bar{\tau}} e^{jr\bar{\tau}} d\bar{\zeta}_1 d\bar{\zeta}_2 d\bar{\tau}.$$

Equation (32) may further be expressed as:

$$\bar{\bar{v}}_3(\bar{x}_1, \bar{x}_2) e^{j\bar{\omega}t} = \int_{-\infty}^{+\infty} \int_{\Omega_W} S^{F_2}(\bar{x}_1 - \bar{\zeta}_1, \bar{x}_2 - \bar{\zeta}_2; r) \bar{\bar{p}}(\bar{\zeta}_1, \bar{\zeta}_2) e^{jr\bar{t}} d\bar{\zeta}_1 d\bar{\zeta}_2 (2\pi)^{-1} \int_{-\infty}^{+\infty} e^{j\bar{\tau}(\bar{\omega}-r)} d\bar{\tau} \quad (33)$$

Finally, by virtue of the well-known integral representation of the Dirac distribution $\delta(t)$ and of its role played in the convolution process, Eq. (33) takes the following form

$$\bar{\bar{v}}_3(\bar{x}_1, \bar{x}_2) = \int_{\Omega_W} S^{F_2}(\bar{x}_1 - \bar{\zeta}_1, \bar{x}_2 - \bar{\zeta}_2, \bar{\omega}) \bar{\bar{p}}(\bar{\zeta}_1, \bar{\zeta}_2) d\bar{\zeta}_1 d\bar{\zeta}_2 \quad (34)$$

where the time factor $\exp(j\bar{\omega}t)$ was suppressed in both sides of (34). Equation (34) shows that $S^{F_2}$ plays the role of the frequency-domain Kernel function. Its expression given by Eq. (23) coincides with the one obtained differentily by Watkins and Berman [5] and Ando and Ichikawa [6]. In addition it is to point out that in (23) the variable r takes the role of the frequency variable $\bar{\omega}$. In light of the above facts and of Eq. (30) it may be concluded that the time-domain kernel function K constitutes the inverse Fourier transform of the frequency-domain kernel function $S^{F_2}$. In the following developments it will be denoted as $\tilde{K}(S^{F_2} \equiv \tilde{K})$. This above relationship between time and frequency-domains supersonic kernels reminds the one established for the unsteady incompressible 2D flow, between the Wagner and Theodorsen functions (see e.g. [10-12]). Examination of the frequency-domain kernel function as given by Eq. (23) allows us the infer that the indicated differentiations under the integral sign could introduce singular terms. It is why the kernel function is to be reduced to a form free of improper integrals. This conversion may be done by following the developments indicated in [5] (see also [13]). Being similar to the expression obtained in [5] this modified form of $\tilde{K}$ will be presented as such, without any proof. It is given by:

$$\tilde{K}(\bar{x}_1, \bar{x}_2; \bar{\omega}) = \frac{\phi}{2\pi C_1} Y(\bar{x}_1 - \frac{B}{\phi}\bar{x}_2) e^{-j\bar{\omega}\bar{x}_1} \frac{\bar{x}_1 e^{-j\frac{\bar{\omega}}{B^2}\bar{x}_1} \cos[(M_\infty \bar{\omega}/B^2)(\bar{x}_1^2 - (B^2/\phi^2)\bar{x}_2^2)^{1/2}]}{(\bar{x}_1^2 - (B^2/\phi^2)\bar{x}_2^2)^{1/2}} \quad (35)$$

$$+ \frac{j\bar{\omega}\bar{x}_2}{2\phi} \int_{A_-}^{A_+} \frac{\sigma}{(1+\sigma^2)^{1/2}} e^{-j\bar{\omega}(\bar{x}_2/\phi)\sigma} d\sigma$$

where

$$A_\pm = \frac{\phi}{B^2|\bar{x}_2|} [\bar{x}_1 \pm (\bar{x}_1^2 - \frac{B^2}{\phi^2}\bar{x}_2^2)^{1/2}],$$

$\sigma$ denoting a dummy variable.

For $M_\infty \to 1$, the specialized counterpart of (35) may be reduced to:

$$\tilde{K}(\bar{x}_1,\bar{x}_2,\bar{\omega})_{M_\infty \to 1} = \frac{\phi^2}{2\pi C_1 \bar{x}_2^2} Y(\bar{x}_1) e^{-j\bar{\omega}\bar{x}_1} \left( e^{j\frac{\bar{\omega}}{2}\bar{x}_1 \left(1 - \frac{\bar{x}_2^2}{\bar{x}_1^2}\right)} \right.$$

$$\left. + j \frac{\bar{\omega}|\bar{x}_2|}{2\phi} \int_a^\infty \frac{\sigma}{(1+\sigma^2)^{1/2}} e^{-j\frac{\bar{\omega}}{\phi}|\bar{x}_2|\sigma} d\sigma \right) \qquad (36)$$

where

$$a = -\phi/(2|\bar{x}_2|) \left[ \bar{x}_1 - \frac{1}{\phi^2} \frac{\bar{x}_2^2}{\bar{x}_1^2} \right].$$

Similar expressions of the modified kernel function may be obtained through appropriate specialization of the results obtained in [12] for electrically-conducting supersonic gas flows in the presence of an ambient magnetic field.

## SEVERAL PROPERTIES OF THE 3D AERODYNAMIC KERNEL

As it was already mentioned, Eq. (28) governing the relationship between the known downwash velocity and the unknown pressure field was obtained without any limitation on the time-dependence of the field variables and on the extent in time of the motion.

However, as is well understandable, we are interested to model such a motion that starts from rest at a certain instant of time $t = t_o$.

In such a case it must be considered that prior to the beginning of motion, i.e. for the system at rest (defined over $-\infty < t < t_o$), all the disturbances are zero being different from zero only on the interval $t_o \leq t < \infty$.

In this case we shall replace $\bar{v}_3$ and $\bar{p}$ appearing in (28) by:

$$\begin{aligned} \bar{v}_3(\bar{x}_1,\bar{x}_2,\bar{t}) &\to \bar{v}_3(\bar{x}_1,\bar{x}_2;\bar{t})Y(\bar{t} - \bar{t}_o) \\ \bar{p}(\bar{\zeta}_1,\bar{\zeta}_2,\bar{\tau}) &\to \bar{p}(\bar{\zeta}_1,\bar{\zeta}_2;\bar{\tau})Y(\bar{\tau} - \bar{t}_o). \end{aligned} \qquad (37)$$

On the other hand, owing to the fact that $K(\bar{x}_1,\bar{x}_2,\bar{t})$ and $\tilde{K}(\bar{x}_1,\bar{x}_2,\bar{\omega})$ are correlated by a Fourier integral (see Eq. (30)) we are interested to know the conditions that $\tilde{K}$ must satisfy in order to be the $(F_2T)$ of the causal function $K$. The corresponding conditions resulting in the causality of $K(\bar{x}_1,\bar{x}_2,\bar{t})$ (i.e. in the requirement that $K(\bar{x}_1,\bar{x}_2,\bar{t}) = 0$ for $\bar{t} < 0$), may be expressed in terms of the following theorem (see e.g. [14]) which, adapted for the present case reads:
If $\tilde{K}$ extended in the complex space $\Omega = \omega + j\sigma$ has no singularities in the lower half of the $\Omega$-plane, then $K$ will be causal.

A cursory inspection of (35) and (36) shows that the terms of this theorem are fulfilled, both in the supersonic and sonic instances.

By virtue of the causality property of the Kernel function K and of the representations (37), one obtains the counterpart of Eq. (28), valid for the case of motion starting from rest at the time $t = t_o$:

$$\bar{v}_3(\bar{x}_1,\bar{x}_2,\bar{t}) = \int_{t_o}^{\bar{t}} \int_{\Omega_w} K(\bar{x}_1 - \bar{\zeta}_1, \bar{x}_2 - \bar{\zeta}_2; \bar{t} - \bar{\tau}) \bar{p}(\bar{\zeta}_1,\bar{\zeta}_2,\bar{\tau}) d\bar{\zeta}_1 d\bar{\zeta}_2 d\bar{\tau} \qquad (38)$$

The left hand-side of Eq. (38) may be expressed in terms of Z. Towards this goal, insertion in Eq. (4) of Z represented as $Z(\bar{x}_1,\bar{x}_2,\bar{t}) Y(\bar{t} - \bar{t}_o)$ yields the l.h.s. of (38) in the form:

$$\bar{v}_3(\bar{x}_1,\bar{x}_2,\bar{t})Y(\bar{t} - \bar{t}_o) \rightarrow (\frac{\partial \bar{Z}}{\partial \bar{x}_1} + \frac{\partial \bar{Z}}{\partial \bar{x}_1})Y(\bar{t} - \bar{t}_o) \bar{Z}(\bar{x}_1,\bar{x}_2,\bar{t}_o)\delta(\bar{t} - \bar{t}_o), \qquad (39)$$

where the last term identifies the initial condition at $\bar{t} = \bar{t}_o$, associated to Z. Equation (38), considered in conjunction with (35), (36) and (29), governs the unsteady pressure distribution in the time domain for 3D supersonic and sonic flows.

UNSTEADY AERODYNAMICS OF 2D LIFTING SURFACES. ARBITRARY TIME-DEPENDENT MOTIONS

In the case of infinite aspect-ratio lifting surfaces, the dependence on the spanwise coordinate $x_2$ is immaterial. Owing to this fact the 2D counterparts of (15) and (16) write as:

$$\frac{L_1F_2}{\bar{v}_3} = S^{L_1F_2} \frac{L_1F_2}{\bar{p}} \qquad (40)$$

where

$$S^{L_1F_2} \equiv S^{L_1F_2}(s,\bar{x}_3=0;r) = \frac{B}{2C_1} \frac{s+j\frac{M_\infty^2}{B^2}r + j\frac{r}{B^2} + \frac{r^2}{B^2(s+jr)}}{\left[(s+j\frac{M_\infty^2}{B^2}r)^2 + \frac{M_\infty^2 r^2}{B^4}\right]^{1/2}} \qquad (41)$$

Making use of the identity

$$L_1^{-1}(\frac{1}{[(s+a)^2 + b^2]^{1/2}}) = e^{-a\bar{x}_1} J_o(b\bar{x}_1)Y(\bar{x}_1)$$

as well as the basic laws of the inverse Laplace transform, we get easily:

$$S^{F_2}(\bar{x}_1,\bar{x}_3 = 0;r) = \frac{B}{2C_1}\left\{e^{-j\frac{M_\infty^2 r}{B^2}\bar{x}_1}\left[\delta(\bar{x}_1) + j\frac{r}{B^2}Y(\bar{x}_1)J_o(\frac{M_\infty r}{B^2}\bar{x}_1)\right.\right.$$

$$\left.\left. - \frac{M_\infty r}{B^2}Y(\bar{x}_1)J_1(\frac{M_\infty r}{B^2}\bar{x}_1)\right] + \frac{r^2}{B^2}e^{-jr\bar{x}_1}\int_0^{\bar{x}_1} e^{-j\frac{r\lambda}{B^2}}Y(\bar{\lambda})J_o(\frac{M_\infty r}{B^2}\bar{\lambda})d\bar{\lambda}\right\} \qquad (42)$$

where $\lambda$ denotes a dummy variable associated to $\bar{x}_1$, while $J_m$ denotes Bessel functions of first kind and of order m. Equation (42) represents the 2D counterpart of Eq. (23). It coincides with the one obtained in a more difficult manner in [5]. By parallelling further the developments performed for the 3D case, one may obtain easily for the 2D case the equation

$$\bar{v}_3(\bar{x}_1, \bar{x}_3=0; \bar{t}) = \int_{-\infty}^{+\infty} \int_0^{\bar{x}_1} K(\bar{x}_1 - \bar{\zeta}_1; \bar{t} - \bar{\tau}) \bar{p}(\bar{\zeta}_1; \bar{\tau}) d\bar{\zeta}_1 d\bar{\tau} \tag{43}$$

Here $K(\bar{x}_1, \bar{t})$ denotes the time-domain 2D supersonic Kernel function correlated to $S^{F_2}(\bar{x}_1; r)$ like in the 3D case, i.e. through Eq. (30), where $S^{F_2}(\bar{x}_1, r)$ given by (42) stands for the 2D frequency-domain Kernel function.

For the sonic flow regime, Eq. (41) reduces to

$$S^{L_1 F_2}_{M_\infty \to 1} = \frac{1}{C_1} \frac{jr(1 - \frac{j}{2}\frac{r}{s+jr})}{(s+j\frac{r}{2})^{1/2}} \tag{44}$$

Making use of the identity

$$L_1^{-1}\left(\frac{1}{(s-a)^k}\right) = \frac{\bar{x}_1 e^{a\bar{x}_1}}{\Gamma(K)} \tag{45}$$

and of the convolution rule, where $\Gamma(K)$ denotes the hypergeometric function ($\Gamma(1/2) = \sqrt{\pi}$), the 2D frequency-domain Kernel function in the sonic range becomes:

$$S^{F_2}_{M_\infty \to 1}(\bar{x}_1, r) = \frac{r}{2C_1}\left[2j\frac{e^{-j\frac{r}{2}\bar{x}_1}}{\sqrt{\pi \bar{x}_1}} + \frac{r}{\sqrt{\pi}}e^{-jr\bar{x}_1}\int_0^{\bar{x}_1}\frac{e^{j\frac{r}{2}\bar{\lambda}}}{\sqrt{\bar{\lambda}}}d\bar{\lambda}\right] \tag{46}$$

This expression agrees with the one obtained in [15] as the borderline case $M_\infty \to 1$, when starting from the subsonic side.

Needless to say that the functional relationship between $K_{M_\infty \to 1}(\bar{x}_1, \bar{t})$ and $S^{F_2}_{M_\infty \to 1}(\bar{x}_1, r)$ remains the same as the one given by Eq. (30). By using the same reasonings as in the 3D case, it may easily be shown that $K(\bar{x}_1, \bar{t})$ and $K_{M_\infty \to 1}(\bar{x}_1, \bar{t})$ enjoy the causality property, so that the integral equation approriate to the 2D case, in the case of the motion starting at the instant $\bar{t} = \bar{t}_o$ reads:

$$\bar{v}_3(\bar{x}_1,\bar{t}) = \int_{\bar{t}_o}^{\bar{t}} \int_0^{\bar{x}_1} K(\bar{x}_1 - \bar{\xi}_1; \bar{t} - \bar{\tau})\bar{p}(\bar{\xi}_1,\bar{\tau})d\bar{\xi}_1 d\bar{\tau}. \tag{47}$$

## CONCLUSIONS

The <u>indirect</u> aerodynamic theory of 3D and 2D lifting surfaces undergoing arbitrary small motions in a supersonic flow field has been considered in this paper which constitutes the generalization of a previous one [16]. As it was shown, the theory involves the determination of appropriate time-domain Kernel function. The relationship between frequency-domain and their corresponding time-domain Kernel-functions was established and the causality property of time-domain Kernel-functions was put into evidence. This property enables to express the integral equation, in the case of the motion starting at a certain time, under the form (38) and (47). The sonic flow instance of 2D and 3D lifting surface theories was also considered in the paper. It should be pointed out that the generalization of simple harmonic motion to growing and decaying oscillatory motion as considered in [15] for subsonic compressible flows could be extended in the low- and high supersonic ranges and for 2D and 3D lifting surfaces, by using some of the present results. It must also be added that alternative approaches to the problem treated in this paper concern: a) the <u>direct</u> approach of 2D and 3D theories of supersonic lifting surfaces undergoing arbitrary small motions as developed in [18], which may be viewed as a complement to the present analysis, and b) the extension in the time-domain [19] of Green's function method of potential aerodynamics developed for harmonic time-dependent motions in [20].

## REFERENCES

1. R. Vepa, Finite State Modeling of Aeroelastic Systems NASA CR-2779, 1977.

2. J. W. Edwards, Unsteady Aerodynamic Modeling and Active Aeroelastic Control, SUDSAAR 504, Stanford University, Februrary 1977.

3. E. H. Dowell, A Simple Method of Converting Frequency-Domain Aerodynamics to the Time-Domain, NASA Technical Memorandum 81844, NASA 1980.

4. G. J. Hancock, Role of Unsteady Aerodynamics in Aircraft Response, in "Special Course on Unsteady Aerodynamics", AGARD 1980.

5. Ch. E. Watkins and Berman, J. H., On the Kernel Function of the Integral Equation Relating Lift and Downwash Distributions of Oscillating Wings in Supersonic Flow, 1956 NACA Rep. 1257.

6. S. Ando and Ichikawa, A., Derivation by a Transform Method of Integral Equations of Unsteady Lifting Surface Theory in Subsonic and Supersonic Flow, The Aernautical Quarterly, Vol. xxx, Nov. 1979, pp. 529-543.

7. L. Librescu, Unsteady Pressure Loads for 3D Flutter Calculations of Planar Configuration Panels in a Supersonic Ionized Gas Flow, Israel J. Technol, Proc. XXIII Israel Annual Conf. Aviation & Astronautics, 18, pp. 37-46.

8. L. Librescu, Unsteady Aerodynamics of Chemically Reacting Flows Past Oscillating Thin Bodies, in "Progress in Astronautics and Aeronautics", Vol. 95, "Dynamics of Flames and Reactive Systems", 1985, pp. 593-609.

9. S. Ando and Ichikawa A., Effect of Forward Acceleration in Aerodynamic Characteristics of Wings in an Inviscid Incompressible Fluid, Trans. Japan Soc., Aerospace Sci., 1979, 22, 56, pp. 57-69.

10. R. L. Bisplinghoff, Ashley, H., and Halfman, R. L., Aeroelasticity, Addison-Wesley, Co. Inc. 1955.

11. R. L. Bisplinghoff and Ashley H., Principles of Aeroelasticity, John Wiley and Sons, Inc., New York, London, 1962.

12. E. H. Dowell, Ed; Curtiss, H. C. Jr., Scanlan R. H., and Sisto, F., A Modern Course in Aeroelasticity, Sijthoff & Noordhoff (Netherlands), 1978.

13. L. Librescu, Non-Classical Developments of the Unsteady Supersonic 3D Lifting Surface Theory, Paper Issued in "Collection of Papers of the 25th Israel Annual Conference on Aviation and Astronautics, Israel, Feb. 23-25, 1983, pp. 237-244.

14. H. Urkowitz, Signal Theory and Random Processes, ARTECH HOUSE Inc., 1983.

15. C. E. Watkins, Runyan, H. L., and Woolston, D. S., On the Kernel Function of the Integral Equation Relating the Lift and Downwash Distributios of Oscillating Finite Wings in Subsonic Flow, NACA Rep. 1234, 1955.

16. L. Librescu, An Exact Formulation of the Unstedy Aerodynamic Theory of Lifting Surfaces Undergoing Arbitrary Small Motions in a Supersonic Flow Field. Paper presented at "The Second International Symposium on Aeroelasticity and Structural Dynamics," Aachen, April 1-3, 1985.

17. H. J. Cunningham and Desmarais, R. N., Generalization of the Subsonic Kernel Function in the s-Plane, with Applications to Flutter Analysis, NASA TP 2292, 1984.

18. L. Librescu, Unsteady Aerodynamic Theory of Lifting Surfaces and Thin Elastic Bodies Undergoing Arbitrary Small Motions in a Supersonic Flow Field, to appear in Quarterly of Mechanics and Applied Mathematics.

19. M. I. Freedman and Tseng, K., A First-Order Time-Domain Green's Function Approach to Supersonic Unsteady Flow, NASA CR 172208, April 1985.

20. L. Morino, A General Theory of Unsteady Compressible Potential Aerodynamics, NASA CR-2464, December, 1974.

# AEROELASTICITY OF VERY LIGHT AIRCRAFT

Ilan Kroo
Department of Aeronautics and Astronautics
Stanford University
Stanford, CA 94305

## ABSTRACT

The design of aircraft with exceptionally lightweight structures is strongly influenced by aeroelastic considerations. This paper addresses some of the problems which have been encountered by this unusual class of low speed aircraft, some of the methods by which they may be analyzed, and some of the ways in which potential aeroelastic difficulties have been turned to advantage.

## INTRODUCTION

Very light aircraft comprise a unique category of flight vehicles including very long endurance platforms, hang gliders and ultralights, human powered aircraft, and other aircraft characterized by very lightweight structures [1-3]. (See figure 1.) While encompassing a diverse set of design goals and missions, these aircraft share many similar aeroelastic characteristics. Although conventional flutter and divergence may be important for some of these designs, in many cases major design considerations are based on unusual aeroelastic phenomena, rarely significant for more conventional aircraft. In this paper some of the aeroelastic effects of possible significance to very light aircraft are identified, methods for predicting these effects are outlined, and examples are given of constructive uses of aeroelasticity.

First, however, the concern with aeroelasticity should be motivated. It is sometimes argued that at the low speeds typical of most very light aircraft, the dynamic pressure is so low that aeroelasticity is not important — that wings under these conditions are very stiff and it is strength, not stiffness, that is the issue. As will be shown, the opposite is true. Aircraft such as these, for which structural weight is kept to a minimum, are often more flexible as limit load factors may be low and aspect ratios are in some cases extremely high. Weight-saving construction techniques such as cable or strut bracing may increase wing bending stiffness but increase torsional flexibility. The inappropriateness of dimensionless parameters such as the ratio of dynamic pressure to Young's modulus, $q/E$, in assessing the importance of aeroelasticity is shown in the following simplified analysis.

Consider a high aspect ratio, cantilevered wing with a structure that may be approximated as a simple beam. With the bending stress given by:

$$\sigma = \frac{M t}{2 I}$$

where M is the local bending moment, t is the structural box thickness, and I is the area moment of inertia, the deflection, z, is given by:

$$E I z'' = M$$

Thus, the deflection may be expressed simply as:

$$z'' = \frac{2}{t}\frac{\sigma}{E}$$

Now, if the wing structure is designed based on strength considerations in such a way that the stress is nearly constant, $\sigma = \sigma_{design}$, then the deflection depends only on the wing exterior geometry.

In the case of a rectangular wing:

$$\bar{z}_{tip} = \frac{\sigma}{E}\frac{AR}{2\,t/c}$$

where $z_{tip}$ is the deflection of the tip in semi-spans, AR is the wing aspect ratio and t/c is the thickness to chord ratio.

The relative bending flexibility of a fully-stressed wing therefore depends on AR, t/c, the material properties, and the ratio of working stress to design maximum stress (which for a linear structure is approximately the ratio of load factor to design limit load factor). The high aspect ratios of very long endurance aircraft and relatively low design load factors make such aircraft very flexible despite their low dynamic pressure operating conditions. Potential aeroelastic problems are, therefore, not restricted to high speed aircraft; however, because of the unconventional configurations and high aspect ratios common to this class of aircraft, some of the most significant effects of aeroelasticity are different from those commonly encountered by more conventional aircraft.

## PHENOMENA ASSOCIATED WITH WING BENDING

The preceding discussion produces the interesting, but perhaps obvious result that the deflection of a fully-stressed wing at its design load is independent of the shape or magnitude of the applied load. This permits simple, conceptual analysis of several static aeroelastic effects on the stability of very light aircraft.

### Static stability of swept wings

The effect of wing bending on the static stability of swept wings is taken into consideration in the early stages of transport aircraft design [4] but is often neglected for lack of an existing database in the advanced design of many very light aircraft. This is often justifiable when the design has little or no sweep; however, a number of moderately-swept, tailless aircraft have been proposed recently [5,6] for which this issue is of critical importance.

Bending of a swept wing along its elastic axis introduces a streamwise twist of magnitude:

$$\theta = \frac{dz}{dy}\,\text{Sin}\,\Lambda \qquad \text{with } \frac{dz}{dy} \text{ the slope of the deflected elastic axis and } \Lambda \text{ the sweep.}$$

For a high aspect ratio, strength-design wing structure, the expression for the curvature may be integrated once to obtain:

$$z' = \frac{dz}{dy} = \frac{\sigma}{E}\frac{AR}{t/c}\,y$$

in the case of an untapered wing. This leads to a corresponding change in twist and a pitching moment increment as the load distribution over the wing varies. If the change in pitching moment due to tip twist is written $C_{m\varepsilon}$, then for a rectangular wing:

$$\frac{\partial C_m}{\partial C_L} = \left(\frac{\partial C_m}{\partial C_L}\right)_{rigid} + \frac{C_{m_\varepsilon}}{C_L} \left.\frac{\sigma}{E}\right|_{1-g} \frac{AR}{t/c} \sin \Lambda$$

This increment is shown in figure 2 as a function of aspect ratio and sweep for moderately tapered wings (taper ratio =.5). It was computed by integrating the expression for beam deflection and evaluating its effect on pitching moment with a vortex lattice aerodynamics code. While the effects on aircraft with aspect ratios in the range of 10-15 are not insignificant, the effect on wings with aspect ratios in the 25-30 range is dramatic. (Note the assumed value of $\sigma/E = .001$.)

Especially in the case of tailless configurations which exhibit low damping in pitch, the pitching moment produced by wing bending may be even more significant than indicated by the above results. Dynamic instabilities involving coupling of the low frequency wing bending modes with the rigid body pitch mode (short-period) are quite possible and easily demonstrated with lightweight models.

### Effects on lateral stability

Lateral-directional characteristics also may be influenced by the large deformations of lightweight, high aspect ratio wings. Such effects may be estimated for preliminary design purposes in a manner similar to that discussed above. The primary result is an increase in dihedral due to deformation. The effect may be quite large as illustrated in the design of the human powered aircraft, Musclair [7]. (See figure 3.) The original design incorporated polyhedral in the wing panels to obtain desirable handling qualities. It was found, however, that the bending deformation of the wings produced adequate dihedral itself and the final configuration jig-shape incorporated no dihedral.

## PHENOMENA ASSOCIATED WITH WING TORSION

While wing bending deformations may have a strong influence on the stability of very light aircraft, wing torsional deflections have, historically, been of greater significance. This is due in part to attention paid to minimize structural weight: surface quality and durability may be sacrificed in order to reduce the weight of minimum gauge skin, for example. For aircraft with very low wing loadings this is an effective method for reducing structural weight — and, of course, torsional stiffness. Resulting aeroelastic phenomena include conventional divergence, flutter, and aileron reversal as well as some less well-known phenomena.

### Conventional configurations

Conventional torsional divergence arises when the elastic axis of the wing lies behind the aerodynamic center. This is often the case in ultralight aircraft designs with one structural member at the leading edge and another well behind the half-chord line. Indeed, torsional divergence of such designs has been demonstrated in full-scale and model tests [8,9]. The dynamic pressure resulting in structural divergence of a uniform wing is given by the well-known expression [10]:

$$q_{Div} = \frac{GJ\,\pi^2}{c\,e\,C_{l_\alpha}\,b^2}$$

with c, the chord; $C_{l_\alpha}$, the lift curve slope; b, the span and e, the distance from the a.c. to the elastic axis in chords.

A similar expression may be derived based on the assumption of a fully-stressed design as in the case of wing bending, but the approximation is less useful since structural design conditions often differ markedly from the flight conditions near torsional divergence speed and are sensitive to airfoil section properties. It should be noted, however, that the conclusion of the previous section applies to torsional stiffness as well: despite the low dynamic pressures, aeroelastic torsional deformations are often important.

An example is found in a number of mysterious accidents involving ultralight aircraft. On several occasions ultralights have entered dives from which recovery was not possible. Torsional flexibility and even conventional divergence of the horizontal tail have been suggested as explanations in some of these cases, but torsional flexibility of the wing may also be a factor [11]. In particular, coupling of the wing torsional deformations with vehicle longitudinal dynamics may produce an instability at a speed substantially lower than the torsional divergence speed. The parameters of importance in this case are revealed in the following simplified model.

If the dimensionless torque on a typical section of the wing is given by:

$$C_\tau = C_L e + C_{m_0}$$

and produces a twisting of the section:

$$\varepsilon = C_\tau q c K_\varepsilon$$

then torsional divergence will occur at the value of q given by:

$$q_{Div} = \frac{1}{c e K_\varepsilon C_{L_\varepsilon}} \quad \text{with } C_{L_\varepsilon} \text{ the change in lift due to unit twist.}$$

Now with $x_w$ the distance from the aircraft center of gravity to the wing a.c. and $x_t$, the distance to the tail and $S_{tail}/S_{wing}$ the tail to wing area ratio:

$$\frac{\partial C_{m_{cg}}}{\partial \alpha} = \bar{x}_w C_{L_{\alpha\,rigid}} (1 + C_{L_\varepsilon} q c K_\varepsilon e) - \bar{x}_t C_{L_\varepsilon} \frac{S_{tail}}{S_{wing}} \quad \text{at a fixed value of q}$$

or:

$$\frac{\partial C_m}{\partial C_L} = \frac{\partial C_m}{\partial C_L}\bigg|_{rigid} + \bar{x}_w \frac{q_{Div}}{q}$$

Thus, the static stability of the aircraft is reduced by wing torsion when the wing aerodynamic center is located behind the airplane center of gravity. This is, in fact, the case for some ultralight aircraft and the effective reduction in stability can be large, producing uncontrollable aircraft at speeds substantially below the clamped divergence speed of the wing. Clear instances of this phenomena have been observed in radio controlled sailplane models with high aspect ratio, lightweight wings.

A related phenomena is apparent from the above expressions when they are cast in dimensional form:

$$\frac{\partial M_{c.g.}}{\partial q} = q c \bar{x}_w C_{L_\varepsilon} e K_\varepsilon C_\tau$$

with $C_\tau = C_L e + C_{m_o}$ the moment change with q may be negative for aircraft with large negative values of $C_{m_o}$ and positive values of $x_w$.

This is not uncommon for very light aircraft designed to fly at high $C_L$'s; it leads to a reduction in "speed stability" and "speed divergence" in extreme cases.

## Tailless configurations

Changes in the twist of swept wings can produce substantial changes in the overall aircraft pitching moment as discussed in the previous section. While wing bending produces an effective twist change, torsional deformations do so more directly and have been responsible for a host of problems with these designs. Nowhere have the results been so apparent as in the case of flexible-wing hang gliders.

On a typical hang glider configuration (see figure 1) a flexible membrane is affixed to a relatively stiff frame in a manner that results in an extremely torsionally-flexible structure. As the angle of attack is reduced, torsional loads on the outer portion of the wing are reduced and the wing twist decreases. A decrease in wing twist reduces the aircraft pitching moment, producing a reduction in static stability. This effect is apparent in figure 4 which shows how the sail shape changes with angle of attack. At smaller angles of attack the geometric stiffness is small; the twist changes rapidly, and the aircraft exhibits a slightly unstable pitching moment curve up to 15° incidence.

These results were obtained from an aeroelastic prediction code developed especially for analyzing and predicting sail deformation and its influence on the stability of these aircraft [12]. Although linear analysis methods may provide adequate results for more conventional structures, the flexible sail over a relatively stiff frame introduces important geometric nonlinearities. A direct energy minimization technique was therefore employed to determine the deformed sail shape as a function of angle of attack. The numerical optimization procedure combined with a vortex lattice representation of the wing permitted the combined elastic strain energy and work done by the aerodynamic forces to be minimized simultaneously, yielding the equilibrium shape of the surface. The addition of flexible battens or constraints imposed by various changes in sail attachment could be easily modeled by changing the expression for total strain energy. The methodology is outlined in figure 5 and some example results showing sail shape changes with angle of attack and pitching moment characteristics are shown in figures 4 and 6. For the case shown in figure 6 the sail shape has been constrained by including a tip batten rigidly attached to the frame. This prevents much of the torsional deformation at low angles of attack and the resulting increase in stability is apparent. These results are confirmed by flight experience and by a series of elastically-scaled wind tunnel model tests [13].

Although a stability reduction is produced by wing bending, for both forward and aft swept wings, the reduction in pitch stability shown here applies only to aft-swept wings. The longitudinal stability of forward-swept wings can be *augmented* by this torsional deformation. It is possible that Nature has exploited this situation. Reference 14 describes how such effects may be used to reduce the control requirements of certain Pterosaurs, creatures with a wing structure bearing some resemblance to flexible-wing hang gliders. Initial reconstructions [15] of the wing planform of the largest of these animals, the *Quetzalcoatlus northropi*, indicated that the wing was substantially swept forward as shown in the center view in figure 7. The above-mentioned analysis program was employed to study the effect of torsional deformations on the longitudinal stability of this creature. The results suggest a sharp increase in stability due to torsional deformation with a sufficiently flexible wing. More recent reconstructions associated with the development of a large flying replica [16] have incorporated much lower sweep angles with smaller aeroelastic stability increments.

Aircraft performance and handling are also affected by torsional flexibility through its influence on aerodynamic damping. Once again, hang gliders provide an extreme example of how important this can be. For both forward and aft-swept wings torsional deformations reduce the pitching moment due to pitch rate, $C_{mq}$, or damping in pitch. Such changes affect the glider's dynamics, changing the character of short-period motion from an overdamped, non-oscillatory response to more conventional damped motions in some cases. The importance is magnified in the analysis of large perturbation, nonlinear dynamics of a very flexible glider. Figure 8, shows the time history of the motion of this glider after an initial condition not far from the trim condition. The behavior is stable and well-damped. When the same vehicle is started in a vertical dive with an initial negative pitching rate a tumbling motion is produced which is not damped. This behavior has been observed on occasion with full-scale gliders. The simulation shows that the addition of increased static stability and damping associated with reduced torsional flexibility eliminates the possibility of this undesirable mode.

## CAMBER CHANGES

Aeroelastically-produced camber changes rarely produce significant changes in the performance or stability of conventional aircraft; this is not true in the case of many very light aircraft. Figure 9 illustrates the importance of aeroelastically-induced changes in both camber and twist of an early "Rogallo-type" hang glider design with a very flexible sail. The pitching moment curve is highly nonlinear with several distinct breaks in the curve corresponding to qualitative changes in the aeroelastic response of the sail. Characteristics such as these can be quite dangerous as evidenced by several instances of unrecoverable full-luffing dives with this sort of design [17]. The possibility of these low angle of attack stable trim points may be seen from the data — the solution was to exercise greater control over changes in wing twist and camber without substantial increases in weight.

One technique for accomplishing this involves the use of wing battens. The changes in wing camber can be controlled by batten bending stiffness and the resulting wing geometry may be calculated using the analysis procedure discussed previously. Figure 10 shows the changes in wing camber, lift, and pitching moment with changes in angle of attack as computed by this code. The slightly nonlinear deflections act in an opposite sense from those produced by wing twist changes, with a small stabilizing influence in the low angle of attack regime.

## CONSTRUCTIVE USES OF AEROELASTICITY

Although the extreme flexibility of very light aircraft has proven to be a serious disadvantage in a number of cases, it has also been used to advantage—not only eliminating aeroelastic problems, but in some instances producing characteristics unobtainable with a rigid vehicle. By necessity, perhaps, aeroelastic tailoring of these aircraft is far more advanced and more frequently exploited than in other aircraft designs or research programs. In this section a variety of examples (by no means comprehensive) are discussed in which aeroelasticity has been constructively exploited in the design of very light aircraft. In most of these cases the effect is not one of improving the stability or performance by a small margin; the effects are absolutely crucial for the operation of the aircraft.

### Performance improvements

Although changes in wing twist with angle of attack may reduce the effective stability of aft-swept wings they may also be used to reduce drag over a range of flight conditions. To prevent undesirable stalling characteristics, the tip chords of aft-swept wings are often made larger than is ideal from the standpoint of achieving minimum drag span loadings. While washout may be used to minimize the drag, a given amount of washout produces an ideal span loading at only one angle of attack. To minimize drag over a range of speeds the twist must be changed, with lower twist desired at lower angles of attack. This may be achieved simply by locating the elastic axis of the wing forward of the aerodynamic center. In the case of flexible wing hang gliders the effective elastic axis is located very far forward and the result is a larger useful speed range. Additional benefits include more benign stalling characteristics and load alleviation as the wing twists to as much as 30° in some cases.

An extreme example of using wing twist changes to improve performance through span loading changes was encountered in the design of a replica of the *Quetzalcoatlus northropi* (and, most likely, in the design of the actual creature) [16]. In order to achieve the desired distribution of lift over the wing in flapping flight, large changes in wing twist are required [18]. In the case of the full size Pterodactyl at typical flight speeds, the required twist change is of order +/- 45°. The replica was constructed with the proper torsional flexibility so that the optimal twist change during the flapping cycle was approximated. Most ornithopter are, in fact, constructed in this way, with very large aeroelastically-induced twist changes required to achieve efficient flight.

The camber changes which were partially responsible for the unappealing pitching moment characteristics of early hang glider designs may be controlled and used to increase the performance of very flexible aircraft. Figure 10 shows the increase in section camber with angle of attack. This deformation is desirable in that the ideal angle of attack increases with the geometric angle, producing an airfoil section with low drag at both low and high speeds. The section is capable of producing higher maximum lift coefficients than would be achievable with a moderately cambered section while retaining attached flow on the lower surface at high speeds, a characteristic not possible with a highly cambered thin wing section. These effects are discussed in references 12 and 19 and are apparent in the previously cited wind tunnel tests [13].

## Stability and control enhancement

Just as some of the most severe problems encountered by very light aircraft are associated with aeroelastic effects on stability and control, some of the most constructive applications of aeroelastic design have been made in this domain.

Aileron reversal proved problematic in the case of the Musculair human powered aircraft and carbon fiber torsional stiffeners were added. This phenomenon was considered in the design of M.I.T.'s long range HPA as a possible mechanism for increased roll authority, relying on aileron reversal and deflecting the ailerons in the "wrong" direction to initiate a turn. The idea was not adopted in that case, but remains an intriguing option for future very light, flexible designs. A closely related idea has been employed in some hang glider designs, however. It was observed that reflexed airfoil sections near the the wing tips, used to increase the zero lift pitching moment of certain swept, tailless designs, actually reduced the pitching moment at high speeds. As the lift coefficient was reduced, these sections produced a positive moment about the elastic axis and reduced the wing's washout. The result was a sort of elevon reversal which led to loss of stability in steep dives. It was therefore discovered that more highly cambered sections near the tips produced increased washout at very high speeds and an improvement (even with respect to rigid designs) in longitudinal stability was realized.

Extreme torsional flexibility is also essential for the lateral control of hang gliders. A rigid wing with the span of contemporary hang gliders cannot be controlled with a small lateral shift in center of gravity position. As spans were increased to improve performance, it became necessary to increase the torsional flexibility in order to improve roll response. Reducing sail tension would produce this effect but would result in unacceptably high washout angles and reductions in longitudinal stability. A solution was found in frame construction and sail attachment methods which permitted anti-symmetric deformations without large deformations under symmetric conditions (figure 11). The result is a highly maneuverable vehicle with roll damping much lower than could be achieved without aeroelastic control enhancement.

The final example discussed here involves the use of camber changes to enhance longitudinal stability. This constructive exploitation of aeroelasticity has made the tailless configuration an acceptable design and has been one of the major contributors to the improved safety of foot-launched gliding. The concept is illustrated in figure 12. It involves the use of flexible battens, connected from their trailing edge to a point above the plane of the wing by flexible "luff lines". At normal flight angles of attack the lines remain slack and the airfoil sections assume efficient cambered shapes. When the angle of attack becomes small or negative the center section of the flexible airfoil deforms downward while the trailing edge is held in place by the luff lines. The result is a highly reflexed airfoil shape with large positive pitching moment which tends to return the glider to normal flight attitudes. A dramatic increase in static longitudinal stability at the critical low angles of attack is seen in the wind tunnel data presented in the figure.

## CONCLUSIONS

Because of the importance placed on low structural weight and despite the very low dynamic pressures at which very light aircraft operate, aeroelasticity is often an important consideration in the design of these vehicles. In this paper simple analyses and more refined techniques have been used to illustrate several aeroelastic phenomena of significance to such aircraft. As this class of aircraft includes many unconventional configurations, a variety of possible instabilities arise in addition to classical flutter and torsional divergence; some of the parameters affecting these modes have been identified. Several examples of the constructive use of aeroelasticity in the design of very light aircraft are given. It has been shown that the unusually flexible character of many of these aircraft affords the designer with great potential — either for constructive improvements or for disastrous errors.

## REFERENCES

1. Proceedings of the Fourth International Symposium on the Science and Technology of Low Speed and Motorless Flight, Feb. 1984.

2. Progress in High Altitude Long Endurance Aircraft, papers presented at the Second Annual HAPP/VLEA Symposium, sponsored by the Association of Unmanned Vehicle Systems, Menlo Park, California, Nov. 1985.

3. Drela, M., Langford, J., "Human-Powered Flight," Scientific American, Nov. 1985.

4. Torenbeek, E., *Synthesis of Subsonic Airplane Design*, Delft Univ. Press, 1978.

5. Pegg, R., "Preliminary Analysis of Flying Wing, High Altitude, Long Endurance Aircraft," presented at the Symposium on Long Endurance Aircraft, 1985.

6. Henderson, C., McQuillen, E., Lehman, L., "High Altitude Long Endurance RPV Design Technology Study," Unmanned Systems, Winter 1986.

7. Gremmer, H., Moulton, R., "Muscle Hustle," Aeromodeller Magazine, 1985.

8. Schörnherr, M., "Investigation of Alternative (Ultralight) Aircraft by the Aerodynamic Test Vehicle Method," Proceedings of the Fourth International Symposium on the Science and Technology of Low Speed and Motorless Flight, Feb. 1984.

9. Uhl, S., "Static Aeroelastic Behavior of an Ultralight Aircraft Wing," Low Speed and Motorless Flight Symposium, Feb. 1984.

10. Bisplinghoff, R., Ashley, H., Halfman, R., *Aeroelasticity*, Addison-Wesley, 1955.

11. National Transportation Safety Board accident reports and statistics related to FAR Part 23, 1983 and personal correspondence with NTSB and Zane Meyers.

12. Kroo, I., *Aerodynamics, Aeroelasticity, and Stability of Hang Gliders*, Stanford University Department of Aeronautics and Astronautics, Ph.D. Dissertation, May 1983.

13. Kroo, I., "Aerodynamics, Aeroelasticity, and Stability of Hang Gliders — Experimental Results," NASA TM-81269, April 1981.

14. Sneyd, A., Bundock, M., Reid, D., "Possible Effects of Wing Flexibility on the Aerodynamics of Pteranodon," The American Naturalist, Oct. 1982.

15. McMasters, J., "Reflections of a Paleoaerodynamicist," Perspectives in Biology and Medicine, Spring 1986.

16. Brooks, A., MacCready, P., Lissaman, P., Morgan, W., "Development of a Wing-Flapping Flying Replica of the Largest Pterosaur," AIAA-85-1446, July, 1985.

17. Jones, R.T., "Dynamics of Ultralight Aircraft — Dive Recovery of Hang Gliders," NASA TM X-73229, 1977.

18. Jones, R.T., "Wing Flapping with Minimum Energy," NASA TM 81174, 1980.

19. Ormiston, R., "Theoretical and Experimental Aerodynamics of an Elastic Sailwing," Ph.D. thesis, Dept. of Mathematical Sciences, Princeton Univ., 1969.

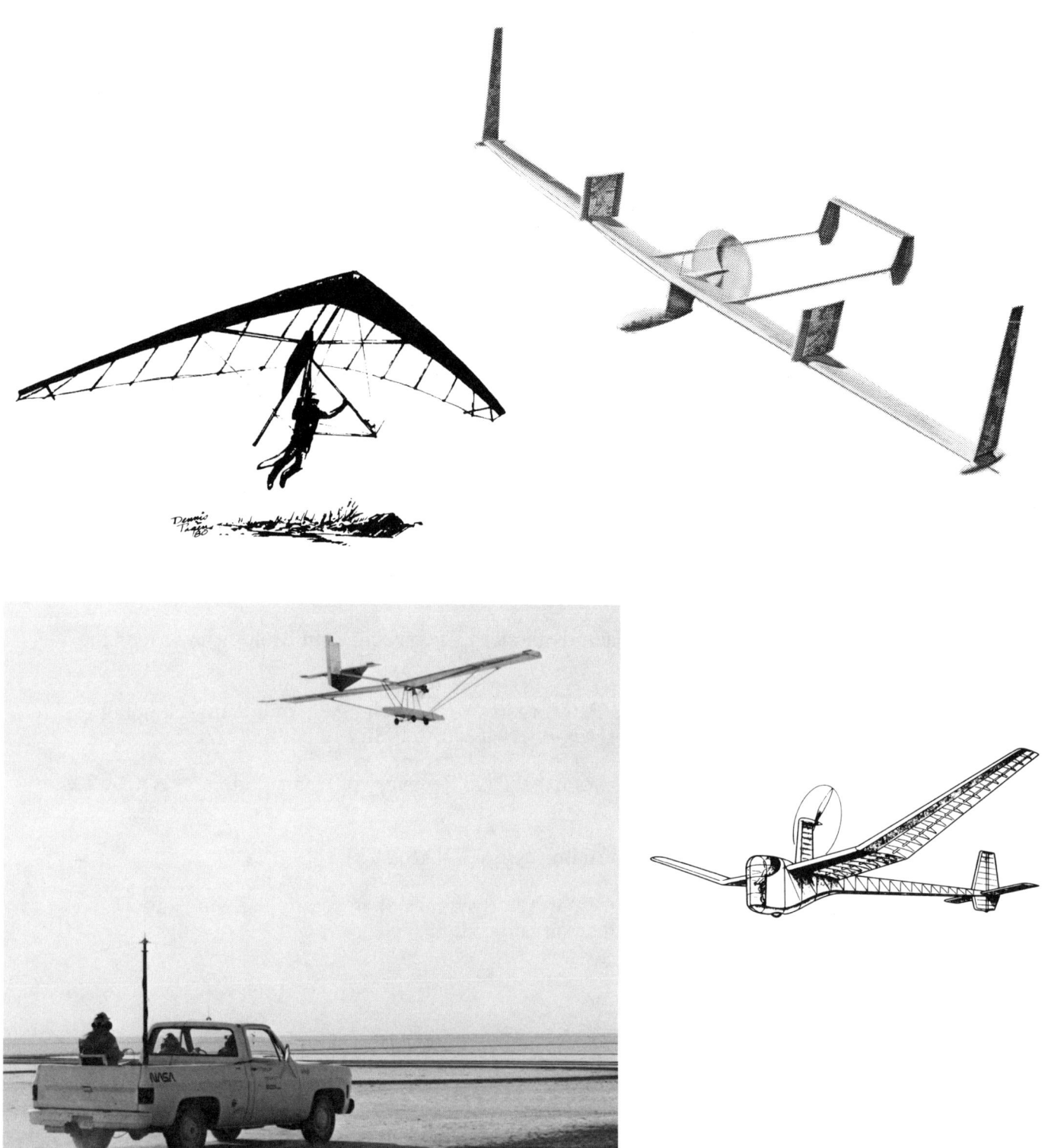

Figure 1.
Some representative very light aircraft.

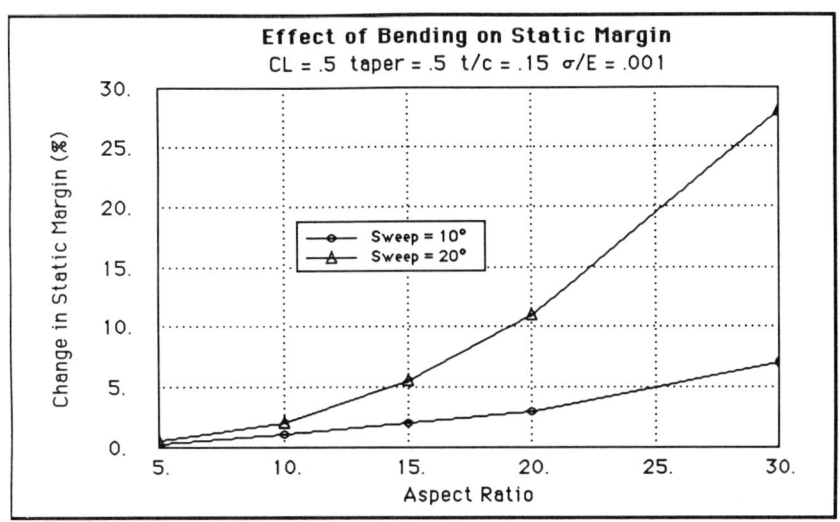

Figure 2.
Effect of wing bending on static stability of swept wings.

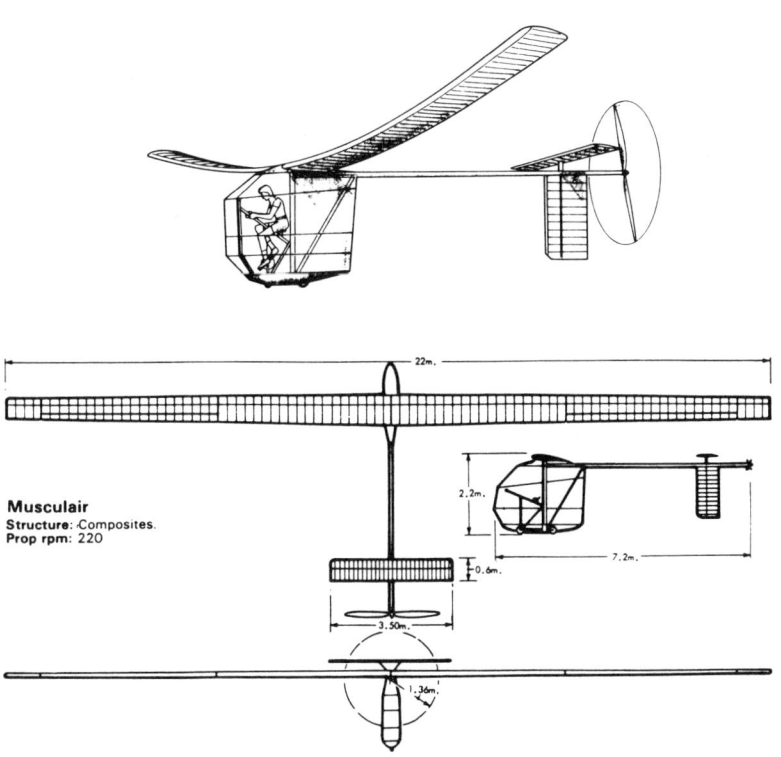

Figure 3.
Musculair, human powered aircraft with composite, cantilever wing.
Note dihedral — the jig-shape is flat.

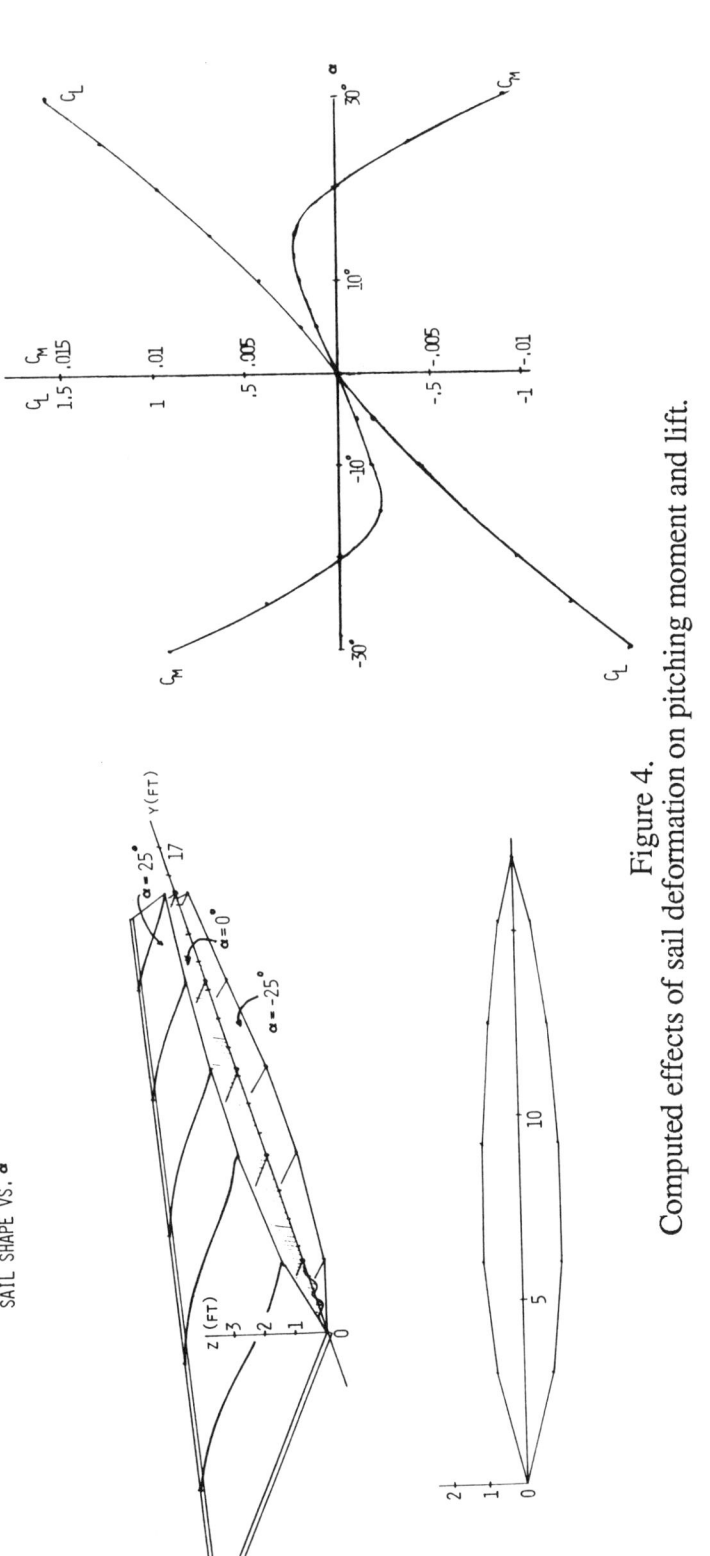

Figure 4.
Computed effects of sail deformation on pitching moment and lift.

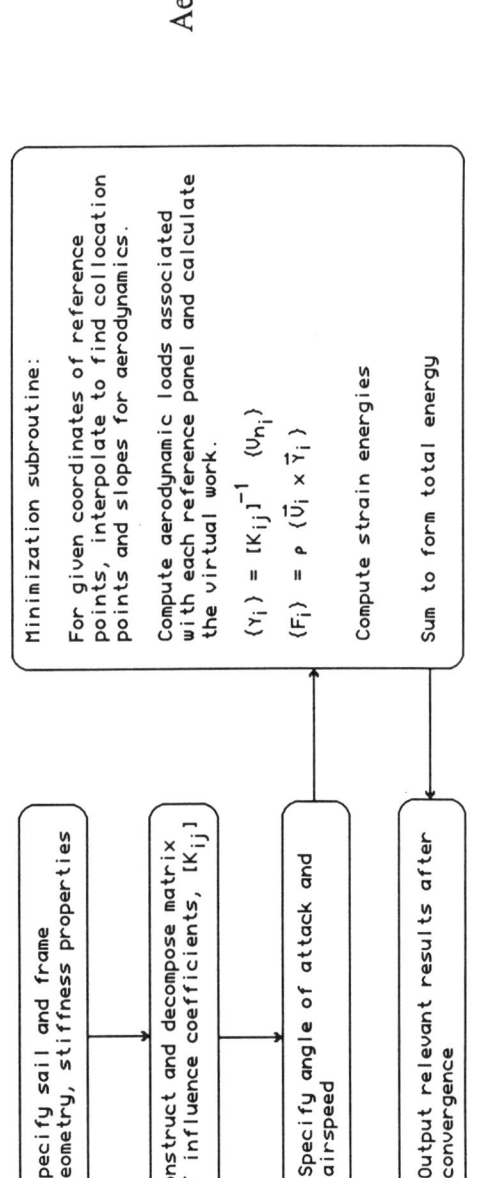

Figure 5.
Aeroelastic Code — Basic Procedure.

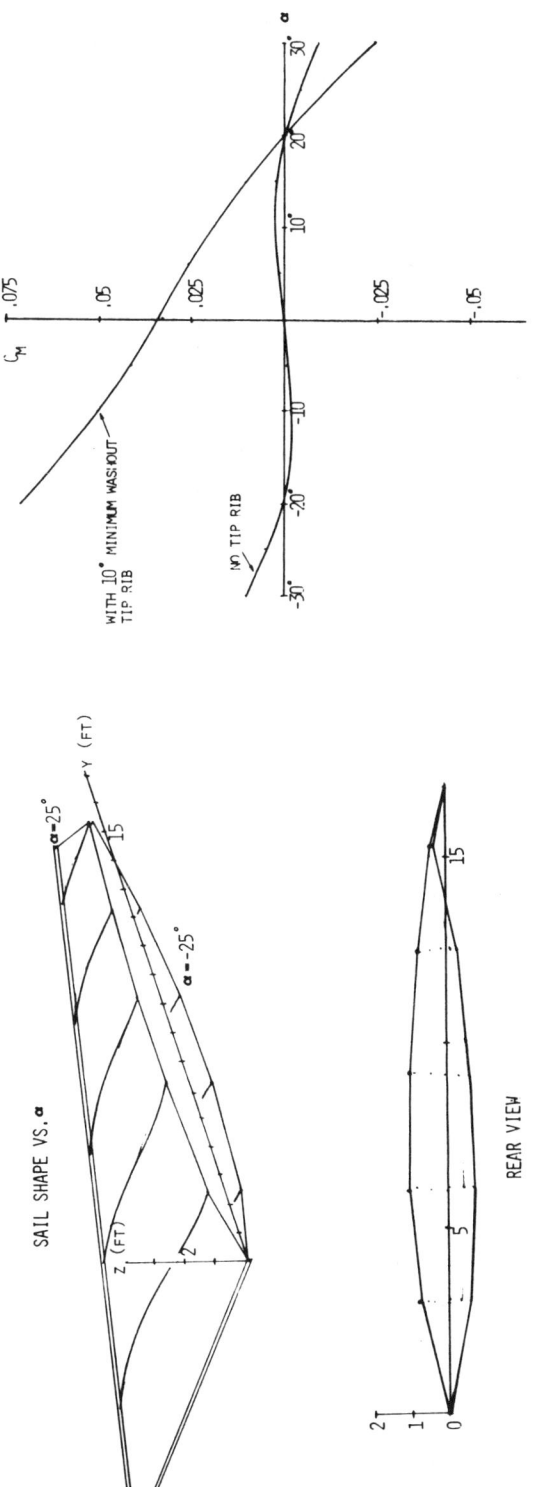

Figure 6.
Effect of wing torsional flexibility on sail shape and pitching moment.

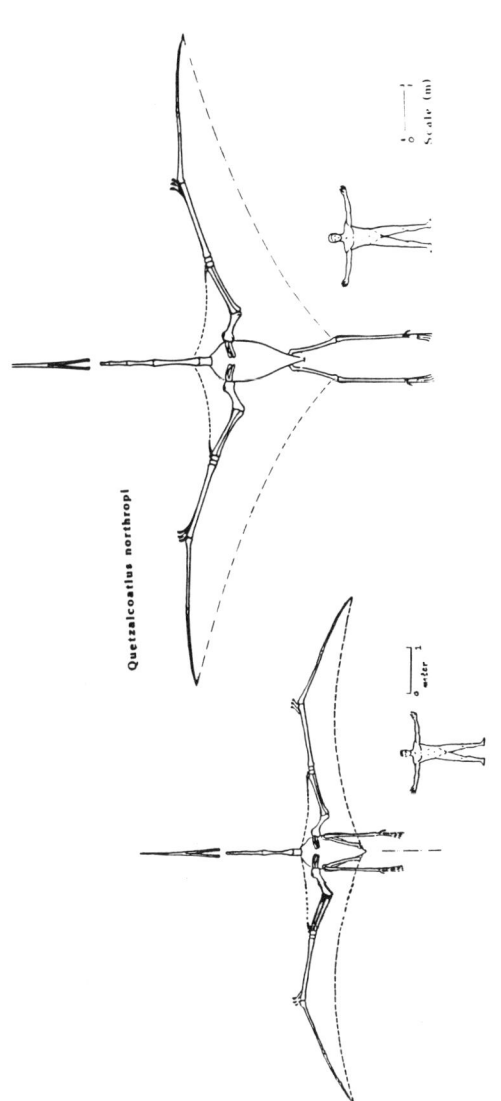

Figure 7.
Reconstructions of the planform shapes of pterosaurs.

200 / PART II

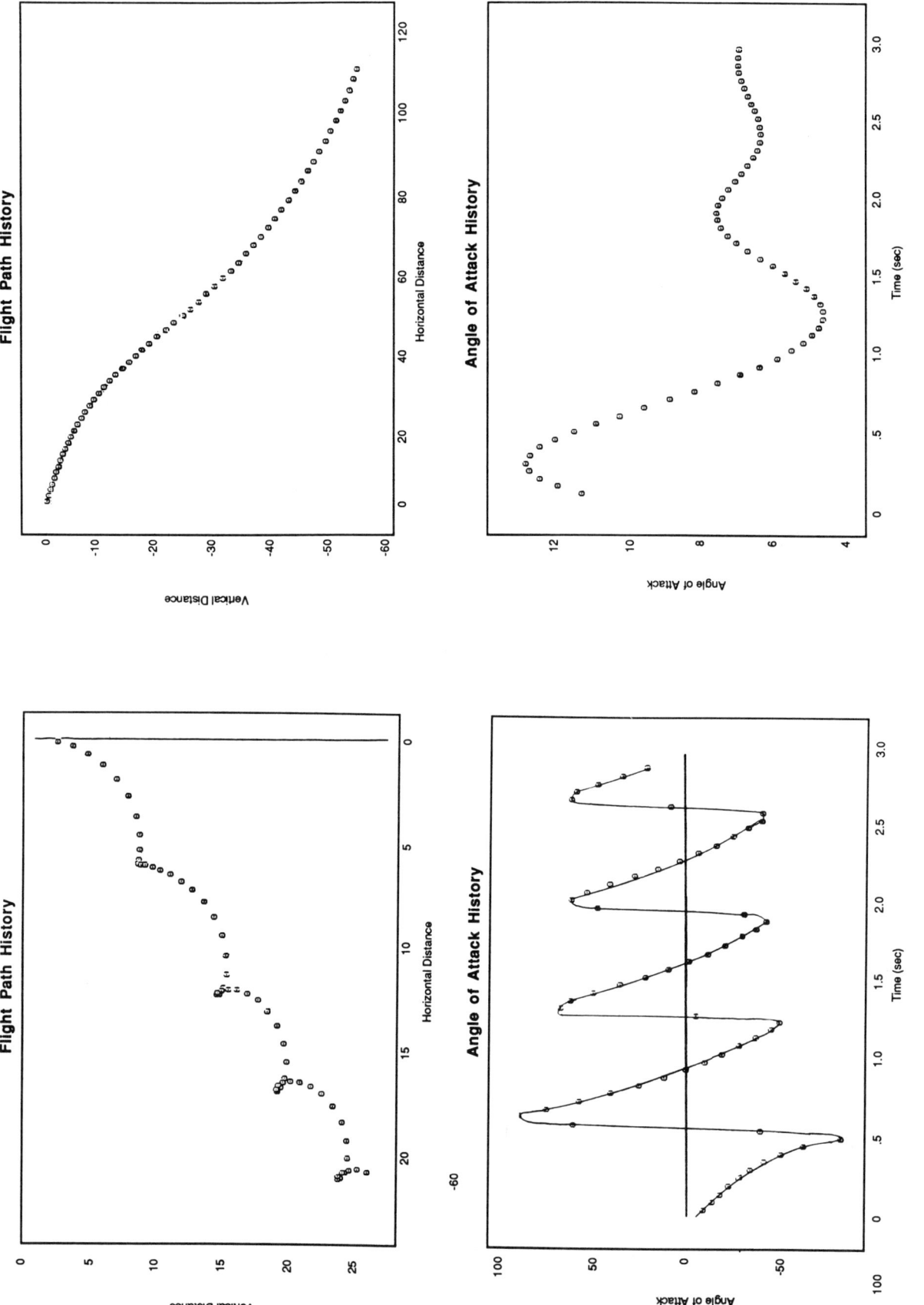

Figure 8.
Time histories of angle of attack and flight path for a glider with low damping and small static margin produced by a very flexible sail.

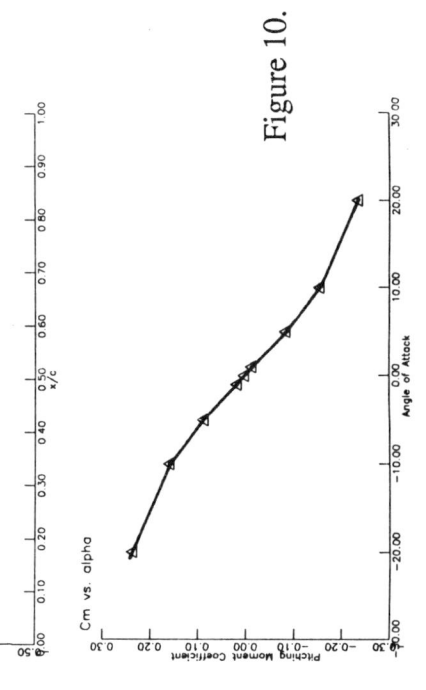

Figure 10.
Computed camber changes with flexible battens and effect on pitching moment.

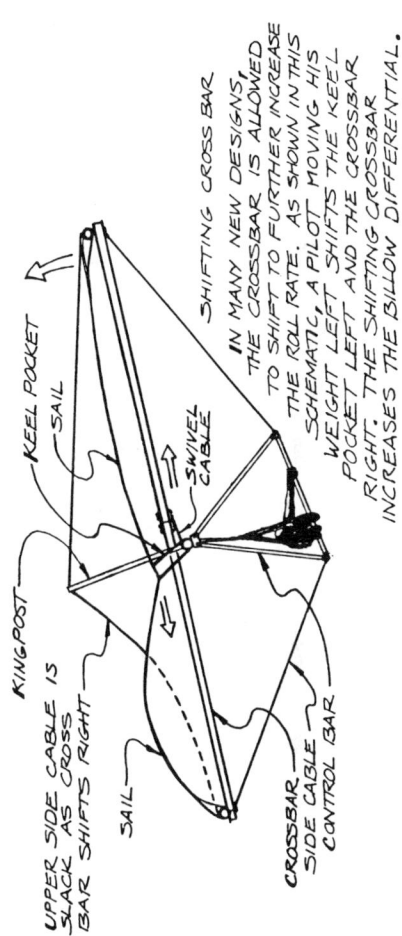

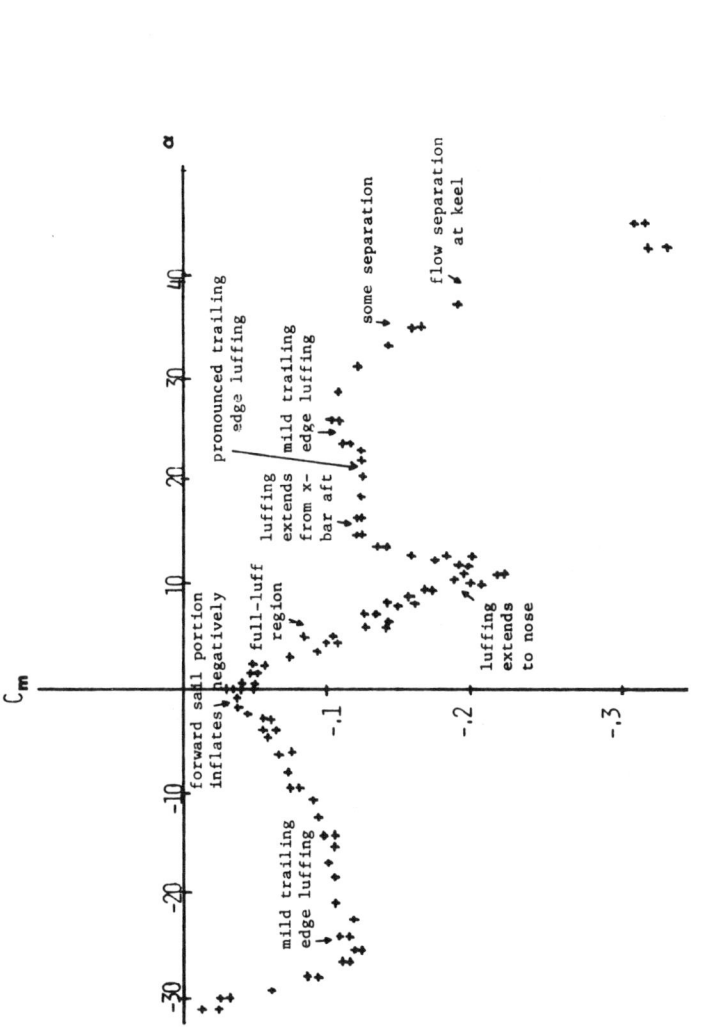

Figure 9.
Pitching moment characteristics of a very flexible early hang glider design.

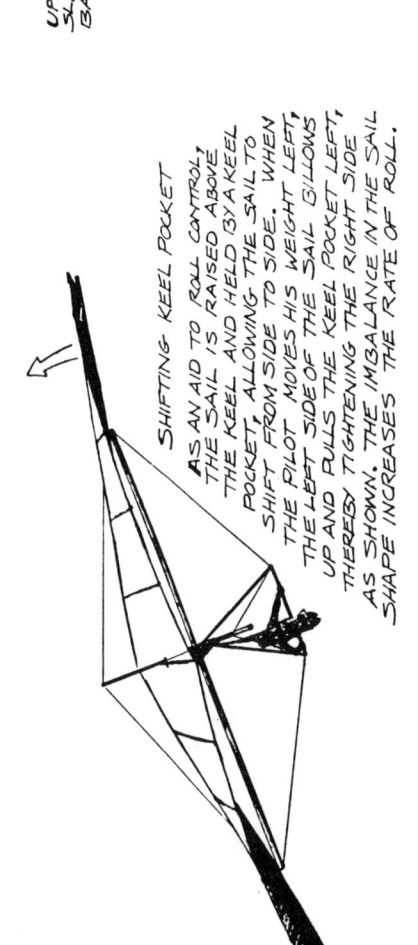

Figure 11.
Use of torsional flexibility to increase maneuverability in roll. (Drawings by Dennis Pagen.)

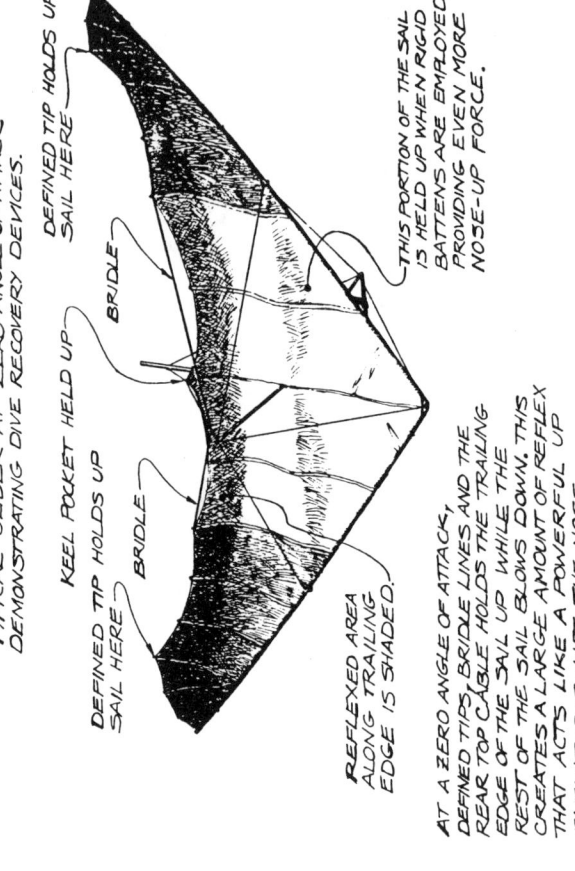

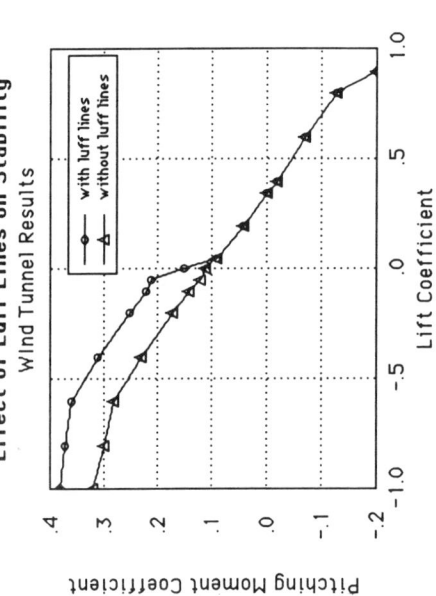

Figure 12.
Aeroelastically-induced camber changes increase pitch stability at low angles of attack.
(Drawings by Dennis Pagen.)

PROBLEMS AND PROGRESS IN AEROELASTICITY FOR INTERDISCIPLINARY DESIGN


E. Carson Yates, Jr.
NASA Langley Research Center
Hampton, VA 23665


ABSTRACT

Some problems and progress in the development of aerodynamic and aeroelastic computational capabilities are reviewed with emphasis on needs for use in current interdisciplinary design procedures as well as for stand-alone analyses. The primary focus is on integral-equation methods which are well suited for general, accurate, efficient, and unified treatment of flow around vehicles having arbitrary shapes, motions, and deformations at subsonic, transonic, and supersonic speeds up to high angles of attack. Computational methods for potential flows and viscous flows are discussed, and some applications are shown. Calculation of steady and unsteady aeroelastic characteristics of aircraft with nonlinear aerodynamic behavior is also addressed briefly.

INTRODUCTION

The need for accurate, efficient methods for steady and unsteady aeroelastic computations is a continuing problem of growing importance in the analysis and design of high-performance aircraft. Moreover, the emergence of computer-aided multidisciplinary design and optimization procedures over the last three decades has enhanced the importance of efficiency in repetitive applications and introduced a new requirement for the computation of the sensitivities of performance aerodynamics, structural loads, dynamic response and flutter characteristics, stability, and control to variations in aircraft geometry or structure. Typically and perennially the development of suitable aerodynamic methods has been one of the primary pace setters for progress.

To highlight these growing and changing needs, consider the evolving scope of aeroelasticity in computer-aided interdisciplinary airframe design and optimization. During the 1950's and 1960's, the objective of computational design processes was to minimize structural mass while satisfying strength requirements for fixed structural geometry with predetermined aerodynamic load--usually for a single load condition. The design variables were generally the material thicknesses. In the late 1960's and the 1970's, the scope was broadened to include both strength and stiffness (e.g., flutter) requirements for fixed structural geometry and with aeroelastic loads for single or multiple load conditions. Linearized aerodynamic methods were used (e.g., [1]). The design variables were still the material thicknesses. From the late 1970's to the present time, computer-aided design optimization methods have been expanded further, for example, to minimize mission or lifetime fuel consumption or cost, etc., while satisfying strength, stiffness, performance, etc., requirements with variable structural and external geometry and with aeroelastic loads for multiple load conditions employing both linear and nonlinear aerodynamics (e.g.,[2]). The design variables include material thicknesses, arrangement of anisotropic materials, structural arrangement, and external shape.

The purpose of this paper is to review some problems and progress in the development of aerodynamic and aeroelastic computational capabilities with emphasis on needs for current interdisciplinary design procedures as well as for stand-alone analyses. Because of space limitations this review is limited to work done at or supported by the NASA Langley Research Center, and the primary focus is on the integral-equations research program for steady and unsteady aerodynamics. This effort is directed toward general, accurate, efficient, and unified treatment of flows around vehicles having arbitrary shapes, motions, and deformations (including control motions) at subsonic, transonic, and supersonic speeds up to high angles of attack. Special attention is given to real-world design and operating conditions (e.g., Mach number, angle of attack, maneuver) as well as to efficient computation for both design and analysis applications. As will be brought out in the subsequent discussion, the integral-equation approach is well suited for these purposes because flow complexities such as viscous effects or transonic flow need to be addressed only in the flow regions where they actually occur, and there is no requirement for patching and matching flow domains or regional solutions. Moreover, for design applications repetitive and nonrepetitive portions of the computations are readily separable. Although the integral-equations research program has been given only limited and intermittent support for the last several years, it has nevertheless produced some significant results.

The concluding portion of this paper addresses requirements for and accomplishments in efficient computation of steady and unsteady aeroelastic characteristics of aircraft especially when aerodynamic behavior is nonlinear.

Finally, note that several technical disciplines interface with the material discussed herein but are beyond the scope of this review. These include the development of optimization methods, aeroelastic tailoring methodology, aeroservoelastic analysis and design, structural analysis, and rotor-wing technology.

INTEGRAL-EQUATION METHODS

Following a long-range plan established a number of years ago, initial efforts addressed the development of surface-panel methods for subsonic ([3] to [7]) and supersonic ([3],[4],[8],[9]) linearized potential flow. Current activities include nonlinear methods implementing the full-potential equation for high-subsonic/transonic/low-supersonic speeds[10]. Although the initial high-subsonic/transonic proof-of-concept codes ([11] to [13]) implemented the small-perturbation potential equation, there is no particular benefit in refining codes for small-perturbation conditions or two-diminsional flow as stepping-stones toward more realistic conditions.

The Euler equations are not addressed in this program. Modification of the full-potential equation to account for entropy changes across shock waves (e.g., as in [14]) should greatly expand the usefulness of potential-flow solutions well into the range of flow conditions that would otherwise require Euler solutions. Consequently, for present purposes we go directly from the modified full-potential method to the Navier-Stokes equations which are being addressed by use of the classical Helmholtz scalar/vector-potential decomposition ([15] to [17]). We are specifically concerned with several types of viscous influences: Thin wakes separating from lifting-surface edges are well represented by inviscid-flow singularities (vortex sheets). Other viscous influences require solution of Navier-Stokes equations or equivalent. These influences include boundary-layer effects, especially on deflected and/or deflecting control surfaces, shock/boundary-layer interaction, and large areas of flow separation in general. Specifics of these problem areas are addressed in the subsequent sections of this paper.

The time-dependent full-potential partial differential equation

$$\nabla^2 \phi - \frac{1}{a_\infty^2} \left( \frac{\partial}{\partial t} + U_\infty \frac{\partial}{\partial x} \right)^2 \phi = F \tag{1}$$

is the governing equation for most of the work described herein. Application of the generalized Green's-function method to this equation yields an equivalent integral equation for the velocity potential $\phi$ at any point P in the flow or on the surface of a body in the flow at any time t [3]

$$\phi(P,t) = \underbrace{\int \int \int \int GF dV_1 dt_1}_{\text{nonlinear terms}}$$
$$+ \underbrace{\int \int \int \left[ \nabla_1 S \cdot (G \nabla_1 \phi - \phi \nabla_1 G) - \frac{1}{a_\infty^2} \frac{dS}{dt_1} \left( G \frac{\partial \phi}{\partial t_1} - \phi \frac{\partial G}{\partial t_1} \right) \right] |\Box S|^{-1} dS dt_1}_{\text{linear terms}} \tag{2}$$

where $\phi$ is the perturbation velocity potential, G is the Green's function, F represents all the nonlinear terms, $a_\infty$ is the freestream speed of sound, $U_\infty$ is the freestream speed, x is the coordinate in the freestream direction, $S(x,y,z,t) = 0$ defines the body surface, and

$$|\Box S| = \sqrt{S_x^2 + S_y^2 + S_z^2 + S_t^2}$$

The exact boundary condition on the body surface is

$$\frac{DS}{Dt} = 0 \tag{3}$$

An important point here is that only the nonlinear terms need to be integrated over a fluid volume. The linear terms are integrated only over the surface of the body and its wake. Note also that the Green's function is a function of freestream Mach number, not local Mach number. Equations (2) and (3) have been formulated and computationally implemented in a moving frame of reference so that they are applicable to problems such as helicopter rotors and maneuvering aircraft as well as aircraft in uniform motion. The time integration with respect to $t_1$ in equation (2) is made trival by choice of a subsonic or supersonic source pulse as the Green's function.

Linearized Theory

If perturbations from freestream velocity are small, and Mach number is not near one nor too high in the supersonic range, the non-linear terms are negligible, and the volume integral can be ignored. The remaining surface integral of the linear terms is discretized by surface paneling [4] (e.g., arbitrary twisted quadrilateral panels [5],[6]). The unsteady-flow solution can then be obtained directly by integration in time domain or a time solution by Laplace transform

([4] to [6]) converts to a complex-frequency domain formulation which is generally more efficient for use in solving linear aeroelastic problems.

The velocity potential on the paneled surface is then found in terms of the normalwash distribution which, in general, is known from the input shape, motion, and deformation.

$$[\tilde{Y}_{jh}] \{\tilde{\phi}_h\} = [\tilde{Z}_{jh}] \{\tilde{\psi}_h\} \tag{4}$$

where $\tilde{\phi}_h$ is the Laplace transform of the perturbation velocity potential, and $\tilde{\psi}_h$ is the Laplace transform of the normalwash

$$\tilde{Y}_{jh} = \delta_{jh} - (C_{jh} + sD_{jh})e^{-s\theta_{jh}} - \sum_n (F_{jn} + sG_{jn})S_{nh}e^{-s(\theta_{jn} + \pi_n)} \tag{5}$$

$$\tilde{Z}_{jh} = B_{jh}e^{-s\theta_{jh}} \tag{6}$$

$\delta_{jh}$ is Kronecker delta, s is the Laplace transform variable (complex frequency), $B_{jh}, C_{jh}, D_{jh}, F_{jn}, G_{jn}$ are integrals over surface panels, $\theta_{jh}, \pi_{jn}$ are lag functions, and $S_{nh} = \pm 1$ for panels adjacent to a trailing edge on upper or lower surface of the body and is zero otherwise. Surface pressures are obtained from the potential by use of Bernoulli's equation.

Several features of equations (4) to (6) are significant. First, the elements of the Y and Z influence matrices are independent of the normalwash and hence independent of the mode of motion or deflection. Moreover, these matrix elements are simple functions of the complex frequency s so that the cost of changing frequency or calculating for multiple frequencies is small. The influence integrals B, C, and D represent integrals of source, doublet, and "ratelet" distributions over each body-surface panel, and integrals F and G are the corresponding doublet and "ratelet" integrals for wake panels. For a given paneling geometry, all of these integrals are functions only of Mach number. If a problem (e.g., dynamic response or flutter) involves multiple modes of normalwash, the normalwash vector in the equation becomes a matrix of modal columns, and the potential distributions for all the modes can be found in a single solution. Similarly, solutions for additional modes or revised modes (as in a structural-design optimization problem) can be obtained without recalculating the Y and Z matrices. For use in design processes, this formulation also provides a general and very efficient means for evaluating sensitivities, i.e., changes in aerodynamic properties caused by changes in external shape. Demonstration calculations have been initiated.

The generality and versatility of this approach is indicated by its use by Rockwell International for flutter analysis of the space shuttle (fig.1). Nearly 800 panels were used on the orbiter, and up to 60 modes of motion were used in both symmetric and antisymmetric flutter analyses. Subsequently, the external tank and solid rocket boosters were added, and the calculations were repeated for the entire launch configuration.

For development purposes equations (4) to (6) have been implemented in a prototype code called SOUSSA P1.1 (Steady, Oscillatory, and Un-steady Subsonic and Supersonic Aerodynamics - Version 1.1) which is applicable to vehicles having arbitrary shapes, motions, and deformations in subsonic flow only. The P1.1 code employs zeroth-order (constant-potential) panels along with the data base and data-handling utilities of the SPAR finite-element structural-analysis program. These were incorporated because SOUSSA P1.1 originally was intended for the calculation of steady-state structural loads and unsteady aerodynamics for flutter and gust-response calculation in multidisciplinary structural-optimization computations employing the SPAR structural analysis. The SPAR components, however, are unnecessary for stand-alone use. More efficient data handling methods for stand-alone operation are available.

Subsequent to the completion of SOUSSA P1.1 several significant improvements have been incorporated and others have been defined [7]. Among the latter are implementation of higher-order panels, elimination of the SPAR components, transposition and revision of the solution algorithm to substantially reduce input/output operations, and improved implementation of the trailing-edge (Kutta) flow condition.

Some program improvements already incorporated in the SOUSSA code include the development of an "out-of-core" solver to permit the use of paneling schemes that lead to coefficient matrices too large to fit in the memory of modest-size computers; the replacement of the paneled wake by an analytical wake (reducing the cost of a typical run by about one-half) but retaining an option to use paneled wakes if needed (e.g., when there is another lifting surface in the wake); and replacing the rectangular integration of pressures by a Gaussian guadrature scheme to improve the accuracy of the calculated generalized aerodynamic forces. These improvements are incorporated in a replacement for the SOUSSA code (called UTSA) which is under development at a low level of effort.

Figure 2 (reproduced from [7]) compares a chordwise distribution of pressure coefficient $C_p$ calculated by the SOUSSA surface-panel method with pressures measured on a clipped delta wing oscillating in pitch. The wing had a six-percent-thick circular-arc airfoil. The agreement is good and is representative of results obtained with this code. Figure 3 compares calculated and measured steady upper-surface pressures at two chordwise locations x on an outboard station (y=0.85) on the same clipped delta wing. Two points are to be made: First, in the range of angle of attack α (-2 deg to +2 deg) where pressure varies linearly, the agreement is excellent. Second, for this sharp-edge wing, the influence of the leading-edge vortex is substantial and begins at a low angle of attack. The latter behavior emphasizes the importance of our treatment of vortex-type flow separation to be discussed below. A phenomenological description of the relation between the vortex development and the pressure variation shown is given in [18] (from which figure 3 was taken) and in Appendix A of [19].

In addition to the subsonic capability of the SOUSSA program, a supersonic proof-of-concept surface-panel code has been written to implement linear-theory algorithms developed in [8] and [9]. The code employs first-order panels and, like SOUSSA, is applicable to vehicles having arbitrary shapes, motions, and deformations. Validation and application of the code have begun.

The only significant difference between subsonic and supersonic formulations is in the expressions for the influence integrals B,C,D, F,G in equations (5) and (6) (see, e.g., [4]). Other portions of the computations, such as paneling geometry and solution alorithms, are common to both. Consequently, it is possible that the computational capability for supersonic flow derived from this proof-of-concept code will subsequently be incorporated into the subsonic code UTSA.

The status and near-term plans for linear-theory surface-panel methods, which are applicable to vehicles having arbitrary shapes, motions, and deformations, may be summarized as follows: As planned the SOUSSA program will be superseded by an improved program UTSA which incorporates first-order panels as well as other improvements indicated by earier work with SOUSSA. Ultimately, the code may include both subsonic and supersonic capabilities. Frequency-domain computations are most efficient for implementing linear theory, but a time-domain version is also retained for evaluation of the surface integral in the nonlinear methods described next. Specific activities include configuring the UTSA code for efficient use in interdisciplinary design processes, incorporating special elements to improve accuracy and efficiency near normalwash discontinuities (e.g., at control surfaces), completing the initial demonstration of the efficient computation of sensitivities of aerodynamic pressures and loads to variations in planform, and general check out and validation.

Nonlinear Theory

When the flow approaches transonic conditions and/or flow perturbations (e.g., angle of attack) become large, the nonlinear terms represented by F in equation (2) are no longer negligible, and the volume integral must be evaluated in combination with the surface-panel evaluation of the linear terms. For nonlinear problems it is important to note (1) that the Green's function depends on freestream Mach number, not local Mach number, and (2) that the integrand of the volume integral diminishes rapidly in magnitude with increasing distance from the body and its wake.

For application to nonlinear problems the integral-equation method has several features which make it particularly attractive for general, efficient computational implementation: (1) Evaluation of an integral is required rather than the numerical solution of a partial differential equation, which is a more sensitive process. (2) The volume integral need be treated only in the limited region of flow in which nonlinear terms are of significant magnitude rather than over an entire computational domain. In fact, as the integration proceeds away from the body, it is terminated when the integrand falls below a preselected threshhold value. (3) Required accuracy can be attained with relatively few computational grid points in the fluid (computational domain of the volume integral). (4) The code is numerically stable even when moderate-to-large time steps are employed. (5) Correct far-field boundary conditions are automatically imposed. This condition is particularly important for unsteady flow. Linear-theory behavior in the far field is inherent in the integral-equation solution. (6) When viscous flows are treated by the scalar/vector-potential decomposition (to be discussed below), interfacing (patching and matching) of regional solutions (e.g., inner viscous solution and outer inviscid solution) is not required.

In this section small-perturbation transonic attached flow will be considered first followed by large-perturbation subsonic and transonic flow conditions involving vortex-type flow separation in the form of thin wakes emanating from lifting-surface edges and finally flow conditions involving significant viscous effects which require solution of Navier-Stokes equations for attached or separated flow for which the scalar/vector-potential method is employed.

Small-Perturbation Transonic Flow: For proof-of-concept demonstration of transonic capability, only the small-perturbation terms were retained in the volume integral of equation (2), and the resulting time-domain computer code [13] was called SUSAN (Steady and Unsteady Subsonic Aerodynamics-Nonlinear). Figure 4 shows chordwise pressure distribution near the root of a rectangular wing as calculated by the SUSAN code and by a transonic small-perturbation finite-difference code. The shock is captured, and the agreement is quite good even though only a few elements were used to evaluate the volume integral, and the domain of integration extended only one chord length from the wing perimeter. Good agreement with measured pressures is shown in figure 5 for a sharp-edge wing under conditions involving supercritical flow over much of the chord.

Evolution of the lifting pressure $\Delta C_p$ on a wing oscillating slowly in pitch about the leading edge is shown in figure 6 at three times during a cycle of motion. Although only ten computational elements along the wing chord were used to evaluate the volume integral of the nonlinear terms, the build-up of lift and the appearance of a shockwave are clearly indicated. The symbols shown are used only to distinguish the curves and do not indicate computational points.

The formulation described here and its implementation in the SUSAN code demonstrated the merits of the integral-equation method for transonic flow. However, no further development of the small-perturbation approximation is planned.

Subsonic/Transonic Flow with Vortex Separation: All of the preceding involved calculation of the velocity potential. For solving nonlinear problems, however, there are advantages in calculating velocities directly, especially when large velocity variations occur, when shocks are present, when thin-wake (vortex-like) flow separation from wing leading or side edges occurs, or even when trailing-edge wake deformations are significant (fig.7). Taking the gradient of the integral equation for the potential (equation (2)) or alternatively applying the Green's-function method to the full-potential equation in the form (for steady state)

$$\nabla^2 \phi = \frac{-1}{\rho} \nabla \rho \cdot \nabla \phi \equiv Q \tag{7}$$

gives [10]

$$\bar{V}(x,y,z) = \bar{E}_\infty + \frac{1}{4\pi} \iint_{BODY} \frac{\bar{\omega} \times \bar{R}}{R^3} dS + \frac{1}{4\pi} \iint_{WAKE} \frac{\bar{\omega} \times \bar{R}}{R^3} dS \tag{8}$$

$$+ \frac{1}{4\pi} \iiint_{VOL.} \frac{Q}{R^2} \bar{E}_R dV$$

where $\rho$ is the fluid density, $\bar{\omega}$ is the vorticity vector, $\bar{R}$ is the vector from "sending" point to "receiving" point, $\bar{E}_R$ is a unit vector in the $\bar{R}$ direction, and $\bar{E}_\infty$ is a unit vector in freestream direction.

Equation (8) is an expression for the velocity field V as the sum of four components: (1) freestream, (2) a surface integral which gives the velocity induced by the flow singularities representing the solid body, (3) a surface integral which gives the velocity induced by the vorticity representing the thin wake, and (4) a volume integral representing the compressibility terms (right-hand side of equation (7)). The integrand of this volume integral decreases more rapidly than the square of the distance from the body or vortex surface, so the domain of integration can be relatively small. The integrands in the three integrals are not independent, and solution is by iteration to satisfy the boundary conditions on the body and to deform the free vortex sheets into a force-free shape [10]. Note that the form of the integrand shown in the body integral indicates the use of a vorticity distribution to represent a thin wing in some proof-of-concept calculations. One of the major generalizations of this method, to be initiated, consists of replacing this body integral with the UTSA surface-panel formulation so that transonic flow over bodies of arbitary shape, including vortex-type separation, can be calculated. Other planned improvements include (1) replacing the vortex-lattice model used in the wake integral for proof-of-concept calculations with the hybrid-vortex formulation [20],[21] in which second-order distributed-vorticity panels are used to compute near-field influence, reducing to zeroth-order (discrete-vorticity) elements for far-field influence, (2) shifting the linear compressibility term $M^2\phi_{xx}$ from volume integral to surface integral by solving

$$\nabla^2\phi - M^2\phi_{xx} = Q - M^2\phi_{xx} \equiv Q_{nonlin} \qquad (9)$$

instead of equation (7), thereby significantly reducing the region over which the volume integral needs to be evaluated, (3) replacing constant source strength with linearly varying source strength in the volume elements and introduing a threshold cutoff value for the integrand of the volume integral to terminate integration when the integrand diminishes to negligible magnitude, (4) accelerating convergence of the solution by possible use of shock fitting [10], (5) accounting for entropy changes across shockwaves (see, e.g.,[14]). Code development for unsteady flow is in progress. Research on suitable configuration of these codes for efficient use in computer-aided interdisciplinary design will be a continuing activity.

Completion of the improvements listed above should provide a powerful tool for calculating transonic and/or free-vortex flows around arbitrary aircraft configurations with sharp leading edges or with assumed separation line locations. Establishing the separation line on a vibrating wing, however, is a tough viscous-flow problem, but may be amenable to treatment by the scalar-vector potential method to be discusssed below. The importance of expediting this activity should be underscored. The ability to calculate accurately the complicated transonic vortical flows around highly swept wings and complete aircraft at high angles of attack is a key problem for the future development of highly maneuverable fighter aircraft and is already needed to improve the assessment and understanding of steady and transient flight loads and flutter problems of current combat aircraft. It should be especially noted that vortex-type flow separations produce typically detrimental effects on structural loads and flutter.

Figure 8 shows the calculated velocity field and shape of the free-vortex surface in a crossflow plane slightly downstream of the trailing edge of a delta wing with vortex sheets representing thin wakes emanating from leading and trailing edges as in figure 7. The volume integral (equation (8)) has not been included for this incompressible-flow calculation. The results compare quite favorably with the low-Mach-number experiments of Hummel even though relatively few vortex elements were used in this exploratory calculation. The leading-edge vortex core is clearly defined as is the incipient deformation of the trailing-edge vortex sheet into a trailing-edge core with rotation opposite to that of the leading-edge core. The corresponding spanwise distributions of lifting pressure $\Delta C_p$ are shown in figure 9 for crossflow planes at 0.7 and 0.9 of the root chord aft of the wing apex. Agreement with measured values is very good.

Inclusion of the volume integral (equation (8)) permits calculation of transonic flow. Figure 10 shows the spanwise distribution of upper-surface pressure $C_{pu}$ and the flow field, including a captured shock, in a crossflow plane at 0.8 of the root chord aft of the apex of a delta wing. In this exploratory calculation the vortex sheet was not allowed to roll up enough to exert its full inductive effect on the wing surface before the vorticity was transferred into the vortex core. If an additional quarter turn of rollup were allowed, the pressure peak would be slightly higher and a little farther outboard, resulting in even better agreement with experiment. In contrast, the pressure peak from the Euler solution is considerably weaker and farther outboard than the experimental peak because of spatial and numerical diffusion in the Euler calculation.

Structural design loads do not occur at small-perturbation conditions but at limit load-factor conditions such as high angle of attack. Aeroelastic deformations are important. Wind-tunnel results may be of questionable accuracy because of large wall effects. The important influence of large perturbation conditions and free-vortex flows on structural design loads is typically detrimental, as is illustrated by the calculations shown in figure 11 (from [22]). Even if the linear and nonlinear spanwise load distributions shown were compared on the basis of same total normal force (same area under the curves), it is evident that the effect of the wing-tip vortex is to shift the load outboard and hence increase wing bending movements.

Linearized aerodynamic theory indicates that there should be no effect of angle of attack on flutter dynamic pressure. However, a detrimental effect typically does occur with increasing angle of attack. If adequate flutter margins are to be maintained when angle of attack is not near zero, the degradation must be predictable. Wind-tunnel testing is not the answer because stiffness-scaled flutter models are typically too weak to sustain more than very small static loads. Figure 12 shows experimental variation of flutter dynamic pressure with angle of attack for a stiff wing that was spring supported [23]. The initial decline in flutter dynamic pressure between 0 and 7 deg is attributed to the effect of the tip vortex. Confirming calculations by methods just described are in early stages. The drastic decline beyond 7 deg is probably caused by flow separation progressing forward from the trailing edge. Prediction of that behavior will require solutions of Navier-Stokes equations as discussed below.

Summarizing the status of integral-equation methods for vortex-type (thin wake) flow separation: The hybrid-vortex method for low-Mach-number steady flow [20], [21] is complete. Computations based on equation (8) for steady transonic flow with vortex-type separation and shockwaves have been demonstrated [10], and the corresponding unsteady code development is in progress. Major generalizations and improvements in efficiency are planned to start in the immediate future. Further developments for transonic flow, with or without vortex-type flow separation, will be based on equation (8).

Scalar/Vector-Potential Method: When viscous influences (other than thin wakes from lifting-surface edges) are important -- for example, boundary-layer effects on control-surface forces, shock/boundary-layer interaction, or flow separation from surfaces -- solution of Navier-Stokes equations in some form is required. The approach taken here is a scalar/vector-potential (SVP) decomposition of the velocity field by use of the classical Helmholtz representation of a vector field as the sum of an irrotational part and a solenoidal part. Thus

$$\bar{v} = \text{grad } \phi + \text{curl } \bar{A} \tag{10}$$

where $\phi$ is again the scalar potential which is evaluated by the methods already described herein, and the vector potential $\bar{A}$ is related to the vorticity $\bar{\omega}$ by

$$\nabla^2 \bar{A} = -\bar{\omega} = -\text{curl } \bar{v} \tag{11}$$

The vorticity, in turn, is governed by the vorticity-dynamics equation

$$\frac{D}{Dt}\left(\frac{\bar{\omega}}{\rho}\right) - \frac{\bar{\omega}}{\rho} \cdot \text{grad } \bar{V} = \frac{1}{\rho} \text{curl } \bar{A}$$

$$= \frac{1}{\rho} \text{grad } T \times \text{grad } S + \frac{1}{\rho} \text{curl } \frac{1}{\rho} \text{div } (\bar{\bar{T}} + p\bar{\bar{I}}) \tag{12}$$

which is obtained by taking the curl of Navier-Stokes equation for general, three-dimensional, unsteady, compressible, viscous, heat-conducting flow [15] to which the present formulation is fully equivalent. In equation (12), T is temperature, S is entropy, and $\bar{\bar{T}}$ is stress tensor.

The formulation in equations (10) to (12) appears to be a computationally attractive alternative to direct solution of the Navier-Stokes equations in primitive variables. Methods of this type have been used for a long time for viscous incompressible flow, but they have not proved to be readily generalizable to compressible flow. The present formulation is quite general and is directly applicable to compressible flow. Since the outer region of the flow about an aircraft is essentially irrotational, an integral-equation implementation is an especially attractive method of solution. The initial proof-of-concept code for two-dimensional incompressible flow has been used to calculate boundary layers on a flat plate, flow over an airfoil, and separated flow around a rectangle -- all with good results [16],[17]. Application to a circulation-control airfoil is being initiated. For applications to turbulent flows this method, of course, requires a good turbulence model just as any other method does. In addition to its computational use, the SVP formulation has also generated considerable insight into the relations between surface boundary conditions, viscosity, vorticity and its diffusion [16],[17].

Current activities under a university grant are extending the proof-of-concept code to three dimensions and to compressible flow. The types of applications planned include viscous flow over lifting surfaces with and without control-surface deflection, lifting surfaces with flow separation from edges in compressible flow, and lifting surfaces with separated flow following a step change in angle of attack.

Summary of Integral-Equation Activities

The activities described here and the computational capabilities summarized in table I indicate that completion of this work will provide efficient and <u>unified</u> treatment of flow over vehicles having arbitrary shapes, motions, and deformations at subsonic, transonic, and supersonic speeds up to high angles of attack. Moreover, the computational capabilities that are emerging appear to be well suited for repetitive use in design applications as well as for stand-alone use. As pointed out previously, the UTSA surface-panel program for attached flow may contain both subsonic and supersonic modules in a single program. Flow complexities, such as transonic nonlinearities, thin wakes, or viscous influences, are addressed only if and where they occur. Thus, if the volume-integral module is included with UTSA, the program implements the full-potential equation for transonic nonlinear attached flow. With modification for shock-generated entropy change, the program can apply also to flows with shocks of finite strength, including supersonic Mach numbers beyond the linear range, as long as shock-generated vorticity is of minor importance. If the hybrid-vortex module representing the free vortex sheets is also included, the code treats transonic flow with vortex-type separation. Finally, combination of the vector potential with these scalar-potential methods (SVP formulation) permits the formal equivalent of Navier-Stokes solutions for high angles of attack where flow separation from surfaces may occur (for example, on advanced fighter aircraft in combat maneuvers and in highly transient conditions) and also even for low angles of attack when control-surface deflections or deflection rates are large enough or shock waves are strong enough to cause significant boundary-layer thickening or separation. The latter conditions are particularly important for generating control forces and for design of active control systems.

COMPUTATION OF AEROELASTIC CHARACTERISTICS

Aerodynamic developments such as those just described exert a dominant influence on what can be done in aeroelastic analysis and design. Moreover, for repetitive design applications it is especially important that both aerodynamic and aeroelastic computations be performed as efficiently as possible. This task is difficult when aerodynamic behavior becomes nonlinear because of the intrinsically higher cost of each calculation and because of the larger number of variables that influence the results. Significant progress has nevertheless been made in achieving efficiency in steady and unsteady aeroelastic computations involving nonlinear aerodynamics. A few of these accomplishments are mentioned here for completeness albeit briefly because of space limitations.

When aerodynamic behavior is nonlinear, load calculations are usually iterative. However, combining aerodynamic and aeroelastic iterations and converging them concurrently permits evaluation of static loads and deformations on an elastic lifting surface at a cost only about 35% higher than that for a rigid-body computation [24]. Figure 13 shows results of such a calcuation made by use of the FLO22 finite-difference code. Deflection contours are shown as well as the spanwise and chordwise load changes caused by static aeroelastic deformation. Computational methods such as these need to be incorporated into iterative design processes.

The use of nonlinear time-domain aerodynamics directly in time-marching calculations is an expensive way to compute flutter characteristics or subcritical dynamic behavior. Considerable improvement in the efficiency of flutter calculations for two-dimensional airfoils has been achieved by Fourier decomposition of the dynamic response to an impulse perturbation in angle of attack [25]. Extraction of the response fundamental for use in classical frequency-domain flutter computations has yielded accurate flutter characteristics with transonic small-perturbation theory. This result, however, may very well be method-dependent, that is, it may not work so well with full-potential, Euler, or Navier-Stokes equations. Moreover, its suitability for general application to three-dimensional problems is uncertain. Other methods for efficiently determining points on flutter boundaries are under current development, and methodology suitable for iterative design applications will be the subject of continuing study.

CONCLUDING REMARKS

Some problems, progress, and plans in steady and unsteady aerodynamics and aeroelastic analyses have been reviewed with emphasis on computational needs for airframe design. The primary focus has been on applications to (1) vehicles having arbitrary shapes, motions, and deformations, (2) appropriate design and operating conditions, especially for transonic speeds and high angles of attack, (3) efficient computation of aerodynamics and aeroelastic behavior for both design and analysis. Current and future activities have been highlighted.

REFERENCES

1. Haftka, Raphael T.; and Yates, E. Carson, Jr.: Repetitive Flutter Calculations in Structural Design. Journal of Aircraft, Vol. 13, No. 7, July 1976, pp. 454-461.

2. Sobieszczanski-Sobieski, J.; Barthelemy, J.-F. M.; and Giles, G. L.: Aerospace Engineering Design by Systematic Decomposition and Mutlilevel Optimization. ICAS-84-4.7.3, 1984.

3. Morino, Luigi: A General Theory of Unsteady Compressible Potential Aerodynamics. NASA CR-2464, 1974.

4. Morino, Luigi; and Chen, Lee-Tzong: Indicial Compressible Potential Aerodynamics Around Complex Aircraft Configurations. In "Aerodynamic Analyses Requiring Advanced Computers." NASA SP-347, Part II, pp. 1067-1110, 1975.

5. Morino, Luigi: Steady, Oscillatory, and Unsteady Subsonic and Supersonic Aerodynamics - Production Version (SOUSSA P1.1) - Vol. I, Theoretical Manual. NASA CR 159130, 1980.

6. Smolka, Scott A.; Preuss, Robert D.; Tseng, Kadin; and Morino, Luigi: Steady, Oscillatory, and Unsteady Subsonic and Supersonic Aerodynamics - Production Version 1.1 (SOUSSA P1.1), Vol. II - User/Programmer Manual. NASA CR 159131, 1980.

7. Yates, E. Carson, Jr.; Cunningham, Herbert J.; Desmarais, Robert N.; Silva, Walter A.; and Drobenko, Bohdan: Subsonic Aerodynamic and Flutter Characteristics of Several Wings Calculated by the SOUSSA P1.1 Panel Method. AIAA Paper 82-0727, 1982.

8. Freedman, Marvin I.; Sipcic, Slobodan; and Tseng, Kadin: A First-Order Green's Function Approach to Supersonic Oscillatory Flow - A Mixed Analytic and Numerical Treatment. NASA CR-172207, 1984.

9. Freedman, Marvin I.; and Tseng, Kadin: A First-Order Time-Domain Green's Function Approach to Supersonic Unsteady Flow. NASA CR-172208, 1985.

10. Kandil, Osama A.; and Yates, E. Carson, Jr.: Computation of Transonic Vortex Flows Past Delta Wings - Integral Equation Approach. AIAA Paper 85-1582, 1985. To appear in AIAA Journal, Sept. 1986.

11. Morino, Luigi; and Tseng, Kadin: Time-Domain Green's Function Method for Three-Dimensional Nonlinear Subsonic Flows. AIAA Paper 78-1204, 1978.

12. Tseng, K. and Morino, L.: Nonlinear Green's Function Method for Unsteady Transonic Flows. In "Transonic Aerodynamics", edited by David Nixon. AIAA Series, Progress in Aeronautics and Astronautics, Vol. 81, 1982, pp. 565-603.

13. Tseng, K.: Nonlinear Green's Function Method for Transonic Potential Flow. Ph.D. Dissertation, Boston University, 1983.

14. Fuglsang, Dennis F.; and Williams, Marc H.: Non-Isentropic Unsteady Transonic Small Disturbance Theory. AIAA Paper 85-0600, 1985.

15. Morino, Luigi: Scalar/Vector Potential Formulation for Compressible Viscous Unsteady Flows. NASA Contractor Report 3921, 1985.

16. Morino, L.: Helmholtz Decomposition Revisited: Vorticity Generation and Trailing Edge Condition. Part 1: Incompressible Flows. Computational Mechanics, Vol. 1, 1986, pp. 65-90.

17. Morino, L.; Bharadvaj, B. K.; and Del Marco, S. P.: Helmholtz Decomposition and Navier-Stokes Equations. In "Proceedings of International Conference on Computational Mechanics", May 25-29, 1986, Tokyo, Japan.

18. Yates, E. Carson, Jr.; and Olsen, James J.: Aerodynamic Experiments with Oscillating Lifting Surfaces - Review and Preview. AIAA Paper 80-0450. Invited Lecture given at AIAA 11th Aerodynamic Testing Conference, Colorado Springs, March 1980.

19. Hess, Robert W.; Cazier, F. W.; and Wynne, Eleanor C.: Steady and Unsteady Transonic Pressure Measurements on a Clipped-Delta Wing for Pitching and Control-Surface Oscillations. NASA TP 2594, 1986.

20. Kandil, O. A.; Chu, L. C.; and Yates, E. C., Jr.: Hybrid Vortex Method for Lifting Surfaces with Free Vortex Flow. AIAA Paper 80-0070, 1980.

21. Kandil, O. A.; Chu, L.; and Turead, T.: A Nonlinear Hybrid Vortex Method for Wings at Large Angle of Attack. AIAA Journal, Vol. 22, No. 3, March 1984.

22. Kandil, Osama A.: Prediction of the Steady Aerodynamic Loads on Lifting Surfaces Having Sharp-Edge Separation. Ph.D. dissertation, Virginia Polytechnic Institute and State University, 1974.

23. Farmer, Moses G.: A Two-Degree-of-Freedom Flutter Mount System with Low Damping for Testing Rigid Wings at Different Angles of Attack. NASA Technical Memorandum 83302, 1982.

24. Whitlow, Woodrow, Jr., and Bennett, Robert M.: Application of a Transonic Potential Flow Code to the Static Aeroelastic Analysis of Three-Dimensional Wings. AIAA Paper No. 82-0689, 1982.

25. Seidel, D. A., Bennett, R. M., and Whitlow, W., Jr.: An Exploratory Study of Finite-Difference Grids for Transonic Unsteady Aerodynamics. AIAA Paper 83-0503, 1983.

Table 1.–Summary of Integral-Equation Activities.

| α Range \ M Range | Subsonic | Transonic | Supersonic |
|---|---|---|---|
| Low (attached flow) w/large control deflection | UTSA SVP | Nonlinear UTSA SVP | UTSA |
| Moderate (vortex separation) w/large control deflection | UTSA + Hybrid vortex SVP | Nonlinear UTSA + Hybrid vortex SVP | |
| Large (separated flow) w/ or w/o control deflection | SVP SVP | SVP SVP | |

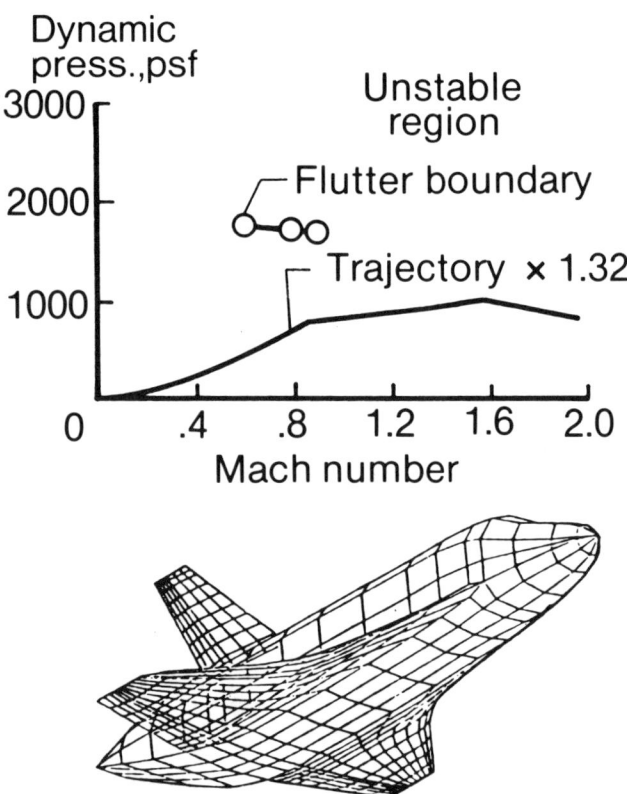

Fig. 1-Shuttle orbiter flutter analysis

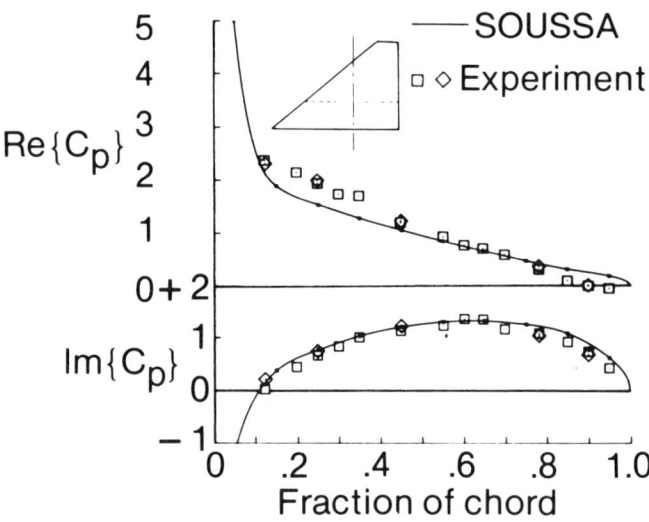

Fig. 2-Unsteady surface pressures on clipped-delta wing at Mach number 0.4, reduced frequency 0.66

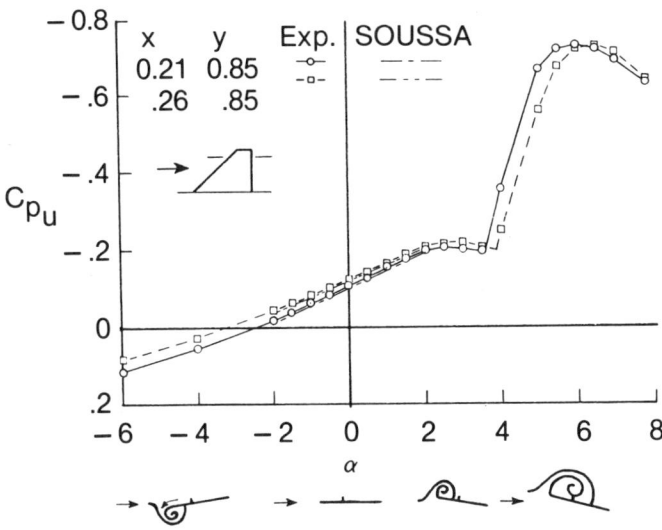

Fig. 3-Steady pressures on clipped-delta wing with 6-percent biconvex airfoil at Mach number 0.4, Reynolds number $2. \times 10^6$

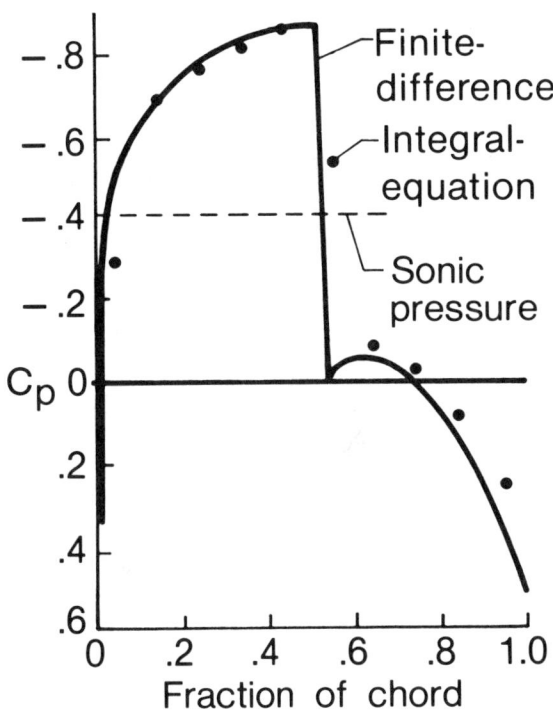

Fig. 4-Steady pressures near root of aspect-ratio 6 rectangular wing with NACA 0012 airfoil at Mach number 0.82, $\alpha=0$

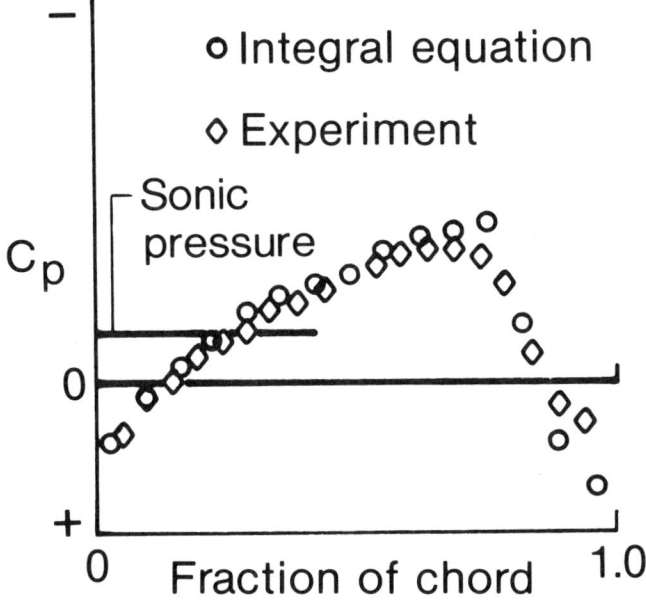

Fig. 5-Steady pressures for aspect-ratio 4 rectangular wing with 6% biconvex airfoil at Mach number 0.908, $\alpha=0$

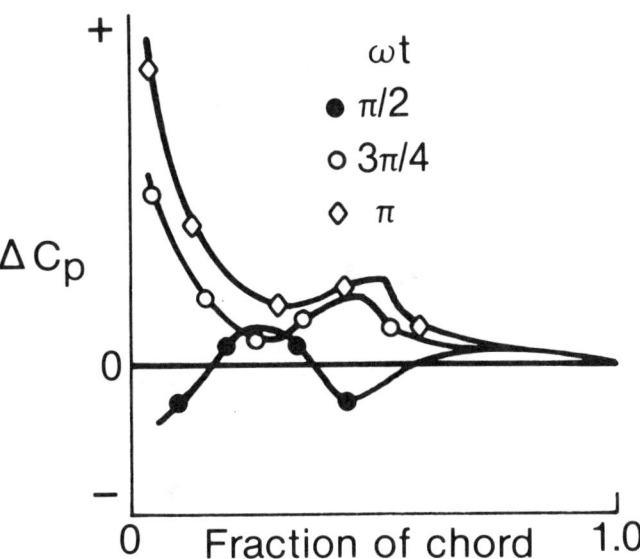

Fig. 6-Unsteady pressures for aspect-ratio 5 rectangular wing with NACA 64A006 airfoil pitching at reduced frequency 0.06, Mach number 0.875

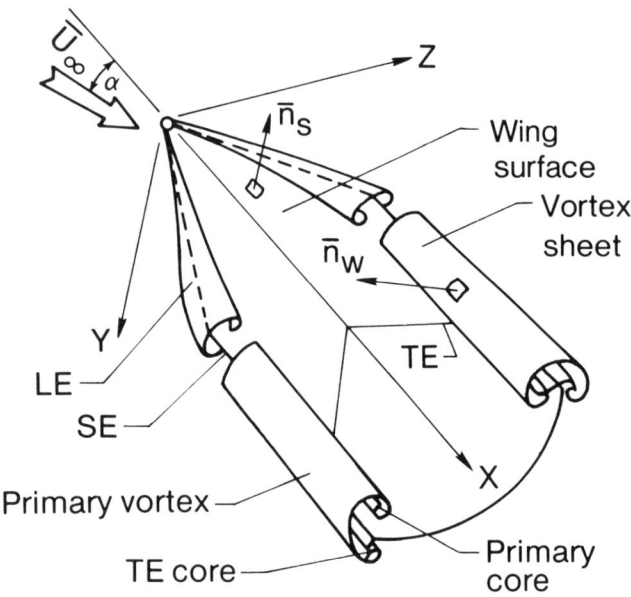

Fig. 7-Sketch of vortex sheets separating from wing edges

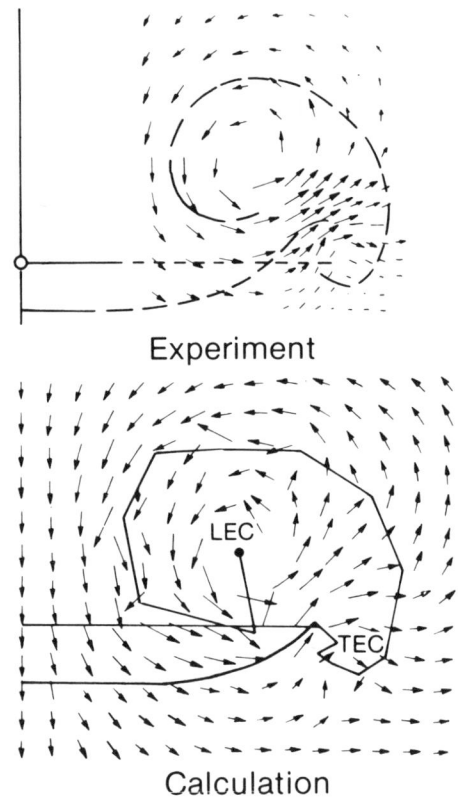

Fig. 8-Flow field behind aspect-ratio 1 delta wing at $\alpha=20.5°$

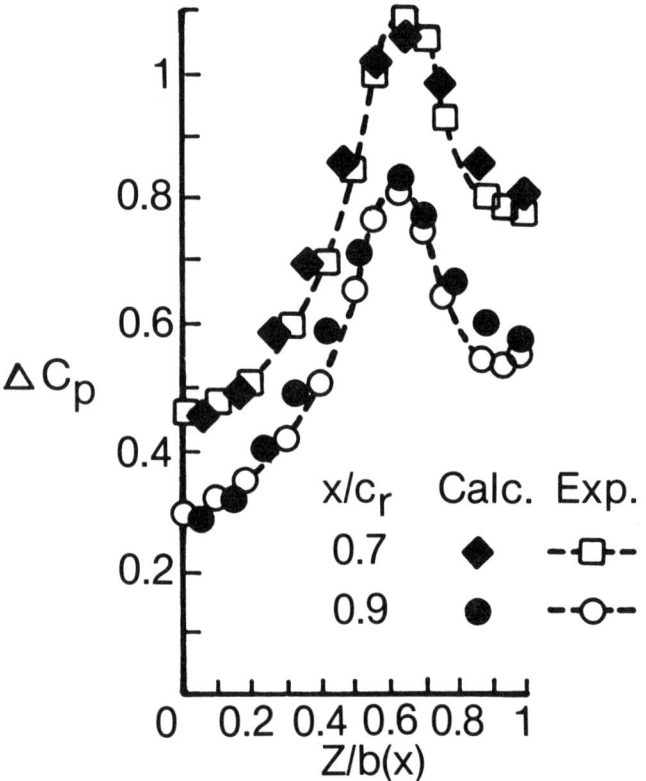

Fig. 9-Steady pressures on aspect-ratio 1 delta wing at $\alpha=20.5°$

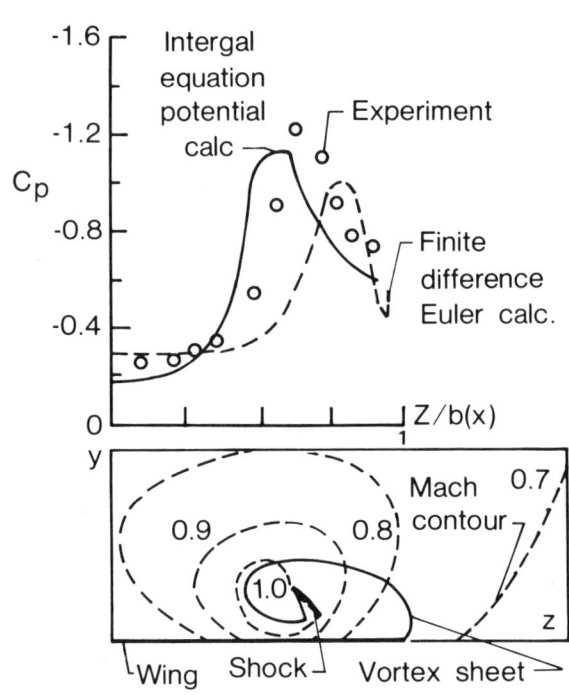

Fig. 10-Steady pressures on aspect-ratio 1.5 delta wing at Mach number 0.7, $\alpha=15.0°$, $x/c_r=0.8$

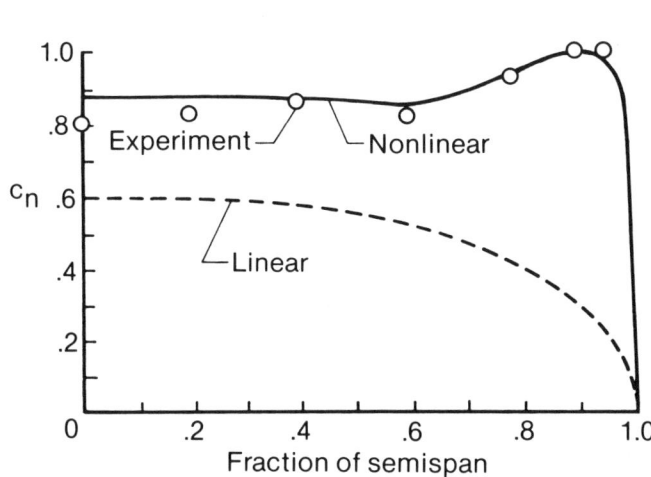

Fig. 11-Steady spanwise variation of local normal-force coefficient for aspect-ratio 1 rectangular wing at $\alpha=19.4°$

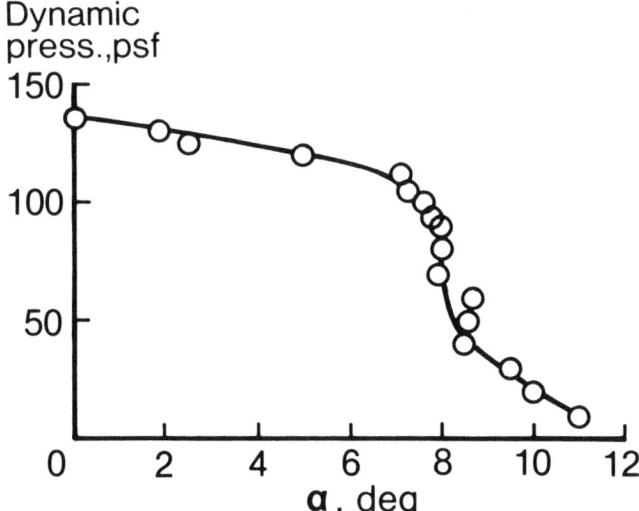

Fig. 12-Effect of angle of attack on flutter of aspect-ratio-6 rectangular wing with NACA 64A010 airfoil

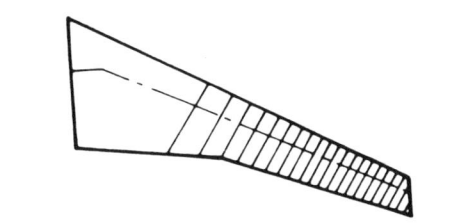

Constant-deflection contours

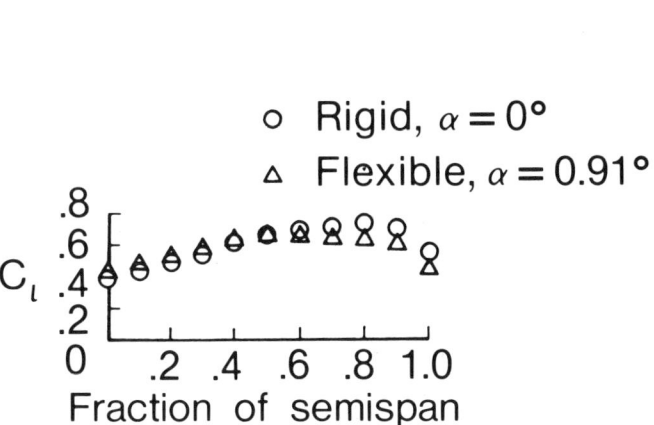

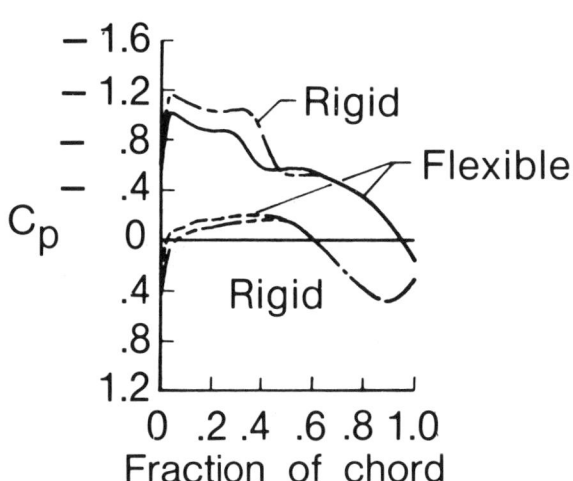

Fig. 13-Loads on elastic supercritical wing at Mach number 0.60, lift coefficient 0.58, dynamic pressure 126.4 psf

III. Structures, Structural Dynamics and
     Active Control

# AN OVERVIEW OF FINITE ELEMENT FORMULATIONS BY GENERALIZED VARIATIONAL PRINCIPLES

Theodore H.H. Pian
Massachusetts Institute of Technology
Cambridge, Massachusetts  02139

## ABSTRACT

In variational formulations of finite element methods in solid mechanics it is often desirable or even necessary to relax the compatibility and equilibrium constraints that are required in the basic principles of minimum energy.  Many generalized and extended variational principles have been constructed especially for finite element applications.  A review is made of the variational formulations for finite elements for matrix displacement methods.

## INTRODUCTION

The finite element method for structural continuum has its origin in the matrix methods for structural analysis.  In the late 1940's, matrix formulations were first adopted by aircraft structural engineers.  The methods used belong mainly to the matrix force method for which redundant internal forces are unknowns, and the matrix displacement method for which the nodal displacements are unknowns.  A structure is discretized into simple structure elements of which the strain energy is expressed exactly in terms of displacements or internal forces based on simplified modeling.  The energy or variational methods are then employed to determine the resulting matrix equations for the solutions.  For the matrix force method, Levy [1], Bisplinghoff and Lang [2], Wehle and Lansing [3], and Denke [4] and others employed the complementary energy principle, which is, of course, applicable under the strict constraint of stress equilibrium conditions.  For the matrix displacement method, Levy [5] and Scheurch [6] used the principle of minimum potential energy, which is applicable only under the constraint of displacement compatibility conditions.

The very early engineering developments of finite element method for solid continuum are based on element stiffness matrices derived from assumed displacement functions [7, 8].  The results are matrix displacement methods based on either the principle of minimum potential energy or the principle of virtual work.  In a survey article [9] it was shown that formulations by generalized variational principles may lead to finite element analyses in solid mechanics in the form of matrix displacement method or matrix mixed method. This paper is to present an updated survey, to indicate that in the finite element formulation it is desirable to relax the constraint equations in the basic variational principles in solid mechanics and to show that many extended forms of generalized variational principles have been employed. The present discussion is limited to finite elements by matrix displacement methods.

RELAXATION OF COMPATIBILITY CONDITIONS

The principle of minimum potential energy in the finite element formulation takes the form

$$\Pi_p(u) = \sum_n \left\{ \int_{V_n} \left[\tfrac{1}{2} \varepsilon^T C \varepsilon - u^T F\right] dV - \int_{S_{\sigma_n}} u^T T dS \right\} = \text{minimum} \quad (1)$$

where $\varepsilon$ are strains in term of displacements u by using differential operators D,

$$\varepsilon - D u = 0, \quad (2)$$

C are elastic stiffness coefficients, F are prescribed body forces, $V_n$ is the n-th element and $S_{\sigma_n}$ is its boundary where the tractions T are prescribed. In the finite element formulation [10], the element displacements u are interpolated in terms of nodal displacements q by shape functions N that are maintaining interelement compatibility,

$$u = N q \quad (3)$$

From this,

$$\varepsilon = (D N)q = B q \quad (4)$$

and $\Pi_p$ takes the form

$$\Pi_p(q) = \sum_n \left\{ \tfrac{1}{2} q^T k\, q - q^T Q_n \right\} \quad (5)$$

where

$$k = \int_{V_n} B^T C\, B\, dV = \text{element stiffness matrix} \qquad (6)$$

and $Q_n$ are the nodal forces due to applied loads on the n-th element.

At the early days of finite element development the task for the constructing shape functions for thin plates and shells to satisfy $C^1$ interelement continuity was found to be very difficult. One of the effective remedies is to introduce independent boundary displacements u that are common to two neighboring elements and to maintain the constraint equations

$$u - u = 0 \qquad \text{on } \partial V_n \qquad (7)$$

by using the Lagrange multipliers which turn out to be the corresponding boundary tractions T for the particular element [11]. The resulting variational principle thus becomes a stationary principle with u, u and T as independent variables. In the finite element formulation, u and T are represented in terms of parameters that are independent from one element to the other while u are interpolated in terms of nodal displacements q. The parameters for u and T thus can be eliminated from the element level and the resulting functional contains only element nodal displacements q in the same form as Eq (5). This variational priniciple has been utilized to derive stiffness matrices of a rectangular flat plate element [11] and of doubly curved quadrilateral elements which are parts of shells of revolution [12]. This principle has been more widely used for the derivation of special elements to be used at the tip of sharp cracks [13 - 15]. With the singular behavior included in the approximation of both u and T, the resulting elements are very effective in the solution of stress intensity factors for both linear and nonlinear fracture analyses.

Another difficulty that exists in finite element methods based on the potential energy principle is the locking phemomena resulted from constraint in strains under certain limiting conditions. Specifically, these are (1) the vanishing of mean normal strain for incompressible materials [16]; (2) the vanishing of transverse shear strain when plane and solid elements or plates and shells under the Reissner-Mindlin theory are used for slender and thin solids under bending action [17]; and, (3) the vanishing of membrane strain when the ratio of shell thickness to the radius of curvature is small such that the inextentional shell theory governs [18]. When an element is derived using the exact strain- displacement relation, the nodal displacements will be governed by constraint conditions near any one of the above limiting conditions. The resulting element may become extremely rigid and lead to the locking phenomena. A remedy for the locking difficulty has been devised for the assumed displacement method by using reduced order of numerical integration. However, such scheme may lead to kinematic instability [19].

An alternative remedy is to modify the functional $\Pi_p$ by including the stress displacement relation given by Eq (2) as a condition of constraint using the Lagrange multiplier method. The Lagrange multipliers are the corresponding stress components $\sigma$. The resulting generalized and multifield variational principle which is the so called Hu-Washizu principle now has three sets of field variables u, ε and σ [20, 21]. In using this functional for the finite element formulation, the element displacements are expressed in terms of nodal displacement q while the strains ε and stresses σ are in terms of parameters in each element. When the number of stress parameters is the same as that of the strain parameters, it is possible to determine the strains in terms of the nodal displacements. With appropriate choice of strain and stress terms the strain-displacement relation Eq (2) is satisfied only in integral sense. Here, since the element strain energy can be expressed in terms of the nodal displacements, the element stiffness matrix can be resulted. For the formulation of a problem that requires $C^1$ continuity this variational functional can also be extended by adding Eq (7) as a condition of constraint. However, the corresponding boundary tractions can be expressed in terms of element stresses. Tang et al originated the so called quasi-conforming element based the satisfaction of Eq (2) in the weighted average sense [22]. It can be easily shown that their method is equivalent to one obtained by the Hu-Washizu principle. Belytschko et al has also used this principle to derived plane elements that can avoid both the shear locking and the incompressibility locking [23].

When the stress-strain relation is satisfied the Hu-Washizu principle is reduced to the Hellinger-Reissner principle which contains only stresses and displacements as field variables and is stated as [20]

$$\Pi_R(u, u, \sigma) = \sum_n \left\{ \int_{V_n} \left[ -\frac{1}{2} \sigma^T S \sigma + \sigma^T (D u) - F^T u \right] dV \right.$$

$$\left. - \int_{\partial V_n} T^T (u - u) dS - \int_{S_{\sigma_n}} T^T u dS \right\} = \text{stationary} \qquad (8)$$

This functional can also be derived by a relaxation of the equilibrium conditions from the principle of minimum complementary energy given in the next section.

Another effective remedy for the locking problem is to use a modified version of the Hellinger-Reissner principle. As an example, for nearly incompressible materials the total strain energy is split into the deviatoric and spherical parts represented by the strains $\varepsilon_1$ and $\varepsilon_2$ respectively. In reducing $\Pi_p$ variational principle to a multifield principle, the strain-displacement relation for only the mean strain is relaxed. The corresponding Hellinger-Reissner functional is then of the form [24]

$$\Pi_R = \sum_n \left[ \int_{V_n} \left\{ \tfrac{1}{2}(D_1 u)^T C_1 (D_1 u) + \sigma_2^T (D_2 u) - \tfrac{1}{2} \sigma_2^T S_2 \sigma_2 - F^T u \right\} dV \right.$$

$$\left. - \int_{S_{\sigma_n}} T^T u \, dS \right] \tag{9}$$

where $(D_1 u)$ and $(D_2 u)$ are respectively the two different strains in terms of u, $\sigma_2$ are the stresses corresponding to $\varepsilon_2$, and $C_1$ and $S_2$ are the respective elastic constants. It is seen that the field variables in this functional are u and $\sigma_2$. In the finite element formulation the displacements u are expressed in terms of nodal displacements q, while $\sigma_2$ are in terms of internal stress parameters that are made independent from one element to the other. The latter thus can be eliminated in the element level and the functional is reduced to the form of Eq (5). When the stress-strain relation is satisfied, an alternative functional in terms of u and $\varepsilon_2$ can be used. This has been used by Lee et al for the formulation of plate and shell elements with C° continuity condition [25, 26]. The idea of splitting strain energy has also been used in conjunction with quasi-conforming elements [27].


RELAXATION OF EQUILIBRIUM CONDITIONS

The principle of minimum complementary energy for a continuum with discretized elements takes the form

$$\Pi_C(\sigma) = \sum_n \left\{ \int_{V_n} \tfrac{1}{2} \sigma^T S \, \sigma \, dV - \int_{S_{u_n}} T^T u \, dS \right\} = \text{minimum} \tag{10}$$

where

$T = \nu \sigma$

$\nu$ = the directional cosine of the surface normal

$S$ = elastic compliance coefficients

$S_{u_n}$ = boundary of the n-th element over which the displacement u are prescribed

The stresses must be equilibrating, i.e.

$$D^T\sigma + F = 0 \quad \text{in } V_n \tag{11}$$

and

$$T^a + T^b = 0 \quad \text{on the boundary between "a"} \tag{12}$$
$$\text{and "b" elements}$$

When the interelement equilibrium condition Eq (12) are included by the Lagrange multiplier method, the interelement displacements u are introduced as independent variables. The modified variational principle is of the form [28].

$$\Pi_{mc}(\sigma, u) = \sum_n \left\{ \int_{V_n} \tfrac{1}{2} \sigma^T S\sigma dV - \int_{\partial V_n} T^T u dS \right.$$

$$\left. + \int_{S_{\sigma_n}} T^T u dS \right\} = \text{stationary} \tag{13}$$

This is the basis for the early version of assumed stress hybrid elements [29]. The stresses $\sigma$ are in a priori equilibrium and are expressed in terms of stress parameters $\beta$, while the boundary displacements u are interpolated in terms of element nodal displacements q. Because the $\beta$-parameters are independent for different elements, the carrying-out of $\delta\Pi_{mc}=0$, will enable the expressing of $\beta$ in terms of q in the element level and hence the functional $\Pi_{mc}$ can be expressed in the same form as that by Eq (5). The assumed stress hybrid element, thus, is again based on the matrix displacement method. The equilibrium model by Fraeijs de Veubeke [30] can also be formulated by the same variational principle [31].

When the stress equilibrium equations Eq (11) are also incorporated in the $\Pi_c$ functional by the Lagrange multiplier method, the modified functional becomes

$$-\Pi_R = \sum_n \left\{ \int_{V_n} \left[ \tfrac{1}{2} \sigma^T S\sigma + (D^T\sigma)^T u + F^T u \right] dV - \int_{\partial V_n} T^T u dS \right.$$

$$\left. + \int_{S_{\sigma_n}} T^T u dS \right\} \tag{14}$$

The stationary condition for this functional is one version of the Hellinger-Reissner principle. By integrating by parts, the other version of $\Pi_R$ as given by Eq (8) can be obtained. In the finite element formulation, $\sigma$ are expressed in terms of stress parameters while both u and u can be interpolated in terms of nodal displacements q. Again because the stress parameters can be eliminated in the element level, the resulting functional is of the same form as that of Eq (5). It should be remarked that in the formulation by $\Pi_R$ the assumed stresses need not satisfy the equilibrium equations a priori. However, if they do the resulting element stiffness matrix will become identical to that by the $\Pi_c$ formulation [32].

If the assumed stresses are entirely unconstrained and are complete polynomials, the resulting element will tend to become that of the assumed displacement method according to the limitation principle of Fraeijs de Veubeke [33]. It turns out that, in the early days of hybrid element formulation, it was thought that the stress equilibrium constraint is necessary. The current thought, however, is that in many applications as listed in the following, it is desirable to relax the stress equilibrium conditions.

[1] For general shells it is difficult to satisfy the exact equilibrium equation a priori.

[2] In order to avoid the locking pheonomena it is sometimes desirable to reduce the coupling between stress terms through the equilibrium equations.

[3] For elements of distorted geometry it is most logical to express the stresses in natural coordinates in order to maintain the invariant behavior of the element and to reduce the deterioration due to geometric distortion. In such case the stress equilibrium condition cannot be maintained in general.

[4] For geometrically nonlinear problems, it is not feasible to satisfy the equation of equilibrium which include the displacements that are still unknowns.

The Hellinger-Reissner principle for finite element formulation can be modified so that the stress-equilibrium conditions can be satisfied in the integral sense in the element level. The scheme is to divide the element displacements u into two parts, with $u_q$ expressed in terms of nodal displacements q, and $u_\lambda$ expressed in terms internal displacement parameters $\lambda$ that can be eliminated in the element level. In the case that $u_q$ are compatible hence are identical to u on the boundary, the variational functional takes the form

$$\Pi_R = \sum_n \Bigg\{ \int_{V_n} \Big[ -\tfrac{1}{2} \sigma^T S \sigma + \sigma^T(Du_q) - (D^T\sigma)^T u_\lambda$$
$$- F^T(u_q + u_\lambda) \Big] dV - \int_{S_{\sigma_n}} T^T(u_q + u_\lambda) dS \Bigg\} \tag{15}$$

Here the internal displacements $u_\lambda$ play the role of Lagrange multipliers to

enforce the stress equilibrium conditions in the element level [34]. In the finite element formulation the stresses σ are initially unconstrained, while the variation with respect to λ will lead to equilibrium constraints to the original assumed stress terms. This scheme can lead to a rational procedure for the choice of stresses that are in balance with the displacement interpolation. The idea is to begin with stresses in complete polynomials. Internal displacements $u_\lambda$ are then added such that the total displacements are

now complete polynomials and the corresponding strains are of the same order as that of the stresses [35]. It has been found that the number of resulting constraint equations are often less than the independent parameters used for $u_\lambda$. In that case a perturbation of element geometry is introduced in order to

obtain additional constraint equations. It has been found that this is an ideal scheme to obtained the stress terms for the hybrid stress elements [35-39].

The idea of using $u_\lambda$ terms to introduce the stress equilibrium

constraints has also been applied to element formulation by the Hu-Washizu principle [34, 40]. The introduction of such constraint by Lagrange multiplier can also be made directly in the complementary energy functional $\Pi_c$ [41].

The above method, however, is not always convenient for obtaining constraint equations for geometrically distorted elements. For example for thin plates under the Kirchhoff theory, the equilibrium equation which contains second derivatives of moments will become very difficult to implement. An alternative is to obtain a desirable stress terms by this approach only for regular rectangular geometry. The same stress expansions are used for tensor stresses based on the natural coordinates. These stresses are then converted to physical components based on Cartesian coordinates. In this transformation, however, it is desirable to replace the Jacobians for the two coordinates system by their values at the centroid of the element [35, 36, 39, 42].

In the assumed displacement finite element formulation it is also possible to introduce additional internal displacement parameters which are statically condensed in the element level [43]. The so called incompatible elements used by Wilson et al to remove shear locking of quadrilateral plane elements and hexahedral elements are based on the introduction of incompatible internal displacements [44]. Chen and Tang also applied this technique for their quasi-conforming element [45].

It is seen that the trend in the formulation of hybrid stress elements is to include the possibility of relaxing the a priori satisfaction of stress equilibrium condition. Thus, the Hellinger-Reissner principle is the preferred variational principle instead of the complementary energy principle. In fact, for problems involving required $C^1$ continuity for displacements, independent boundary displacements u can be added in the $\Pi_R$ functional [35]. One exception in the application is the development of special elements at tips of plane cracks for which stress expansions can be obtained that satisfy both equilibrium and compatibility conditions. In that case the functional $\Pi_c$ can be converted to a form that contains only boundary integrals [46]. Special singular elements obtained by this variational principle are, with no question, the most efficient ones for linear and plane fracture analyses.

The inter-connections of the various variational principles that have been utilized to the formulation of element stiffness matrices are indicated by a flow chart given in Figure 1. At the left hand column are variational principles that were developed prior to the age of finite element methods. At the right hand column are modified variational principles by relaxed continuity or reciprocity requirements along the inter-element boundaries. At the middle column are variational principles that are modified by additional internal displacements that are the Lagrange multipliers to maintain the stress equilibrium conditions in the element level.

## CONCLUSIONS

The two basic minimum energy principles in solid mechanics can be generalized and extended for the finite elment formulations by relaxing, in various degrees, the corresponding constraint conditions. The broad nature in the possibilities of such extensions plays a very important role in the advances in finite element methods.

## REFERENCES

[1] Levy, S. "Computation of Influence Coefficients for Aircraft Structures with Discontinuities and Sweepback", J. Aero Sci. Vol. 14, pp. 547-560, 1947.

[2] Bisplinghoff, R.L. and Lang, A. "An Investigation of Deformations

and Stresses in Sweptback and Tapered Wings with Discontinuities", M.I.T. Aero-Elastic and Structures Research Report, July 1949.

[3] Wehle L.B. and Lansing, W. "A Method for Reducing the Analysis of Complex Redundant Structures to a Routine Procedure", J. Aero Sci, Vol. 19, pp. 677-684, 1952.

[4] Denke, P.H. "A Matrix Method of Structural Analysis", Proc. Second U.S. National Congress of Applied Mechanics ASME, pp. 445-451, 1954.

[5] Levy, S. "Structural Analysis and Influence Coefficients for Delta Wings", J. Aero Sci., Vol. 20, pp. 449-454, 1953.

[6] Scheurch, H.U. "Delta Wing Design Analysis", S.A.E. National Aeronautics Meeting Prepring, No. 141, 1953.

[7] Turner, M.J., Clough, R.W., Martin, H.C. and Topp, L.J. "Stiffness and Deflection Analysis of Complex Structures", J. Aero Sci, Vol. 23, pp. 805-823, 854, 1956.

[8] Argyris, J.H. _Energy Theorems and Structural Analysis_, Butterworths Scientific Publications, London, 1960.

[9] Pian, T.H.H. "Variational and Finite Element Methods in Structural Analysis", Israel J. of Technology, Vol. 16, pp. 23-33, 1978, also in RCA Review, Vol. 39, pp. 648-664, 1978.

[10] Melosh, R.J. "Basis for Derivation of Matrices for the Direct Stiffness Method", AIAA J., Vol. 1, pp. 1631-1637, 1963.

[11] Tong, P. "New Displacement Hybrid Finite Element Model for Solid Continua", Int. J. Num. Meth. Engng, Vol. 2, pp. 78-83, 1970.

[12] Atluri, S. and Pian, T.H.H. "Finite-Elment Analysis of Shells of Revolution by Two Doubly Curved Quadrilaterial Elements", J. Structural Mechanics, Vol. 1, p. 393, 1973.

[13] Atluri, S.N., Kobayashi, A.S., and Nakagaki, M. "An Assumed Displacement Hybrid Finite Element Model for Linear Fracture Mechanics", Int. J. of Fracture, Vol. 11, pp. 257-271, 1975.

[14] Atluri, S.N. and Kathiresan, K. "An Assumed Displacment Hybrid Finite Element Model for Three-Dimensional Linear Fracture Mechanics Analysis," Proc. 12th Annual Meeting Soc. of Engng. Sci., Univ. of Texas, Austin, pp. 77-87, Oct. 1975.

[15] Atluri, S.N. and Nakagaki, M., "Post-Yield Analysis of a 3-Point Bend Fracture Test Specimen: An Embedded Singularity Finite Element", Development in Theoretical and Applied Mech., Vol. 8, Virginia Polytechnic Inst., Blacksburg, VA., pp. 206-224, 1976.

[16] Nagtegaal, J.C., Parks, D.M. and Rice, J.R. "On Numerically

Accurate Finite Element Solutions in the Fully Plastic Range" Computer Methods in Applied Mechanics and Engineering, Vol. 4, pp. 153-177, 1974.

[17] Tsach, U. "Locking of Thin Plate/Shell Elements", Int. J. Num. Meth. Engng, Vol. 17, pp. 633-644, 1981.

[18] Ashwell, D.G. and Gallagher, R.H. (Ed.) <u>Finite Elements for Thin Shells and Curved Members</u>, John Wiley and Sons, London, 1976.

[19] Bicanic, N. and Hinton, E. "Spurious Modes in Two-Dimensional Isoparametric Elements", Int. J. Num. Meth. Engng, Vol. 14, pp. 1545-1557, 1979.

[20]. Washizu, K. <u>Variational Methods in Elasticity and Plasticity</u>, 3rd Ed., Pergamom Press, Oxford, 1982.

[21] Hu, H.C. "On Some Variational Principles in the Theory of Elasticity and Plasticity" Scintia Sinica, Vol. 4, pp. 33-54, 1955.

[22] Tang, L. Chen, W. and Liu, Y. "Quasi-Conforming Element and Generalized Variational Illegalities", Proc. Symposium on Finite Element Method, Hefei, China, May 19-23, 1981, Science Press, Beijing, pp. 353-389, 1982.

[23] Belytschko, T. and Bachrach, W.E. "Simple Quadrilaterlas with High Coarse-mesh Accuracy" in <u>Hybrid and Mixed Finite Element Methods</u>, ASME AMD, Vol. 73, Ed. by R.L. Spilker and K.W. Reed, pp. 39-56, 1985.

[24] Key, S.W. "A variational Principle for Incompressible and Nearly Incompressible Anisotropic Elasticity", Int. J. Solids and Structures, Vol. 5, pp. 951-964, 1969.

[25] Lee, S.W. and Pian, T.H.H. "Improvement of Plate and Shell Finite Elements by Mixed Formulations", AIAA J, Vol. 16, pp. 29-34, 1978.

[26] Lee, S.W., Wong, S.C. and Rhiu, J.J. "Study of a Nine-node Mixed Formulation Finite Element for Thin Plates and Shells", Computers and Structures, Vol. 21, pp. 1324-1334, 1985.

[27] Tang, L and Liu, Y. "Quasi-Conforming Element Techniques for Penalty Finite Element Methods", Finite Elements in Analysis and Design, Vol. 1, pp. 25-33, (1985).

[28] Tong, P. and Pian, T.H.H. "A Variational Principle and the Convergence of a Finite Element Method Based on Assumed Stress Distribution", Int. J. Solids and Structures, Vol. 5, pp. 463-472, 1969.

[29] Pian, T.H.H. "Derivation of Element Stiffness Matrices by Assumed

Stress Distributions", AIAA J. Vol. 2, pp. 1333-1336 (1964).

[30] Fraeijs de Veubeke, B. "Upper and Lower Bounds in Matrix Structural Analysis" in Matrix Methods of Structural Analysis, Ed. by B. Fraeijs de Veubeke, MacMillian, New York, pp. 165-201, 1964.

(31] Pian, T.H.H. "On Hybrid and Mixed Finite Element Methods", Proc. Invitational Sympoisum on Finite Element Method, Hefei, Anhui, China, May 19-23, 1981. Science Press, pp. 1-19, 1982.

[32] Pian, T.H.H. "Finite Element Methods by Variational Prionciples with Relaxed Continuity Requirement" in Variational Methods in Engineering, Ed. by C.A. Brebbia and H. Tottenheim, Southampton University Press, pp. 3/1-3/24, 1973.

[33] Fraeijs de Veubeke, B. "Displacement and Equilibrium Models in the Finite Element Method" in Stress Analysis, Ed. by O.C. Zienkiewicz and G.S. Holister, John Wiley & Sons, London, pp. 145-197, 1969.

[34] Pian, T.H.H. and Chen, D.P. "Alternative Ways for Formulation of Hybrid Stress Elements", Int. J. Num. Meth. Engng, Vol. 18, pp. 1679-1684, 1982.

[35] Pian, T.H.H. "Finite Elements Based on Consistently Assumed Stresses and Displacements", in Finite Elements in Analysis and Design, Vol. 1, pp. 131-140, 1985.

[36] Pian, T.H.H. and Sumihara, K. "Rational Approach for Assumed Stress Finite Elements", Int. J. Num. Meth. Engng, Vol. 20, pp. 1685-1695, 1984.

(37] Pian, T.H.H., Kang, D. and Wang C. "Hybrid Plate Elements Based on Balanced Stresses and Displacements" to be published in State-of-the-Art Texts on Finite Elements Methods in Plate and Shell Structural Analysis, Ed. by T.J.R. Hughes and E. Hinton, Pineridge Press Ltd., Swansea, Wales, 1986.

[38] Tian, Z.S., and Pian, T.H.H., "Axisymmetric Solid Elements by a Rational Hybrid Stress method", Computers and Structures, Vol. 20, pp. 141-149, 1985.

[39] Pian, T.H.H., Li, M.S. and Kang, D. "Hybrid Stress Elements Based on Natural Isoparametric Coordinates" paper for presentation at Workshop on Finite Element Methods, Beijing Institute of Aeronautics and Astronautics, June 1986.

[40] Pian, T.H.H., and Sumihara, K. "Hybrid SemiLoof Elements for Plates and Shells Based Upon a Modified Hu-Washizu Principle", Computers and Structures, Vol. 19, pp. 165-173, 1984.

[41] Tong, P. "A Family of Hybrid Plate Elements", Int. J. Num. Meth. Engng, Vol. 18, pp. 1455-1468, 1982.

[42] Punch, E.F. and Atluri, S.N. "Applications of Isoparametric Three Dimensional Hybrid-stress Finite Element with Least Order Stress Fields", Computers and Structures, Vol. 19, pp. 409-430, 1984.

[43] Pian, T.H.H. "Derivation of Element Stiffness Matrices", AIAA J., Vol. 2, pp. 576-577, 1964

[44] Wilson, E.L., Taylor, R.L., Doherty, W.P. and Ghaboussi, J. "Incompatible Displacement Models" in Numerical and Computer Methods in Structural Mechanics, Ed. by S.J. Fenves et al., Academic Press, pp. 41-57, 1973.

(45] Chen, W., and Tang, L. "Isoparametric Quasi-conforming Element" (in Chinese), Journal of Dalian Institute of Technology, Vol. 20, No. 1, pp. 63-74, 1981.

[46] Tong, P., Pian, T.H.H. and Lasry, S. "A Hybrid-element Approach to Crack Problems in Plane Elasticity", Int. J. Num. Meth. Engng, Vol. 7, pp. 297-308, 1973.

## FIGURE 1: VARIATIONAL FORMULATIONS OF FINITE ELEMENTS
### (BASED ON MATRIX DISPLACEMENT METHOD)

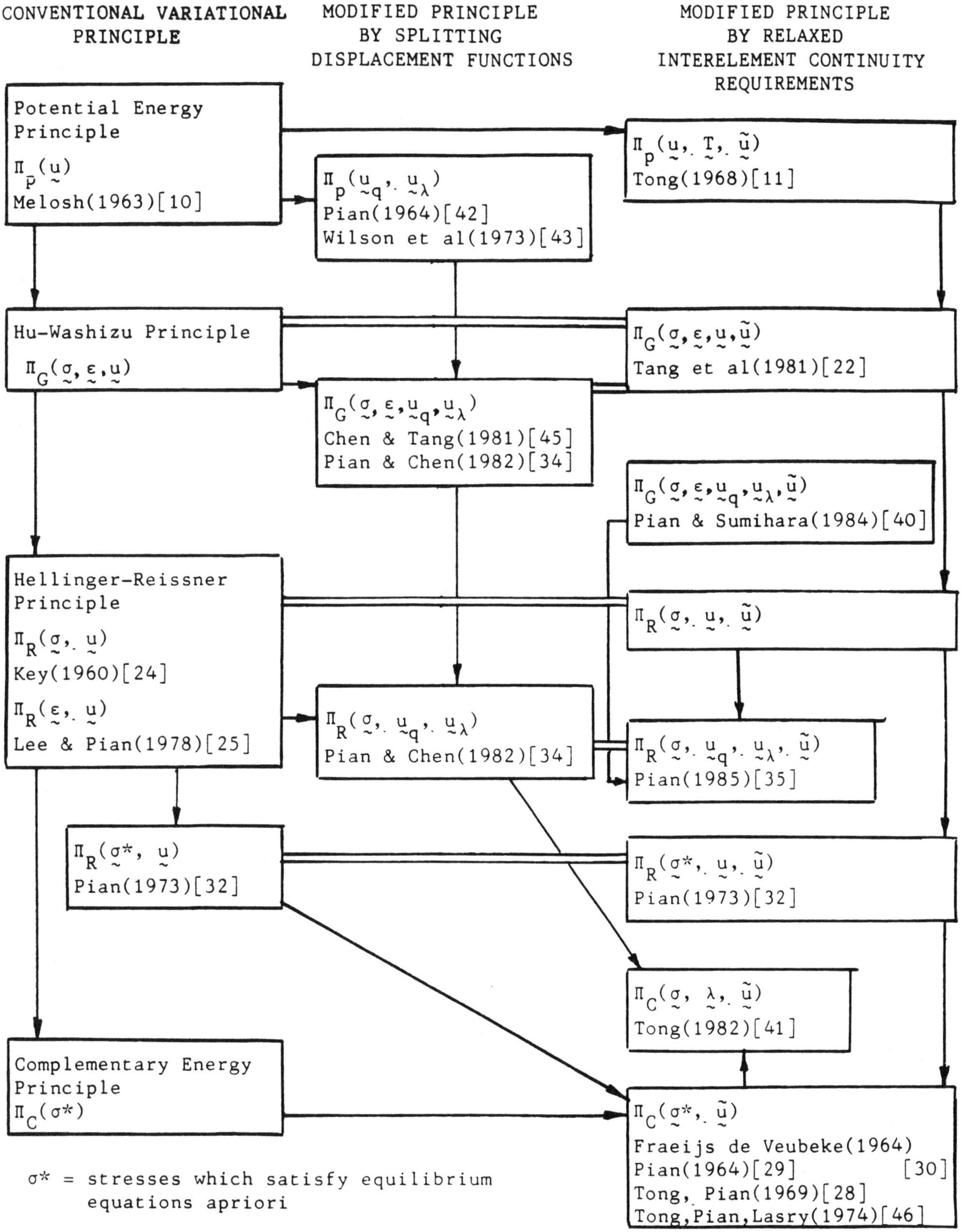

**CONVENTIONAL VARIATIONAL PRINCIPLE**

- Potential Energy Principle
  $\Pi_p(u)$
  Melosh(1963)[10]

- Hu-Washizu Principle
  $\Pi_G(\sigma, \epsilon, u)$

- Hellinger-Reissner Principle
  $\Pi_R(\sigma, u)$
  Key(1960)[24]
  $\Pi_R(\epsilon, u)$
  Lee & Pian(1978)[25]

- $\Pi_R(\sigma^*, u)$
  Pian(1973)[32]

- Complementary Energy Principle
  $\Pi_C(\sigma^*)$

**MODIFIED PRINCIPLE BY SPLITTING DISPLACEMENT FUNCTIONS**

- $\Pi_p(u_q, u_\lambda)$
  Pian(1964)[42]
  Wilson et al(1973)[43]

- $\Pi_G(\sigma, \epsilon, u_q, u_\lambda)$
  Chen & Tang(1981)[45]
  Pian & Chen(1982)[34]

- $\Pi_R(\sigma, u_q, u_\lambda)$
  Pian & Chen(1982)[34]

**MODIFIED PRINCIPLE BY RELAXED INTERELEMENT CONTINUITY REQUIREMENTS**

- $\Pi_p(u, T, \tilde{u})$
  Tong(1968)[11]

- $\Pi_G(\sigma, \epsilon, u, \tilde{u})$
  Tang et al(1981)[22]

- $\Pi_G(\sigma, \epsilon, u_q, u_\lambda, \tilde{u})$
  Pian & Sumihara(1984)[40]

- $\Pi_R(\sigma, u, \tilde{u})$

- $\Pi_R(\sigma, u_q, u_\lambda, \tilde{u})$
  Pian(1985)[35]

- $\Pi_R(\sigma^*, u, \tilde{u})$
  Pian(1973)[32]

- $\Pi_C(\sigma, \lambda, \tilde{u})$
  Tong(1982)[41]

- $\Pi_C(\sigma^*, \tilde{u})$
  Fraeijs de Veubeke(1964)
  Pian(1964)[29]         [30]
  Tong, Pian(1969)[28]
  Tong, Pian, Lasry(1974)[46]

$\sigma^*$ = stresses which satisfy equilibrium equations apriori

CATASTROPHIC FAILURE OF LAMINATED CYLINDERS UNDER INTERNAL PRESSURE

James W. Mar, Jerome C. Hunsaker Professor of Aerospace Education
Department of Aeronautics and Astronautics
Massachusetts Institute of Technology

ABSTRACT

The damage tolerance of cylindrical shells fabricated of graphite/epoxy filamentary composite materials has been studied. The damage, called "flaws", was in the form of longitudinal slits, slits at angles to the longitudinal axis, colinear slits, circular holes, long holes and holes with slits. The original series of cylinders used six plies of unidirectional prepreg in the laminate while the remainder used four plies of fabric. In one set of experiments the damage was inflicted by a guillotine which dropped a knife onto a cylinder under pressure. In another series, the flaw was pre-cut, covered with a non-intrusive patch and then monotonically pressurized to failure. It has been determined that the catastrophic failure, i.e., rapid fracture of these cylinders, can be correlated to the fracture of flat coupons of the same laminate under uni-axial tension.

INTRODUCTION

Pressure vessels fabricated of filamentary composite materials offer a large weight saving because the tensile strength to weight ratio of the advanced composite materials are significantly higher than the metals. In transport category aircraft, the fuselage is a pressure vessel which experiences pressures close to limit design values for each takeoff and landing. The airplane wing, if their pilot is skilled and bad weather is avoided, will experience stresses which are a relatively much smaller fraction of the limit design values. Thus, an understanding of the behavior of the failure modes of the fuselage under internal pressure is important. Other structures which will benefit from this understanding include the oxygen tanks used by scuba divers and the cases of solid rocket boosters.

The most recent addition to design philosophies for structural integrity is "damage tolerance". This requires the designer to design in the presence of flaws. In metals, the flaw is a crack which is well characterized mathematically and analytically. There is no analagous flaw for filamentary composite materials although there is data which shows that holes and slits affect the strength of filamentary composite materials in the same manner as do cracks in metals [1]. Designing for damage tolerance, therefore, requires data on the strengths of a material in the presence of flaws. As should be appreciated, this is more difficult to obtain than the unflawed strength data such as can be found in handbooks.

This paper presents a summary of research directed toward the rapid (sometimes referred to as "catastrophic") fracture of pressurized cylinders whose wall thicknesses are laminates of graphite/epoxy advanced composite materials. Details can be found in references 2 through 9.

THE MATERIAL SYSTEMS

The basic material used in these experiments is the Hercules AS1 fiber in the

3501-6 epoxy resin and is designated as AS1/3501-6. Two different prepreg forms were used. One was the uni-directional tape with a nominal per ply thickness of .00525 inches and was obtained in 36 inch wide broadgoods. The other was a five harness satin weave (4.3 yarns/cm) fabric with a nominal per ply thickness of .0135 inches and was obtained in widths of one meter. This fabric was also manufactured by Hercules and is designated as A370-5H/3501-6. One set of cylinder specimens, [2], was fabricated with a laminate consisting of a layup of $[\pm 45/0]_s$ in which the 0 degree directional is along the longitudinal axis of the cylinder and the 90 direction coincides with the hoop direction. Other sets of specimens used the fabric, [4], in two different layups of $(0,45)_s$ and $(45,0)_s$. The nomenclature of $(0,45)_s$ means the outside and inner plies of the cloth have their fiber directions coinciding with the longitudinal and hoop directions; the two inner plies have their fiber directions oriented at plus and minus 45 degrees to the axis of the cylinder. The other layup, $(45,0)_s$, reverses the order and has the outer plies with the fibers at plus and minus 45 degrees to the longitudinal axis.

The elastic moduli and tensile strength properties of the basic material, uni-ply and fabric, as well as the laminates are shown in Table 1.

## FABRICATION OF THE CYLINDERS

All the cylinders were laid up by hand on a 12 inch diameter, .25 inch thick wall, circular aluminum mandrel 48 inches in length (see figure 1). The first layer on the mandrel was a release layer of nonporous teflon. Next the plies of the prepreg were wrapped around the assembly in the desired sequence and angular orientation. On top of these plies which constitutes the desired laminate, there overwrapped a peel ply. Finally, other layers of porous teflon, bleeder and fiberglass air breather also overwrapped the previously described layers. This assembly was then vacuum-bagged before it was placed into an autoclave for curing in a two part cycle (see figure 2). The first part is at a temperature of 240 degree F and a pressure of 100 psi for one hour and the second is at 350 degrees for 2 hours at the same pressure. After removal from the mandrel, the cylinders were post cured in an oven at 350 degrees F for eight hours (see figure 3).

End-caps were bonded onto the graphite/epoxy thin-walled cylinders to create a pressure vessel for burst testing. The end-caps were one inch in thickness with a one-half inch deep circular groove to accept the wall of the cylinder. These end-caps were removed from each failed specimen by heating to 500 degrees F for five hours and reused.

Six cylinders of the $(0,45)_s$ fabric laminates were reinforced with 3 inch wide "belly bands" consisting of 4 uni-directional plies placed 12 inches off the centers. Three different reinforcing schemes were used. One scheme placed all four of the reinforcing plies on the outer surface of the cylinder. A second placed two plies on the outer and two plies on the inner surface. The third scheme interleaved the 4 plies such that at the "belly band location, the stacking sequence of uniply and fabric resulted in a laminate which was symmetric as follows: 0 fabric, uni-ply, 45 fabric, uni-ply, uni-ply, 45 fabric, uni-ply, 0 fabric. The intent of the reinforced specimens was to determine if the fracture could be arrested.

## METHODS OF TESTING

Two methods of testing have been used. The initial series of tests, [2] & [3], dropped a guillotine with a given size of blade onto the cylinder while it was under pressure. If failure did not occur, the slit caused by the guillotine was patched, the specimen was re-pressurized and the next size of guillotine was used, This was continued until failure occured. Figure 4 is a photograph of some of the blade sizes which were used. Figure 5 shows the testing set-up with the original guillotine mechanism. This was later modified to look like the French-style of guillotine (see figure 6).

The guillotine method of testing required several tests to obtain the failure pressure for rapid fracture and as can be seen in figure 4, the increment in blade size means there is an uncertainty of .25 inches in the length of the flaw at failure. A second method of testing was devised in which flaws were pre-cut through the wall of the cylinder with jeweler's saws and a patch (see figures 7 and 8) was placed over the flaw on the inner surface of the cylinder. As can be seen in figures 7 and 8, the patch consists of several thin sheets of aluminum about .015 inches thick arranged into an overlapping pattern such that the flaw became pressure tight with sealant applied to the edges of the patch. It was determined that the patch did not significantly affect the local stiffnesses in the vicinity of the flaw so that the results obtained with the pre-cut flaws and guillotine-inflicted flaws are deemed to be the same. This method of testing permitted the montonic increase of pressure until rapid fracture occurred.

## THE FLAWS

The flaws inflicted by the guillotine are called "slits" and these were mostly oriented along the longitudinal axes of the cylinders. The pre-cut flaws included slits along the longitudinal axes, slits at angles to the longitudinal axes, collinear slits along the longitudinal axes, circular holes, long holes and holes with with slits. Sketches of these flaws will be shown with the experimental results.

## TEST PROCEDURES

Testing was accomplished in a large blast chamber which is designed to accommodate the energy release of equivalent to that from the explosion of two pounds of TNT. Pressures used in these tests have been as high as 230 psi and at this pressure, the 12 inch diameter end caps are subjected to a load of 26,000 pounds. The catastrophic failure of the specimens releases a lot of stored energy from the compressed nitrogen gas which was used as the medium for pressurization. Data acquisition was performed by a PDP-11 computer from pressure and strain gage transducers. An important step in the test procedure was the post mortem determination of the fracture paths; this required a painstaking retrieval and re-assembly of the failed pieces of each cylinder (see figures 9 and 10).

## THE EXPERIMENTAL RESULTS

The experimental data are shown in two tables. Table 2 summarizes the data obtained from experiments using the "guillotine" procedure and Table 3 summarizes the data from cylinders with pre-cut flaws.

Sketches of the typical failure paths initiated by the various flaw configurations are shown in figures 11 through 18.

Strains in the laminates are typical of those being used in aircraft structures. In the $[\pm 45/0]_s$ laminates, the hoop and longitudinal strains at an internal pressure of 100 psi are approximately + .0025 and - .0002, respectively. The negative value of the longitudinal strain is a consequence of the large poisson ratio (see table 1) of this particular laminate and the two-to-one ratio of the hoop to longitudinal stresses in a pressurized cylinder. The hoop and longitudinal strains in the (0,45)s and (45,0)s laminates at a pressure of 100 psi are approximately +.0012 and +.00025, respectively.

METHODOLOGY AND CORRELATION

The methodology used for correlating the experimental results is based upon the observation that holes and slits cause rapid fracture in filamentary composite materials in the same manner as do cracks in metals. This has led to the following expression for the fracture of graphite/epoxy laminates [5, 10, 11]:

$$\sigma_{fp} = H_c (2a)^{-.28} \qquad (1)$$

where
$\sigma_{fp}$ is the flat plate fracture stress

$H_c$ is the fracture parameter experimentally determined

$2a$ is the length of the flaw

.28 is the exponent for AS1/3501-6 material

It should be noted that equation 1 has the same mathematical form as the expression used to predict the fracture stress of metals in the presence of cracks of length 2a except that the exponent for graphite/epoxy is .28 instead of .5 as it is for isotropic materials. The $H_c$ fracture parameter is akin to the fracture toughness of isotropic materials; note that the dimension of $H_c$ is (stress) x (length)$^{.28}$. The fracture parameter, $H_c$, is obtained by determining the fracture stresses of coupon specimens which have flaws of different sizes. For the $(0,45)_s$ and $(45,0)_s$ specimens (these are the laminates fabricated with fabric), the value of $H_c$ is 44.7 ksi (inch)$^{.28}$ [4]. Experiments have not been conducted to determine $H_c$ for the cylinders which used the $[\pm45/0]_s$ laminate. Thus, the correlation to be described applies only to the specimens fabricated of the cloth fabric, i.e., the cylinders using the laminates designated as $(0,45)_s$ and $(45,0)_s$.

There are local "stress intensifications" at the tips of slits and other forms of flaws which go through the walls of circular cylindrical shells. The term "stress intensifications" is used because if the cylinders were constructed of isotropic materials and if the flaws were cracks, then it would be proper to use the terminology of linear elastic fracture mechanics to say that the stress intensity at the tip of a crack is increased by the curvature of the cylinder [12]. Physically, the presence of a flaw in a pressurized cylinder interrupts the membrane state of stress which results in local bending. Thus, equation 1 must be modified to account for the "stress intensification" due to the local bending caused by the presence of the flaw. This has been accomplished by introducing a factor K, [5], according to the equation

$$\sigma_{fc} = \sigma_{fp} / K \qquad (2)$$

where $\sigma$ is the fracture stress in the hoop direction and K is the factor which is dependent upon the geometry of the flaw. There will be a different K for slits,

slits at an angle, colinear slits, circular holes, long holes and holes with slits.

$$K_{slit} = (1 + .317 \lambda^2)^{\frac{1}{2}} \tag{3}$$

$$\lambda^2 = \frac{a^2 [12(1-\nu^2)]^{\frac{1}{2}}}{Rh} \tag{4}$$

$$K_{angle} = K_1 + K_2 - 1 \tag{5}$$

$$K_1 = (1 + .317 \lambda_1^2)^{\frac{1}{2}} \tag{6}$$

$$K_2 = (1 + .05 \lambda_2^2)^{\frac{1}{2}} \tag{7}$$

$$\lambda_2^1 = \frac{R}{\rho_1} \lambda^2 \tag{8}$$

$$\lambda_2^2 = \frac{R}{\rho_2} \lambda^2 \tag{9}$$

$$\rho_1 = \frac{R}{\sin^2(\theta + \frac{\pi}{2})} \tag{10}$$

$$\rho_2 = \frac{R}{\sin^2 \theta} \tag{11}$$

$$K_{two} = 2.18 \quad \text{for} \quad \frac{a}{c} \quad .5 \, , \, \lambda = 3.1 \tag{12}$$

$$K_{two} = -7.916 + 48.455 \left(\frac{a}{c}\right) - 75.788 \left(\frac{a}{c}\right)^2$$
$$+ 38.529 \left(\frac{a}{c}\right)^3 \text{ for } .5 \leq \frac{a}{c} \leq .89, \lambda = 3.1 \tag{13}$$

$$K_{hole} = K_{slit} \text{ for } \lambda \leq 3.677 \tag{14}$$

$$K_{hole} = .532 + .696 \lambda - .0586 \lambda^2$$
$$\text{for } 3.677 < \lambda \leq 8.202 \tag{15}$$

$$K_{long} = K_{slit} \text{ for } \lambda \leq 3.677 \tag{16}$$

$$K_{long} = K_{slit} - \left(\frac{d}{a}\right)\left(K_{slit} - K_{hole}\right)$$
$$\text{for } .\lambda > 3.677 \tag{17}$$

$$K_{plus} = K_{slit} \text{ for } \lambda \leq 3.677 \tag{18}$$

$$K_{plus} = K_{slit} - \left(\frac{r}{a}\right)\left(K_{slit} - K_{hole}\right)$$
$$\text{for } \lambda > 3.677 \tag{19}$$

where R is the radius of the cylindrical shell (R = 6.028 inches); h is the wall thickness of the shell (.05512 inches); is poisson's ratio for the layup (.315), $K_{slit}$ is the factor for longitudinal slits in the cylindrical wall; $K_{angle}$ is the factor for slits oriented at an angle θ to the longitudinal axis; $K_{two}$ is the factor for two colinear slits each of length 2a spaced 2c on centers so that the distance between adjacent tips is 2(c-a); $K_{hole}$ is the factor for circular holes of diameter D; $K_{long}$ is the factor for a flaw with dimensions of 2d in the circumferential direction and 2a in the longitudinal direction; $K_{long}$ is the factor for a flaw consisting of two longitudinal slits emanating from a circular hole of radius r with an overall length from tip-to-tip of the slits of 2a.

The calculated failure pressures according to the cited formulae are given by the expression

$$p_{theory} = \frac{h}{R} \frac{H_c (2a)^{-.28}}{K} \tag{20}$$

Results are shown in table in the columns marked "p theory" in tables 2 and 3.

The measured failure pressures can be used to back-calculate an experimental value of the $\sigma_{fp}$ stresses by using the K factors. These calculations are shown in figure 19 against the theoretical values.

Figure 20 shows an empirical correlation of the cylinders with the $[\pm 45/0]_s$ laminates. The equation of the empirical correlation is as follows:

$$p_{theory} = 95(2a)^{-.46} \tag{21}$$

## DISCUSSION

The data from the tests of cylinders with colinear slits are not shown in figure 19 because the other parameter, a/c, cannot be included in the same plot. Of special interest is the size of the ligament between the two slits such that the two slits behave as one slit [7,9].

None of the three schemes of reinforcement were able to arrest the fracture or to increase the failure presure. Two of the schemes did significantly change the path of the fracture (figures 21 and 22) [4].

## ACKNOWLEDGEMENTS

It was my very good fortune to have been associated with Ray Bisplinghoff both as a professional colleague and friend. He brought me into the Department of Aeronautical Engineering and provided the initial stimuli and the challenges which have shaped my career. My wife and I saw him periodically during the past few years and we were both heartened and saddened while he battled against his illness. He will be sorely missed.

The author wishes to acknowledge the support of the United States Airforce AFSC/Aeronautical Systems Division, Wright Patterson Air Force Base. Dr. Stephen W. Tsai and Dr. James Whitney were the contract monitors.

## REFERENCES

1. Mar, J.W. and Lin, K.Y., "Fracture Mechancis Correlation for Tensile Failure of Filament Composites with Holes," J. of Aircraft, Vol. 14, July 1977, pp. 703-704.

2. Rogers, J.D., "An Investigation of the Damage Tolerance Characteristics of Graphite/Epoxy Pressure Vessel," Master's Thesis, Dept. of Aeronautics and Astronautics, Massachusetts Institute of Technology, Cambridge, September, 1981.

3. Graves, M.J., "The Catastrophic Failure of Pressurized Graphite/Epoxy Cylinders," Ph.D. Thesis, Dept. of Aeronautics and Astronautics, Massachusetts Institute of Technology, Cambridge, September, 1982.

4. Graves, M.J., and Lagace, P.A., "Damage Tolerance of Composite Cylinders," Composite Structures, Vol.4, (1985), pp.75-91.

5. Chang, S.G. and Mar, J.W., "The Catastrophic Failure of Pressurized Graphite/Epoxy Cylinders Initiated by Slits at Various Angles," J. of Aircraft, Vol. 22, No. 6, June 1985.

6. Chang, S.G., and Mar, J.W., "The Catastrophic Failure of Pressurized Graphite/Epoxy Cylinders Initiated by Slits at Various Angles," TELAC REPORT 84-5, Dept. of Aeronautics and Astronautics, Massachusetts Institute of Technology, May, 1984.

7. Chang, S.G., and Mar, J.W., "The Catastrophic Failure of Pressurized Graphite/Epoxy Cylinders Flawed with Slits and Holes," TELAC REPORT 84-13, Department of Aeronautics and Astronautics, Massachusetts Institute of Technology, March, 1984.

8. Trop, D.W., "Damage Tolerance of Internally Pressurized Sandwich Walled Graphite/Epoxy Cylinders," Masters Thesis, Department of Aeronautics and Astronautics, Massachusetts Institute of Technology, February, 1985.

9. Zhang, X. and Mar, J.W., "The Catastrophic Failure of Pressurized Graphite/Epoxy Cylindrical Shells Flawed with two Colinear Slits," Presented at the International Symposium on Composite Materials and Structures in Beijing, China, June, 10-13, 1986.

10. Lin, K.Y., "Fracture of Filamentary Composite Materials," Ph.D. Thesis, Dept. of Aeronautics and Astronautics, Massachusetts Institute of Technology, Cambridge, February, 1977.

11. Lagace, P.A., "Static Tensile Fracture of Graphite/Epoxy," Ph.D. Thesis, Massachusetts Institute of Technology, April, 1982.

12. Folias, E.S., "Asymtotic Approximation to Crack Problems in Shells," Mechanics of Fracture, Vol. 3, Noordhoff International, Leiden, the Netherlands, 1977, pp. 117-160.

FIGURE 1    ALUMINUM MANDREL

FIGURE 2    ASSEMBLY IN AUTOCLAVE

FIGURE 3    CURED CYLINDER

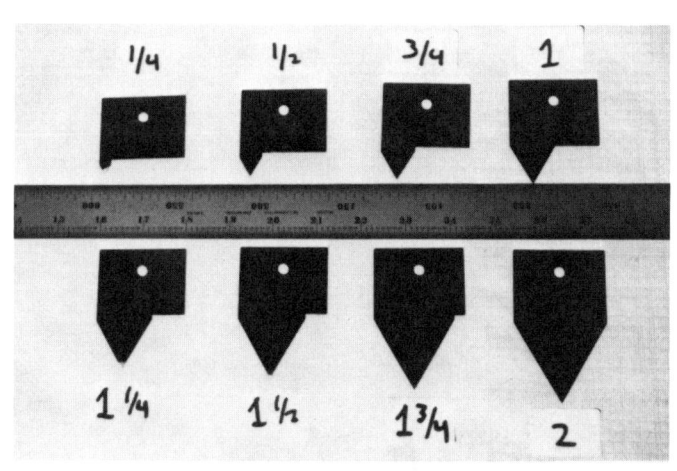

FIGURE 4    BLADE SIZES

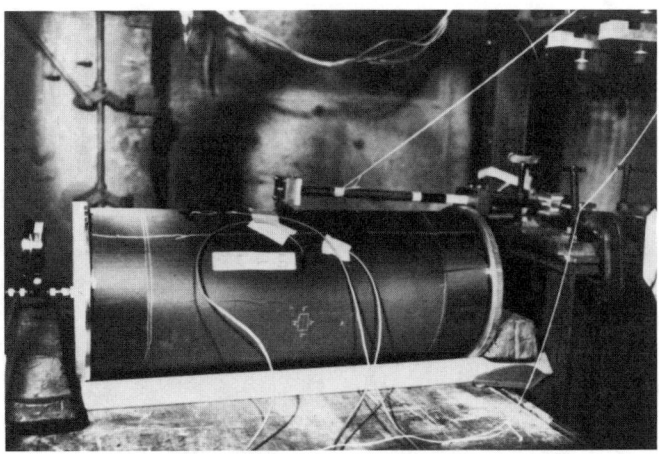

FIGURE 5    TEST SET UP

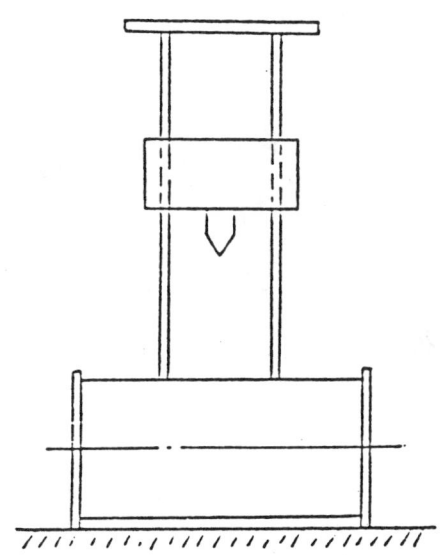

FIGURE 6    GRAVE'S GUILLOTINE

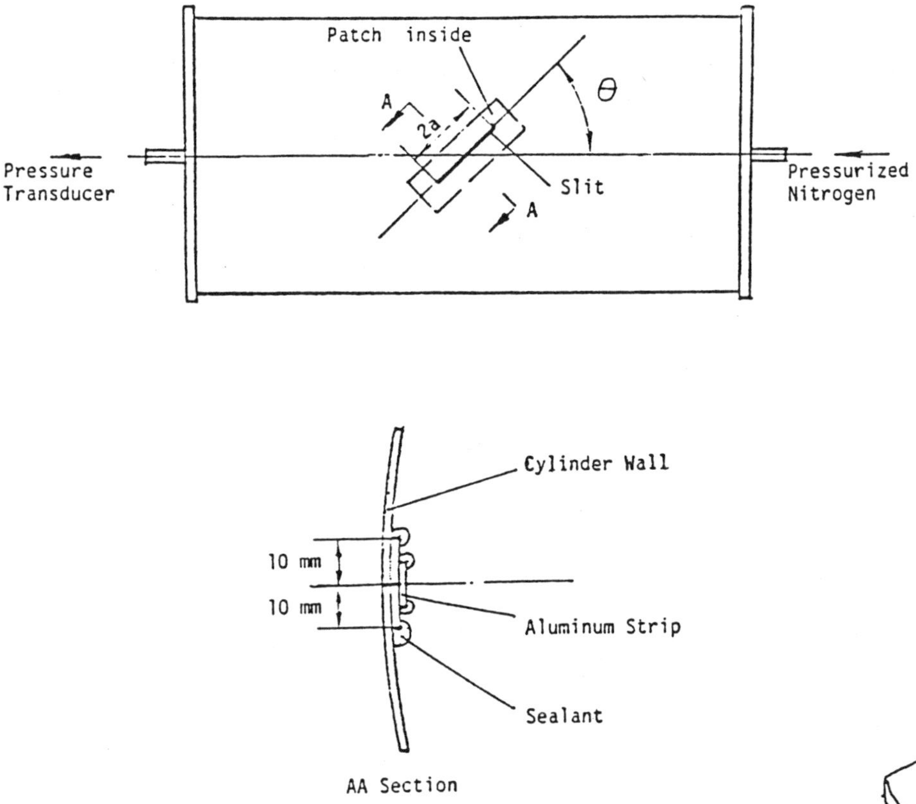

FIGURE 7    PATCH DETAILS

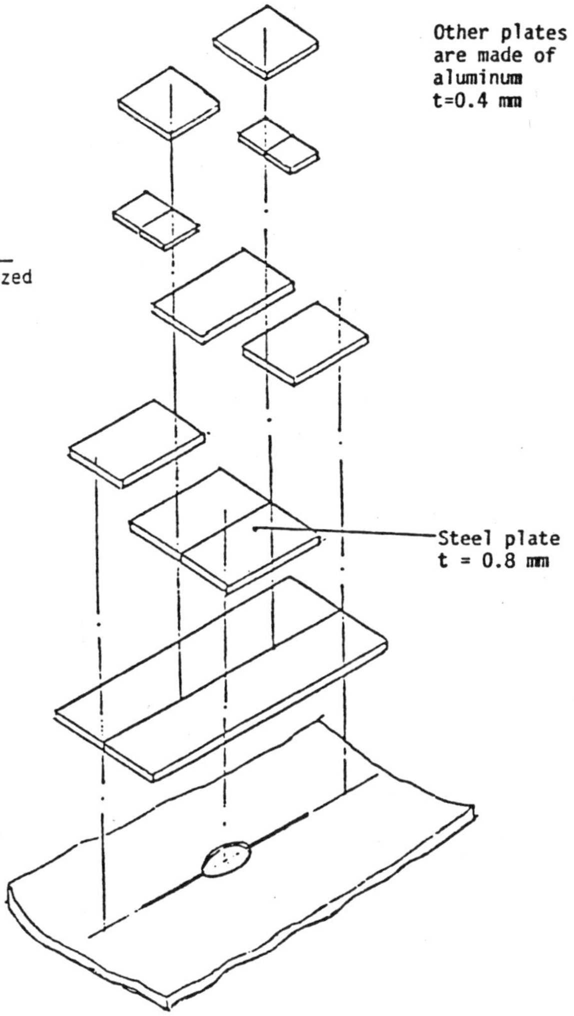

FIGURE 8    PATCH FOR HOLE AND SLITS

FIGURE 9   FAILURE OF (0,45)s CYLINDER

FIGURE 10   FAILURE MODE WITH SLIT

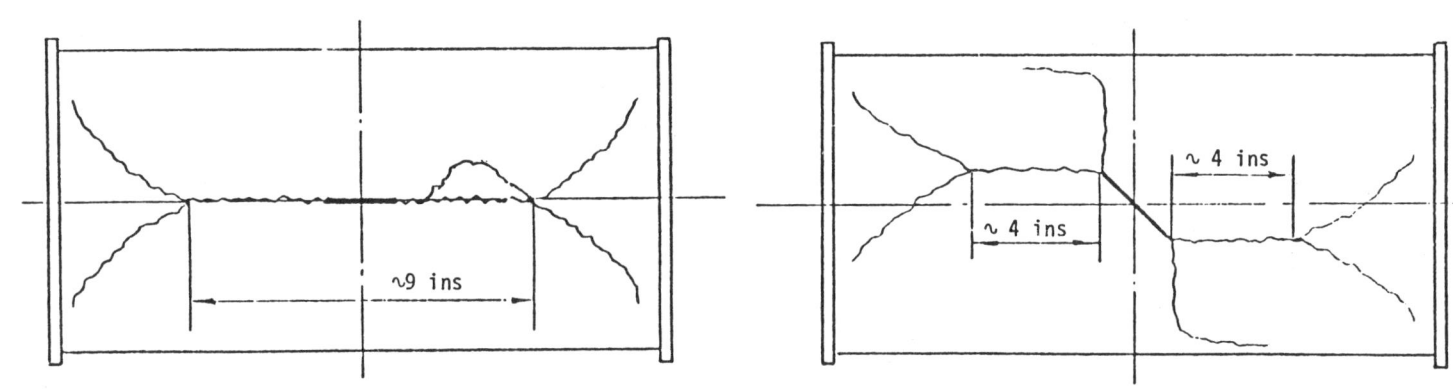

FIGURE 11   FAILURE MODE SLIT AT 0 DEGREES

FIGURE 12   FAILURE MODE SLIT AT 45 DEGREES

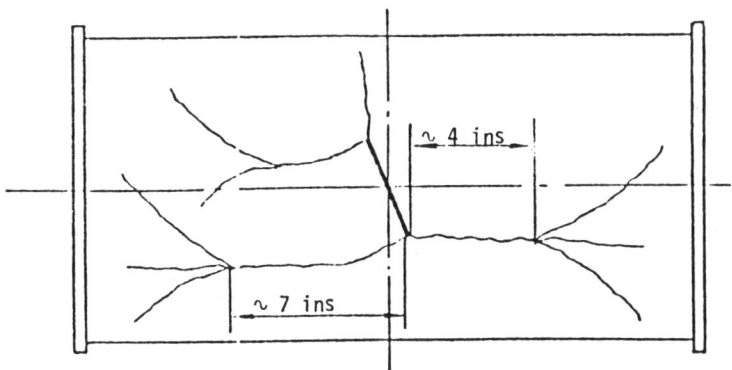

FIGURE 13    FAILURE MODE SLIT AT 67.5 DEGREES

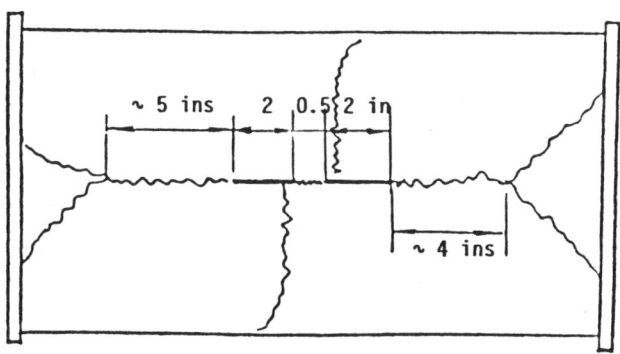

FIGURE 14    FAILURE MODE COLINEAR SLITS

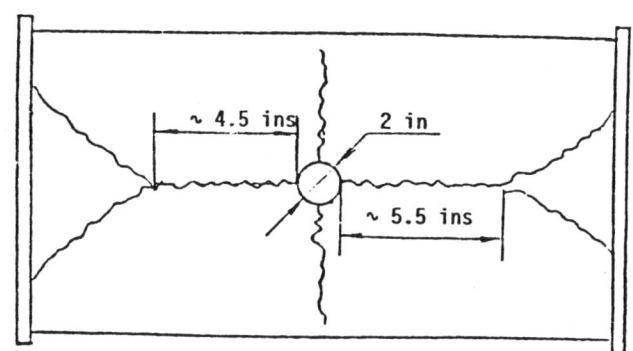

FIGURE 15    FAILURE MODE CIRCULAR HOLE

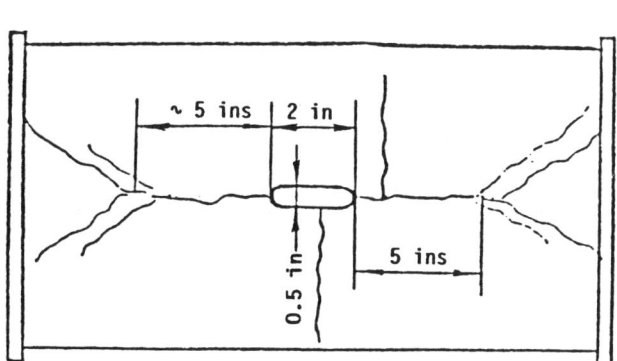

FIGURE 16    FAILURE MODE LONG HOLE

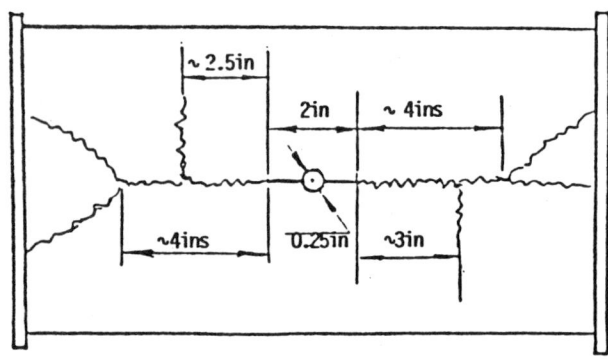

FIGURE 17  FAILURE MODE HOLE WITH SLITS

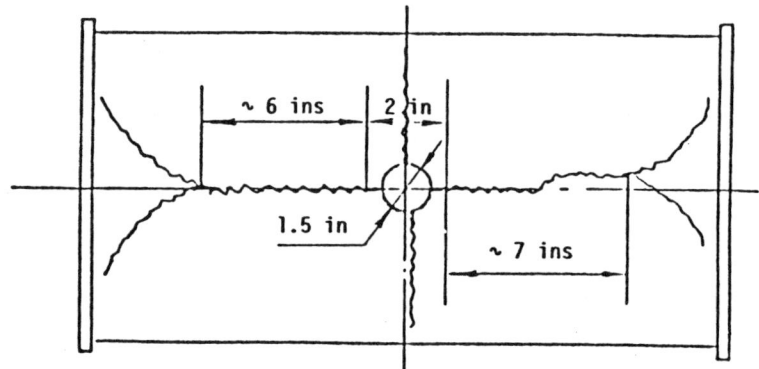

FIGURE 18  FAILURE MODE HOLE WITH SLITS

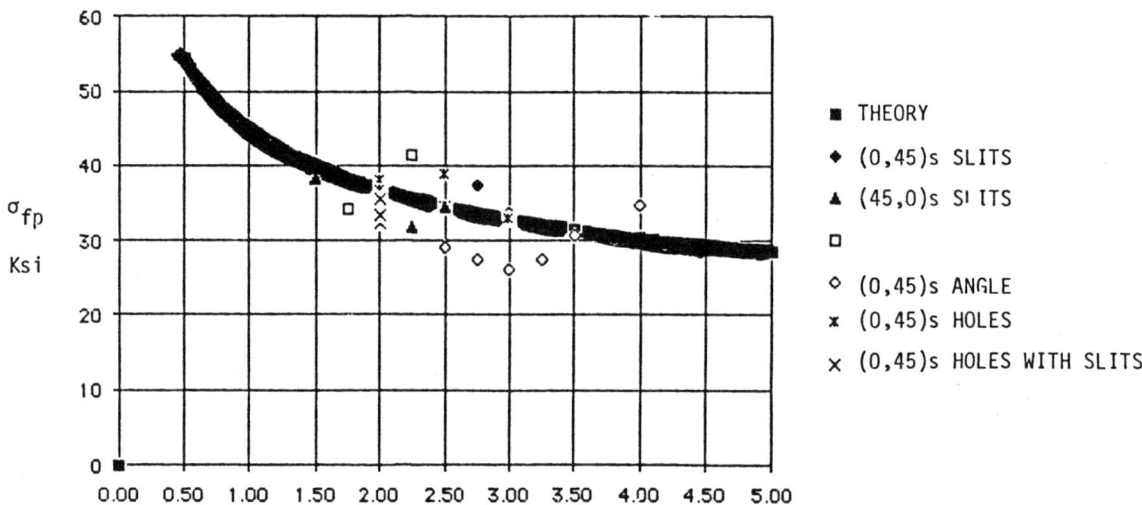

FIGURE 19  FLAW SIZE INCHES

- ■ THEORY
- ♦ (0,45)s SLITS
- ▲ (45,0)s SLITS
- □
- ◊ (0,45)s ANGLE
- × (0,45)s HOLES
- × (0,45)s HOLES WITH SLITS

$\sigma_{fp}$ Ksi

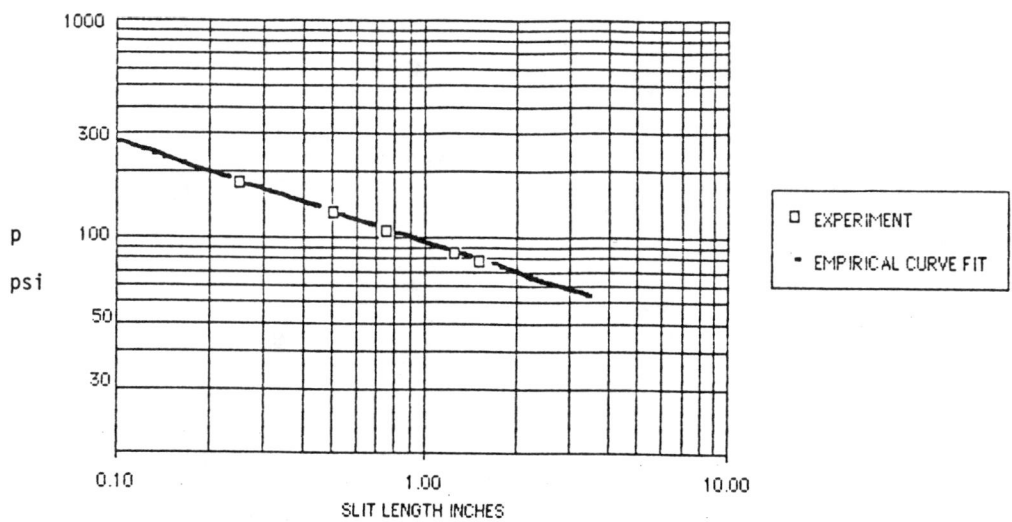

FIGURE 20  CORRELATION OF [+-45/0] EXPERIMENTS

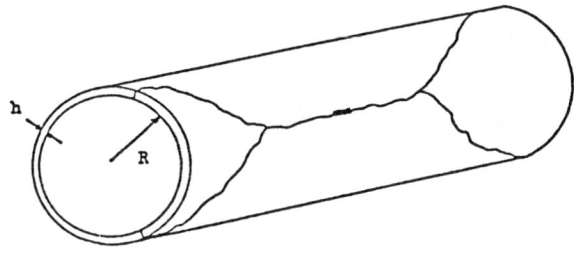

CYLINDER FAILURE MODE -- $(0,45)_s$ LAMINATE

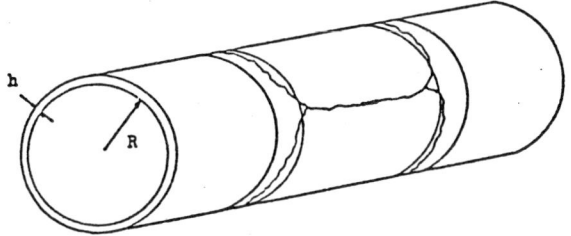

CYLINDER FAILURE MODE -- $(0,45)_s$ LAMINATE
REINFORCED: CASE 1 and CASE 2

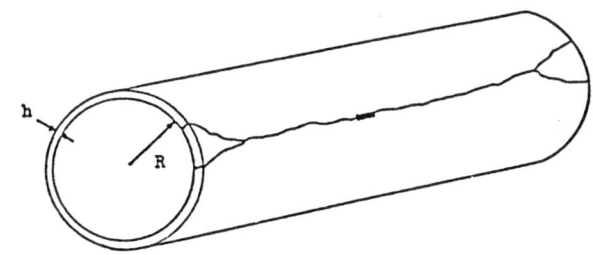

CYLINDER FAILURE MODE -- $(45,0)_s$ LAMINATE

FIGURE 21  FAILURE MODES FOR $(0,45)_s$ and $(45,0)_s$

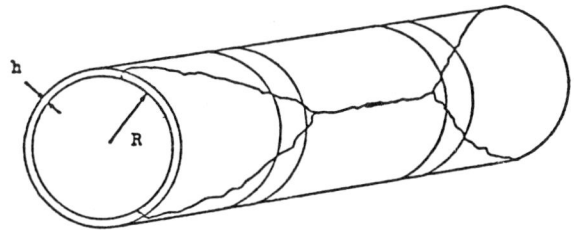

CYLINDER FAILURE MODE -- $(0,45)_s$ LAMINATE
REINFORCED: CASE 3

FIGURE 22  FAILURE MODES FOR REINFORCED CYLINDERS

| SYSTEM TYPE LAMINATE | | AS1-3501-6 UNI-PLY [0] | AS1-3501-6 UNI-PLY [±45/0]s * | AS1-3501-6 A370-5H FABRIC "(0) | AS1 3501-6 A370-5H FABRIC "(0,45) | AS1 3501-6 A370-5H FABRIC "(45,0) |
|---|---|---|---|---|---|---|
| E11 | Msi | 18.9 | 8.33 | 10.4 | 7.75 | 7.75 |
| E22 | Msi | 1.52 | 3.7 | 10.4 | 7.75 | 7.75 |
| NU12 | | 0.28 | 0.67 | 0.06 | 0.32 | 0.32 |
| G12 | Msi | 0.87 | 3.54 | 0.65 | | |
| FTU,long | Ksi | 241 | 31.8 | 128 | 90 | 90 |
| FTU,transv | Ksi | 7.8 | | 115 | | |

*calculated from uni-ply values

TABLE I - DATA ON MATERIAL SYSTEMS

| MATERIAL SYSTEM AND NOTES | LAY-UP | FLAW CONFIGURATION | FAILURE PRESSURE PSI | FLAW SIZE INCHES | P THEORY PSI |
|---|---|---|---|---|---|
| AS1/3501-6 UNI PREPREG Reference | [+-45/0]s " " " " " " | TRANSVERSE SLIT LONGITUDINAL SLIT " " " " " | 110.5 78.5 87.3 106.0 133.4 178.5 | 1.50 1.50 1.25 0.75 0.50 0.25 | |
| A370-5H/3501-6 FABRIC PREPREG References 3 and 4 " " " " " " " " " " " " " | (0,45)s " " " " " (45,0)s " " " " " " | LONGITUDINAL SLIT " " " " " LONGITUDINAL SLIT " " " " " " | 130 130 170 170 210 210 130 130 170 170 170 210 210 | 2.75 2.50 2.00 2.00 1.50 1.50 2.25 2.50 2.00 2.00 2.00 1.50 1.50 | 117 130 166 166 220 220 146 130 166 166 166 220 220 |
| ply reinforced 4 plies out " ply reinforced 2 plies out " ply reinforces 4 plies inter " | (0,45)s " " " " " | LONGITUDINAL SLIT " " " " " | 170 170 170 170 170 170 | 1.75 1.75 2.25 2.20 2.00 2.25 | 190 190 146 166 166 146 |

TABLE II - DATA FROM "GUILLOTINE" EXPERIMENTS

| FLAW CONFIGURATION | FAILURE PRESSURE PSI | FLAW SIZE INCHES | NOTES | P THEORY PSI |
|---|---|---|---|---|
| SLITS AT ANGLES " " " " " " " | 145 122 175 153 135 131 137 239 | 2@0 3@22.5 2.5@45 2.75@45 3@45 3.25@45 3.5@45 4@67.5 | FIRST NUMBER IS LENGTH OF SLIT SECOND NUMBER IS ANGLE IN DEGREES | 166 119 208 186 169 154 141 209 |
| COLINEAR SLITS " " " " " " | 145 150 160 105 70 140 | 2 2-.5-2 2-1-2 2-.25-2 2-.125-2 2-2-2 | THIS IS SINGLE SLIT *FIRST NUMBER IS LENGTH OF SLIT **SECOND NUMBER IS LENGTH OF LIGAMENT ***THIRD NUMBER IS LENGTH OF SLIT | 166 163 159 136 127 155 |
| CIRCLUAR HOLES " " " | 170 165 120 150 | 2 2 3 2.5 | NUMBER IS DIAMETER OF HOLE | 162 162 120 135 |
| LONG HOLES " " | 185 189 150 | 2-.5 2-1 2.5-2 | *FIRST NUMBER IS OVERALL LENGTH *SECOND NUMBER IS WIDTH OF LONG HOLE | 166 165 135 |
| HOLES WITH SLITS " " | 157 150 160 | .25-2 1-2 1.5-2 | *FIRST NUMBER IS DIAMETER OF HOLE **SECOND NUMBER IS OVERALL LENGTH FORM TIP-TO-TIP- OF SLITS | 166 166 166 |

TABLE III - DATA FROM "PRE-CUT" EXPERIMENTS

# NATURAL FREQUENCIES OF TRIANGULAR PLATES USING CHARACTERISTIC ORTHOGONAL POLYNOMIALS IN RAYLEIGH-RITZ METHOD

R.B. Bhat
Department of Mechanical Engineering
Concordia University
Montreal, Quebec, Canada

## ABSTRACT

Natural frequencies of triangular plates are obtained employing a set of characteristic orthogonal polynomials involving two variables in Rayleigh-Ritz method. These orthogonal polynomials are constructed using Gram-Schmidt orthogonalization procedure. The first member of the set is constructed so as to satisfy the geometric boundary conditions of the triangular plate. The natural frequencies are compared with those obtained previously by finite element methods and are found to agree very well.

## INTRODUCTION

Vibration of rectangular and circular plates have been studied extensively in literature and well documented [1]. In some cases solution is separable in terms of two coordinates describing the domain, such as in the case of rectangular plates with atleast two opposite sides simply-supported. Even when that is not the case, it is possible to assume a deflection function as the product of two functions, one in each of the two variables, and solve using either Rayleigh-Ritz or Galerkin's techniques.

Studies on triangular plates and polygonal plates are rather limited since the two variables are implicitly related and are not separable. Earlier work on the triangular plates are restricted to some regular shapes such as delta (right angle) plates [2-3], isosceles and equilateral plates [4-6]. Experimental determination of natural frequencies of swept back plates was carried out by Gustafson and Stokey [7]. Cowper et al [8] used high precision triangular plate bending elements to study triangular plates. Mirza and Bijlani [9] investigated triangular plates, with one edge fixed and the other two free, using finite element mehtods. They studied the plate behavior for different plate configurations and provided mode shapes for one case.

Pairs of characteristic orthogonal polynomials in single variables were used by Bhat [10,11] to study the vibrations of rectangular plates following a Rayleigh-Ritz approach. He also used the method of characteristic orthogonal polynomials to study the vibrations of a tapered cantilever beam [12] and the plate deflections under various types of static loads [13].

In the present study a set of orthogonal polynomials in two variables are constructed in the triangular domain by applying Gram-Schmidt orthogonalization process [14] and subsequently using this orthogonal set as deflection functions in Rayleigh-Ritz Method to obtain the natural frequencies of the plate. Results obtained by applying the method on triangular plates are presented for different configurations of the triangular plate and are compared with those published in literature.

## CHARACTERISTIC ORTHOGONAL POLYNOMIALS

An ordered polynomial in two variables can be constructed by using a sequence of monomials

$$1, x, y, x^2, xy, y^2, x^3, x^2y, xy^2, \ldots x^n, x^{n-1}y, \ldots, x^{n-k}y^k \quad (1)$$

In a triangular domain there are six geometrical boundary conditions and the first polynomial in the orthogonal set will be chosen such that it satisfies all the six geometrical boundary conditions. The higher members of the orthogonal set are constructed by applying Gram-Schmidt orthogonalization. Let $\phi_1(x,y)$ be the first member polynomial satisfying the geometrical boundary conditions. If, for example, $\phi_1(x,y)$ is of the form

$$\phi_1(x,y) = c_0 + c_1 x + c_2 y + c_3 x^2 + c_4 xy + c_5 y^2 + c_6 x^3 \quad (2)$$

the next member in the orthogonal set, $\phi_2(x,y)$ can be constructed by expressing $\phi_2$ as

$$\phi_2(x,y) = x^2 y + a_{21} \phi_1(x,y) \quad (3)$$

Note that the term $x^2 y$ appears after $x^3$ in the sequence of monomials in Eq. (1). Orthogonality condition between $\phi_1(x,y)$ and $\phi_2(x,y)$ requires that

$$\iint \phi_i(x,y) \cdot \phi_j(x,y) \, \gamma(x,y) \, dx \, dy \quad (4)$$

$$= 0 \text{ if } i \neq j$$

$$= \mu_i \text{ if } i = j$$

where $\gamma(x,y)$ is a weight function.

Making use of this property, the constant $a_{21}$ in Eq. (3) is obtained as

$$a_{21} = \frac{-\iint x^2 y \, \gamma(x,y) \, \phi_1(x,y) \, dx \, dy}{\iint \gamma(x,y) \, \phi_1^2(x,y) \, dx \, dy} \quad (5)$$

The third member $\phi_3$ can be expressed as

$$\phi_3(x,y) = xy^2 + a_{32} \phi_2(x,y) + a_{31} \phi_1(x,y) \quad (6)$$

The polynomial $\phi_3$ must be orthogonalized to both $\phi_2$ and $\phi_1$ to obtain the constants $a_{32}$ and $a_{31}$. This process is continued until required number of orthogonal polynomials are constructed. Unlike in the case of polynomials in single variables where such orthogonal polynomials have a recurrence relation involving only three polynomials [15], in the case of polynomials in two variables, orthogonalization must be carried out with every other member in the set.

If the plate is uniform throughout, the weight function $\gamma(x,y)$ in Eq. (4) can be chosen as unity. If the plate thickness varies in any fashion, such a variation can be included in the weight function in constructing the orthogonal set.

## EIGENVALUE PROBLEM

The geometry of the triangular plate is shown in Fig. 1. The deflection of a triangular plate undergoing free flexural vibration can be expressed in terms of the characteristic orthogonal polynomials as

$$W(x,y) = \sum_{i=1}^{n} c_i \phi_i(x,y) \tag{7}$$

where $x = \xi/u$, $y = \eta/a$, a and b are the sides of the triangle as in Fig. 1 and $\xi$ and $\eta$ are the cartesian coordinate system. The kinetic and strain energies of the plate are given by

$$T_{max} = \frac{1}{2} \rho h a^2 \omega^2 \iint W^2(x,y) \, dx \, dy \tag{8}$$

$$U_{max} = \frac{1}{2} D a^2 \iint [W_{xx}^2 + W_{yy}^2 + 2\nu W_{xx} W_{yy} + 2(1-\nu) W_{xy}^2] \, dx \, dy \tag{9}$$

where
- $\rho$ is density of the plate material
- h is thickness of the plate
- D is flexural rigidity of the plate
- $\nu$ is the Poisson's ratio

and the subscripts x and y refer to differentiation with respect to the subscript as many number of times as the subscripts appear.

Substituting the deflection function into the kinetic and strain energy expressions and minimizing the Rayleigh quotient with respect to the coefficients $c_i$ yields the eigenvalue problem

$$\sum_i [E_{ij} - \lambda F_{ij}^{(0,0,0,0)}] c_i = 0 \tag{10}$$

where

$$E_{ij} = F_{ij}^{(2,0,2,0)} + F_{ij}^{(0,2,0,2)} + \nu [F_{ij}^{(0,2,2,0)} + F_{ij}^{(2,0,0,2)}]$$
$$+ 2(1-\nu) [F_{ij}^{(1,1,1,1)}]$$

$$F_{ij}^{(m,n,r,s)} = \iint [d^m \phi_i(x,y)/dx^m] \cdot [d^n \phi_i(x,y)/dy^n]$$
$$\cdot [d^r \phi_j(x,y)/dx^r][d^s \phi_j(x,y)/dy^s] \, dx \, dy$$

where
- $\lambda = \rho h \omega^2 a^4/D$
- $i,j = 1,2,3,\ldots$
- $m,n,r,s = 0,1,2$

Solution of the eigenvalue problem will provide the natural frequencies and the mode shapes.

The integration of expressions in the triangular domain is facilitated by transformation into area coordinates $L_1$, $L_2$, $L_3$. The transformation relations are

$$\begin{Bmatrix} x \\ y \\ 1 \end{Bmatrix} = \begin{bmatrix} x_1 & x_2 & x_3 \\ y_1 & y_2 & y_3 \\ 1 & 1 & 1 \end{bmatrix} \begin{Bmatrix} L_1 \\ L_2 \\ L_3 \end{Bmatrix} \qquad (11)$$

Integration of expressions over the triangular domain in terms of the area coordinates are given by

$$\iint L_1^{t_1} L_2^{t_2} L_3^{t_3} \, dx \, dy = \frac{t_1! \, t_2! \, t_3! \cdot 2\Delta}{(t_1 + t_2 + t_3 + 2)!} \qquad (12)$$

where $\Delta$ is the area of the triangle and is given by

$$\Delta = \iint dx \, dy = \frac{ab}{2} \begin{vmatrix} 1 & x_1 & y_1 \\ 1 & x_2 & y_2 \\ 1 & x_3 & y_2 \end{vmatrix} \qquad (13)$$

## NUMERICAL RESULTS

Natural frequencies of triangular plates, shown in Fig. 1, with the edge AB clamped and edges AC and BC free, are computed using the above approach, for different plate configurations keeping the edge lengths a and b same. Table 1 shows the convergence of the natural frequencies with increasing number of terms used. A value of 0.3 has been used for the Poisson's ratio in the numerical computations. Table 2 provides the first six natural frequencies for different angles θ, and are compared with values available in literature. 21 terms have been used in obtaining the results of Table 2. The finite element results in Ref. [9] have considered 45 degrees of freedom compared to the 21 degrees of freedom in the present study. This is the reason for the discrepencies in the present study results for higher modes.

## ACKNOWLEDGEMENTS

This work was supported by the Natural Sciences and Engineering Research Council of Canada Grant A1375. The author wishes to acknowledge the help of Robert Heimbach in obtaining the numerical results.

# REFERENCES

1. A.W. Leissa, "Vibration of Plates", NASA SP-160, 1969.

2. R.M. Christensen, "Vibration of a 45° right triangular cantilever plate by a gridwork method", AIAA Journal, Vol. 1, 1963, pp. 1790-1795.

3. P.W. Hanson and W. Tuovila, "Experimentally determined natural vibration modes of some cantilever wing flutter models by using an acceleration method", NACA TN 4010, 1957.

4. M.P. Kumaraswamy and V. Cadambe, "Experimental study of the vibration of cantilevered isosceles triangular plates", J. of Scientific and Industrial Research (India), Vol. 15B, No. 2, 1956, pp. 54-60.

5. T. Ota, M. Hamada and T. Tarumoto, "Fundamental frequency of an isosceles triangular plate", Bulletin of the Japan Society of Mechanical Engineers, Vol. 4, No. 15, 1961, pp. 478-481.

6. R. Williams, Y.T. Yeow and H.F. Branson, "An analytical and experimental study of vibrating equilateral triangular plates", SESA Spring Meeting Proceedings, Chicago, May 1975.

7. P.N. Gustafson, W.F. Stokey and C.F. Zorowski, "An experimental study of natural vibrations of cantilevered triangular plates", J. of the Aerospace Sciences, Vol. 20, 1953, pp. 331-337.

8. G.R. Cowper, E. Kosko and G.M. Lindberg, "Static and dynamic application of high precision triangular plate bending elements", AIAA Journal, Vol. 7, 1969, pp. 1957-1965.

9. S. Mirza and M. Bijlani, "Vibration of triangular plates", AIAA Journal, Vol. 21, 1983, pp. 1472-1475.

10. R.B. Bhat, "Natural frequencies of rectangular plates using characteristic orthogonal polynomials in Rayleigh-Ritz Method", J. of Sound and Vibration, Vol. 102, No. 3, 1985.

11. R.B. Bhat, "Vibration of rectangular plates using beam characteristic orthogonal polynomials in Rayleigh-Ritz method", 3rd International Modal Analysis Conference Proceedings, Orlando, Florida, Jan. 28-31, 1985.

12. R.B. Bhat, "Vibration of Structures using characteristic orthogonal polynomials in Rayleigh-Ritz method", 10th Canadian Congress of Applied Mechanics, London, Ontario, June 2-7, 1985.

13. R.B. Bhat, "Plate deflections using orthogonal polynomials", Engineering Mechanics, Trans. ASCE, J. of Engineering Mechanics, Vol. 111, 1985, pp. 1301-1309

14. R.A. Askey (Ed.), "Theory and application of special functions", Academic Press, New York, 1975.

15. T.S. Chihara, "An introduction to orthogonal polynomials", Gordon and Breach Science Publishers, London, 1978.

## Table I

Convergence of Natural Frequencies with the Number of Terms

$$\lambda = (\omega/2\pi \cdot (\rho\ a^4 h/D)^{1/2}$$

| Mode No. | No. of terms | | | | |
|---|---|---|---|---|---|
| | 3 | 6 | 10 | 15 | 21 |
| 1 | 1.0510 | 0.9910 | 0.9851 | 0.9837 | 0.9825 |
| 2 | 4.6535 | 4.2189 | 3.8707 | 3.7406 | 3.7365 |
| 3 | 9.4123 | 5.8991 | 5.3151 | 5.2271 | 5.2069 |
| 4 | | 12.394 | 10.467 | 9.5132 | 8.9772 |
| 5 | | 21.574 | 14.441 | 13.205 | 12.447 |
| 6 | | 44.203 | 23.084 | 18.478 | 17.155 |

## Table II

### Natural Frequencies of Triangular Plates for Different Configurations

$$l = (\omega/2\pi \cdot (\rho\, a^4 h/D)^{1/2}$$

| $\theta$ in Deg. | Mode No. | Present Method | Ref. [9] | Ref. [8] | Ref. [7] |
|---|---|---|---|---|---|
| 0 | 1 | 0.9825 | 0.9802 | 0.9801 | 0.9238 |
|  | 2 | 3.7365 | 3.6703 | 3.7302 | 3.6418 |
|  | 3 | 5.2069 | 5.2981 | 5.1976 | 5.0878 |
|  | 4 | 8.9772 | 8.8992 | 8.9385 | 8.7028 |
|  | 5 | 12.447 | 11.855 | 12.195 | 11.809 |
|  | 6 | 17.155 | 15.831 | 15.885 | 15.478 |
| +30 | 1 | 0.9099 | 0.9178 | --- | --- |
|  | 2 | 3.4259 | 3.3583 |  |  |
|  | 3 | 5.9612 | 5.7216 |  |  |
|  | 4 | 8.9225 | 8.6171 |  |  |
|  | 5 | 11.877 | 10.731 |  |  |
|  | 6 | 19.202 | 16.323 |  |  |
| -30 | 1 | 1.4200 | --- | --- | --- |
|  | 2 | 5.5915 |  |  |  |
|  | 3 | 6.1283 |  |  |  |
|  | 4 | 14.418 |  |  |  |
|  | 5 | 14.885 |  |  |  |
|  | 6 | 17.521 |  |  |  |

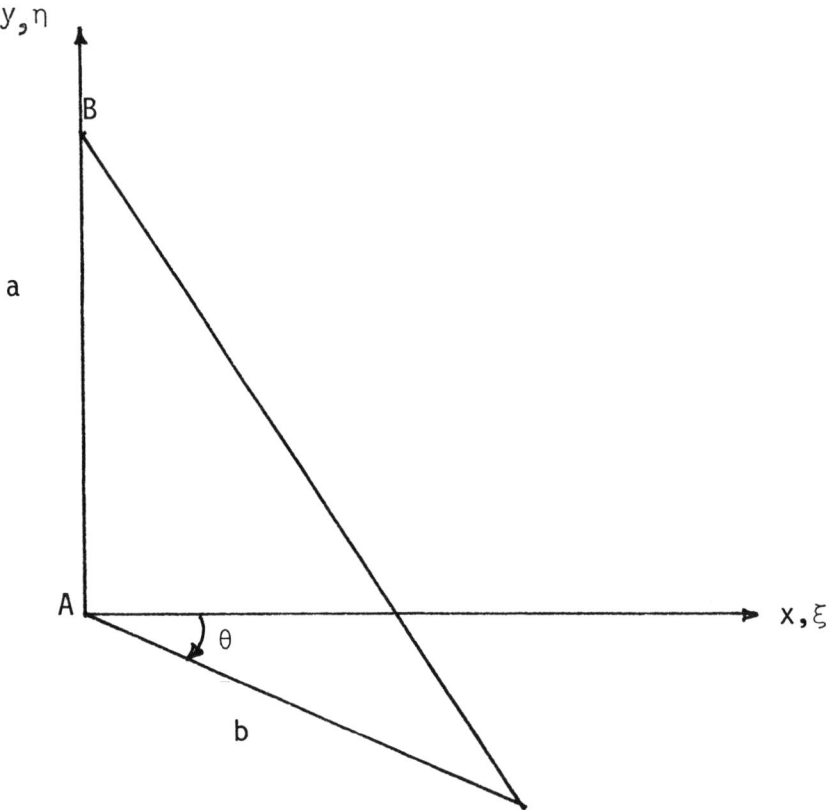

Fig. 1. Triangular Plate Geometry

STRUCTURAL STABILITY IN TURBULENT FLOW

Y. K. Lin
Florida Atlantic University, Boca Raton, FL  33431

## ABSTRACT

Turbulence in the ambient flow is shown to affect the dynamic stability of a structure. It can be either stabilizing or de-stabilizing depending on the type of structure and the way in which the structure interacts with the surrounding flow. In particular, turbulence always de-stabilizes a single degree of freedom linear system, but its effect is likely to be small in the case of gusty wind exciting earth-bound structures. Under favorable conditions turbulence can increase the stability of a linear structure of more than one degree of freedom by providing a conduit to feed energy from the least stable mode to other more stable modes, but the effect is again likely small. Greater effects, either beneficial or otherwise, may be expected of a nonlinear system, as examplified by the coupled flap-lag motion of a helicopter rotor blade.

## INTRODUCTION

The effect of turbulence in the ambient flow is often neglected in the stability analysis of an airborne vehicle in flight or an earth-bound structure exposed to gusty wind. This can either be conservative or unconservative depending on many interacting factors. If turbulence can be modeled as a stochastic process, then the problem belongs to the general subject of stochastic stability for which substantial progress has been made during the past twenty some years. In the present paper a review is given of recent theoretical investigations involving several structural systems using the mathematical tools of stochastic stability. Considerable insight can be gained from these studies which help to point the direction for future works.

To set the stage for the review, a brief comment on the concepts of stochastic stability is in order. Let $X(t)$ be the structural response in question. Then stability means that $X(t)$ is bounded in some sense, and asymptotic stability means additionally that $X(t)$ approaches to a well-defined limit (which may be taken to be zero without loss of generality). When $X(t)$ is a random function, both boundedness and convergence must be interpreted in terms of probability or statistics. Two modes of stochastic stability are of special interest, and they are defined as follows [1,2]:

(i) Stability with probability one -- for every pair of $\varepsilon_1$, $\varepsilon_2 > 0$, there exists a $\delta(\varepsilon_1, \varepsilon_2, t_0) > 0$ such that

$$\text{Prob}\left[\sup_{t > t_0} |X(t)| \geq \varepsilon_1\right] \leq \varepsilon_2, \quad \text{provided } |X(t_0)| \leq \delta. \tag{1}$$

(ii) Stability in the nth moment -- for every $\varepsilon > 0$, there exists a $\delta(\varepsilon, t_0) > 0$, such that

$$|E[X^n(t)]| \leq \varepsilon, \quad t > t_0, \quad \text{provided } |X(t_0)| \leq \delta, \tag{2}$$

where $E[\ ]$ denotes a statistical average.

Stability with probability one is also known as sample stability. For some time it was thought to be a more suitable design criterion. Unfortunately, results obtained for sample stability have been limited to rather simple systems, and these are mostly sufficient conditions, generally much too conservative. On the other hand, necessary and sufficient conditions for moment stabilities are easier to obtain. Kozin and Sugimoto [3] have investigated the relation between the two types of stochastic stability and have shown that for linear systems, sample stability is assured if

$$\lim_{n \downarrow 0} \frac{1}{n} \log \{E[|X(t)|^n]\} < 0 \tag{3}$$

Inequality (3) also suggests that stability in the second moment (the meansquare stability) may, in fact, be a stronger criteron for design than sample stability. Therefore, the ensuing discussion will restrict to moment stability alone.

LINEAR STRUCTURES

Consider first a single pitching degree of freedom of a structure in a two-dimensional flow, governed by the following linear equation [4]

$$\ddot{\alpha} + 2\omega_\alpha \zeta_\alpha \dot{\alpha} + \omega_\alpha^2 \alpha = \frac{1}{2}\rho(2B^2)\frac{\partial C_M}{\partial \alpha}\left\{X_\alpha U^2(t)\dot{\alpha}(t) + U^2(t)\alpha \right.$$

$$\left. + \int_{-\infty}^{t} U^2(\tau)[X_{M\alpha}(t-\tau) - 1]\frac{d\alpha(\tau)}{d\tau}\right\} \tag{4}$$

where $\alpha$ = pitch angle, $\omega_\alpha$ = the natural frequency, $\zeta_\alpha$ = ratio of structural damping to the critical damping, $\rho$ = air density, B = scale of structure, U = total flow velocity (the average velocity plus turbulent fluctuation), $X_{M\alpha}$ = aerodynamic indicial function, and $X_\alpha$ and $\partial C_M/\partial \alpha$ are aerodynamic constants. The forcing terms on the right hand side of Eq.(4) depend on the structural response variables $\alpha$ and $\dot{\alpha}$. These are called self-excited loads which would disappear if the structure were held motionless at all times. The so-called buffeting loads, independent of structural motion, have been dropped from Eq.(4), since they do not affect the structural stability.

The right hand side of Eq.(4) differs from the form given in Ref. [5] in two respects: Firstly, the motion is assumed to have begun in the infinite past; therefore, it has lost the memory of initial conditions. Secondly, the "quasi-steady" term, $U^2\alpha$, has been taken out of the Duhamel integral. If the quasi-steady approximation were used, the indicial function $X_{M\alpha}$ would take the constant value of unity, and the integral in Eq.(4) would vanish. Then the self-excited load would depend only on the instantaneous configuration, not on its past history, between the structure and the flow.

Let $U(t) = u[1 + \xi(t)]$ in which u is the mean wind velocity, and $u\xi(t)$ represents the turbulent fluctuation. For small random perturbation from the mean, the squared velocity may be approximated by $U^2 \approx u^2[1 + 2\xi(t)]$. Then Eq.(4) can be simplified to

$$\ddot{\alpha} + 2\omega_0\zeta_0\dot{\alpha} + \omega_0^2\alpha = 2a_6\xi(t)\alpha + 2\omega_0 a_7\xi(t)\dot{\alpha}$$

$$+ a_6 \int_{-\infty}^{t} [1 + 2\xi(\tau)][X_{M\alpha}(t - \tau) - 1]\dot{\alpha}(\tau)dt \qquad (5)$$

with

$$a_6 = \frac{\rho u^2 B^2}{I} \frac{\partial C_M}{\partial \alpha}, \quad \omega_0 = (\omega_\alpha^2 - a_6)^{1/2}, \quad a_7 = a_6 X_\alpha/\omega_0,$$

$$\zeta_0 = (2\omega_\alpha\zeta_\alpha - a_6 X_\alpha)/2\omega_0$$

We digress to comment on whether or not Eq.(5) can be a realistic mathematical model for a structural stability study. Indeed it is known that linear flutter-type instability cannot occur to a single degree of freedom airfoil; yet Scalan [6] has found that motion of many a suspension bridge is dominated by torsion and that the indicial function $X_{M\alpha}(t)$ can be fitted in the following form

$$X_{M\alpha}(t) = 1 + c_1 \exp(-\gamma_1 t) + c_3 \exp(-\gamma_3 t) \qquad (6)$$

in which $\gamma_1 = a_2 u/B$, $\gamma_3 = a_4 u/B$, and $c_1$, $c_3$, $a_2$, $a_4$ are constants. It can be shown that in the ideal case of no turbulence, the onset of instability occurs at the flow velocity and frequency satisfying the following equations

$$\omega_0^2 - \omega^2 - a_6 \left( \frac{c_1\omega^2}{\gamma_1^2 + \omega^2} + \frac{c_3\omega^2}{\gamma_3^2 + \omega^2} \right) = 0 \qquad (7)$$

$$2\omega_0\zeta_\alpha - a_6 \left( X_\alpha + \frac{c_1\gamma_a}{\gamma_1^2 + \omega^2} + \frac{c_3\gamma_3}{\gamma_3^2 + \omega^2} \right) = 0 \qquad (8)$$

It is clear that the three terms in Eq.(8), with a common factor $a_6$, represent negative aerodynamic damping. In a numerical example in [6] the following aerodynamic parameters are selected:

$$\rho = 1.226 \text{ kg/m}^3, \quad X_\alpha = 1.52, \quad \partial C_M/\partial\alpha = 0.93 \quad c_1 = 1.64,$$

$$c_3 = -51.61, \quad \gamma_1 = \frac{0.38u}{B}, \quad \gamma_3 = \frac{19.74u}{B}$$

Since the signs for $c_3$ and $\gamma_3$ are opposite in the example, the third term within the parentheses in Eq.(8) actually contributed to <u>positive</u> aerodynamic damping, while the other two terms contributed to negative aerodynamics damping. Thus, even without turbulence this linear mathematical model is capable of single mode instability when the deterministic wind velocity exceeds a certain critical value $u_{cr}$.

We now substitute Eq.(6) into Eq.(5) to obtain

$$\ddot{\alpha} + 2\omega_0\zeta_0\dot{\alpha} + \omega_0^2\alpha = 2a_6\xi(t)\alpha + 2\omega_0 a_7\xi(t)\dot{\alpha}$$
$$+ a_6 \int_{-\infty}^{t} [1 + 2\xi(\tau)][c_1 e^{-\gamma_1(t-\tau)} + c_3 e^{-\gamma_3(t-\tau)}]\dot{\alpha}(\tau)d\tau \quad (9)$$

As shown in [4], this equation can be replaced by four first order equations as follows:

$$\dot{Y}_1 = Y_2$$
$$\dot{Y}_2 = -\omega_0^2 Y_1 - 2\omega_0\zeta_0 Y_2 + a_6(Y_3 + Y_4) + 2a_6\xi(t)Y_1 + 2\omega_0 a_7\xi(t)Y_2$$
$$\dot{Y}_3 = -\gamma_1 Y_3 + c_1[1 + 2\xi(t)]Y_2$$
$$\dot{Y}_4 = -\gamma_3 Y_4 + c_3[1 + 2\xi(t)]Y_2 \quad (10)$$

The second moment stability requires that all the second moments $E[Y_j Y_k]$ be bounded where j and k can have values from 1 to 4. Equations for these second moments can be obtained if $\xi(t)$ is a wide-band stochastic process and the state vector $\{Y_1, Y_2, Y_3, Y_4\}$ can be approximated by a random Markov vector [4]. These are ten linear first order differential equations with constant coefficients. The system is meansquare stable if the real part of every eigenvalue of the coefficient matrix is negative, and the stability boundary is reached when one eigenvalue crosses the imaginary axis in the complex plane. For details, the reader is referred to [4].

The stability boundaries for the first and second moments computed from the above numerical data are plotted in Fig. 1 in which $u/u_{cr}$ is the ratio between the mean wind velocity u and the critical velocity $u_{cr}$ computed from Eqs.(7) and (8), and $D_0$ is the non-dimensional spectral density of the turbulence. It shows clearly that turbulence destabilizes the system. We emphasize particularly the second moment stability which should be assured for structural safety. However, in normal strong wind conditions possibly experienced by a bridge, the value for $D_0$ is in the order of 0.01. Therefore, for this particular model, the presence of turbulence reduces the stability margin by about 1 to 2 percent which is not very significant.

An obvious question is whether or not turbulence can stabilize a multi-degree-of-freedom linear system. Intuitively, this is possible if turbulence can help feeding energy from the least stable mode to more stable modes. Extension of the above analysis to more degrees of freedom is straight forward except for much greater amount of computations. For example, to add a plunging mode to the pitching mode, the number of first order linear equations in (10) will increase to twelve if each indicial function can be modeled in the same form as in Eq.(6). Then the total number of equations for the second order statistical moments will increase to seventy-eight. However, using a quasi-steady approximation, only ten second moment equations are needed for a two-degree-of-freedom system, and the results so obtained may still shed some light on the possibility of stabilizing of a linear system by turbulence. The results from such a simplified analysis [7] are shown in Fig. 2 where $\zeta_\alpha$ and $\zeta_L$ are the damping ratios for the torsional mode and the lateral deflection mode, respectively, and $u_{cr}$ is the deterministic critical wind velocity when turbulence is ignored. We see now that turbulence can indeed stabilize a two-degree-of-freedom linear system under certain conditions. However, by suppressing either the torsional or the lateral mode, we reached an important conclusion in [7] that turbulence is <u>always</u> de-stabilizing for the remaining degree of freedom of the linear system. Clearly, the stabilizing effect of turbulence can occur in a linear system only if energy interchange is permissible among different modes. However, the increase in the margin of stability of the order of 5% or less, as shown in Fig. 2, is again rather small.

## HELICOPTER ROTOR BLADES

Even under the ideal smooth flow assumption, a helicopter rotor blade is a time-variant system since forward motion of the vehicle introduces periodically varying coefficients in the equations of motion. The system is both structurally and aerodynamically complicated. Drastic simplification is requried to obtain analytical solutions. Usual aerodynamic assumptions include incompressible and sectionally two-dimensional flow, steady or quasi-steady fluid forces on the blade, etc. Structurally, coupling between flap and torsional motions and coupling between flap and lag motions are often treated separately. These assumptions have also been made in the investigations of turbulence effects on the stability of blade motions [8,9,10,11,12].

Particularly interesting results have been obtained for coupled flap-lag motion [10,11,12] which is nonlinear and also the least stable. The following equations of motion for a spring-restrained rigid blade under ideal no-turbulence condition have been derived by Peters [13]:

$$\ddot{\beta} + \sin\beta \cos\beta(1 + \dot{\zeta})^2 + (P - 1)(\beta - \beta_{pc}) + 2\zeta = \int_0^1 \bar{F}_\beta \bar{r} \, d\bar{r} \quad (11)$$

$$(\cos^2\beta)\ddot{\zeta} - 2\sin\beta\cos\beta(1+\dot{\zeta})\dot{\beta} + W\zeta + Z(\beta - \beta_{pc})$$
$$= \cos\beta \int_0^1 \bar{F}_\zeta \bar{r}\, d\bar{r} \qquad (12)$$

where an over-dot denotes one differentiation with respect to the non-dimensional time $\psi = \Omega t$; $\beta$ = flap angle; $\beta_{pc}$ = pre-cone angle; $\zeta$ = lag angle; P, W, Z = elastic constants; $\bar{r}$ = non-dimensional coordinate along the blade; $\bar{F}_\beta$, $\bar{F}_\zeta$ = normalized airloads given by

$$\bar{F}_\beta = \pm \frac{\gamma}{2}\left[\bar{U}_t^2 \sin\theta - \bar{U}_t \bar{U}_p \left(\cos\theta + \frac{c_{d_o}}{a}\right)\right] \qquad (13)$$

$$\bar{F}_\zeta = \pm \frac{\gamma}{2}\left[\bar{U}_p^2 \left(\cos\theta - \frac{1}{2}\frac{c_{d_o}}{a}\right) - \bar{U}_p \bar{U}_t \sin\theta - \bar{U}_t^2 \frac{c_{d_o}}{a}\right] \qquad (14)$$

$$\bar{U}_t = (1 + \dot{\zeta})\bar{r}\cos\beta + \mu\sin(\psi + \zeta) \qquad (15)$$

$$\bar{U}_p = \bar{r}\dot{\beta} + \lambda\cos\beta + \mu\sin\beta\cos(\psi + \zeta) \qquad (16)$$

In Eqs.(13) to (16), $\gamma$ = Lock number; $\theta$ = pitch angle; $\lambda$ = inflow ratio; $\mu$ = advance ratio; $c_{d0}$ = profile drag coefficient; a = slope of lift curve; $\bar{U}_t$ and $\bar{U}_p$ are nondimensional relative flow velocities tangent and perpendicular to the blade, respectively.

To include turbulence, we modify Eqs. (15) and (16) as follows

$$\bar{U}_t = (1 + \dot{\zeta})\bar{r}\cos\beta + (\mu + \eta)\sin(\psi + \zeta) + \xi\cos(\psi + \zeta) \qquad (17)$$

$$\bar{U}_p = \bar{r}\dot{\beta} + (\lambda + \nu)\cos\beta + (\mu + \eta)\sin\beta\cos(\psi + \zeta)$$
$$- \xi\sin\beta\sin(\psi + \zeta) \qquad (18)$$

where $\xi$, $\eta$ and $\nu$ are nondimensional turbulence components of the air flow in the lateral, longitudinal and vertical directions, respectively.

To derive a set of linearized equations for stability study, let

$$\beta = \bar{\beta} + \delta\beta$$
$$\zeta = \bar{\zeta} + \delta\zeta \qquad (19)$$
$$\theta = \bar{\theta} + \theta_\beta \delta\beta + \theta_\zeta \delta\zeta$$

where $\bar{\beta}$, $\bar{\zeta}$ and $\bar{\theta}$ are equilibrium flap, lag and pitch angles, respectively, and $\theta_\beta$ and $\theta_\zeta$ are pitch-flap and pitch-lag coupling parameters which account approximately the changes in the blade pitch angle due to changes in flap and lag angles. The equilibrium flap angle $\bar{\beta}$ is assumed of the form

$$\bar{\beta} = \beta_o + \beta_s \sin\psi + \beta_c \cos\psi \tag{20}$$

and the equilibrium pitch angle of the form:

$$\bar{\theta} = \theta_o + \theta_s \sin\psi + \theta_c \cos\psi + \theta_\beta(\bar{\beta} - \beta_{pc}) + \theta_\zeta \bar{\zeta} \tag{21}$$

where $\theta_o$ is the collective pitch, $\theta_s$ and $\theta_c$ are the cyclic pitch coefficients.

Substituting (19) into (17) and (18), substracting out the equilibrium terms, and collecting linear terms in $\delta_\beta$, $\delta\zeta$ and their derivatives, we obtain

$$\begin{Bmatrix} \ddot{\delta\beta} \\ \ddot{\delta\zeta} \end{Bmatrix} + [C] \begin{Bmatrix} \dot{\delta\beta} \\ \dot{\delta\zeta} \end{Bmatrix} + [K] \begin{Bmatrix} \delta\beta \\ \delta\zeta \end{Bmatrix} = \begin{Bmatrix} 0 \\ 0 \end{Bmatrix} \tag{22}$$

where elements of C and K are periodic functions of angle $\psi$, the periodic equilibrium values $\bar{\beta}$ and $\bar{\theta}$, and random turbulence components $\xi$, $\eta$, and $\nu$. Again, the inhomogeneous forcing terms on the right hand side have been dropped for the present stability analysis.

Since turbulence velocities are expected to be small compared with the tip speed of the rotor, the nondimensional $\xi$, $\eta$ and $\nu$ are much smaller than one. It is then reasonable to approximate [C] and [K] as being linearly dependent on $\xi$, $\eta$ and $\nu$. Write

$$\begin{aligned} [C] &= [C_D] + \xi[C_\xi] + \eta[C_\eta] + \nu[C_\nu] \\ [K] &= [K_D] + \xi[K_\xi] = \eta[K_\eta] + \nu[K_\nu] \end{aligned} \tag{23}$$

The elements of matrices $[C_D]$ and $[K_D]$ on the right hand sides are given by Peters in [13] and the other matrices are given in [10].

Eq.(22) is now replaced by four first order equations. Under the assumption that turbulence components are wide-band stochastic processes, the theory of Markov process is applied to obtain moment equations for the response state variables. The coefficients in these moment equations are complicated periodic functions of $\psi$; their solution can only be obtained by numerical integration. In particular, these moment equations are integrated over one period to obtain the so-called Floquet transition matrix which relates the statistical moments of a given order (say the second order) at successive periods. The moments are stable if the real part of every eigenvalue of the Floquet transition matrix is negative, and the stability boundary is reached when one eigenvalue crosses the imaginary axis in the complex plane.

An order-of-magnitude study reveals that in the case of flap-lag coupling, the effect of the vertical turbulence component dominates those of the longitudinal and lateral components. The numerical solution can be greatly simplified by retaining only $\nu$ in Eqs.(23).

Fig. 3 shows the stability boundaries for second moments in terms of the flap-mode and lag-mode frequencies, p and $\omega_\zeta$, neglecting the structural coupling between the two modes. As shown in Fig. 3a, the instability region for a turbulence level $\Phi_\nu^* = 5.7 \times 10^{-4}$ is smaller than the one without turbulence. This is a surprising result. As the turbulence level increases the instability region becomes even smaller, and at $\Phi_\nu^* = 1.71 \times 10^{-3}$, it collapses to point A and disappears entirely. The system remains stable for further increasing of $\Phi_\nu^*$ until the value reaches approximately 19 whereupon a small instability region begins to reappear around point A.

Fig. 4 shows the second moment stability boundaries in terms the lag-mode frequency and the ratio of the thrust coefficient $C_T$ and the blade solidity $\sigma_s$, using an intermediate structural coupling parameter R = 0.4. Again, turbulence appears to stabilize the system significantly.

A partial physical explanation for the stabilizing effect of vertical turbulence on flap-lag blade motion has been given in [11] in terms of a necessary condition for the first moment stability. Although the first moment stability (which is itself a necessary condition for the second moment stability) is too weak for design purposes, it can be examined much more simply. In fact, a closed form analysis is possible for the present case. It can be shown [11] that after higher order terms are neglected, the equations for the first moments can be expressed as

$$\ddot{E}\{X\} + [\bar{C}] \dot{E}\{X\} + [K]\{X\} = \{0\} \qquad (24)$$

where $\{X\} = \{\delta\beta, \delta\zeta\}$, and $[\bar{C}]$ differs from $[C]$ in Eq.(22) only in the (2,2) element. Neglecting structural coupling between the flap mode and the lag mode, and restricting to the hovering case this (2,2) element is found to be

$$\bar{C}_{22} = C_{22} + 2h\,\Phi_\nu^* = h[2(c_{do}/a) + \theta_o \phi] + 2h\,\Phi_\nu^* \qquad (25)$$

where $h = \gamma/8$ and $\phi$ is an inflow parameter. Since the spectral density $\Phi_\nu^*$ must be positive, the vertical turbulence is seen to increase the effective profile drag by an amount $a\Phi_\nu^*$. A parallel analysis for the second moments appears impossible because of large number of equations.

The case of flap-torsion coupling is somewhat simpler since it is essentially linear. Thus, the equations for the perturbed motion $\delta\beta$ and $\delta\alpha$ are the same as those for the original equations for $\beta$ and $\alpha$. Adding turbulence terms in the equations obtained by Sissign and Kuczynski [14], stability analysis can also be carried out in a straight-forward manner. The longitudinal and lateral turbulence components are found to affect the system stability adversely [8,9]; however, their effects are significant only at very high advance ratios. For practical purposes, therefore, the in-plane turbulence may be neglected in the stability analysis of helicopter blades.

CONCLUDING REMARKS

It is clear that proper mathematical modeling is essential in order to obtain a valid theoretical prediction of the turbulence effect. The theoretical investigations discussed herein were based on equations originally derived for ideal smooth flows without turbulence. The implication is that the structure-fluid interaction mechanism remains essentially unchanged with or without turbulence. This is a much simplified assumption, but a logical one to make when the scale of turbulence is greater than the scale of the structure. Improvements in this respect must come from further theoretical and experimental research on turbulent flow fields around different shapes of structures both streamlined and unstreamlined.

The assumption that the structural response state vector is approximately a random Markov vector implies that the correlation time of the turbulence field is short compared with the relaxation time of the dynamic system. This assumption is not critical if we are interested only in motion stability [for reasons see Ref. 15]; however, it must be justified if the numerical values of the statistical moments of the response are also of concern. Response statistics of a stable motion may be computed by retaining the inhomogeneous terms in the equations of motion, and they are needed in the assessment of fatigue damage and thus service life of a structure. If the correlation time of the turbulence is not short enough compared with the relaxation time of the structural system, a suitable linear filter can be added to the system so that the Markov vector assumption remains valid, although the dimensionality of the Markov vector is enlarged to include the dimension of the filter.

REFERENCES

[1] Kozin, F., "A Survey of Stability of Stochastic Systems," Automatika, Journal of International Federation of Automatic Control, Vol. 5, 1969, pp. 95-112.

[2] Lin, Y. K. and Prussing, J. E., "Concepts of Stochastic Stability in Rotor Dynamics," Journal of the American Helicopter Society, Vol. 27, No. 2, 1982, pp. 73-74.

[3] Kozin, F. and Sugimoto, S., "Relations between Sample and Moment Stability for Linear Stochastic Differential Equations," in D. Mason, ed. Proceedings of Conference on Stochastic Differential Equations, Academic Press, 1977, pp. 145-162.

[4] Lin, Y. K. and Ariaratnam, S. T., "Stability of Bridge Motion in Turbulent Winds," Journal of Structural Mechanics, Vol. 8, 1980, pp. 1-15.

[5] Bisplinghoff, R. L. and Ashley, H., *Principles of Aeroelasticity*, John Wiley and Sons, Inc., New York, 1962.

[6] Scanlan, R. H., Belivean, J. G. and Budlong, K. S., "Indicial Aerodynamic Functions for Bridge Decks," Journal of Engineering Mechanics, ASCE, Vol. 100(EM 4), 1974, pp. 657-672.

[7] Lin, Y. K. and She, K., "Can Turbulence Be Good for Structural Stability?" Second ASME/ASCE Mechanics Conference, Albuquerque, New Mexico, June 24-26, 1985.

[8] Lin, Y. K., Fujimori, Y. and Ariaratnam, S. T., "Rotor Blade Stability in Turbulent Flows -- Part I," AIAA Journal, Vol. 17, 1979, pp. 545-552.

[9] Fujimori, Y., Lin, Y. K. and Ariaratnam, S. T., "Rotor Blade Stability in Turbulent Flows -- Part II," AIAA Journal, Vol. 17, 1979, pp. 673-678.

[10] Prussing, J. E. and Lin, Y. K., "Rotor Blade Flap-Lag Stability in Turbulent Flows," Journal of the American Helicopter Society, Vol. 27, 1982, pp. 51-57.

[11] Prussing, J. E. and Lin, Y. K., "A Closed-Form Analysis of Rotor Blade Flap-Lag Stability in Hover and Low-Speed Forward Flight in Turbulent Flow," Journal of American Helicopter Society, Vol. 28, 1983, pp. 42-46.

[12] Prussing, J. E., Lin, Y. K. and Shiau, T. N., "Rotor Blade Flap-Lag Stability and Response in Forward Flight in Turbulent Flows," Journal of American Helicopter Society. Vol. 29, 1984, pp. 81-87.

[13] Peters, D. A., "Flap-Lag Stability of Helicopter Rotor Blade in Forward Flight," Journal of American Helicopter Society, Vol. 20, 1975, pp. 2-13.

[14] Sissign, G. J. and Kuczynski, W. A., "Investigations on the Effect of Blade Torsion on the Dynamics of the Flapping Motion," Journal of American Helicopter Society, Vol. 15, 1972, pp. 2-9.

[15] Lin, Y. K., "Some Observations on the Stochastic Averaging Method," Probabilistic Engineering Mechanics, Vol. 1, 1986, pp. 23-27.

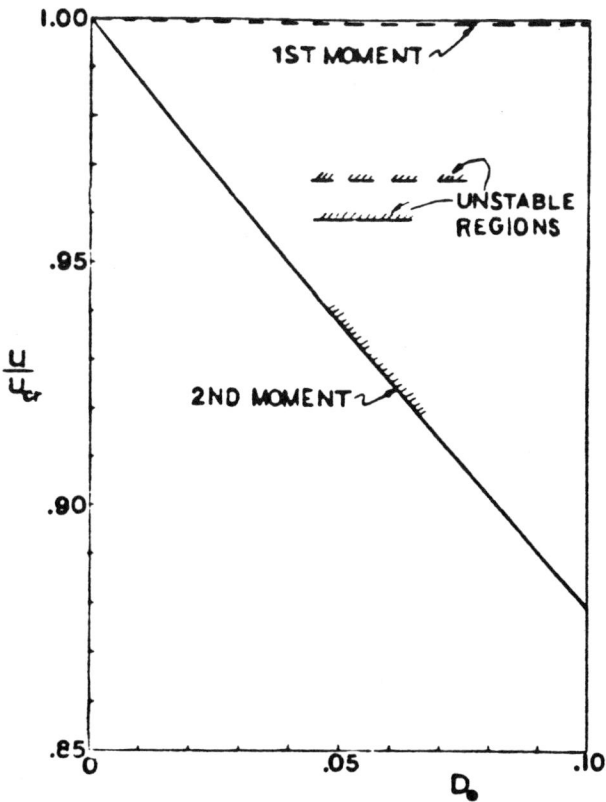

Fig. 1. Stability boundaries of torsional bridge motion in turbulent wind------based on SDF linear model and unsteady aerodynamics,

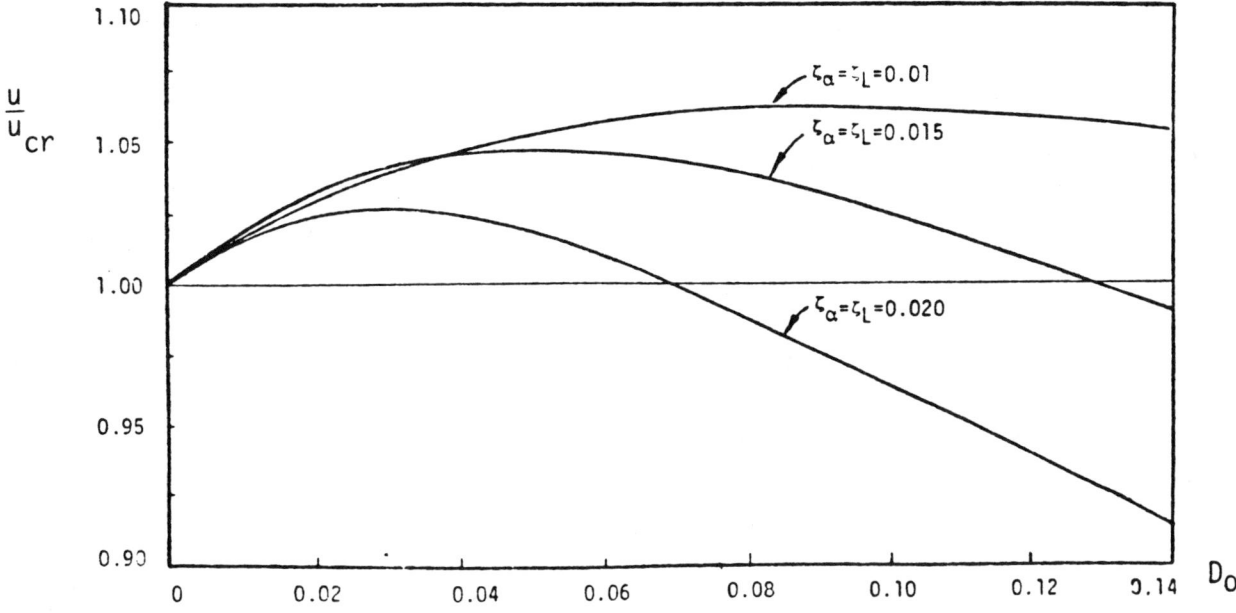

Fig. 2. Second moment stability boundaries of bridge motion in turbulent wind ------ based on 2DF linear model and quasi-steady aerodynamics.

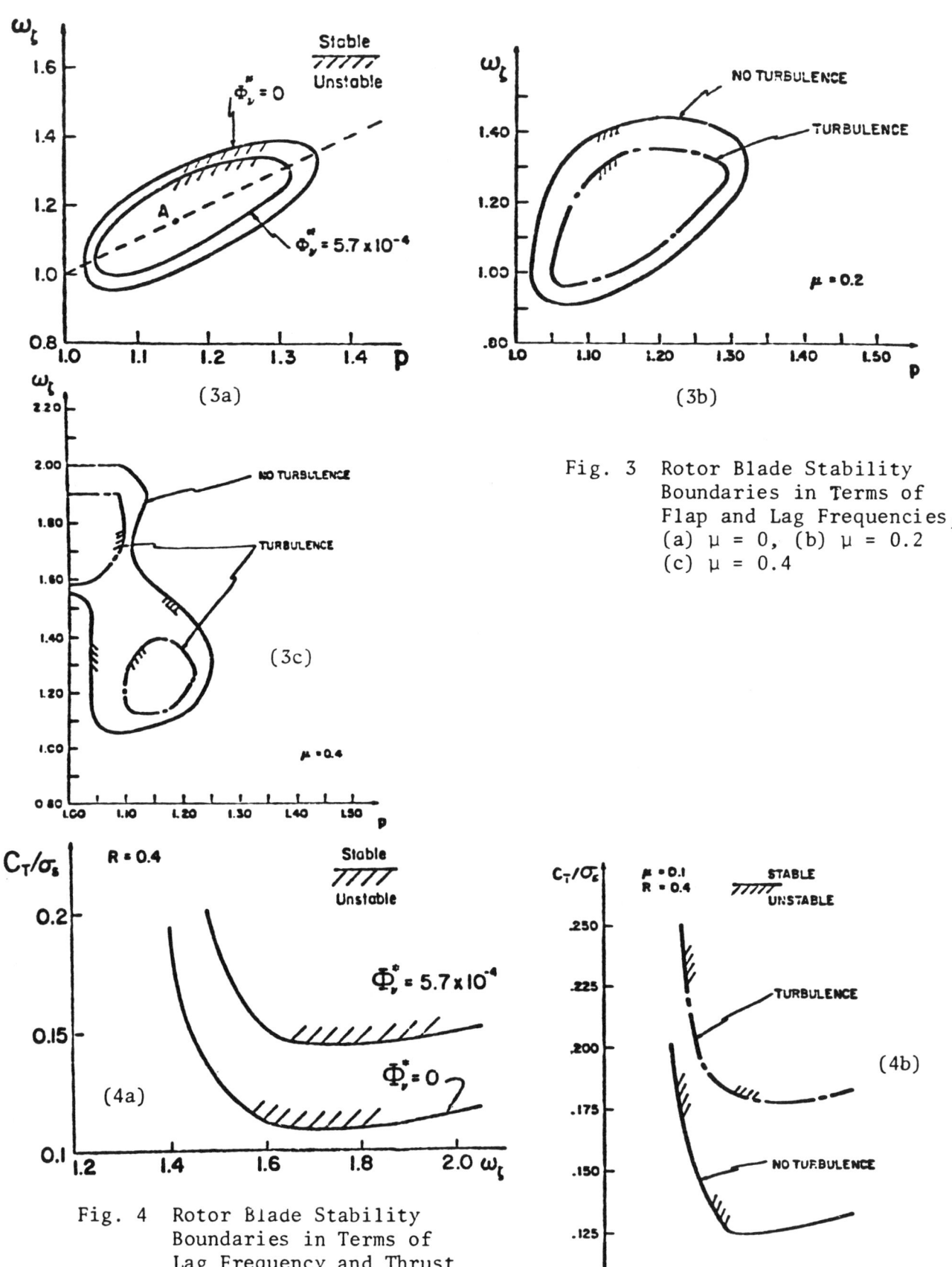

Fig. 3 Rotor Blade Stability Boundaries in Terms of Flap and Lag Frequencies, (a) $\mu = 0$, (b) $\mu = 0.2$ (c) $\mu = 0.4$

Fig. 4 Rotor Blade Stability Boundaries in Terms of Lag Frequency and Thrust Coefficient, (a) $\mu = 0$, (b) $\mu = 0.1$

ON THE TRANSIENT DYNAMICS DURING THE ORBITER BASED
DEPLOYMENT OF FLEXIBLE MEMBERS

V. J. Modi, Professor*
Department of Mechanical Engineering
The University of British Columbia
Vancouver, B. C., Canada V6T 1W5

ABSTRACT

Using a relatively general formulation procedure, the paper reviews complex interactions between deployment, attitude dynamics and flexural rigidity for configurations representing deployment of beam and tether type appendages. The governing nonlinear, nonautonomous and coupled hybrid set of equations are extremely difficult to solve even with the help of a computer, not to mention the cost involved. Results suggest substantial influence of the flexibility, deployment velocity, initial conditions, and appendage orientation on the response, and under critical combinations of parameters the system can become unstable. The information has relevance to the design of control systems for: (i) the next generation of communication satellites; (ii) the Orbiter based experiments such as SAFE (Solar Array Flight Experiments), SCOLE (Structural COntrol Laboratory Experiment), STEP (Structural Technology Experiment Program), and the NASA/CNR tethered subsatellite system; as well as (iii) the evolutionary transient and postconstruction operational phases of the proposed space station.

INTRODUCTION

Flexibility effects on satellite attitude motion and its control have become topics of considerable importance. Over the years, a large body of literature pertaining to the various aspects of satellite system response, stability and control has appeared. A recent issue of the Journal of Guidance, control, and Dynamics published by the AIAA (American Institute of Aeronautics and Astronautics) contains a series of articles reviewing the state of the art in the general area of large space structures [1].

Attention is also directed towards planning of in-orbit experiments such as SCOLE, the Orbiter Mounted Large Platform Assembler Experiment, NASA/Lockheed Solar Array Flight Experiment (SAFE) and a host of others to check, calibrate and improve algorithms. It is generally concluded that in-orbit information acquired during the construction phase of a space station is the only dependable procedure for its overall design. Obviously, this promises to open up an exciting area of in-flight measurements of structural dynamics, stability and control parameters necessary for design. With the U.S. commitment to a space station in early 1990's, the need for understanding structural response and control characteristics

---

*Presently Visiting Professor, The Institute of Space and Astronautical Science, Tokyo, Japan

of such time varying, highly flexible systems is further emphasized. This paper briefly touches upon representative results concerning dynamics of spacecraft having flexible deploying members with the attention focused on two classes of problems of contemporary interest:

(i) the Orbiter based deployment of a beam-type appendage in and out of the orbital plane;
(ii) deployment and retrieval of tethered subsatellite systems.

Details of the mathematical formulations and analyses being extremely lengthy are omitted here, however, appropriate references are cited. Emphasis throughout is on the analysis of results and corresponding conclusions.

## THE ORBITER BASED DEPLOYMENT OF A BEAM IN ARBITRARY ORIENTATION

A relatively general formulation of the nonlinear, nonautonomous and coupled equations of motion applicable to a large class of spacecraft with flexible plate/beam-type of members has been discussed by the authors earlier [2]. It has been extended to account for membrane and tether type of appendages with viscous/structural damping and momentum/reaction wheels. Essential features of the general formulation may be summarized as follows:

- spacecraft of an arbitrary inertia distribution in a general orbit undergoing three-axis librations;
- arbitrary number and orientation of flexible appendages (tether, membrane, beam, plate) deploying independently at an arbitrary velocity and acceleration;
- the appendage is permitted to have variable mass density, flexural rigidity and cross-sectional area along its length;
- governing equations account for gravitational effects, shifting center of mass, changing rigid body inertia, and appendage offset together with transverse oscillations;
- modified Eulerian rotations $\gamma$, $\beta$, $\alpha$ (roll, yaw, pitch, respectively) are so chosen as to make the governing equations applicable to both spin stabilized and gravity gradient orientations;
- the equations are programmed in nonlinear as well as linearized forms to permit the study of:
  (i) large angle manoeuvers;
  (ii) nonlinear effects.

The configuration selected for study corresponds to the Orbiter Mounted Large Platform Assembler Experiment once proposed by Grumman Aerospace Corporation (Fig. 1). Its objective is to establish capability of manufacturing beams in space which would serve as one of the fundamental structural elements in construction of the future space station. The assembler is fully collapsible and automatically deployed.

For analysis, the flexibility and deployment rate parameters were taken to be of the same order of magnitude as used or likely to be employed in practice. In the diagrams e represents orbital eccentricity; $\Psi$, $\Lambda$, $\Phi$ (roll, yaw, pitch, respectively) are the librational angles; EI is the beam flexural rigidity, assumed constant over the length in this particular example; and $\dot{L}$ corresponds to the deployment rate. $\lambda_{in}$ and $\lambda_{out}$ denote beam inclinations to the local vertical in and normal to the orbital plane, respectively. The perigee was taken to be 331 km. The truss or beam vibrations were represented by a maximum of the first four modes, $\psi_i$, of a cantilever. $P_\ell$, $Q_\ell$ represent generalized coordinates associated with the admissible functions used to represent beam-type appendage oscillations in the $\ell$ th mode in $\zeta$ and $\eta$ directions, respectively. $\bar{P}_\ell$ and $\bar{Q}_\ell$ represent transverse generalized coordinates normalized with respect to the total length.

Numerical values for some of the more important parameters used in the computation are given below:

Orbiter:
- Mass = 79,710 kg
- $I_{xx}$ = 8,286,760 kg m$^2$     $I_{xy}$ = 27,116 kg m$^2$
- $I_{yy}$ = 8,646,050 kg m$^2$     $I_{yz}$ = 328,108 kg m$^2$
- $I_{zz}$ = 1,091,430 kg m$^2$     $I_{zx}$ = -8,135 kg m$^2$

Here x,y,z are the principal body coordinates of the Orbiter with the origin coinciding with the center of mass. In the nominal configuration x is along the orbit normal, y coincides with the local vertical and z is aligned with the local horizontal in the direction of motion, $\gamma$ (roll), $\beta$ (yaw), and $\alpha$ (pitch) refer to rotations about the local horizontal, local vertical and orbit normal, respectively.

Beam:
- Mass ($M_b$) = 129 kg
- Length (L) = 33 m
- Flexural Rigidity (EI) = 436 kg m$^2$

Figure 2 shows tip response of the beam for two different orientations in the plane defined by the local vertical and the orbit normal, $\lambda_{out}$ = 20° and 90°. Note, the two transverse motions, $\zeta$ and $\eta$, are completely coupled with the plane of vibration precessing, due to the Coriolis force, at a uniform speed which is governed by the beam inclination angle $\lambda_{out}$. For the case of $\lambda_{out}$ = 0, the uncoupled motion showed no precession. On the other hand, the precessional velocity increased with $\lambda_{out}$ and reached a maximum value at $\lambda_{out}$ = 90°. The plane of vibration of the beam precessed in one direction only (in this case clockwise) for a given $\lambda_{out}$.

Effect of beam deployment on the tip dynamics is studied in Fig. 3. Initial tip deflection is the same as before. Two time histories with the same duration of deployment are considered. As can be expected, the frequency of oscillation in and out of the orbital plane gradually decreases with deployment finally attaining a steady state value upon its termination. It is of interest to recognize that they reach the same steady state amplitude, although it is much larger during deployment compared to the deployed case.

In practice the Orbiter's librations will be controlled to a specified tolerance limit. A typical time history [3] of the controlled Space Shuttle librations during an orbit is shown in Fig. 4. In the following results attention is focused on response of the deployed beam during such forced excitation of the Orbiter in the Lagrange configuration.

Figure 5 shows the forced tip response as well as the first two modes contributing to it for a beam deployed along the orbit normal with the Orbiter in the Lagrange configuration. At the outset it should be recognized that, for this out-of-plane configuration of the beam, the out-of-plane motion $\zeta$ and inplane response $\eta$ are coupled as seen before (Fig. 2). Hence one would expect the Orbiter's yaw and roll to be reflected in both $\eta$ (inplane) and $\zeta$ (out-of-plane) motions. The response shown in Fig. 5 precisely reveals these trends. However, the roll disturbance being at a higher frequency, and hence with a higher acceleration, appears to be dominant as apparent from the amplitude modulation of the response at the roll frequency (around 13 cycles per orbit).

A word concerning accuracy of the results presented here would be appropriate. To this end permissible error during numerical integration of the governing nondimensional differential equations was varied systematically to assure reliable data. The case corresponding to the Lagrange configuration with $\lambda_{out} = 90°$ was studied using error tolerance of $10^{-4}$, $10^{-6}$ and $10^{-8}$. Both components of tip deflections and generalized coordinates were compared. The tolerance level of $10^{-4}$ was found to be inadequate and gave misleading response. However, the results obtained using the tolerance levels of $10^{-6}$ and $10^{-8}$ were essentially the same. Hence during the numerical integration the error tolerance of $10^{-8}$ or lower was used.

## THE ORBITER BASED TETHERED SUBSATELLITE SYSTEM

A vast potential of the Shuttle based tethered systems (Fig. 6) has led to many investigations concerning their dynamics during operational (station-keeping), deployment and retrieval phases. In its utmost generality, the problem is indeed quite challenging as the system dynamics is governed by a set of ordinary and partial, nonlinear, nonautonomous and coupled equations which account for:

- three dimensional rigid body dynamics (librational motion) of the Shuttle and the subsatellite;
- swinging inplane and out-of-plane motions of the tether, of finite mass and elasticity, with longitudinal and transverse vibrations superimposed on them;
- offset of the tether attachment point from the Shuttle's center of mass;
- aerodynamic drag in a rotating atmosphere.

Over the years, investigators have attempted to obtain some insight into the complex dynamics of the system using a variety of models which have been summarized by Misra and Modi [4]. In general, the studies show that the dynamics of the system during deployment is stable, however, the retrieval dynamics is basically unstable. The system involves a negative damping approximately proportional to $\dot{L}/L$, where L represents the unstretched tether length. This suggests a need for an active control strategy particularly to limit inplane ($\theta$) and out-of-plane ($\phi$) swing motions of the tether.

Figure 7 shows inplane and out-of-plane response of a 100 km long tether during retrieval of a 170 kg subsatellite to the Shuttle in a circular equatorial orbit at a height of 220 km. Physical properties of the tether, subsatellite, atmosphere as well as the exponential retrieval procedure are taken to be the same as those used by Baker and others [5]. Both the inplane and out-of-plane tether librations become unstable even in the presence of dissipation ($C_\theta' = C_\phi' = 0.1$) due to the negative damping introduced by the terms proportional to $\dot{L}/L$. Depending upon the retrieval rate used, one can easily calculate the damping level required to guarantee stability, however, it may not always be possible in practice to provide the required level of damping. Clearly this suggests a need to evolve an effective control procedure.

Attension was now turned to assess effectiveness of a number of nonlinear control strategies. A procedure where the tether tension is controlled as

$$T = K_\ell \ell + K_{\ell'} \ell' + K_{\phi'} \phi'^2 \qquad (1)$$

appeared to be the most promising. Here: T = tether tension level; and $\ell$ = non-dimensional difference between the actual and the commanded tether length. Primes denote differentiation with respect to the non-dimensional time. Figure 8 presents representative results for the tether response during retrieval from 100 km to 500 m using the control strategy mentioned in equation (1). Note, the amplitudes in both pitch and roll are substantially reduced. The controller continued to remain effective under a wide rage of diverse situations involving different initial conditions, orbital elements, physical properties of the system, retrieval rate, etc.

In the actual practice the tether material is indeed elastic causing a longitudinal stretch ($\xi$) of the tether. Any realistic analysis must account for it. Hence it was essential to check the controller's effectiveness when the tether is elastic. To this end appropriate stretch equation was added to the system of equations representing $\theta$ and $\phi$ degrees of tether rotations. Figure 9 summarizes controlled retrieval dynamics of such a tethered subsatellite system when the Space Shuttle is in a circular polar orbit at a height of 220 km. The high frequency longitudinal oscillations of the tethered system made the numerical integration of the governing nonlinear coupled equations quite expensive. Clearly the nonlinear control strategy remains effective even when the longitudinal stretch is accounted for.

Transverse vibrations of the tether would further complicate the problem. They are excited by the Coriolis forces during deployment and retrieval as well as the aerodynamic forces when a part of the tether is at a relatively low altitude. The transverse and longitudinal vibration are strongly coupled, particularly at the terminal phase of retrieval. The transverse vibrations are also unstable during uncontrolled retrieval. Although the growing transverse vibrations have a negligible effect on the tension at the beginning of retrieval, it can make the tether slack towards the end [6]. This can be avoided in two ways: one is to speed up the retrieval when the length is small, to avoid very slow retrieval rate due to an exponential character. Alternatively, thrusters may be used in conjunction with a tension control or a length rate law. Using three mutually perpendicular thrusters, one along the nominal tetherline and the other two opposing inplane and out-of-plain rotations, Xu et al. [7] have shown a remarkable success in controlling all the degrees of freedom.

CONCLUDING COMMENT

With the relatively general formulations in hand and the programs operational, efforts are in progress to develop a comprehensive data bank for spacecraft with flexible appendages. Not only will it prove useful to design engineers involved in planning of future communications satellites but also help in assessing dynamical, stability and control considerations associated with the Orbiter based construction of space-platforms.

The entire field is wide open to innovative contributions. Dynamics and control of such nonlinear, time-dependent systems accounting for damping and environmental forces remains virtually untouched. Their application to the construction of a space station has received attention only recently. Development of an algorithm to predict the effect of mass, inertia and stiffness of the station as it evolves on the dynamics and control parameters represents an exciting challenge. Application of the tether and associated dynamics presents an area of enormous potential.

ACKNOWLEDGEMENT

The investigation reported here was supported by the Natural Sciences and Engineering Research Council of Canada, Grant No. 67-1547.

REFERENCES

[1] Special Section, "Large Space Structure Control: Early Experiments," Journal of Guidance, Control, and Dynamics, Vol. 7, No. 5, September-October 1984, pp. 513-562.

[2] V. J. Modi and A. M. Ibrahim, "A General Formulation for Librational Dynamics of Spacecraft with Deploying Appendages," Journal of Guidance, Control, and Dynamics, Vol. 7, No. 5, September-October 1984, pp. 563-569.

[3] R. J. Budica and K. L. Tong, "Shuttle On-Orbit Attitude Dynamics Simulation," AIAA/AAS Astrodynamics conference, San Diego, California, August 1982, Paper No. AIAA-82-1452.

[4] A. K. Misra and V. J. Modi, "Dynamics and Control of Tether Connected Two-Body System," Invited Address, 33rd Congress of the International Astronautical Federation, Paris, France, September 1982, Paper No. IAF-82-316; also Space 2000, Selected Papers from the 33rd IAF Congress, Editor: L. G. Napolitano, AIAA Publisher, pp. 473-514.

[5] P. L. Baker, et al., "Tethered Subsatellite Study," NASA TM X-73314, March 1976.

[6] V. J. Modi, D. M. Xu and A. K. Misra, "Influence of Dynamical Modeling of a Tethered Satellite System on its Control System," AIAA 23rd Aerospace Sciences Meeting, Reno, Nevada, January 1985, Paper No. AIAA-85-0025.

[7] D. M. Xu, A. K. Misra and V. J. Modi, "On Thruster Augmented Active Control of a Tethered Subsatellite System During its Retrieval," AIAA/AAS Astrodynamics Conference, Seattle Washington, August 1984, Paper No. 84-1993.

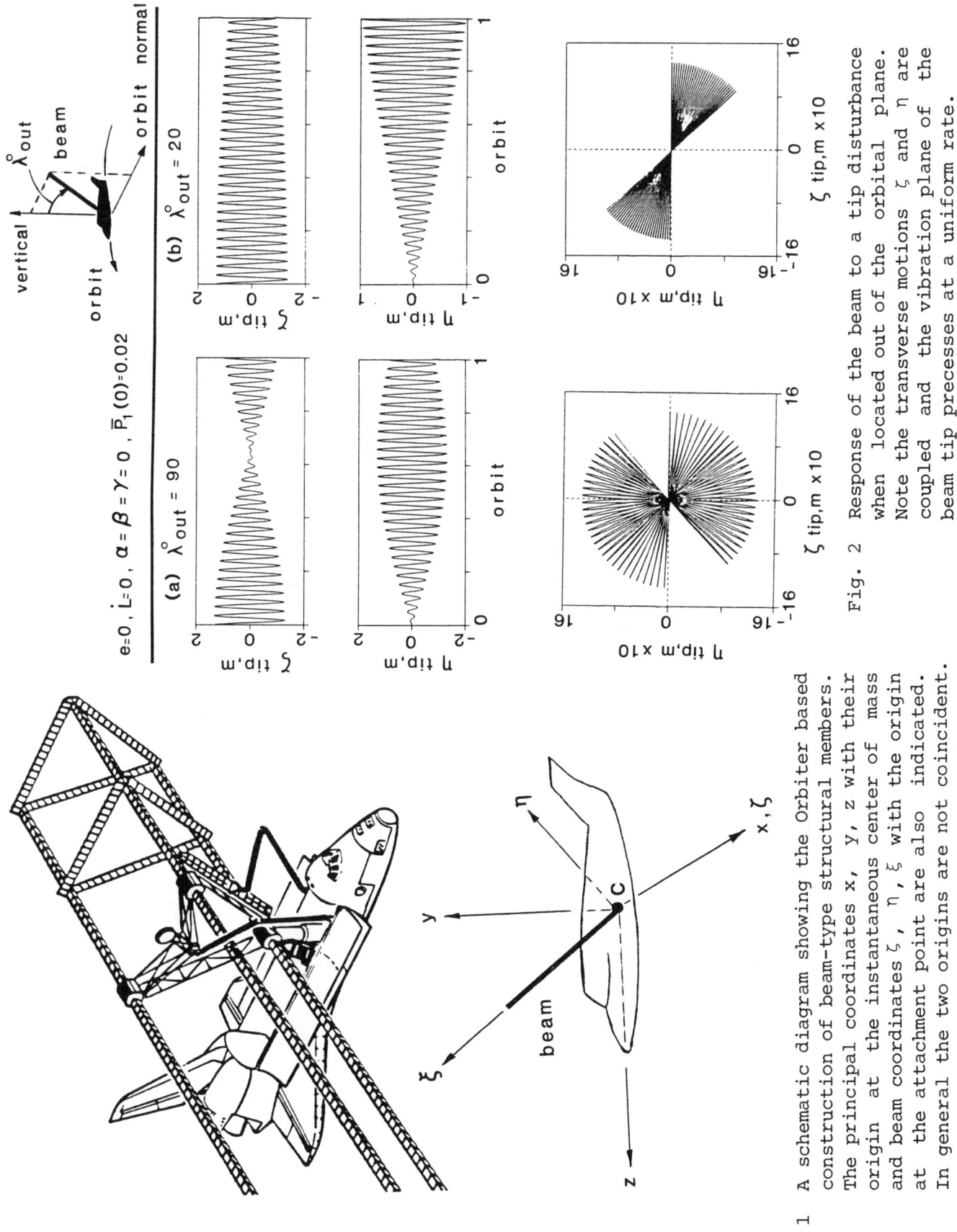

Fig. 1 A schematic diagram showing the Orbiter based construction of beam-type structural members. The principal coordinates $x$, $y$, $z$ with their origin at the instantaneous center of mass and beam coordinates $\zeta$, $\eta$, $\xi$ with the origin at the attachment point are also indicated. In general the two origins are not coincident.

Fig. 2 Response of the beam to a tip disturbance when located out of the orbital plane. Note the transverse motions $\zeta$ and $\eta$ are coupled and the vibration plane of the beam tip precesses at a uniform rate.

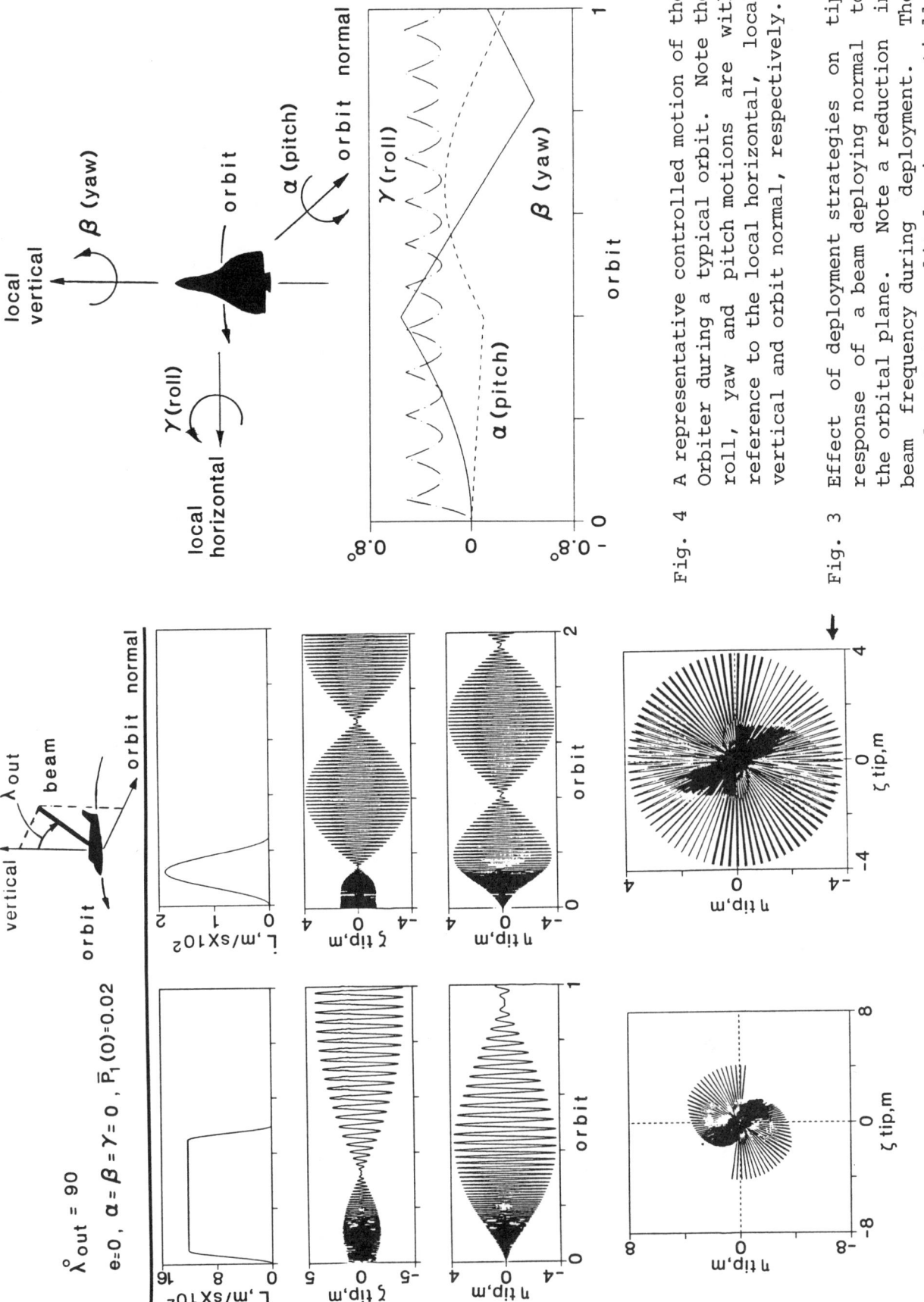

Fig. 4 A representative controlled motion of the Orbiter during a typical orbit. Note the roll, yaw and pitch motions are with reference to the local horizontal, local vertical and orbit normal, respectively.

Fig. 3 Effect of deployment strategies on tip response of a beam deploying normal to the orbital plane. Note a reduction in beam frequency during deployment. The steady state amplitude is essentially independent of the strategy for a given time of deployment.

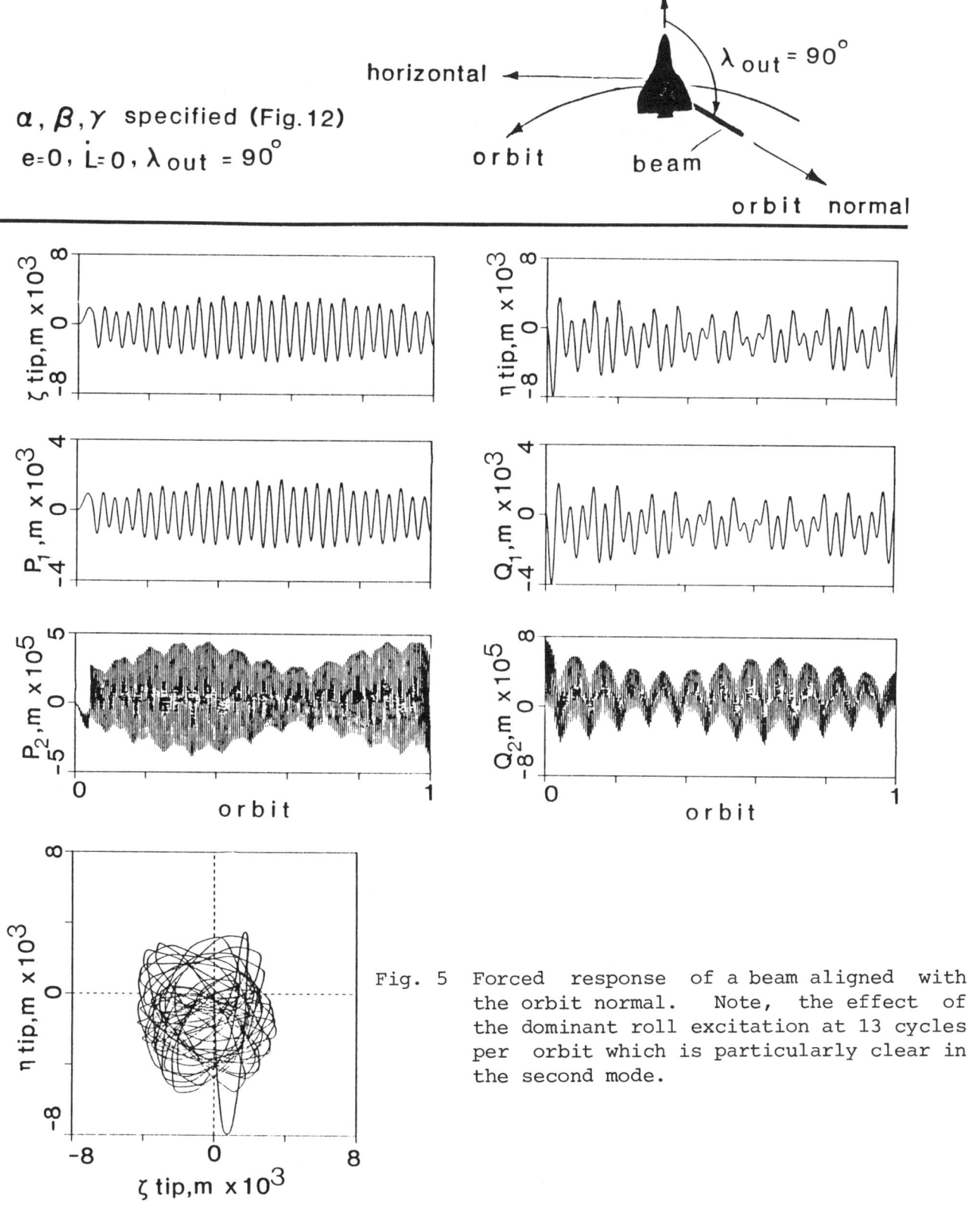

Fig. 5 Forced response of a beam aligned with the orbit normal. Note, the effect of the dominant roll excitation at 13 cycles per orbit which is particularly clear in the second mode.

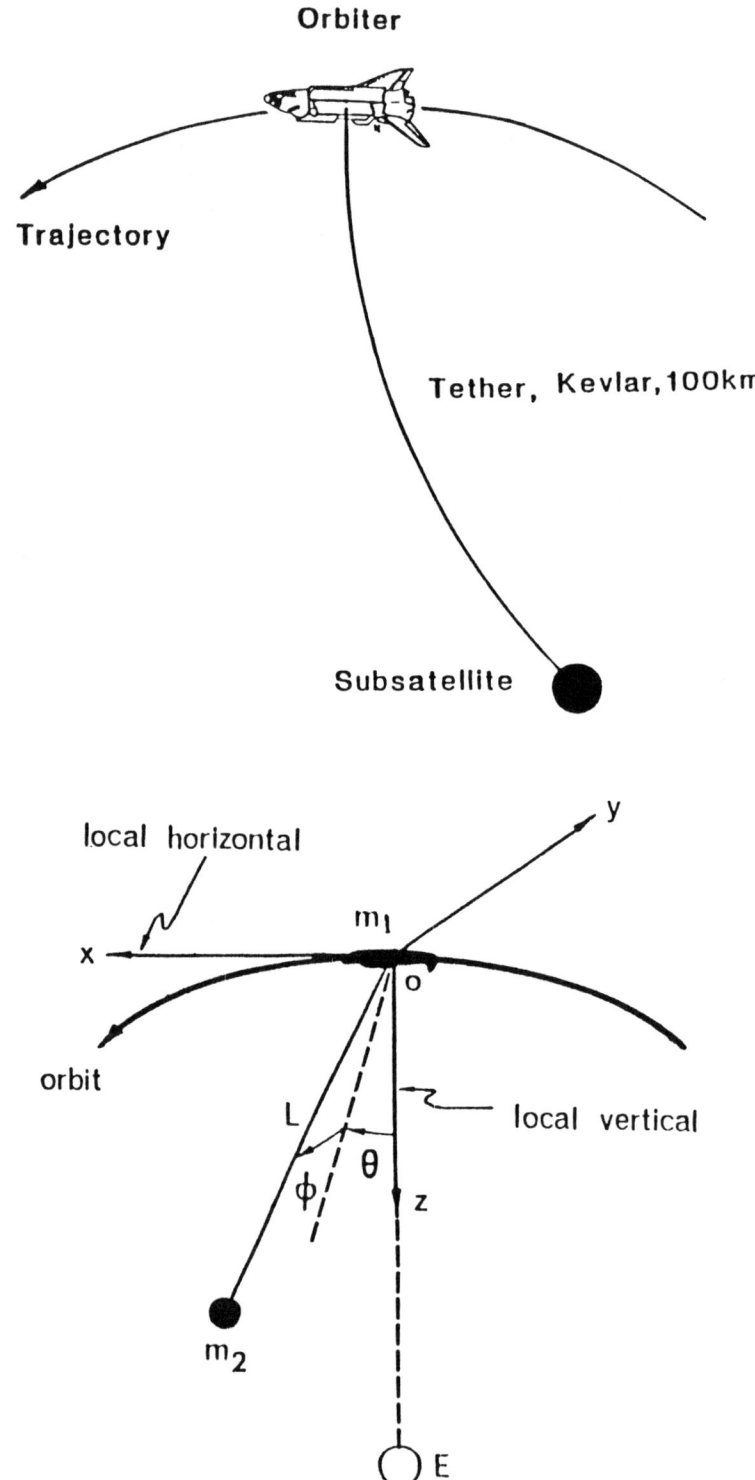

Fig. 6 The NASA/CNR proposed Orbiter based tethered subsatellite system.

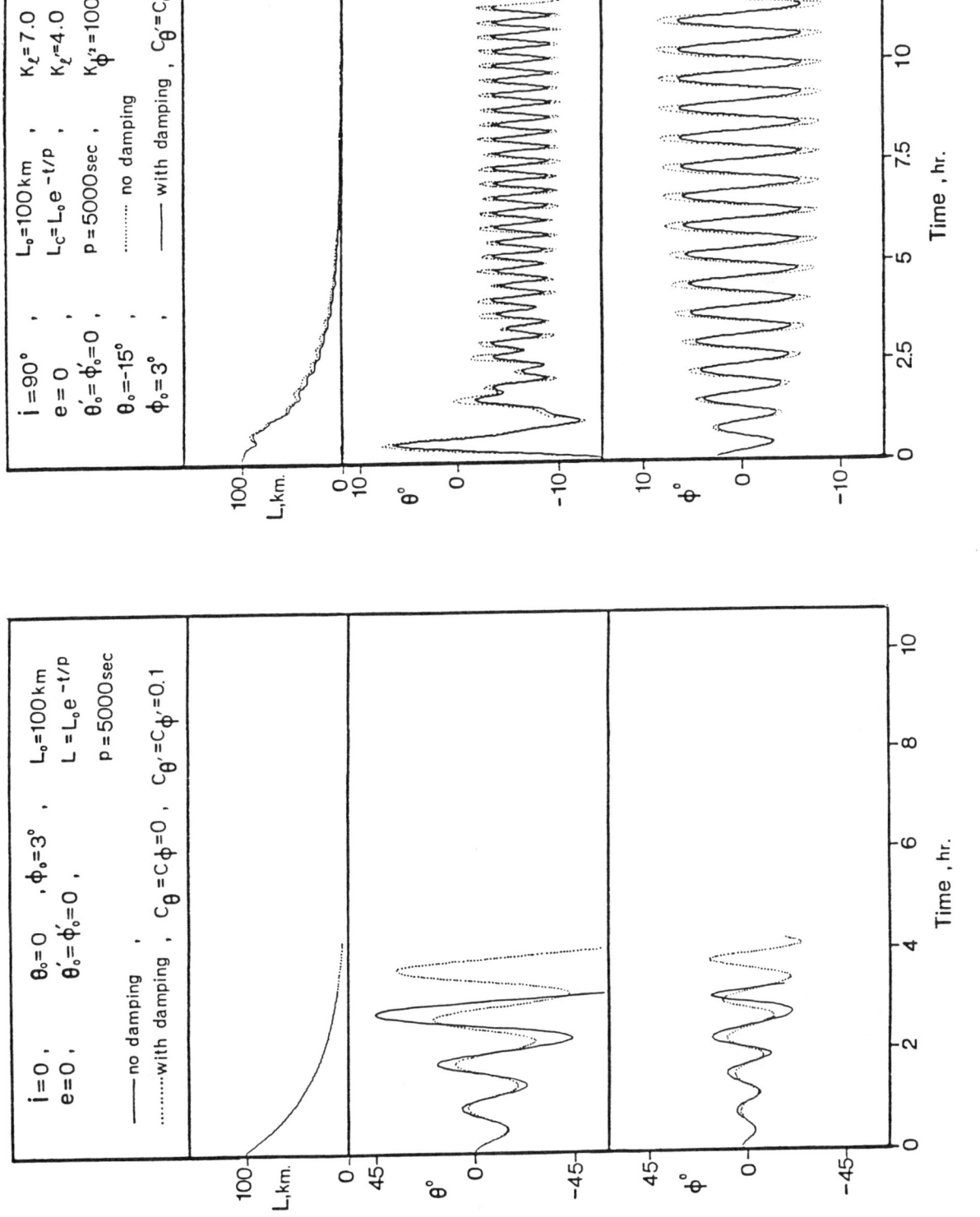

Fig. 7 Typical unstable response of the tethered system during uncontrolled retrieval.

Fig. 8 Retrieval dynamics in presence of the proposed nonlinear control showing dramatic reductions in limit cycle amplitudes.

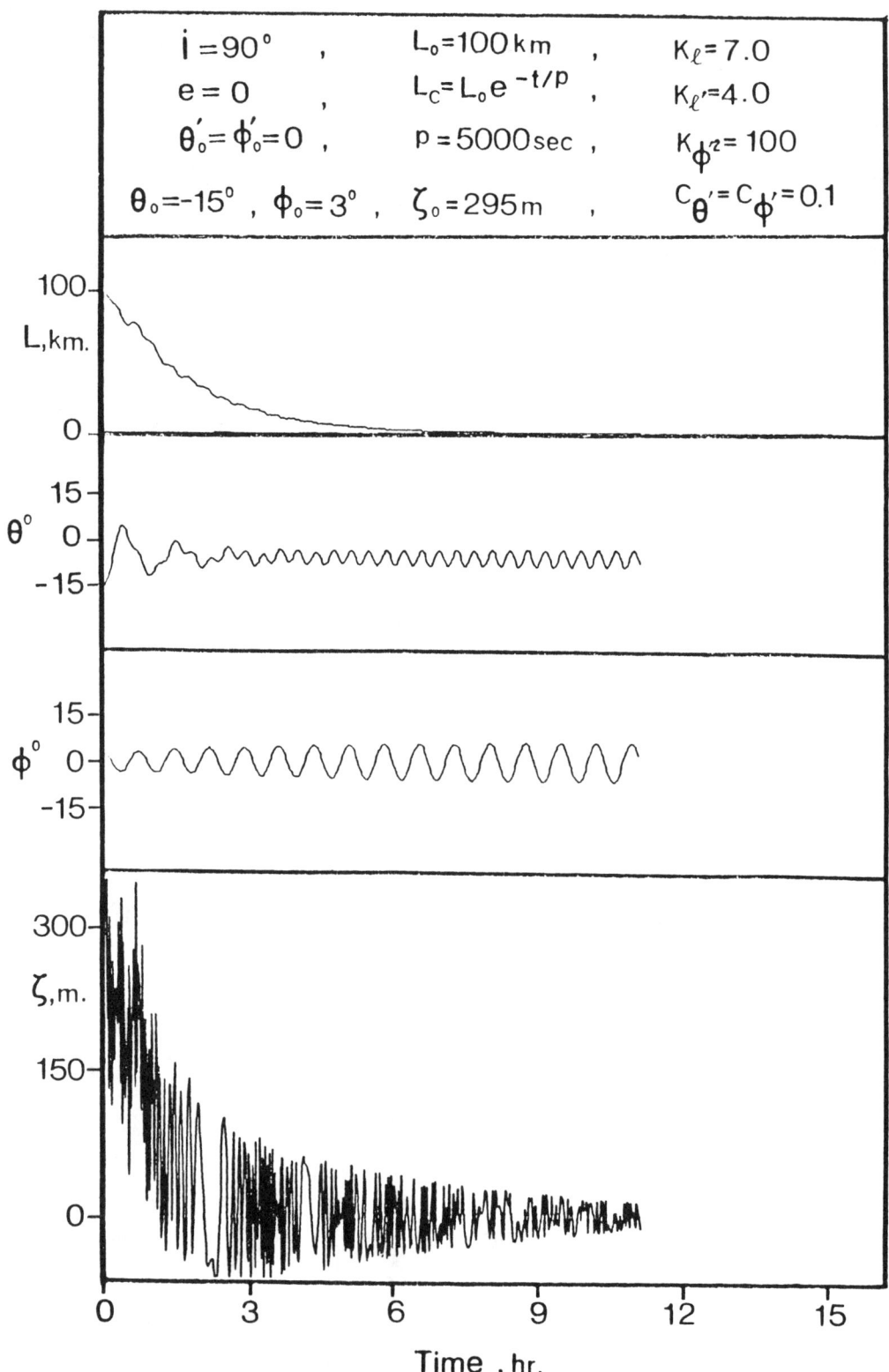

Fig. 9 Plots showing effectiveness of the proposed nonlinear control strategy accounting for longitudinal flexibility of the tether.

# INTERACTIVE STRUCTURAL AND CONTROLLER SYNTHESIS FOR LARGE SPACECRAFT

S.C. McIntosh, Jr.
McIntosh Structural Dynamics, Inc.
Palo Alto, CA

M.A. Floyd
Integrated Systems, Inc.
Palo Alto, CA

## ABSTRACT

A technique is developed for least-weight optimal design of a tubular-truss space structure, subject to constraints on its natural frequencies and its open-loop disturbance-rejection properties. The disturbance-rejection properties of the structure are measured by disturbance-to-regulated-variable grammians. It is shown how this technique can be embedded in a model-reduction scheme based on internal balancing. The procedure is applied to a simple "dumbbell" model and CSDL Model No. 1.

## PROBLEM DESCRIPTION

Large, flexible spacecraft have ambitious performance requirements that must be met in the presence of a variety of disturbances. The natural-frequency spectrum of such spacecraft is typically quite dense, with a number of modes below 1.0 Hz, and damping levels are small. Active controls, whether for maneuvering or for disturbance rejection, interact strongly with the flexible structural modes. The design of controllers for these spacecraft involves such problems as the selection of a reduced-order structural model, the choice of sensor and actuator types and locations, and the control strategy itself. Traditionally, this has been preceded by the design of the structure, with very little interaction between the two processes, as is illustrated in Figure 1a. In view of the advances that have been made separately in the optimal design of structures and controls, great interest has arisen in integrating the two into a single optimal design problem.

The controllers considered fall naturally into two classes: those for maneuver control and those for vibration regulation. References [1] and [2] are recent examples of work with the former type. The integrated problem involves solving for both structural parameters and an optimal control for a specified maneuver. Both [1] and [2] use modal truncation in order to reduce the size of the problem.

Approaches to integrated design for vibration regulation have for the most part considered a system with a linear regulator. The model is not reduced, and the cost function involves both structural parameters and the elements of a feedback gain matrix. In [3] and [4], the gains are determined so as to minimize a quadratic cost function that is combined with a structural cost function. Reference [5] has a more general formulation, in that the gains are controlled by eigenvalue placement, and more freedom is allowed in selecting the cost function. More specialized approaches include those of [6] and [7]. In [6], lattice plate finite elements based on a continuum model of a large space structure are used; these elements permit an evaluation

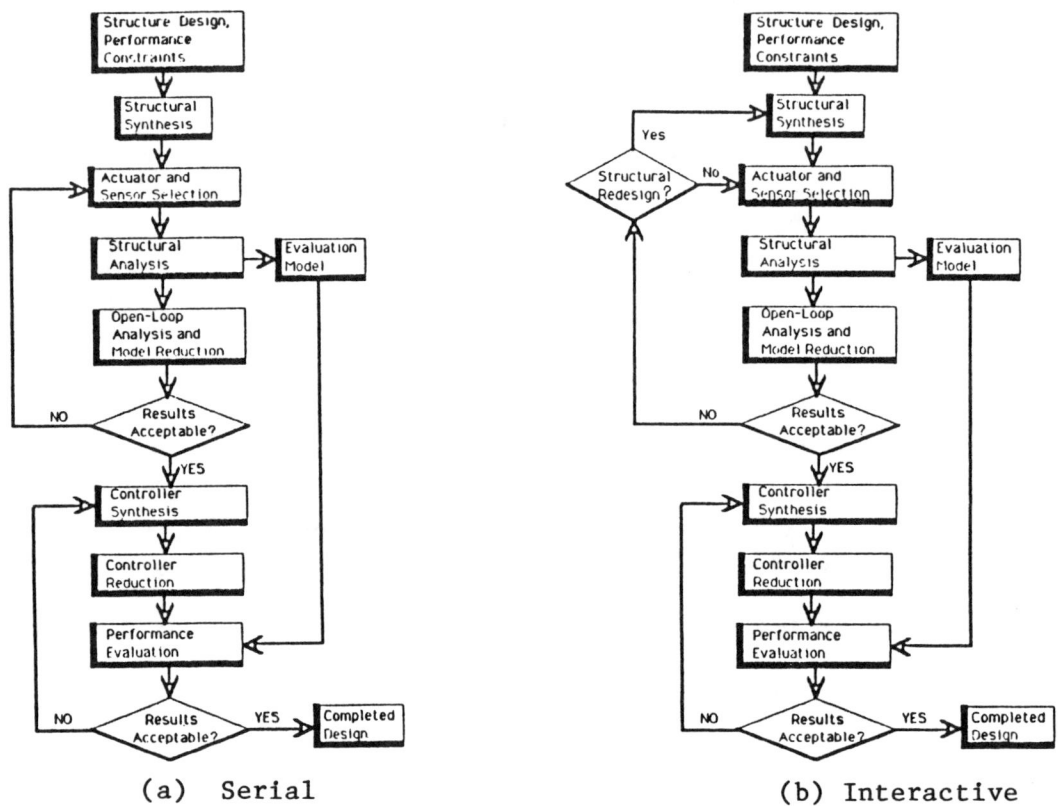

Figure 1. Structural and Controller Synthesis Procedures for Large Spacecraft.

of the effects of structural parameter variations on the performance of reduced-order linear quadratic regulators. In [7], an algorithm to obtain maximal reduction in required control strength for a minimal structural modification is developed and demonstrated for a colocated force-actuator velocity-sensor pair. A completely different approach to integrated design with set-theoretic methods is proposed in [8].

With the exception of [1], [2], and [6], where model-reduction schemes are employed, none of the work cited above deals with one of the principal problems that arises in the design of controllers for truly large spacecraft--the need to adopt a lower-order model both for evaluation purposes and for implementing a practical controller. This is understandable, since the simultaneous design of a structure and a full-order controller is conceptually an enormous task for large systems, without the added complication of including an estimator for a lower-order controller. Furthermore, the model-reduction task itself is far from simple, and there are many different schemes for accomplishing it.

Consider now the generic problem of reducing the effects of disturbances so as to achieve a desired level of performance in a large, flexible spacecraft. Clearly, the role of the controller is to augment whatever disturbance-rejection qualities the spacecraft structure may inherently exhibit. This implies that the controller must be effective where the structure is least effective. And, once the reduced-order model has been selected, filtering is one of the principal steps taken to avoid problems from spillover effects. These considerations lead to exploring the use of automated structural redesign for two purposes:

(1) To enhance as much as possible the inherent disturbance-rejection qualities of the structure, and
(2) To shape the frequency spectrum, if needed, in order to provide adequate spacing for frequency isolation of the controller.

As will be seen, some control over the frequency spectrum is helpful in any event.

This is the controller/structure interaction problem that is considered in this paper. Since the structural redesign takes place before the loop is closed, this is not a truly integrated approach. It is, however, a necessary first step in the development of an integrated design capability for realistic large-scale structures. Moreover, it will be shown that the techniques developed here are readily extendable to the integrated structure/controller design problem.

TECHNICAL APPROACH

The framework for the approach to structural redesign is provided by a model-reduction scheme based on internal balancing [9].

A linear, time-invariant, asymptotically stable system has equations of motion that can be written in the familiar form

$$\{\dot{x}\} = [A]\{x\} + [B]\{u\}, \quad \{y\} = [C]\{x\} \tag{1}$$

Here $\{x\}$ is a column of system states, $[A]$ is the system matrix, $\{u\}$ is a column of control inputs, $[B]$ is the control distribution matrix, $\{y\}$ is a column of outputs, $[C]$ is the corresponding distribution matrix, and the dot symbol ($\cdot$) denotes differentiation with respect to time t. The controllability grammian $[W_C]^2$ and observability grammian $[W_O]^2$ are given by

$$[W_C]^2 = \int_0^\infty e^{[A]t}[B][B]^T e^{[A]^T t} dt$$
$$[W_O]^2 = \int_0^\infty e^{[A]^T t}[C]^T[C] e^{[A]t} dt \tag{2}$$

The system represented by Eqs. (1) is said to be internally balanced if

$$[W_C]^2 = [W_O]^2 = [\Sigma]^2 \tag{3}$$

where the elements of the diagonal matrix $[\Sigma]$ are $\sigma_1^2, \sigma_2^2, \ldots, \sigma_n^2$, with n being the order of the system, and $i < j$ implying $\sigma_i^2 \leq \sigma_j^2$. It is shown in [9] that any model of the form (1) can be taken to internally balanced form by a similarity transformation on the states $\{x\}$. It is also shown that if a balanced model is partitioned into, say, two subsystems, then the subsystems are also asymptotically stable (special steps must be taken if the $\sigma_i^2$ are not distinct).

The ability to quantify controllability and observability rankings and the stability property of subsystems are attractive characteristics for a model-reduction scheme. However, the need for similarity transformation and the concomitant loss of the original model states is undesirable, particularly for large-scale systems. This undesirable effect can be avoided if the structure is lightly damped and has decoupled dynamics; one is therefore led naturally to a modal state-space representation of a large-spacecraft structural-dynamics model in the form

$$\{\ddot{\eta}_i \atop \dot{\eta}_i\} = \begin{bmatrix} -2\zeta_i\omega_i & -\omega_i^2 \\ 1 & 0 \end{bmatrix} \{\dot{\eta}_i \atop \eta_i\} + \begin{bmatrix} b_i \\ 0 \end{bmatrix} \{u\} + \begin{bmatrix} d_i \\ 0 \end{bmatrix} \{w\}, \quad i = 1, 2, \ldots, n \quad (4)$$

$$\{z\} = \sum_{i=1}^{n} [m_{1_i} m_{2_i}] \{\dot{\eta}_i \atop \eta_i\}, \qquad \{y\} = \sum_{i=1}^{n} [c_{1_i} c_{2_i}] \{\dot{\eta}_i \atop \eta_i\}$$

There are n modal coordinates $\eta_i$, with frequency $\omega_i$ and damping ratio $\zeta_i$. The $i^{th}$ row of [B] is $[b_i]$, and the $i^{th}$ row of the distribution matrix [D] is $[d_i]$. The disturbance inputs are $\{w\}$ and the measurements are $\{z\}$. The measurement matrix [M] and regulated-variable matrix [C] are partitioned into blocks multiplying $\{\dot{\eta}_i\}$ and $\{\eta_i\}$, as shown above. The performance measure for the controller is assumed to be

$$J = \lim_{t \to \infty} E(\{y\}^T \{y\}) \quad (5)$$

Let this model be the evaluation model (see Fig. 1a). In [9] it is shown that the internal-balancing transformation simplifies, for $\zeta_i \ll 1$ and certain conditions on the frequency separation, to a simple diagonal scaling transformation. The modal decoupling is thus preserved in the internally balanced form of the model, and in fact it is not necessary to deal explicitly with the transformed model. The elements of the (equal) controllability and observability grammians are given by

$$\sigma_i^2 = \frac{1}{4\zeta_i \omega_i}([b_i][b_i]^T(\{c_{1_i}\}^T\{c_{1_i}\} + \frac{1}{\omega_i^2}\{c_{2_i}\}^T\{c_{2_i}\}))^{1/2} \quad (6)$$

The actual modal-selection process of [9] applies the internal-balancing concept to four separate input-output pairs:

- Disturbability to regulated variable: $[\Sigma]_{DC}^2$
- Controllability to regulated variable: $[\Sigma]_{BC}^2$
- Disturbability to sensors: $[\Sigma]_{DM}^2$
- Controllability to sensors: $[\Sigma]_{BM}^2$

The subscripts correspond to the various matrices that are involved in the computations of the grammians. Thus the second grammian, $[\Sigma]_{BC}^2$, corresponds to the example given above. The elements of these grammians will be identified with the same subscripts, and expressions for all of these resemble the expression (6) above for $\sigma_{BC_i}^2$.

The elements of $[\Sigma]_{DC}^2$ give modal rankings based on open-loop performance. Large values indicate relatively large disturbance propagation to the output. These measures are similar, but not equivalent, to the modal costs of [10].

The elements of $[\Sigma]_{BC}^2$ permit rankings based on the modal controllability of the performance. A small value of the $i^{th}$ element indicates that the given actuator configuration cannot directly affect the contribution of the $i^{th}$ mode to the performance.

The elements of $[\Sigma]_{DM}^2$ rank the observability of disturbances in the sensors. A mode whose corresponding element in this matrix is relatively small may be difficult to estimate on-line.

The elements of $[\Sigma]_{BM}^2$ provide a mode-by-mode measure of potential controller authority for the given set of actuators and sensors.

A mode-selection process involving these four rankings is proposed in [9]. This process involves selecting the modes with the largest disturbance propagation to the input and then including other modes with relatively high rankings based on the other grammians; see [9] for details.

As is indicated in the foregoing discussion, the process of model reduction can be complicated, and combining it <u>ab initio</u> wih an automated structural redesign procedure is not warranted. A more useful first step is to consider the open-loop problem and deal solely with the effects of disturbances on the performance. It is also useful to exert some control over the frequency spectrum. These considerations lead to the following problem statement:

Minimize

$$W(t_i), \; i = 1,2,\ldots,N_d$$

subject to

(7)

$$\sigma_{DC_i}^2 \leq \bar{\sigma}_i, \; i = 1,2,\ldots,N_g; \quad \underline{\omega}_i \leq \omega_i \leq \bar{\omega}_i, \; i = 1,2,\ldots,N_F; \quad \underline{t}_i \leq t_i \leq \bar{t}_i, \; i = 1,2,\ldots,N_d$$

The objective function is the weight W, parameterized as a function of $N_d$ design variables $t_i$. The behavioral constraints are represented by upper bounds on $N_g$ grammians, and upper and lower bounds on the frequencies of $N_F$ modes. Side constraints can be imposed on the design variables. The scenario envisaged here is one in which a designer is seeking to improve a design judged deficient in its disturbance-rejection capabilities, while at the same time tailoring the structure's natural-frequency spectrum to improve filtering or to avoid frequency excursions into an undesirable band. Making weight the objective serves to ensure, at the least, a minimum weight addition to achieve the desired behavior, and the designer can trade off weight against structural capabilities by varying the constraint bounds. This scenario is illustrated in Fig. 1b.

As posed above, this structural redesign problem is amenable to solution by a number of optimal search techniques. Virtually all of these techniques require sensitivity information in the form of derivatives of the objective and constraints with respect to the design variables. While these derivatives can be obtained by finite differencing, it is generally more efficient to compute them from analytical expressions.

First, it is necessary to parameterize the structure itself. The structure is modeled with finite elements, and nodal locations are assumed fixed (i.e., shape changes are excluded). Design variables are typically member sizes, such as bar cross-sectional areas or plate thicknesses. Since space structures are very commonly made up from trusses, the elements to be used are restricted to tubular bar elements. Three resizing options are permitted:

(1) Vary only the tube thickness, with side constraints to ensure that the thin-wall assumption is not violated.

(2) Vary only the radius; side constraints are also applicable here to preserve the thin-wall assumption.

(3) Scale the complete cross section--i.e., both the radius and thickness vary in the same proportion.

For any combination of these options, the discrete mass and stiffness matrices of the structure can be parameterized as follows:

$$[K] = [K_o] + \sum_{i=1}^{N_d} (t_i)^{h_i} [K_i] \quad , \quad [M] = [M_o] + \sum_{i=1}^{N_d} (t_i)^{k_i} [M_i] \tag{8}$$

The matrices $[K_o]$ and $[M_o]$ represent nonactive structure, or structure that is not to be resized (payload, for example). Note that once the geometry of the structure is fixed, the matrices $[K_o]$, $[K_i]$, $[M_o]$, and $[M_i]$ are fixed. Therefore computation of derivatives, as well as updating the mass and stiffness matrices, is particularly simple.

In a manner consistent with Eqs. (8), the weight can be expanded as

$$W = W_o + \sum_{i=1}^{N_d} (t_i)^{m_i} W_i \tag{9}$$

To transform to modal coordinates, the system eigenvector matrix $[\phi]$ is needed. At this point, another important assumption is made. In order to reduce the size of the analysis problem during redesign, the eigenvectors of the initial design are used as basis vectors for the transformation to modal coordinates during the redesign process. Thus the order of the model during redesign remains that of the evaluation model, rather than that of the original discrete model. (Some accuracy may possibly be sacrificed, particularly if the design changes significantly. However, if accuracy is unduly compromised, $[\phi]$ can easily be updated.) With $[GK] = [\phi]^T[K][\phi]$, $[GM] = [\phi]^T[M][\phi]$, Eqs. (8) transform to

$$[GK] = [GK_o] + \sum_{i=1}^{N_d} (t_i)^{h_i} [GK_i] \quad , \quad [GM] = [GM_o] + \sum_{i=1}^{N_d} (t_i)^{k_i} [GM_i] \tag{10}$$

The modal coordinates $\{\eta\}$ in Eqs. (4) are related to the discrete coordinates $\{q\}$ as

$$\{q\} = [\phi][\psi]\{\eta\} \tag{11}$$

Here $[\psi]$ is a square eigenvector matrix in modal coordinates that must be updated at each redesign step. The disturbance-to-regulated-variable grammians are

$$\sigma^2_{DC_i} = \frac{1}{4\zeta_i \omega_i} \left( [d_i][d_i]^T (\{c_{1_i}\}^T \{c_{1_i}\} + \frac{1}{\omega_i^2} \{c_{2_i}\}^T \{c_{2_i}\}) \right)^{1/2} \tag{12}$$

The discrete regulated-variable distribution matrix $[\overline{C}]$ and the disturbance distribution matrix $[\overline{D}]$ are related to their modal counterparts through the transformation (11). This leads to

$$[D_i] = \{\psi_i\}^T [\phi]^T [\overline{D}], \quad \{c_{1_i}\} = [\overline{C}_1][\phi]\{\psi_i\}, \quad \{c_{2_i}\} = [\overline{C}_2][\phi]\{\psi_i\} \tag{13}$$

where $\{\psi_i\}$ is the $i^{th}$ column of $[\psi]$. The problem of computing the gradient of the grammians therefore reduces to that of computing the derivatives of $\{\psi_i\}$ and $\omega_i$. These are obtained as in [11] from the undamped modal representation of the system with a normalization condition on the eigenvectors. A linear system of equations is solved for the derivatives of the eigenvector and the frequency.

Equation (12) can be simplified somewhat by making use of Eqs. (13). The products in Eq. (12) become

$$\delta_i = [d_i][d_i]^T = \{\psi_i\}^T[\phi]^T[\overline{D}][\overline{D}]^T[\phi]\{\psi_i\} = \{\psi_i\}^T[\overline{GD}]\{\psi_i\}$$

$$\sigma_{1_i} = \{c_{1_i}\}^T\{c_{1_i}\} = \{\psi_i\}^T[\phi]^T[\overline{C}_1]^T[\overline{C}_1][\phi]\{\psi_i\} = \{\psi_i\}^T[G\overline{C}_1]\{\psi_i\} \quad (14)$$

$$\sigma_{2_i} = \{c_{2_i}\}^T\{c_{2_i}\} = \{\psi_i\}^T[G\overline{C}_2]\{\psi_i\}$$

Like the matrices $[GK_i]$ and $[GM_i]$, the matrices $[\overline{GD}]$, $[G\overline{C}_1]$, and $[G\overline{C}_2]$ are invariant during redesign. They are also symmetric, so that, for example,

$$\frac{\partial \delta_i}{\partial t_j} = 2\{\psi_i\}^T[\overline{GD}]\{\psi_{i,j}\} \quad (15)$$

Chain-rule differentiation of Eq. (12) produces the desired derivatives:

$$\frac{\partial \sigma^2_{DC_i}}{\partial t_j} = \frac{\partial \sigma^2_{DC_i}}{\partial \omega_i}\frac{\partial \omega_i}{\partial t_j} + \frac{\partial \sigma^2_{DC_i}}{\partial \delta_i}\frac{\partial \delta_i}{\partial t_j} + \frac{\partial \sigma^2_{DC_i}}{\partial \sigma_{1_i}}\frac{\partial \sigma_{1_i}}{\partial t_j} + \frac{\partial \sigma^2_{DC_i}}{\partial \sigma_{2_i}}\frac{\partial \sigma_{2_i}}{\partial t_j} \quad (16)$$

Before the actual optimization strategy is discussed, it is desirable to recast the problem statement of Eqs. (7) into a generic form. The behavioral constraints are rewritten in dimensionless form as follows:

$$(\sigma^2_{DC_i}/\overline{\sigma}_i) - 1 \leq 0, \quad i = 1, 2, \ldots, N_g$$

$$(\omega_i/\overline{\omega}_i) - 1 \leq 0, \quad 1 - (\omega_i/\underline{\omega}_i) \leq 0, \quad i = 1, 2, \ldots, N_f \quad (17)$$

If necessary, the objective function as expressed in Eq. (9) can also be scaled in order to avoid numerical difficulties. Similarly, it is useful to view the design variables $t_i$ as scale factors on the actual physical parameters. Thus the initial design is always characterized by $t_i = 1.0$ for all $i$. The optimization problem becomes

Minimize

$$W(t_i), \quad i = 1, 2, \ldots, N_d \quad (18)$$

subject to

$$G_j(t_i) \leq 0, \quad j = 1, 2, \ldots, N_g + N_f, \quad \underline{t}_i \leq t_i \leq \overline{t}_i, \quad i = 1, 2, \ldots, N_d$$

Here the constraint functions $G_j$ take the appropriate forms as given in Eqs. (17).

The optimal search strategy is based on the method of feasible directions [12], as implemented in the general purpose optimization code CONMIN [13].

## EXAMPLES

The first system to be analyzed, a low-order beam model with point masses, is shown in Fig. 2. Two five-foot aluminum tubes support a 10-lb mass at the center and 5-lb masses at each end. Free-free motion in the y-z plane is allowed. The disturbances are a torque about the z axis and a load in the z direction at node 2, and the measurements combine the difference in z-direction displacements at nodes 1 and 3, and its rate. Six design variables, representing scalings on cross-section properties for each beam segment, were allowed. Modal damping of 0.002 was assumed for the grammian calculations.

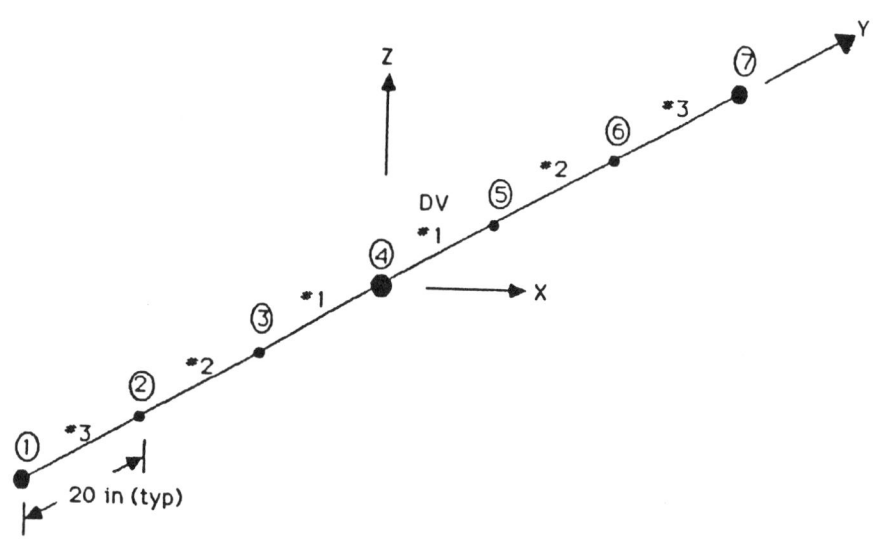

Figure 2. Layout of Dumbbell Model.

The elastic natural frequencies of this model are given in Table I. Modes 6, 8, and 14 are the modes that propagate the disturbance into the measurement.

TABLE I. Elastic Natural Frequencies of the Dumbbell Model

| Mode No. | 5 | 6 | 7 | 8 | 9 | 10 | 11 | 12 | 13 | 14 |
|---|---|---|---|---|---|---|---|---|---|---|
| Frequency, Hz | 35.05 | 138.6 | 237.2 | 510.6 | 513.6 | 553.2 | 680.0 | 880.2 | 1062 | 1235 |

To test the redesign procedure, an optimization problem was set up with this model of the dumbbell. The goal of this optimization problem was to reduce the grammian for mode 6 to 0.2200, with lower bounds on the frequencies up to mode 13 and an upper bound on the frequency of mode 8 as well.

To assess the impact of using fixed mode shapes, normal modes were computed for the final design from the first run, and the optimization procedure was restarted. These results are given in Table II. Comparison of the initial design of Table II with the final design from the first run indicates that frequencies were in general slightly lower and the grammians slightly higher when the computations were done with normal modes. The final design variables in Table II indicate that the optimization process of the first run went slightly too far; the overall final design, given by the products of the final design values from the two runs, is given at the bottom of Table II. In principle, one could continue this process by updating the modes again, but the results would not change significantly.

TABLE II. Results of Redesign of the Dumbbell Model with Updated Normal Modes

| Mode | Initial Frequency (rad/sec) | Design Constraint (rad/sec) | Initial Grammian $\sigma^2_{DC_i}$ | Design Constraint (rad/sec) | Final Frequency (rad/sec) | Final Grammian $\sigma^2_{DC_i}$ |
|---|---|---|---|---|---|---|
| 5  | 268.1 | $\geq 150.0$ |         |                | 280.5 |         |
| 6  | 790.6 | $\geq 750.0$ | 0.2325  | $\leq 0.2200$  | 850.2 | 0.2202  |
| 7  | 1629  | $\geq 1200$  |         |                | 1585  |         |
| 8  | 3069  | $3210 \geq \omega \geq 2500$ | 0.04579 | $\leq 0.2200$ | 3210 | 0.04596 |
| 9  | 3724  | $\geq 3220$  |         |                | 3796  |         |
| 10 | 4281  | $\geq 3220$  |         |                | 4427  |         |
| 11 | 4977  | $\geq 3220$  |         |                | 4845  |         |
| 12 | 5213  | $\geq 3220$  |         |                | 5447  |         |
| 13 | 6696  | $\geq 3220$  |         |                | 7308  |         |
| 14 | 8000  |              |         |                |       |         |

No. of Iterations 14.    Initial Mass (lbs) 40.28.    Final Mass (lbs) 40.65.
Final Design Variables:   0.9559    1.084    1.028
Final Design Variables (Overall):   1.227    1.046    0.8280

The next example is a generic model of a large-antenna feed horn, first proposed by the Charles Stark Draper Laboratory and known as CSDL Model No. 1. A layout of this model is shown in Fig. 3; the data for this model were taken from [14]. The base of the horn is assumed fixed to ground; this is equivalent to neglecting any coupling between antenna motion and horn motion. The regulated quantity is the relative displacement (line-of-sight error, LOS) of node 1 in the x-y plane, expressed as the x and y displacements of node 1; the disturbances are forces in all three coordinate directions at nodes 2, 3, and 4. There are 12 design variables,

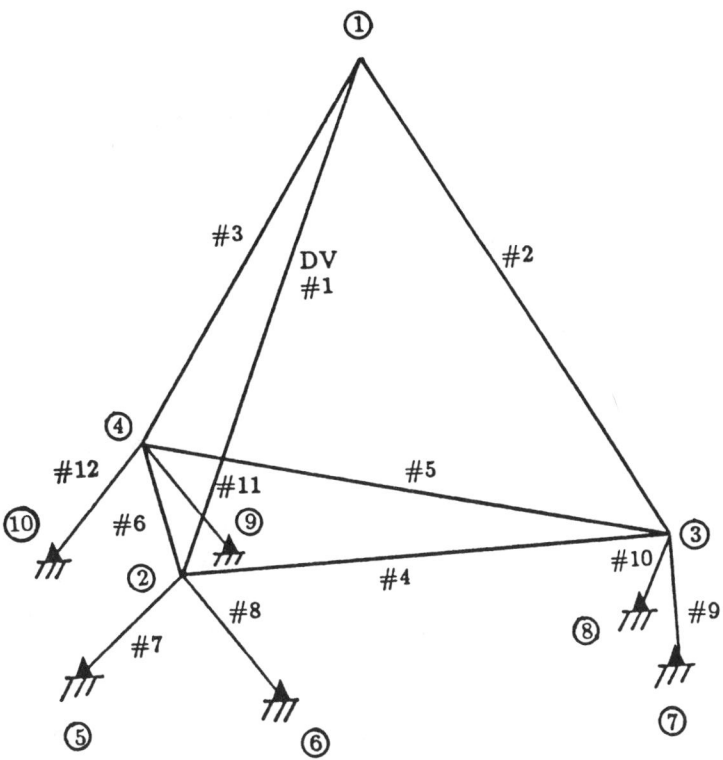

Figure 3. Layout of CSDL Model No. 1.

one for each truss member in the structure. Because the nonactive inertia in this model is so much greater than the inertia in the truss members, the nonactive inertia was not included in the weight calculation during the optimization, and the weight coefficients were scaled in order to avoid numerical problems. Frequencies and grammians for this model are given in Table III. As can be seen, there are two groups of modes, for the most part isolated quite conveniently both spatially and temporally. Note that the grammians for modes 1 and 2 are much greater than those for the other modes.

A number of structural redesigns were run for the CSDL Model 1 structure. Results from two of these runs are shown here. The objective of the first run was to reduce the grammians for modes 1 and 2 while retaining the desirable frequency separation between modes 12 and 13. All of the modes and grammians were constrained, and the desired frequency separation was a full decade, from 20.00 to 200.0 rad/sec. The results are shown in Table IV.

TABLE III. Natural Frequencies and Grammians $\sigma^2_{DC_i}$ of CSDL Model No. 1

| Mode No. | Frequency, Hz | Grammian $\sigma^2_{DC_i}$ |
|---|---|---|
| 1 | 0.2152 | 166.1 |
| 2 | 0.2675 | 203.1 |
| 3 | 0.4701 | 19.47 |
| 4 | 0.4855 | 64.25 |
| 5 | 0.5493 | 59.77 |
| 6 | 0.6750 | $0.7488 \times 10^{-4}$ |
| 7 | 0.7487 | 6.536 |
| 8 | 0.7630 | 22.98 |
| 9 | 1.362 | 13.85 |
| 10 | 1.476 | 0.2267 |
| 11 | 0.640 | 8.907 |
| 12 | 2.057 | 2.220 |
| 13 | 40.88 | $0.1732 \times 10^{-4}$ |
| 14 | 44.86 | $0.1956 \times 10^{-4}$ |
| 15 | 52.63 | $0.3624 \times 10^{-4}$ |
| 16 | 55.52 | $0.5126 \times 10^{-5}$ |
| 17 | 55.79 | $0.2861 \times 10^{-5}$ |
| 18 | 73.31 | $0.2503 \times 10^{-5}$ |

For the second run, an attempt was made to reduce the grammians for modes 1-5 to 50.00, and the design variables were allowed to move between limits of 0.500 and 2.000. All other constraints were the same as in Run 1. The results for this run are given in Table V. The constraints were satisfied, but at the cost of an increase in structural weight of almost 30%. The active constraints were the grammians for modes 2, 3, and 12. The design strategy here was similar to that in Run 1, in that the horn member sizes were reduced and the support member sizes were increased. For this case, however, the upper-bound limits on the support-member sizes were encountered. Also, many more iterations were required. The constraints were satisfied after four iterations, with the structural weight almost doubled, and the remaining iterations were taken up with an agonizingly slow but ultimately successful effort to reduce this weight.

Several control analyses were run to determine the effects of changes in the structure. The top three struts in the structure were considered to be active control elements capable of generating axial forces only.

TABLE IV. Results of Run 1 for CSDL Model No. 1

| Mode | Initial Frequency (rad/sec) | Design Constraint (rad/sec) | Initial Grammian $\sigma^2_{DC_i}$ | Design Constraint (rad/sec) | Final Frequency (rad/sec) | Final Grammian $\sigma^2_{DC_i}$ |
|---|---|---|---|---|---|---|
| 1  | 1.352 | $\geq 1.2$ | 166.1 | $\leq 100$ | 1.200 | 99.41 |
| 2  | 1.681 | $\geq 1.2$ | 203.1 | $\leq 100$ | 1.607 | 100.00 |
| 3  | 2.954 | $\geq 1.2$ | 19.47 | $\leq 100$ | 3.272 | 47.47 |
| 4  | 3.050 | $\geq 1.2$ | 64.25 | $\leq 100$ | 3.774 | 32.81 |
| 5  | 3.451 | $\geq 1.2$ | 59.77 | $\leq 100$ | 4.447 | 53.49 |
| 6  | 4.241 | $\geq 1.2$ | $0.7488 \times 10^{-4}$ | $\leq 100$ | 5.498 | 0.1575 |
| 7  | 4.704 | $\geq 1.2$ | 6.536 | $\leq 100$ | 6.051 | 6.112 |
| 8  | 4.794 | $\geq 1.2$ | 22.98 | $\leq 100$ | 6.325 | 0.2241 |
| 9  | 8.558 | $\geq 1.2$ | 13.85 | $\leq 100$ | 8.021 | 13.16 |
| 10 | 9.277 | $\geq 1.2$ | 0.2267 | $\leq 100$ | 8.444 | 0.5151 |
| 11 | 10.30 | $\geq 1.2$ | 8.907 | $\leq 100$ | 8.856 | 8.448 |
| 12 | 12.93 | $1.2 \geq \omega \geq 75$ | 2.220 | $\leq 100$ | 10.47 | 4.069 |
| 13 | 256.8 | $\geq 200$ | $0.173 \times 10^{-4}$ | | 286.5 | |

No. of Iterations 20. Initial Mass (Scaled) 5.24. Final Mass (Scaled) 3.61.
Final Design Variables: 0.8  0.8  0.8332  0.8  0.8  0.8  1.376  1.374
                       1.229  1.487  1.488  1.232

TABLE V. Results of Run 2 for CSDL Model No. 1

| Mode | Initial Frequency (rad/sec) | Design Constraint (rad/sec) | Initial Grammian $\sigma^2_{DC_i}$ | Design Constraint (rad/sec) | Final Frequency (rad/sec) | Final Grammian $\sigma^2_{DC_i}$ |
|---|---|---|---|---|---|---|
| 1  | 1.352 | $\geq 1.2$ | 166.1 | $\leq 50$ | 1.244 | 42.76 |
| 2  | 1.681 | $\geq 1.2$ | 203.1 | $\leq 50$ | 16.87 | 50.16 |
| 3  | 2.954 | $\geq 1.2$ | 19.47 | $\leq 50$ | 4.708 | 44.80 |
| 4  | 3.050 | $\geq 1.2$ | 64.25 | $\leq 50$ | 5.585 | 9.769 |
| 5  | 3.451 | $\geq 1.2$ | 59.77 | $\leq 50$ | 6.792 | 37.96 |
| 6  | 4.241 | $\geq 1.2$ | $0.7488 \times 10^{-4}$ | $\leq 10$ | 8.640 | $0.4083 \times 10^{-3}$ |
| 7  | 4.704 | $\geq 1.2$ | 6.536 | $\leq 10$ | 8.887 | 1.481 |
| 8  | 4.794 | $\geq 1.2$ | 22.98 | $\leq 10$ | 8.906 | 0.6374 |
| 9  | 8.558 | $\geq 1.2$ | 13.85 | $\leq 10$ | 10.64 | 0.5152 |
| 10 | 9.277 | $\geq 1.2$ | 0.2267 | $\leq 10$ | 11.37 | 0.1674 |
| 11 | 10.30 | $\geq 1.2$ | 8.907 | $\leq 10$ | 11.67 | 4.509 |
| 12 | 12.93 | $1.2 \geq \omega \geq 75$ | 2.220 | $\leq 10$ | 16.15 | 9.968 |
| 13 | 256.8 | $\geq 200$ | $0.173 \times 10^{-4}$ | $\leq 10$ | 422.7 | $0.2205 \times 10^{-4}$ |
| 14 | 281.9 | | | $\leq 10$ | 469.5 | $0.2324 \times 10^{-4}$ |
| 15 | 330.7 | | | $\leq 10$ | 726.7 | 0 |
| 16 | 348.9 | | | $\leq 10$ | 761.3 | $0.1192 \times 10^{-4}$ |
| 17 | 350.5 | | | $\leq 10$ | 762.8 | $0.3576 \times 10^{-5}$ |
| 18 | 460.6 | | | $\leq 10$ | 1344 | 0 |

No. of Iterations 87. Initial Mass (Scaled) 5.24. Final Mass (Scaled) 6.81.
Final Design Variables: 1.439  0.8073  0.7797  1.010  0.8762  1.007
                       2.0  2.0  2.0  2.0  2.0  2.0

Figure 4 shows the open-loop frequency response of the original structure from the actuators to the two LOS outputs, LOS-X and LOS-Y, for frequencies less than 20 rad/sec. Modes 13-18 are not significant in the response. Figure 5 shows the open-

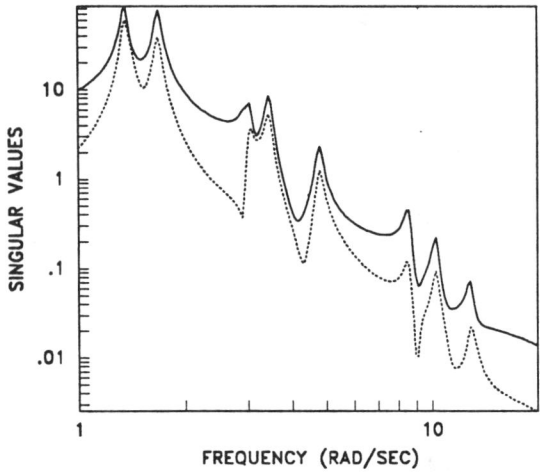

Figure 4. Open-Loop Frequency Response for Original CSDL No. 1 Model.

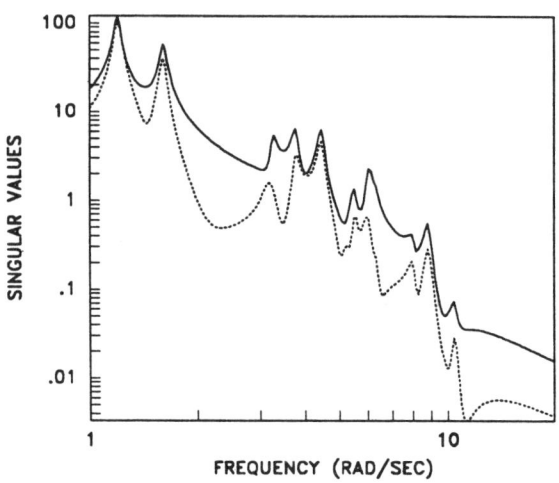

Figure 5. Open-Loop Frequency Response for CSDL No. 1 Model - Run 1.

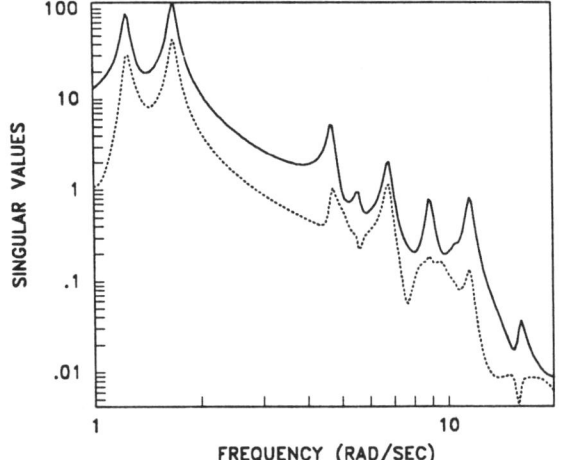

Figure 6. Open-Loop Frequency Response for CSDL No. 1 Model - Run 2.

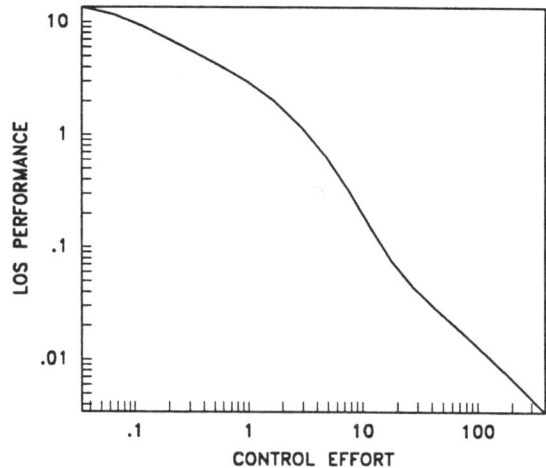

Figure 7. LOS Performance vs Control Effort for Original CSDL No. 1 Model.

loop frequency response of the structure created in Run 1. The differences with the original design are not dramatic. The same holds true for the structure created in Run 2, which is shown in Fig. 6.

To determine what effect the structural changes had on control effort and LOS performance, a series of optimal regulator designs was performed. The optimal regulator design minimized LOS and control effort. Full-state feedback was assumed, since it was not desired to raise the issue of estimation here. Figure 7 shows the results of this analysis for the original model. LOS performance was defined to be the maximum RMS error in LOS-X and -Y, whereas control effort was defined to be the maximum RMS force in the three actuators.

Figures 8 and 9 show performance vs effort for Runs 1 and 2. There are not significant differences with the original model. It is good to note, however, that for the Run 1 model, similar performance is achieved with a much lighter structure (3.61 vs 5.24 mass units, scaled).

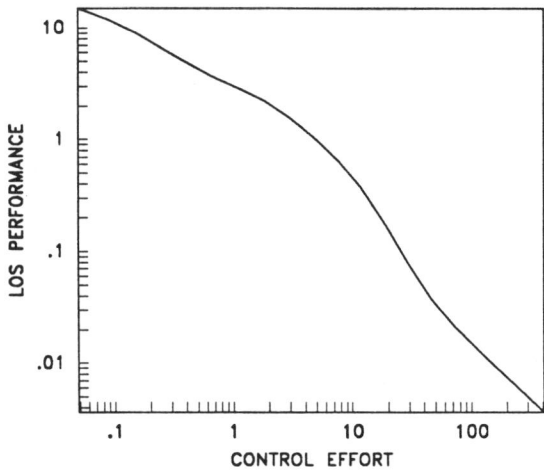

Figure 8. LOS Performance vs Control Effort for CSDL No. 1 Model – Run 1.

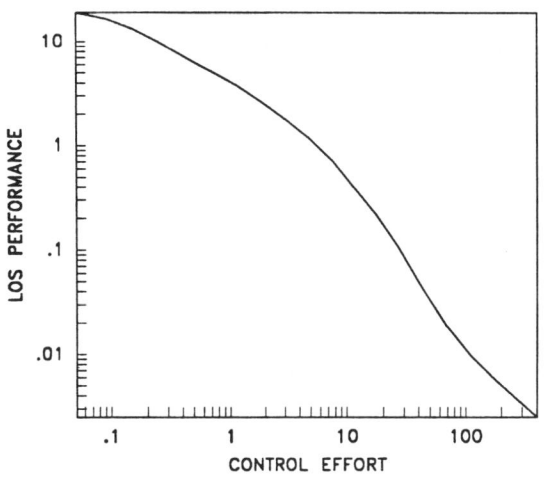

Figure 9. LOS Performance vs Control Effort for CSDL No. 1 Model – Run 2.

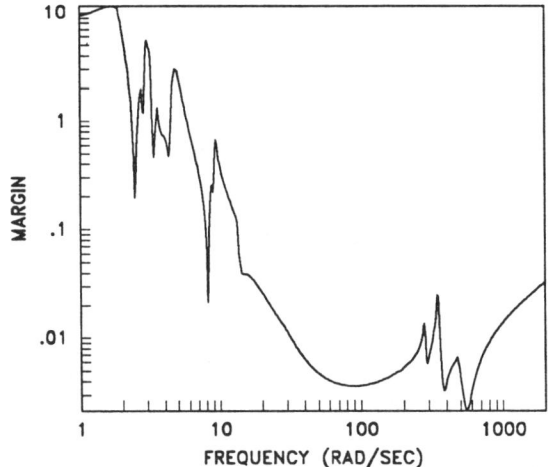

Figure 10. Additive Stability Margin for Original CSDL No. 1 Model.

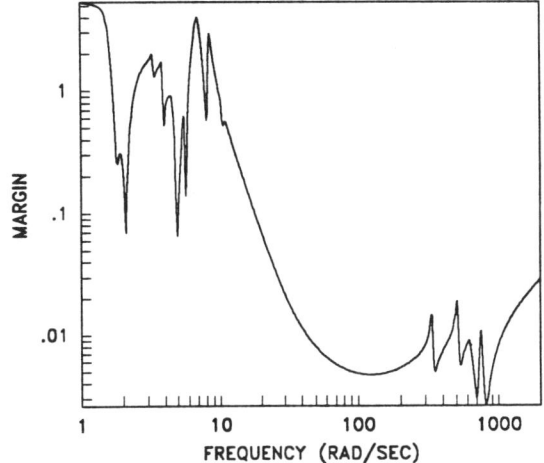

Figure 11. Additive Stability Margin for CSDL No. 1 Model – Run 1.

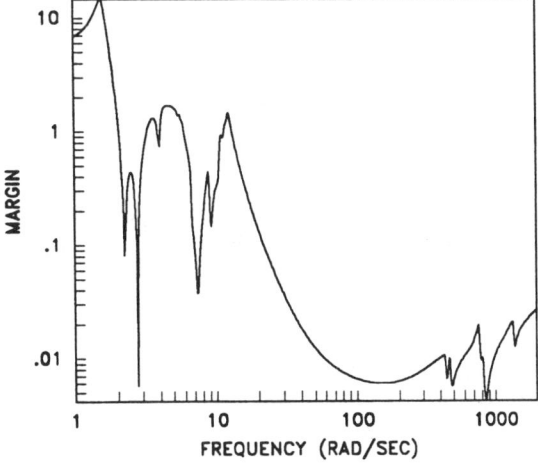

Figure 12. Additive Stability Margin for CSDL No. 1 Model – Run 2.

Finally, the effects of structural changes on the robustness of LQG compensator design was investigated. Three position sensors were placed on the structure, colocated with the three actuators. A Kalman filter was then designed to estimate the states of the first 13 modes. Control designs were used that had equivalent LOS performance and control effort. Figure 10 shows the stability margin for the original design. Figures 11 and 12 show the margins for the Run 1 and Run 2 models. Again there are no significant differences.

CONCLUDING REMARKS

In the examples given above, the ability has been demonstrated to exert control, in some optimal sense, of the natural frequencies and the disturbance-rejection qualities of a structure. Although the structural models used here have only employed up to 12 design variables and 37 constraints, the techniques used have been successfully applied to much larger models in other types of optimization problems, with on the order of 100 design variables and constraints. There is therefore no reason to doubt that the problems treated in this work can be applied to models of similar complexity.

The results of the control analyses performed on CSDL Model 1 are inconclusive. Changes in the structure to reduce measures of its disturbance-rejection capabilities have not significantly altered its characteristics from a control viewpoint. CSDL Model 1 is a structure that is inherently easy to control. There is a large natural break in the frequency response, and the response rolls off almost monotonically after the first two modes. Because the structure is so simple, it does not display all of the pathologies of large space structures that the structural optimization techniques presented in this report are intended to address. Clearly, more complex test cases are required.

In terms of calculating the sensitivities that are required, everything is in place for including the other grammians in the optimization problem, so that other issues that arise in controlling flexible space structures can easily be addressed. Similarly, the extension to integrated controller and structural design with full state feedback, as presented in [2], is at least conceptually a simple matter. For practical applications, however, reduced-order controllers are required, and including state estimation in the integrated design problems remains an open question. For the present, therefore, it appears advisable to continue this work along the following lines:

1. Continue the current applications with more complex models.

2. Incorporate, as appropriate, other grammians into the design problem, and evaluate suitable complex models.

3. As an intermediate step, formulate and evaluate the integrated design problem with full state feedback.

4. Investigate the formulation of the integrated design problem with state estimation included.

## ACKNOWLEDGMENTS

This research was sponsored by the Air Force Office of Scientific Research (AFSC), under Contract F49620-84-C-0025. The United States Government is authorized to reproduce and distribute reprints for governmental purposes notwithstanding any copyright notation hereon.

## REFERENCES

1. Lisowski, R.J. and Hale, A.L.: Optimal Design for Single Axis Rotational Maneuvers of a Flexible Structure. AIAA Paper 84-1041, presented at the AIAA Dynamics Specialists Conference, Palm Springs, CA, May 1984.
2. Messac, A., Turner, J.D., and Soosaar, K.: An Integrated Control and Minimum Mass Structural Optimization Algorithm for Large Space Structures. Paper presented at the JPL Workshop on Identification and Control of Flexible Space Structures, San Diego, CA, June 1984.
3. Salama, M., Hamidi, M., and Demsetz, L.: Optimization of Controlled Structures. Paper presented at the JPL Workshop on Identification and Control of Flexible Space Structures, San Diego, CA, June 1984.
4. Khot, N.S., Eastep, F.E., and Venkayya, V.B.: Optimal Structural Modification to Enhance the Optimal Active Vibration Control of Large Flexible Structures. AIAA Paper 85-0627, presented at the AIAA/ASME/ASCE/AHS 26th Structures, Structural Dynamics and Materials Conference, Orlando, FL, April 1985.
5. Bodden, D.S. and Junkins, J.L.: Eigenvalue Optimization Algorithms for Structural/Controller Design Iterations. Paper presented at the 1984 American Control Conference, San Diego, CA, June 1984.
6. Lamberson, S.E. and Yang, T.Y.: Optimization Using Lattice Plate Finite Elements for Feedback Control of Space Structures. AIAA Paper 85-0592, presented at the AIAA/ASME/ASCE/AHS 26th Structures, Structural Dynamics and Materials Conference, Orlando, FL, April 1985.
7. Haftka, R.T., Martinovic, Z.N., and Hallauer, W.L.: Enhanced Vibration Controllability by Minor Structural Modification. AIAA Journal, Vol. 23, No. 8, August 1985, pp. 1260-1266.
8. Hale, A.L.: Integrated Structural/Control Synthesis via Set-Theoretic Methods. AIAA Paper 85-0806, presented at the AIAA/ASME/ASCE/AHS 26th Structures, Structural Dynamics and Materials Conference, Orlando, FL, April 1985.
9. Gregory, C.Z.: Reduction of Large Flexible Spacecraft Models Using Internal Balancing Theory. Journal of Guidance, Control and Dynamics, Vol. 7, No. 6, November-December 1984, pp. 725-732.
10. Skelton, R.E., Hughes, P.C., and Hablani, H.: Order Reduction for Models of Space Structures Using Model Cost Analysis. Journal of Guidance, Control and Dynamics, Vol. 5, No. 4, July-August 1982, pp. 351-357.
11. Cardani, C. and Mantegazza, P.: Calculation of Eigenvalue and Eigenvector Derivatives for Algebraic Flutter and Divergence Eigenproblems. AIAA Journal, Vol. 17, No. 4, April 1979, pp. 408-412.
12. Zoutendijk, G.: <u>Methods of Feasible Directions</u>. Elsevier Publishing Co., Amsterdam, 1960.
13. Vanderplaats, G.N.: CONMIN--A Fortran Program for Constrained Function Minimization; User's Manual. NASA TM X-62,282, August 1973; Addendum, May 1978.
14. Preston, R.B.: Pareto Optimal Vibration Damping in Large Space Structure Control. M.S. Thesis, Dept. of Aeronautics and Astronautics, Massachusetts Institute of Technology, Cambridge, MA, December 1979.

DAMAGE IN COMPOSITE LAMINATES FROM CENTRAL IMPACTS AT SUBPERFORATION SPEEDS

L.E. Malvern and C.T. Sun
Engineering Sciences Dept., University of Florida, Gainesville, FL 32611
and D. Liu, Metallurgy Mechanics and Materials Science Dept.
Michigan State University, East Lansing, MI 48824

ABSTRACT

An experimental research program at the University of Florida has shown that in central impacts at normal incidence by small hard impactors at subperforation speeds on continuous-filament fiber-reinforced laminated composite plates the major damage is delamination at the interfaces between unidirectionally reinforced layers of different orientation. This is accompanied by some matrix cracking between the fibers of the layers.

Composites with three kinds of fibers have been investigated (glass, Kevlar® and graphite), each in a brittle epoxy matrix. This paper reports some new results on graphite and reviews some previously published results on glass.

Relationships between total and projected delamination areas and the kinetic energy imparted to the target plate are reported for each impactor/target combination at subperforation speeds sufficiently low that delaminations do not extend all the way to the boundaries.

Strength and stiffness retention factors, based on three-point bend tests of damaged specimens and undamaged control specimens have been determined for a range of subperforation speeds on different stacking sequences of each of the composite systems. The percentage strength reduction is greater than the stiffness reduction, as determined by this method.

INTRODUCTION

The ability of structural components to withstand short-duration impact loadings is related to the way in which the stress and deformation are transmitted through the material. This transmission may involve elastic stress waves, propagating plastic deformation zones, and/or propagating cracks. The transmitted stress waves can also produce considerable damage at locations far removed from the impact site, and propagating damage zones can spread the energy absorption over a large volume of the impacted body.

The prediction and control of the failure zones in material systems subjected to transient loading requires understanding of both the basic dynamic materials properties and of the mechanisms by which stress and damage propagate through the material. These problems are interrelated, since the dynamic materials properties must be determined experimentally under conditions of transient loading where the stress and deformation are not uniform through the test specimen.

The introduction of high-strength filamentary types of composite materials as primary and secondary members in structural applications has focused attention on the need to characterize their response to a wide variety of loadings. Because of the anisotropy and heterogeneity of these materials, the observed response can vary considerably from that of monolithic materials for a given set of loadings. To gain understanding of the response of filamentary composite materials to dynamic loadings a program investigating the failure mechanisms in partially penetrated centrally impacted composite plates has been conducted.

The overall objective of the initial program was to obtain from systematic studies on geometrically controlled glass/epoxy composite systems information on the fracture mechanisms occurring in centrally impacted composite plate configurations, which would be useful for establishing design guidelines. The investigation is currently being extended to Kevlar®/epoxy and graphite/epoxy systems. All three of these composite systems have become important for use in aircraft and space vehicle structures because of their high strength-to-weight ratio and, especially for graphite/epoxy, high stiffness-to-weight ratio. The high stiffness is important for keeping the natural frequencies high in order to minimize aeroelastic effects. It is appropriate to dedicate this paper to the memory of Professor R.L. Bisplinghoff, who was renowned for his work in aeroelasticity, even though the paper does not deal explicitly with aeroelasticity.

This research on impact failure mechanisms was motivated by the fact that brittle-matrix laminates are susceptible to impact damage, which has limited applications of these composites in situations where impact is a design consideration, for example in the turbine fan blading of aircraft jet engines. During take-off and flight, ingestion of stones, ice balls, birds and other foreign objects may lead to impact damage on the blades [1]. Small hard impactors produce an effect similar to the effect on a stationary plate produced by a projectile fired from a gas gun at subperforation speeds, as in the experiments reported here.

EXPERIMENTAL PROCEDURE

In each test series, several laminated plates of the same type were subjected to impact at a sequence of increasing subperforation speeds, and the damage produced was correlated with the impact energy imparted to the plates. The impacts were produced by a projectile fired with the gas gun assembly shown schematically in Figure 1. The system differs from that used previously at the University of Florida [2,3,4] only in the method of recording the impactor velocities. The impactor is a short metal cylinder, flat-ended or with a hemispherical or other shaped nose.

The new velocity sensor uses two fiber-optic-transmitted light paths through a cylindrical holder near the impacted plate. The two light paths are 2 inches apart; the first is 5.5 inches from the muzzle of the gas gun, and the second is 2.25 inches from the plate. The signals from the impactor-interrupted light paths are fed into a microprocessor, which reads out times from which it is possible to deduce: (1) the average inbound velocity through the sensor, (2) the arrival time at the plate, (3) the average rebound velocity through the sensor and (4) the time at which the rebounding impactor left the plane of the initial front face of the target plate. Note that the apparent contact time as determined by the difference between (2) and (4) is not necessarily the actual contact time; independent direct measurements [3,4] and calculations [5] both show that the impactor often loses contact with the plate before the maximum plate deflection occurs and makes contact again during the plate spring-back. The microprocessor can also provide a trigger signal for transient recorders.

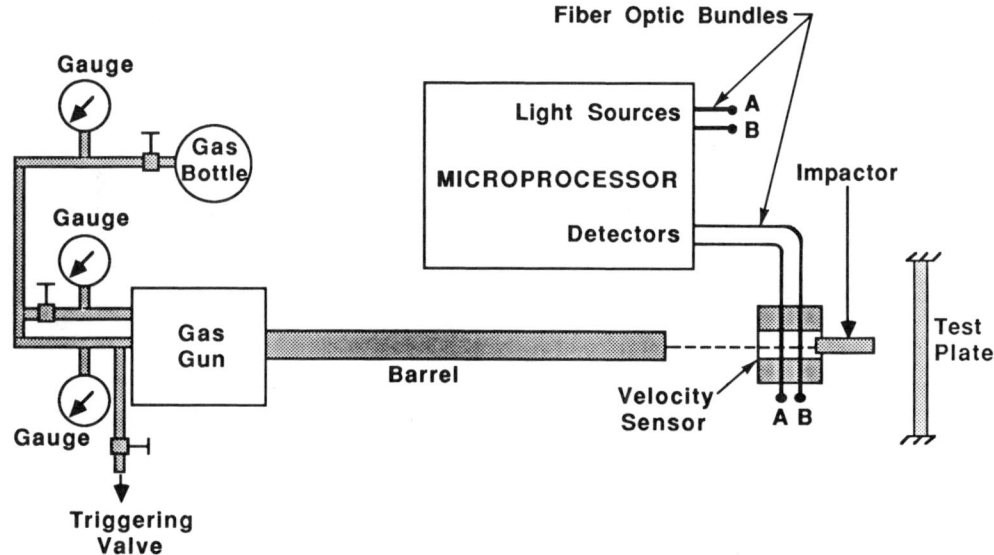

Figure 1. Schematic of Gas Gun and Velocity Measuring System

The rebound velocity is used to determine the rebound kinetic energy (as much as 10 to 20 percent of the initial impactor kinetic energy at these subperforation speeds) in order to determine the energy imparted to the specimen for correlation with the amount of impact damage.

The target plates in most of the investigation were nominally 152 mm (6 inches) square by 3.4 mm thick. The sides of the plates are clamped to steel frames that reduce the exposed plate side length to 140 mm. The edge clamping is far from perfect, so that the actual support provided is somewhere between simply supported and clamped. Elastic plate deflection calculations have obtained close agreement with experimentally determined transient central deflections by treating the 152-mm plate as a clamped plate with an effective side length of 170 mm [5,6]. A few tests have been made on larger plates, nominally 229 mm (9 inches) square, in order to verify that the boundary constraint of the smaller plates did not significantly affect the damage pattern near the center at low-speed impact.

Before impact each plate is weighed, and its thickness is measured at several different points to check uniformity in the sample. Thickness variations between samples are taken into account in the residual strength and stiffness determinations after impact. The plates are also subjected to ultrasonic C-scan examination to ensure that there are no significant initial defects. In the current program, each plate is subjected to a vibration frequency and damping test before impact to establish a baseline for changes in frequency and damping caused by the impact damage.

During impact several different kinds of transient measurements have been made on some of the specimens in order to gain information about the bending wave propagation and the damage development. Transient measurements have included deflection at one or two points by non-contacting eddy-current probes, strain gage measurements, surface crack propagation parallel to the fibers adjacent to the impactor, measured by interrupted electrical paths, and delamination crack propagation in glass/epoxy by high-speed photography. Post-impact nondestructive examination procedures have included visual and photographic recording of delamination areas in semitransparent glass and Kevlar® systems, ultrasonic C-Scan and X-Ray examination, and vibration frequency and damping. Some damaged plates have been cut into strips, and the polished cut surfaces have been examined by an edge replication technique to determine the extent of internal delamination and matrix cracking. Surface matrix cracks have also been revealed by flaw enhancement fluid. In the current program about half of the damaged plates have been subjected to three-point bend tests to compare the retained stiffness and strength to the stiffness and strength of an undamaged control specimen cut from the same fabricated plate.

PLATE SPECIMEN FABRICATION

Several different stacking sequences have been examined [3,4,7], but most of this report will focus on results obtained with 0°/90° layups of epoxy-matrix systems, with unidirectional glass, Kevlar® or graphite continuous filament reinforcement in each ply. All the plates for the research at the University of Florida since 1979 have been fabricated in our laboratory by autoclave curing of laminates prepared from unidirectional prepreg tapes obtained from the manufacturers. Details of the procedure with glass/epoxy were given by Takeda [3]. Prepreg tapes are cut into 12-inch squares at various orientations to the fiber direction and stacked on an aluminum tool plate, with Teflon®-coated vent cloth above and below the tape stack, topped with glass cloth breather layers, a Mylar® film and another aluminum plate, and enclosed in a vacuum bag. Figure 2 shows the package being placed inside the Baron-Blakeslee BAC-24 autoclave with the fitting on top of the package for attaching to the vacuum tube. During cure a vacuum of 27 inches of mercury is provided for the interior of the vacuum bag, while the exterior is subjected to pressure of 40 psi and a suitable temperature cycle including 1 hour at a maximum temperature of 320° F for the glass/epoxy. Heating and cooling rates are carefully controlled following procedures recommended by the prepreg tape manufacturer.

Figure 2. Prepreg Package Being Placed in Autoclave

After cure, specimens are cut from the plates with a diamond saw. Usually four specimens are cut from each plate, three to be impacted at three different speeds and one for a control specimen to be tested to failure for comparison to determine the strength and stiffness retention factors for the damaged specimens.

Before 1979 plates had been fabricated in our lab by a technique of filament winding on a mandrel and matrix impregnation. Some of the earliest tests on 0°/90° glass/epoxy plates showed that somewhat better impact energy absorption was provided by plates where more than one ply was wound in the same direction in each layer. For example, with 15-ply laminates the $[0_5/90_5/0_5]$ laminate with two interfaces and the $[0_3/90_3/0_3/90_3/0_3]$ laminate with four interfaces absorbed somewhat more impact energy without perforation than did the alternating-ply laminate. This motivated a research program to try to understand why this occurred and to optimize the stacking sequence in glass/epoxy. Similar investigations on the other two systems are still in progress, but it already appears that in graphite/epoxy the multiple-ply layer advantage does not exist; the alternating-ply arrangement is more impact resistant. The investigation on glass/epoxy led to the identification of delamination as the major energy absorbing mechanism at subperforation speeds.

DELAMINATION

Some perforated and some partially penetrated glass/epoxy 0°/90° laminates, tested by Ross and Sierakowski [2], were later examined with a bright light behind them, which revealed extensive delamination in the laminates with multiple ply laminas; the delamination occurred at each interface where there was a change in fiber direction. Based on this post-impact examination a <u>sequential delamination mechanism</u> initiated by a <u>generator strip</u> was proposed [8], as follows.

At these moderate impact speeds, the first event was conjectured to be the formation of two through-the-thickness cracks in the first lamina, parallel to the fibers and separated approximately by the diameter of the impactor. This freed a strip of the first lamina from edge constraint; this <u>generator strip</u> was pressed down on the second lamina until the strip was perforated or until rebound. The loading of the second lamina by the generator strip aided the initiation of the first delamination at the interface between the first and second laminas. The maximum extent of the first delamination in the direction of the generator strip was equal to the length of the strip. The first delamination propagates away from the generator strip in the direction of the fibers of the second lamina until the available energy to drive the propagating delamination crack is exhausted. A second delamination occurs at the interface between the second and third laminas, possibly initiated by a second generator strip in the second lamina. The original sequential delamination model [8], based entirely on post-impact examination, envisioned the delaminations as occurring sequentially in time, with the one nearest the impact surface occurring first. From later measurements by high-speed photography [9] in plates with two interfaces, it appears that the two delaminations begin at about the same time. The one in the first interface stops first; the second continues and produces a delamination larger than the first one.

Figure 3 is a schematic illustrating two delamination areas $A_1$ and $A_2$ and the generator strip bounded by the cracks AA and BB. This schematic diagram matches well the post-impact photographs [8] of filament-wound glass/epoxy plates. In many more recent studies of plates fabricated from prepreg tapes, the generator strip is less prominent. For example, it is hardly visible in the photograph of the two-interface glass/epoxy plate shown in Figure 4, where the fibers in the first lamina are horizontal. The first delamination is small and peanut-shaped, appearing to originate from the circular impact area instead of from a generator strip. It propagates in the vertical direction of the fibers of the second lamina.

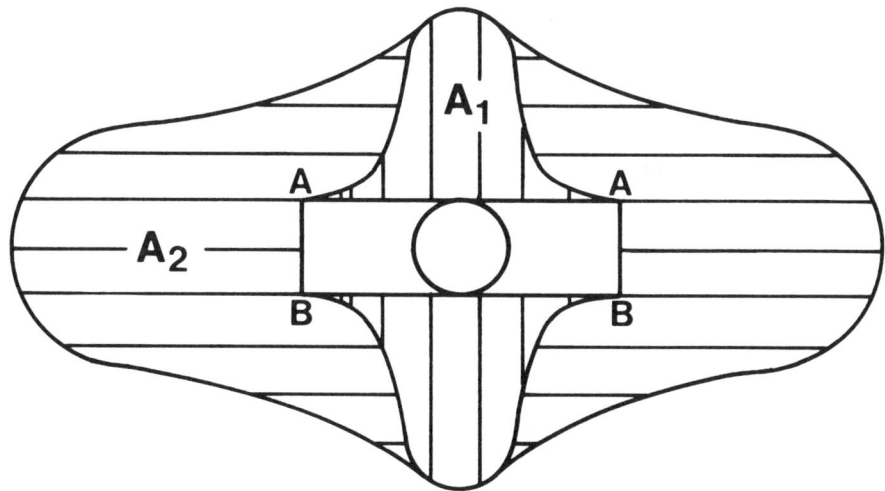

Figure 3. Delamination at Two Interfaces, with Generator Strip (Schematic).

The second (larger) delamination is also peanut-shaped and propagates in the horizontal direction. A number of (horizontal) matrix cracks are also visible in the photo. This photographic technique with a bright light behind the plate has been able to show as many as four overlapping delaminations in a glass/epoxy plate with four interfaces [8].

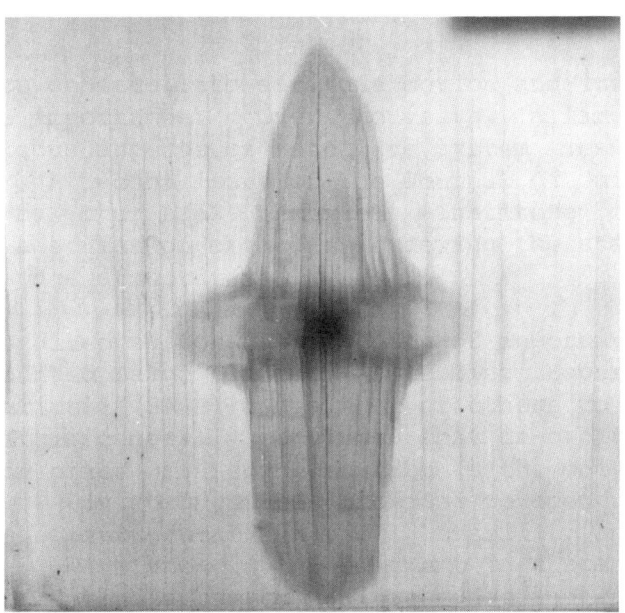

Figure 4. Photo of Two Delaminations in Glass/Epoxy Plate with Two Interfaces

Post-impact determination of delamination in opaque materials has been done by two non-destructive techniques (ultrasonic C-scan and X-ray) and by one destructive technique. For the X-Ray method, it is necessary to inject a penetrant opaque to the X-rays in order to make the delamination visible. Iothalamate sodium was injected with a syringe through a 1.4-mm-diameter hole drilled at the center of the impacted area. After the penetrant had dried, X-ray photography was able to reveal overlapping delaminations in as many as four interfaces, as in the graphite/epoxy plate of Figure 5. In this case, as in many of the visually examined glass/epoxy plates, the separate delamination areas are fairly well delineated, even where they overlap, so that the total delamination area can be determined as the sum of the individual delaminations. In many other cases, including all with more than four interfaces, only the outer boundary of the delaminated area can be determined. The area inside this outer boundary is referred to as the projected delamination area.

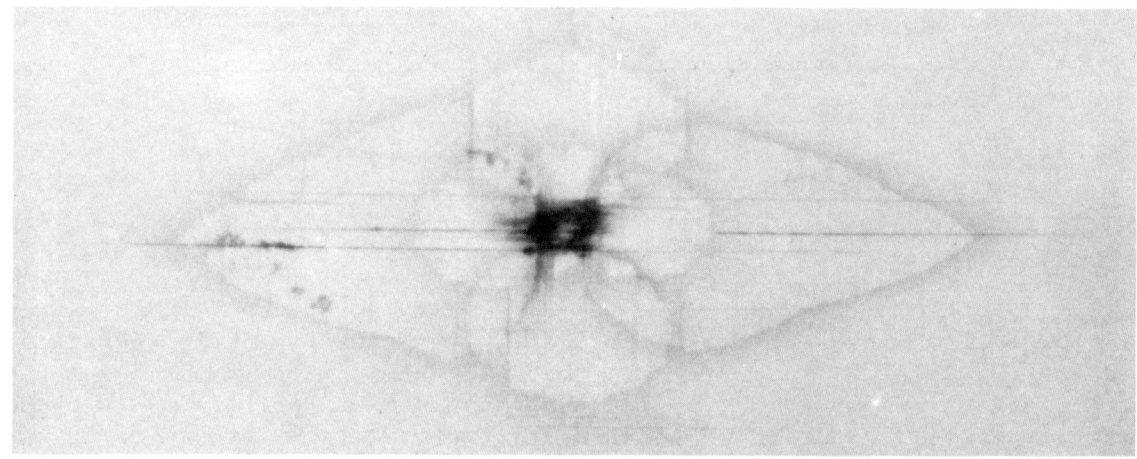

Figure 5. X-Ray Showing Four Delaminations in Graphite/Epoxy Plate.

Figure 6. C-Scan Probe and Graphite/Epoxy Plate

Conventional ultrasonic C-Scan methods determine only the projected delamination. Figure 6 shows the C-Scan probe above a black graphite/epoxy plate lying on an aluminum plate at the bottom of the water tank of the Automation Industries Model US-461 C-Scan system as used in the pulse-echo mode. The probe contains an ultrasonic transducer that sends pulses and receives echoes back from the top of the graphite/epoxy plate and also from the interface with the aluminum plate and/or from internal delaminations as the arm carrying the probe scans along a line parallel to one edge of the plate, and then indexes a short distance perpendicular to the scan direction before scanning along another line. Timed gates and receiver sensitivity can be adjusted, so that an output DC signal is either on or off, depending on whether a sufficient reflection is received within a preset time interval. Meanwhile an electric pen is automatically scanned over a suitably coated piece of paper to reproduce the scan over the target plate. When the DC voltage is on, a black line is drawn on the paper. The result is a series of parallel black lines, except in the areas where there is an internal flaw. This outlines the projected delamination area, as shown in the example of Figure 7.

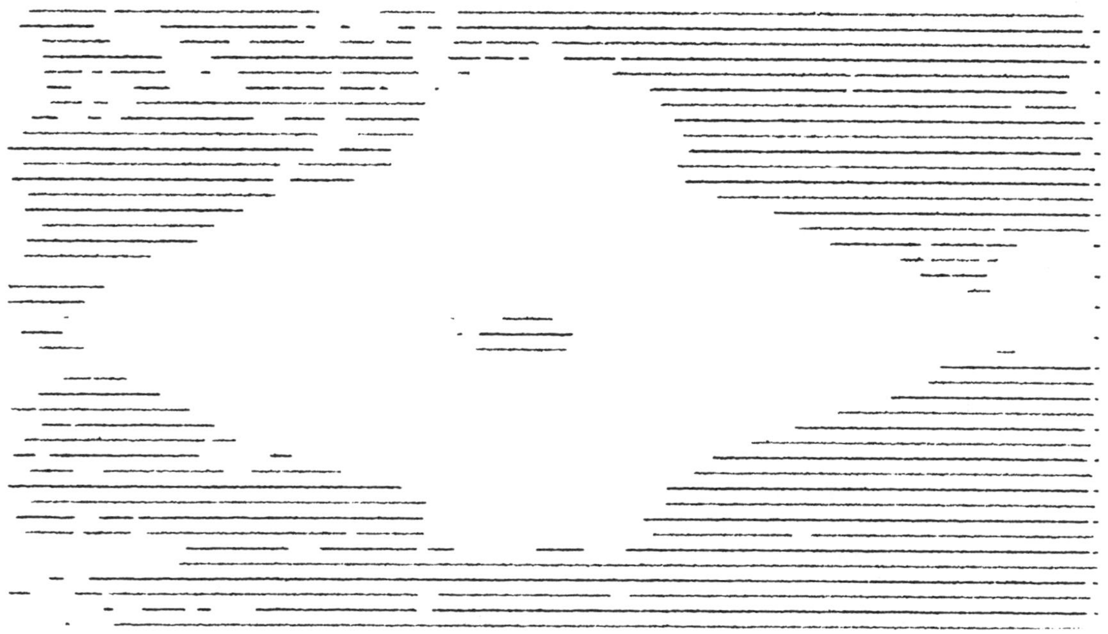

Figure 7. Projected Delamination Area Outlined by C-Scan

Recently B. Ho at Michigan State University [10] has developed an ultrasonic imaging system which displays and records the B-Scan, that is the entire reflected signal from each C-Scan pulse, including multiple reflections from overlapping delaminations. The system has a range resolution of approximately 0.4 mm at an operating frequency of 2.25 MHz. This system was able to display separately the four delaminations [11] in the plate whose projected delamination was shown in Figure 7. This promises to be a useful technique, but is still limited by the range resolution to a maximum of four interfaces in the types of laminates considered in this study.

The only method that we have used for determining more than four overlapping delamination areas is a destructive method in which the damaged plate is sliced into strips about 5 mm wide with a diamond saw. One edge of each strip was polished with emery paper and then successively with three aluminum oxides with grain sizes of five, three and one micron and finally with diamond paste. Then acetone was uniformly applied to the polished surface with a syringe. A section of replicating tape was pressed onto the edge with a piece of rubber to print the damaged area onto the tape. It takes about a minute for the acetone to react with the tape and dry.

The details of the damage pattern, including delamination cracks (and through-the-thickness matrix cracks) intersecting the edge of the strip remain on the edge replica and can be viewed with a microfiche reader. Manual correlation and plotting of the delamination intersections with all the cut sections then gives an outline of each separate delamination. The data reduction could be automated, but it is still a tedious process, although it is easy to do and requires no expensive special equipment. It has given satisfactory results for the resolution of as many as 14 overlapping delaminations [11] and is easier to use than the deply technique [12], which has been used for determining internal damage, including fiber breakage.

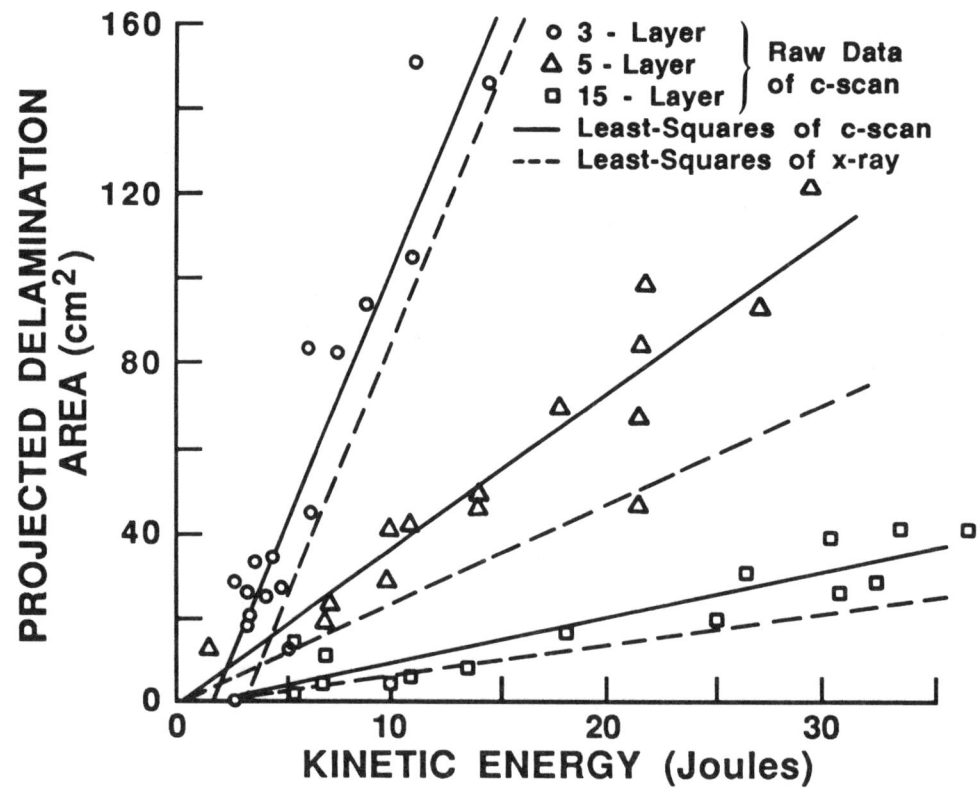

Figure 8. Projected Delamination Area Versus Imparted Kinetic Energy in Three Types of Graphite/Epoxy Laminates.

RESULTS OF DELAMINATION VERSUS IMPARTED KINETIC ENERGY

Graphite/epoxy plates fabricated with Hercules AS4/3501-6 prepreg tape gave the results for projected delamination versus kinetic energy imparted to the plate shown in Figure 8 for 15 tests at various speeds on each of three different layups of 30-ply plates: 3-layer plates with 2 interfaces, 5-layer plates with 4 interfaces, and 15-layer plates with 14 interfaces. The plotted points are from the C-Scan data, while two fitted lines are shown for each layup, one based on the C-Scan data and the other based on X-Ray data

Despite the C-Scan data scatter and some disagreement with the results from the two methods, the results indicate a linear relationship between imparted kinetic energy and projected delamination. If there is really a relationship between delamination area and imparted kinetic energy, the results should fall on one curve when total delamination area is plotted instead of projected delamination.

Figure 9 shows such a plot for 8 three-layer plates and 11 five-layer plates, based on X-Ray data. This preliminary result seems to support the assumption of a single linear relationship, although it is based on limited data. In particular, data for the 3-layer plates was only available at low impact energies, because at higher energies the delaminatinons would extend to the plate boundary. The slope of the fitted line gave a coefficient of about 3440 $J/m^2$ for the energy absorbed by delamination in the three-layer plates.

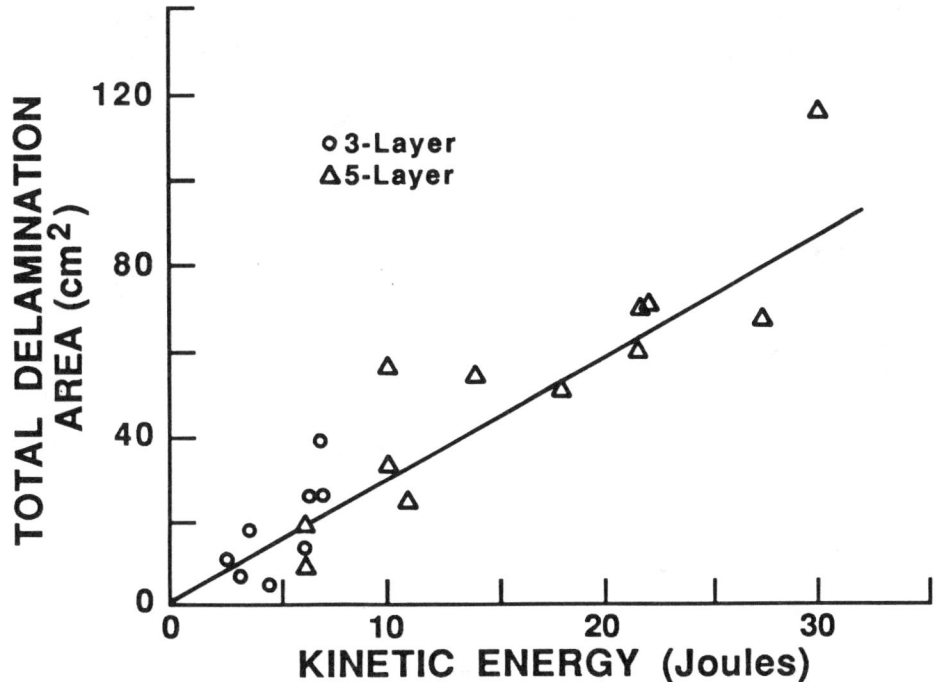

Figure 9. Total Delamination Area Versus Imparted Kinetic Energy in Two Types of Graphite/Epoxy Laminates.

The results on total delamination area versus imparted kinetic energy illustrated in Figure 9 for graphite/epoxy are qualitatively consistent with previously published results [7,13] for glass/epoxy, which showed a single linear relationship between total delamination and initial kinetic energy for each impactor/plate configuration observed in plates with two or four interfaces. In these earlier experiments the rebound velocity was not measured. Another series on glass/epoxy laminates with two interfaces [3,4] gave similar results versus initial kinetic energy, but showed that a different linear relationship was obtained when the mass (and length) of the impactor was changed. The numerical values of the coefficient (energy per unit delamination area produced) varies considerably with the specific materials used, even within the glass/epoxy system. The earlier tests on filament-wound plates fabricated with glass fibers of Owens-Corning Type ECG four-end rovings and Shell Epon 828 epoxy with Magnolia Plastic Curing Agent D [8] gave a coefficient of 3150 $J/m^2$ for plates with two or four interfaces [7]. The linear plots for plates fabricated with Scotchply Type 1003 glass/epoxy prepreg tapes and autoclave cured [3,4] gave a coefficient of 7500 $J/m^2$, more than double the earlier value with the same impactor. An impactor twice as long (and twice as heavy) [3,4] gave a still higher value, 9750 $J/m^2$. An extensive study at NASA [14] of impact resistance of graphite/epoxy laminates has also shown large differences for different resins.

The earlier value of 3150 J/m$^2$ represents an apparent fracture surface energy of 1580 J/m$^2$, since two surfaces are formed by each delamination; this is an order of magnitude higher than fracture surface energies measured in static tests on pure epoxy double cantilever beam specimens [15], even though delamination is usually considered to be primarily a matrix failure. Several reasons have been suggested for the difference (in addition to possible differences between epoxies). An important factor is the crack velocity; a number of investigators [16] have reported fracture surface energy as a function of crack velocity, with dynamic values as much as an order of magnitude above the static values. A second factor is the fracture mode. In the plate impacts the flexural shear stress is believed to be the principal driver, leading to a combination Mode II and Mode III crack, although near the generator strip there may be some crack-opening or Mode I type of loading. The double cantilever beam tests give Mode I cracks. The delaminations in our plate experiments are also close to fiber-matrix interfaces, which may give apparent fracture surface energies different from bulk epoxy material, since these apparent fracture surface energies include deformation energies in regions at some distance from the surface formed and, as in metals, are three or more orders of magnitude higher than theoretical energies required to form surfaces. Some data from static double cantilever beam tests on the cleavage at E-glass/epoxy laminate interfaces [17] have shown apparent fracture surface energies double those of unreinforced epoxy specimens.

The only results available on delamination crack speeds in these impact tests were obtained by high-speed photography [3,9] on glass/epoxy laminates with two interfaces. Examination of photos taken at approximately 25 microsecond intervals permitted velocity determination in two to five of the intervals. Figure 10 summarizes the results of four tests. The actual time of beginning for the delaminations was not determined precisely, but it was less than 25 microseconds before the first plotted point in each case. Note that the boundary support was 70 mm from the center, so that any speeds at distances greater than about 50 mm might be affected by boundary constraint.

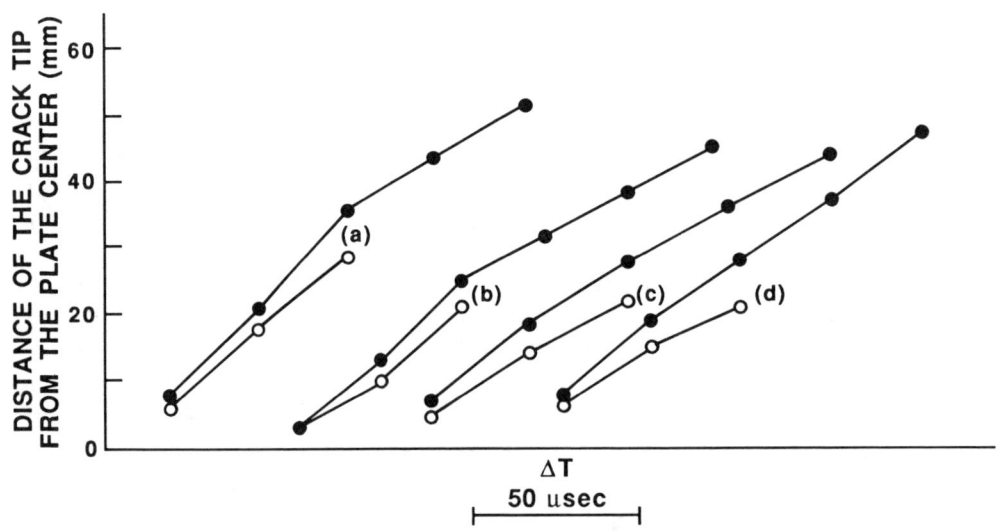

Figure 10. Delamination Propagation in Four Tests (a) to (d). Open circles are for first interface, solid circles for the second. Zero time is not shown. It differs for the different tests.

The delaminations at the two interfaces appear to start at about the same time, but the one at the first interface (open circles) stops first. The delamination speeds slow to 60 to 75 percent of their initially observed values during the period of observation. The highest speed observed was about half the calculated elastic shear wave speed in the epoxy. Delamination speeds during the first interval of observation were in the range of 300-400 m/s in the first interface and 400-500 m/s in the second inteface, slowing to 200-300 m/s in the first interface and 270-400 m/s in the second interface. Approximate times from impact to the stop of the delamination propagation were around 100 microseconds in the first interface and 300 microseconds in the second interface. The delamination speeds vary by less than a factor of three over all the observations, so that the dynamic enhancement of the fracture surface energy should be about the same in all the tests.

A few measurements of generator strip propagation speeds in glass/epoxy, made by observing the breaking times of successive silver paint conducting strips on the surface [3,9], have shown generator strip speeds starting at from 600-700 m/s (some two-thirds the matrix elastic shear wave speed) and slowing to 400-500 m/s during the period of observation, with the complete generator strip propagation completed in less than 100 microseconds.

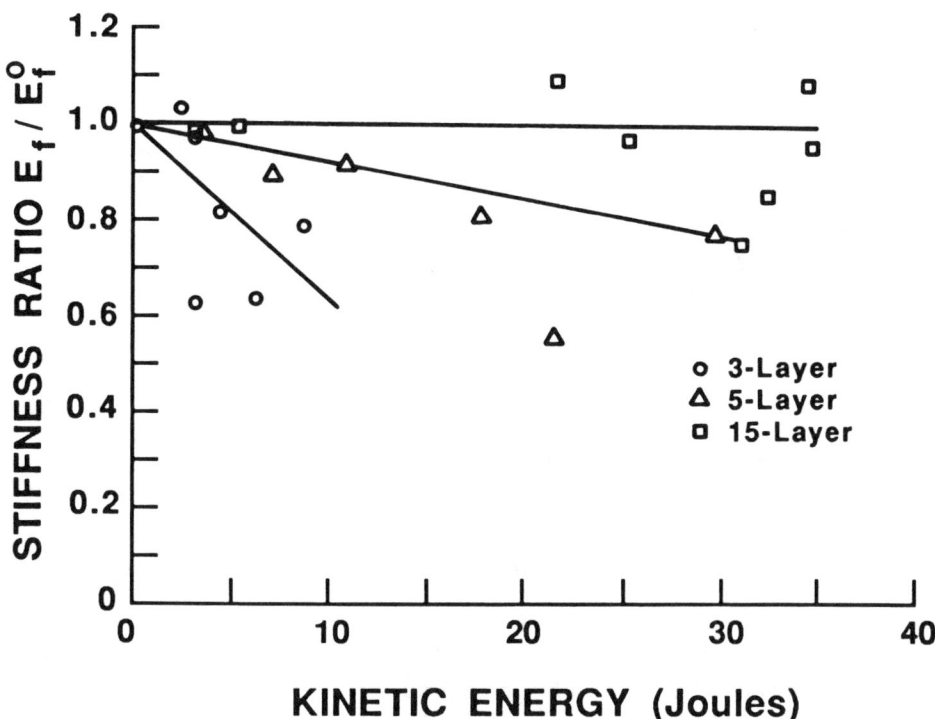

Figure 11. Stiffness Retention Fractions in Graphite/Epoxy Plates.

STRENGTH AND STIFFNESS RETENTION IN DAMAGED PLATES

Three-point bend tests were conducted on damaged plates and on undamaged control specimens cut from the same fabricated plate. Because the compression response is more sensitive to delamination than the tension response, the plates were tested with the non-impacted side in compression, which gave the greater degradation, since the larger delamination was near the non-impacted side.

The load P versus deflection w curve was determined; the strength was determined from the maximum load $P_{max}$, and the stiffness from the initial slope $dP/dw$ of the linear portion of the curve. To take account of the thickness variations between plates, the results were expressed as apparent elastic flexural modulus $E_f$ and maximum strength $\sigma_{max}$, calculated as though the response were linear elastic up to failure, by the following equations.

$$\text{Apparent } E_f = \frac{(dP/dw)L^3}{4bh^3} \qquad \text{Apparent } \sigma_{max} = \frac{3P_{max}L}{2bh^2}$$

where L is the length between supports, b is the width and h is the thickness of the plate. Figures 11 and 12 show fractions of the stiffness and strength retained, plotted versus imparted kinetic energy. $E_f^o$ and $\sigma_{max}^o$ denote stiffness and strength of undamaged control specimens from the same plate.

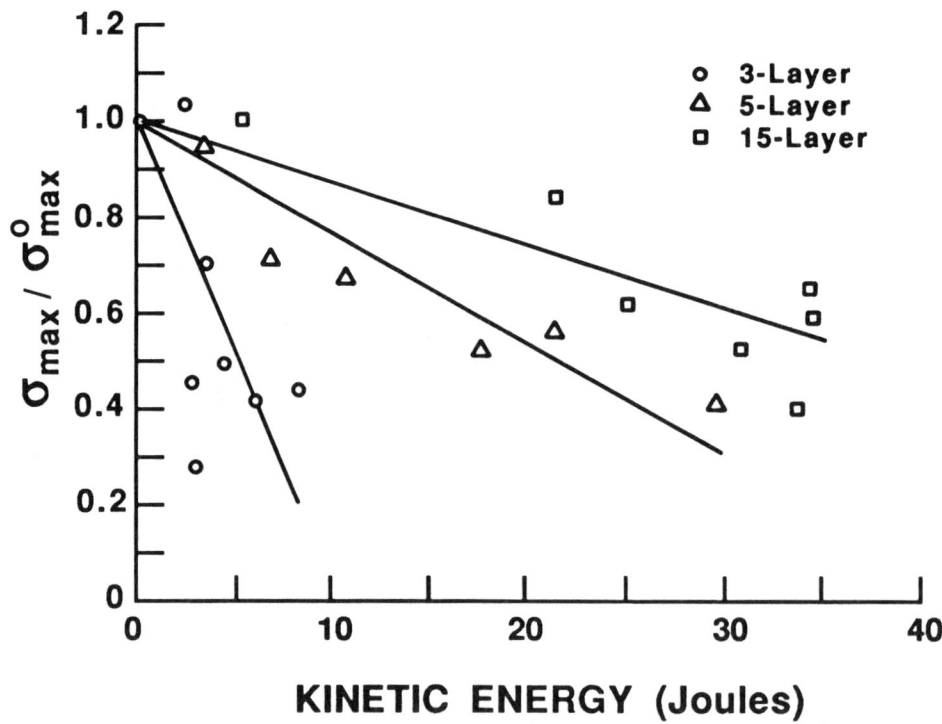

Figure 12. Strength Retention Fractions in Graphite/Epoxy Plates.

The 3-lamina plate has the most degradation and the 15-lamina one has the least degradation of both stiffness and strength. The stiffness reduction is much less than the strength reduction in all cases. Therefore evaluation procedures based on stiffness measurements are likely to underestimate the damage. Pre and post impact vibration tests, for example, show little variation in the natural frequencies of the plates, supported as cantilevers. Damping measurements were much more sensitive to the damage, but the results showed too much scatter to be useful, perhaps because of large and variable frictional damping in the supports.

## SUMMARY AND CONCLUSIONS

This paper has concentrated on research at the University of Florida. Many of the papers cited give references to other work. The 1973 ASTM Symposium volume containing references 1 and 8 also contains many other papers as does a recent NASA Langley Workshop Proceedings [18]. See also papers by C.T. Sun and his colleagues at Purdue [19-22] including representations of the force law between the impactor and target plate or beam and numerical calculations of the propagating elastic waves.

The experimental research program at the University of Florida has shown that in central impacts at normal incidence by small hard impactors at subperforation speeds on continuous-filament fiber-reinforced laminated composite plates the major damage is delamination at the interfaces between unidirectionally reinforced layers of different orientation, accompanied by matrix cracking between the fibers of the layers.

For each impactor/target combination studied, the total delamination area produced by the impacts at subperforation speeds appears to be a linear function of the kinetic energy imparted to the plate by the impactor. From the slopes of these linear functions, apparent fracture surface energies varying from 1580 $J/m^2$ to 4875 $J/m^2$ in glass/epoxy were calculated. Even the lowest of these is an order of magnitude higher than the static double-cantilever beam test measurements in pure epoxy. Some differences were noted between the values found with impactors of different masses, but the impactor nose shape did not appear to affect the total area in the subperforation speed impacts. Large differences (1580 $J/m^2$ to 4875 $J/m^2$) were observed between two glass/epoxy materials used for plate fabrication, which is consistent with other studies on the effect of resin properties on the impact damage resistance of graphite/epoxy laminates [14]. Our preliminary studies give a value of 1720 $J/m^2$ for the graphite/epoxy we used.

Transient measurements by high-speed photography of delamination propagation in glass/epoxy 0°/90° laminated plates with two interfaces showed that the delaminations in the two interfaces began at about the same time, but that the one in the second interface continued longer and propagated faster, with initial speed up to half the elastic shear-wave speed of the epoxy matrix. Although the speed slowed during each crack propagation, it was of the same order of magnitude during the period of observation, so that the dynamic enhancement of the fracture surface energy might be expected to be approximately constant.

Measurement of the retained stiffness and strength of damaged plates by three-point-bend tests indicated that the strength degradation was much greater than the stiffness degradation, so that evaluation procedures based on stiffness measurements are likely to underestimate the damage. In each composite system, the degradation was approximately linear with the imparted energy. The values shown in Figures 11 and 12 of course depend on the size of the test panel and the type of test, and do not give the local degradation near the impact. Greater degradation was observed in the plates with fewer interfaces (for the same total number of plies), where the delaminations extended far from the impact point.

## ACKNOWLEDGMENT

Most of the results reported here were obtained with support from the U.S. Army Research Office, Research Triangle Park, NC., currently under Contract No. DAAG 29-83-K-0107. Facilities and personnel of the University of Florida Center of Excellence Program in New Materials were used.

## REFERENCES

1. Preston, J.L.,Jr. and Cook, T.S., "Impact Response of Graphite/Epoxy Flat Laminates Using Projectiles That Simulate Aircraft Engine Encounters," *Foreign Object Impact Damage to Composites*, ASTM STP 568, 49-71, 1975.
2. Ross, C.A. and Sierakowski, R.L., "Studies on the Impact Resistance of Composite Plates," *Composites*, 4, 157-161, 1973.
3. Takeda, N., *Experimental Studies of the Delamination Mechanisms in Impacted Fiber-Reinforced Composite Plates*, Ph.D. Dissertation, U. of Florida, 1980.
4. Takeda, N., Sierakowski, R.L. and Malvern, L.E., "Studies of Impacted Glass Fiber-Reinforced Composite Laminates," *SAMPE Quarterly*, 12, 9-16, 1981.
5. Petersen, B.R., *Finite Element Analysis of Composite Plate Impacted by a Projectile*, Ph. D. Dissertation, U. of Florida, 1985.
6. Liou, W.-J., *Stress Analysis of Impacted Laminated Composite Plates with a Hybrid Stress Finite Element Method*, Ph. D. Dissertation, U. of Florida, 1986.
7. Sierakowski, R.L., Malvern, L.E. and Ross, C.A., "Dynamic Failure Modes in Impacted Composite Plates," *Failure Modes in Composites III*, ed. T.T. Chiao, AIME, 73-88, 1976.
8. Cristescu, N., Malvern, L.E. and Sierakowski, R.L., " Failure Mechanisms in Composite Plates Impacted by Blunt-Ended Penetrators," *Foreign Object Impact Damage to Composites*, ASTM STP 568, 159-172, 1975.
9. Takeda, N., Sierakowski, R.L., Ross, C.A. and Malvern, L.E., "Delamination-crack Propagation in Ballistically Impacted Glass/Epoxy Composite Laminates," *Experimental Mechanics*, 22, 19-25, 1982.
10. Personal communication with Prof. B. Ho at Michigan State University.
11. Liu, D., Lillycrop, L., Malvern, L.E. and Sun, C.T., "The Evaluation of Delamination - An Edge Replication Study," submitted to *Experimental Techniques*, 1986.
12. Guynn, E.G. and O'Brien, T.K., "The Influence of Lay-Up and Thickness on Composite Impact Damage and Compression Strength," AIAA/ASME/SAE 26th Structures, Structural Dynamics and Materials Conference, Orlando, FL, 1985.
13. Malvern, L.E., Sierakowski, R.L., Ross, C.A. and Cristescu, N., "Impact Failure Mechanisms in Fiber-Reinforced Composite Plates," *High Velocity Deformation of Solids*, IUTAM Symposium Tokyo 1977, Eds. K. Kawata and J. Shiori, Springer-Verlag Berlin and Heidelberg, 120-131, 1979.
14. Williams, J.G. and Rhodes, M.D., *The Effect of Resin on the Impact Damage Tolerance of Graphite/Epoxy Laminates*, ASTM STP 787, 450-480, 1983.
15. Bascom, W.D., Jones, R.L., and Timmons, C.O., "Mixed Mode Fracture of Structural Adhesives," *Adhesion Science and Technology*, Vol. 9-B, ed. L.H. Lee, New York, Plenum Press, 501, 1975.
16. *Dynamic Crack Propagation*, ed. G.C. Sih, Noordhoff, Leyden, 1973.
17. Yeung, P. and Broutman, L.J., "The Effect of Glass-Resin Interface Strength on the Impact Strength of Fiber Reinforced Plastics," *Polymer Engineering and Science*, 18, 62-72, 1978.
18. *Tough Composite Materials, Recent Developments*, (1983 NASA Langley Workshop) Noyes Publications, Park Ridge, NJ, 1985.
19. Yang, S.H. and Sun, C.T., "Indentation Law for Composite Laminates," ASTM STP 787, 425-449, 1982.
20. Tan, T.M. and Sun, C.T., "Use of Statical Indentation Laws in the Impact Analysis of Laminated Composite Plates," *ASME TRANS*, 107, J. Appl. Mech., 52, 6-12, 1985.
21. Sankar, B.V. and Sun, C.T., "Low-Velocity Impact Response of Laminated Beams Subjected to Initial Stress," *AIAA Journal*, 23, 1962-1969, 1985.
22. Joshi, S.P. and Sun, C.T., "Impact Induced Fracture in a Laminated Composite," *J. Composite Materials*, 19, 51-66, 1985.

# DYNAMIC RESPONSES OF ORTHOTROPIC PLATES UNDER MOVING MASSES

O. P. Agrawal[1] and Sunil Saigal[2]

## ABSTRACT

The problem considered is that of heavy masses moving on lightweight rectangular plates of orthotropic materials, slated for use in space structures. The dynamic equation of motion for orthotropic plates which contains singularities in both space and time variables is first presented. The response is expressed as a summation of double series of eigenfunctions. The equation of motion is transformed into an integro-differential equation for modal amplitudes using the Green's function. The Green's function is chosen to satisfy the initial conditions, the boundary conditions, and the transient conditions due to the moving masses. The solution series exhibits a good convergence. The effect of orthotropicity on natural frequencies and dynamic responses is demonstrated.

---

[1]Assistant Professor, Southern Illinois University at Carbondale, Carbondale, Illinois

[2]Assistant Professor, Worcester Polytechnic Institute, Worcester, Massachusetts

## INTRODUCTION

Extensive use of orthotropic materials in space structures is envisaged on account of the high strength to weight ratio of these materials. The problem of determining the dynamic responses of these plates under moving masses is of fundamental importance and arises from considerations of motion of robots or other such devices on such plates in an automated space station environment. Previous studies have considered the dynamic responses of structures of isotropic materials under moving masses for analysis and design of highway bridges. Similar studies for orthotropic structures have not been reported in the literature.

The complexity of the problem of moving masses on structures, caused by the presence of singular Dirac-delta function in spatial and time variables, as a coefficient of the differential equations of motion, has been pointed out [1]. A detailed survey of research efforts in this area may be found in [2,3]. The solution techniques suggested in earlier works were approximate in nature and involved complex mathematical manipulations.

In a recent paper, Stanišić [1] obtained the dynamic responses of a beam under moving masses by expressing the solution as an expansion of eigenfunctions in a series. The eigenfunctions satisfying the initial, boundary, and transient conditions due to the moving masses were determined by performing the integration in the Stieltjes sense. The amplitude coefficients were determined by solving integro-differential equations obtained by transforming the equations of motion using Green's function. This method has since been used by Saigal [4] for dynamic response of isotropic beams with various boundary conditions and by Saigal et al. [3] for isotropic plates under moving masses.

The present paper further extends the theory proposed by Stanišić [1] to study the dynamic responses of orthotropic plates under masses moving parallel to a side of the plate. The effect of degree of orthotropicity on the natural frequencies as well as the dynamic response is presented. The present developments are of vital interest in the design of lightweight space structures.

## THEORY

The differential equation of motion for a rectangular plate of anisotropic construction under a moving mass of magnitude M at instantaneous location $(X_0(t), y_0(t))$ on the plate is written as

$$D_{11} \frac{\partial^4 W}{\partial x^4} + 4 D_{16} \frac{\partial^4 W}{\partial x^3 \partial y} + 2(D_{12} + D_{66}) \frac{\partial^4 W}{\partial x^2 \partial y^2} + 4 D_{26} \frac{\partial^4 W}{\partial x \partial y^3}$$

$$+ D_{22} \frac{\partial^4 W}{\partial y^4} + [\rho h + M \cdot \delta(x-x_0) \cdot \delta(y-y_0)] \frac{d^2 W}{dt^2} \qquad (1)$$

$$= M g \, \delta(x-x_0) \, \delta(y-y_0)$$

where $D_{ij}$ (i,j = 1,2,6) are the modulus of rigidity coefficients for the anisotropic material, $\rho$ is the density, h is the thickness, W is the normal displacement, x and y are the Cartesian coordinates, g is the acceleration due to gravity, t is the time, and $\delta(.)$ is the Dirac-delta function.

For an orthotropic plate, equation (1) reduces to

$$D_1 \frac{\partial^4 W}{\partial x^4} = 2 D_3 \frac{\partial^4 W}{\partial x^2 \partial y^2} + D_2 \frac{\partial^4 W}{\partial y^4} + [\rho h + M \delta(x-x_0) \delta(y-y_0)] \frac{d^2 W}{dt^2}$$

$$= M g \, \delta(x-x_0) \, \delta(y-y_0) \qquad (2)$$

where $D_1$, $D_2$, and $D_6$ are the modulus of rigidity coefficients and can be obtained from the corresponding coefficients for anisotropic materials [5].

Neglecting the convective part of acceleration, equation (2) is expressed in the non-dimensional form as

$$\frac{\partial^4 \bar{W}}{\partial \xi^4} + 2p\gamma_1 \frac{\partial^4 \bar{W}}{\partial \xi^2 \partial \eta^2} + p^2 \gamma_2 \frac{\partial^4 \bar{W}}{\partial \eta^4} + [1 + \varepsilon \delta(\xi-\xi_0) \delta(\eta-\eta_0)] \frac{\partial^2 \bar{W}}{\partial \tau^2}$$
$$= P \delta(\xi-\xi_0) \delta(\eta-\eta_0) \quad (3)$$

where $\bar{W} = W/h$, $\xi = x/a$, $\eta = y/b$, $\varepsilon = \frac{M}{\rho h a b}$, $p = a^2/b^2$, $\tau = t/\alpha$,

$\alpha^2 = \frac{\rho h a^4}{D_1}$, $P = \frac{\varepsilon \alpha^2 g}{h}$, $\delta(\xi-\xi_0) = a \delta(x-x_0)$, $\delta(\eta-\eta_0) = b \delta(y-y_0)$,

$\gamma_1 = \frac{D_3}{D_1}$, and $\gamma_2 = \frac{D_2}{D_1}$

$\gamma_1$ and $\gamma_2$ determine the orthotropicity of the plate material

The plate is assumed to be simply supported on all its edges. The boundary and initial conditions are the same as those given in Reference [3].

The solution for equation (3) is assumed of the form

$$\bar{W}(\xi, \eta, \tau) = \sum_{m=1}^{\infty} \sum_{n=1}^{\infty} a_{mn}(\tau, \xi_0) W_{mn}(\xi, \eta, \xi_0) \quad (4)$$

where $W_{mn}$ are eigenfunctions orthonormalized with respect to the weighting function $\gamma(\xi_0, \eta_0) = 1 + \varepsilon \delta(\xi-\xi_0) \delta(\eta-\eta_0)$ and satisfy the homogeneous part of equation (3), i.e.,

$$\frac{\partial^4 W_{mn}}{\partial \xi^4} + 2p\gamma_1 \frac{\partial^4 W_{mn}}{\partial \xi^2 \partial \eta^2} + p^2 \gamma_2 \frac{\partial^4 W_{mn}}{\partial \eta^4} = \Omega_{mn}^2 \gamma(\xi_0, \eta_0) W_{mn} \quad (5)$$

and the orthonormalization of the eigenfunctions is carried out in the Stieltjes sense as explained in detail in Reference [3], $a_{mn}$ are the amplitude coefficients.

The equation of motion (3) is now transformed into integro-differential equations for the amplitude coefficients $a_{mn}$ by using the Green's function.

Additionally, the highly nonlinear expression for acceleration $\frac{\partial^2}{\partial \tau^2}(a_{mn} X_n Y_m)$ is approximated by the term $\frac{\partial^2 a_{mn}}{\partial \tau^2} X_n Y_m$ and numerical justification for this simplification was given in Reference [1]. Substituting this assumption and the equality in equation (5) in equation (4); multiplying both sides by $W_{rs}$; integrating with respect to $\xi$ and $\eta$ in the domain [0,1] for each variable; and using the property of orthonormality of the eigenfunction leads to

$$\frac{\partial^2 a_{mn}}{\partial \tau^2} + \Omega_{mn}^2 a_{mn} = P X_n(\xi_1 \xi_0) Y_m(\eta_0) \tag{6}$$

with $a_{mn}(0, \xi_0) = \dot{a}_{mn}(0, \xi_0) = 0$ being obtained from initial conditions.

From equation (5), the integer-differential equation for $a_{mn}$ is obtained as

$$a_{mn}(\tau, \xi_0) = \int_0^{\tau_f} G(\phi, \tau) P W_{mn}(\xi_0, \eta_0) d\phi \tag{7}$$

where $G(\phi, \tau)$ is the Green's function satisfying:

$$\frac{\partial^2 G(\phi, \tau)}{\partial \phi^2} + \Omega_{mn}^2 G(\phi, \tau) = \delta(\phi - \tau), \quad G(\tau_f, \tau) = \frac{\partial G(\tau_f, \tau)}{\partial \phi} = 0,$$

and the transient conditions

$$G(\phi, \tau) \Big|_{\tau_-}^{\tau_+} = 0 \text{ and } \frac{\partial G(\phi, \tau)}{\partial \phi} \Big|_{\tau_-}^{\tau_+} = 1$$

## SOLUTION FOR EIGENFUNCTIONS

Except at the instantaneous position $(\xi_0, \eta_0)$ of the mass M, the homogeneous equation given by

$$\frac{\partial^4 W}{\partial \xi^4} + 2p\gamma_1 \frac{\partial^4 W}{\partial \xi^2 \partial \eta^2} + p^2 \gamma_2 \frac{\partial^4 W}{\partial \eta^4} - \Omega^2 W = 0 \qquad (8)$$

should be satisfied everywhere along with the boundary conditions

$W = W_{\eta\eta} = 0$ at $\eta = 0, 1$

$W = W_{\xi\xi} = 0$ at $\xi = 0$ and

$W = W_{\xi\xi} = 0$ at $\xi = 1$

where $W(\xi, \eta) = X(\xi) Y(\eta)$.

Let $\lambda = (\bar{p} + \bar{q})^{1/2}$, $\beta = (\bar{p} - \bar{q})^{1/2}$

where

$$\bar{p} = (p^2 n^4 \pi^4 (\gamma_1^2 - \gamma_2^2) + \Omega^2)^{1/2},$$

$$\bar{q} = p\gamma_1 n^2 \pi^2$$

Three cases, namely, i) $\bar{p} > \bar{q} > 0$, ii) $\bar{q} > \bar{p} > 0$, and iii) $\bar{p}$ imaginary arise. However, from a physical point of view, a solution that contains sine and cosine terms also, given by case (i) of $\bar{p} > \bar{q} > 0$ is of interest.

The general solution for equation (8) is then expressed as

$W(\xi, \eta) = X(\xi) Y(\eta)$,

$X(\xi) = C_1 \cosh \lambda\xi + C_2 \sinh \lambda\xi + C_3 \cos \beta\xi + C_4 \sin \beta\xi$

$Y(\eta) = \sin m\pi\eta \qquad (9)$

m is the wave number.

Five independent conditions, are required to determine the unknowns $\Omega$, $C_1$, $C_2$, $C_3$, and $C_4$ in equation (9). These are obtained by considering three continuity conditions across the instantaneous positions of the mass and a transient condition due to the moving mass that causes a jump in the shear force across the instantaneous position of the mass. These conditions can be derived as [3],

i) Continuity of displacements

$$X_L(\xi_0) = X_R(\xi_0)$$

ii) Continuity of slopes

$$X_L^1(\xi_0) = X_R^1(\xi_0)$$

iii) Continuity of moments

$$X_L^{11}(\xi_0) = X_R^{11}(\xi_0)$$

iv) Jump in shear force

$$\left[\left[\frac{\partial^3 X(\xi)}{\partial \xi^3}\right]\right]_{\xi=\xi_0} = \epsilon \, \Omega^2 \, X(\xi_0)$$

This condition concerning the jump in shear force yields the frequency equation for determining $\Omega$.

The fifth condition is obtained from the orthonormalization of the eigenfunctions which proceeds on the same lines as given in References [1,3]. The weighting function used in the orthonormalization is $Y(\xi_0, \eta_0)$ and the integrations are performed in the Stieltjes sense.

## NUMERICAL RESULTS

The dynamic responses are obtained by summation of the series

$$\sum_{m=1}^{k} \sum_{n=1}^{k} a_{mn}(\tau, \xi_0) W_{mn}(\xi, \eta, \xi_0)$$

Adequate convergence in results is obtained for k=3 [3]. The results presented below were obtained for the motion of the mass along $\eta_0 = 0.25$. Other numerical data used are: p = 4, P = 1, and non-dimensional velocity v α/a = 1.

Natural Frequency and Orthotropic Properties:

The natural frequencies were obtained by solving the condition of jump of shear force due to transients of the moving mass as mentioned earlier. The frequency equation obtained is solved in an iterative fashion. The effect of variation of the parameter $\gamma_1 = D_3/D_1$ on the first natural frequencies is shown in Table 1 for various locations of the moving mass. Similar results for variation of $\gamma_2 = D_2/D_1$ are shown in Table 2. The magnitude of natural frequencies increases with an increase in either of the parameters $\gamma_1$ and $\gamma_2$. The increase in frequencies, however, is higher for an increase in $\gamma_2$ compared to an increase in $\gamma_1$.

Figures 1 and 2 show the effect of mass location on natural frequencies of the system. It should be noted that the frequency of the system is minimum when the mass is in the middle of two nodes, and there is no change in the frequency, when the mass is at a node. Similar results are observed for higher modes also.

Dynamic Response and Orthotropic Properties:

Dynamic responses of a plate for $\gamma_1 = 1.0$ and $\gamma_2 = 0.9$ are shown in Figures 3 and 4. Figure 3 shows the variation of amplitudes for different nodes. It is clear that the combination of each node gets lower as the frequency gets higher. Figure 4, the response of the point (0.5, 0.25) as mass moves from one end to the other end. The nature of the response is similar as that of isotropic plate [3].

An important advantage of orthotropic materials is that parameters, $\gamma_1$ and $\gamma_2$ can be suitably changed to avoid the undesirable dynamic response including undesirable frequency bandwidth.

## CONCLUSION

The dynamic response of a simply supported orthotropic plate under a moving mass is presented by means of operational calculus. The response of the plate is expressed as a double series of eigenfunctions, with time dependent amplitude coefficients. The eigenfunctions satisfy the initial conditions, the boundary conditions, and the transient conditions due to moving masses. Numerical results show that the undesirable frequency band can be avoided by selecting the materials of suitable orthotropic parameters.

Table 1:  Effect of Orthotropic Parameter $\gamma_1 = D_3/D_1$ on the First Natural Frequencies of Plate Under Moving Mass $\varepsilon = 0.01$, $\gamma_2 = 0.1$

| Position of Mass $\xi_0$ | DIMENSIONLESS NATURAL FREQUENCY | | |
|---|---|---|---|
| | $\gamma_1 = 0.1$ | $\gamma_1 = 0.25$ | $\gamma_1 = 0.5$ |
| 0, 1 | 18.199 | 21.168 | 25.355 |
| 0.1, 0.9 | 18.181 | 21.168 | 25.331 |
| 0.2, 0.8 | 18.136 | 21.095 | 25.268 |
| 0.3, 0.7 | 18.080 | 21.030 | 25.190 |
| 0.4, 0.6 | 18.036 | 20.979 | 25.129 |
| 0.5 | 18.019 | 20.960 | 25.105 |

Table 2: Effect of Orthotropic Parameter $\gamma_2 = D_2/D_1$ on the First Natural Frequencies of Plate Under Moving Mass $\varepsilon = 0.01$, $\gamma_1 = 0.25$

| Position of Mass $\xi_0$ | DIMENSIONLESS NATURAL FREQUENCY | | |
|---|---|---|---|
| | $\gamma_2 = 0.10$ | $\gamma_2 = 0.15$ | $\gamma_2 = 0.2$ |
| 0, 1 | 18.199 | 22.935 | 26.575 |
| 0.1, 0.9 | 18.181 | 22.913 | 26.552 |
| 0.2, 0.8 | 18.136 | 22.856 | 26.490 |
| 0.3, 0.7 | 18.080 | 22.785 | 26.415 |
| 0.4, 0.6 | 18.036 | 22.730 | 26.355 |
| 0.5 | 18.019 | 22.709 | 26.333 |

## REFERENCES

1. Stanišić, M. M., "On a New Theory of the Dynamic Behavior of the Structures Carrying Moving Masses," <u>Ing. - Arch.</u>, 55, (1985), 176.

2. Stanišić, M. M., Euler J. A., and Montgomery, S. T., "On a Theory Concerning the Dynamical Behavior of Structures Carrying Moving Masses," <u>Ing. - Arch.</u>, 43, (1974), 22.

3. Saigal, S., Agrawal, O. P., and Stanišić, M. M., "Influence of Moving Masses on Rectangular Plate Dynamics," <u>Ing. - Arch.</u>, to appear.

4. Saigal, S., "Dynamic Behavior of Beam Structures Carrying Moving Masses," <u>J. of Applied Mech.</u>, 53, (1986), 222.

5. Jones, R. M., "Mechanics of Composite Materials," McGraw-Hill, New York, 1975.

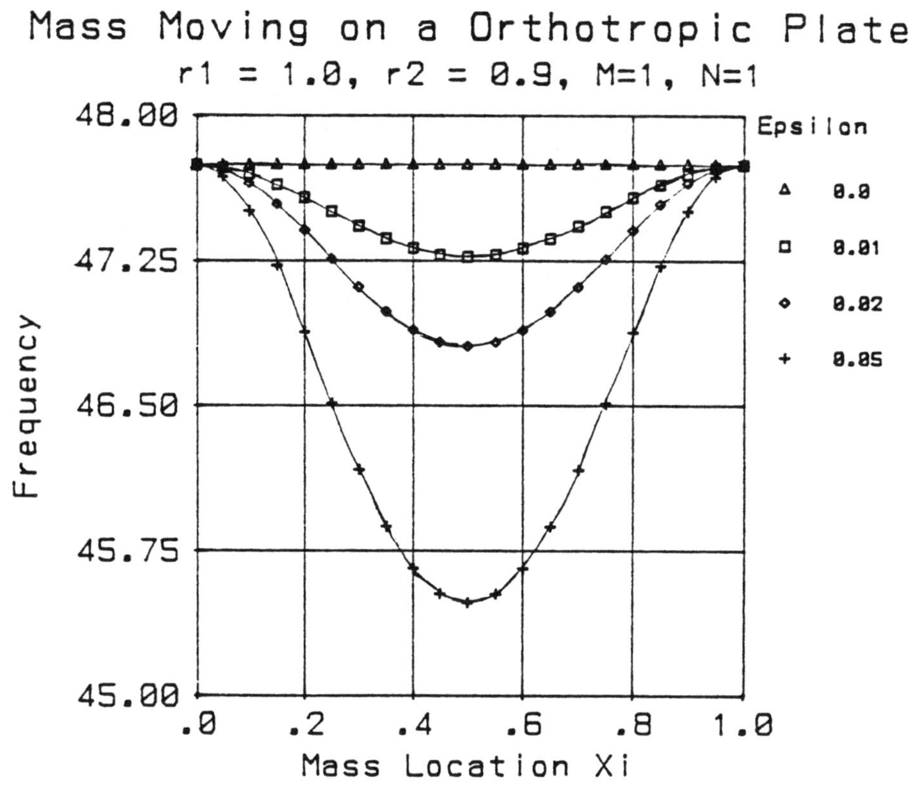

Fig. 1. Nondimensional frequency $\Omega$ vs. position of the moving mass (M=1, N=1)

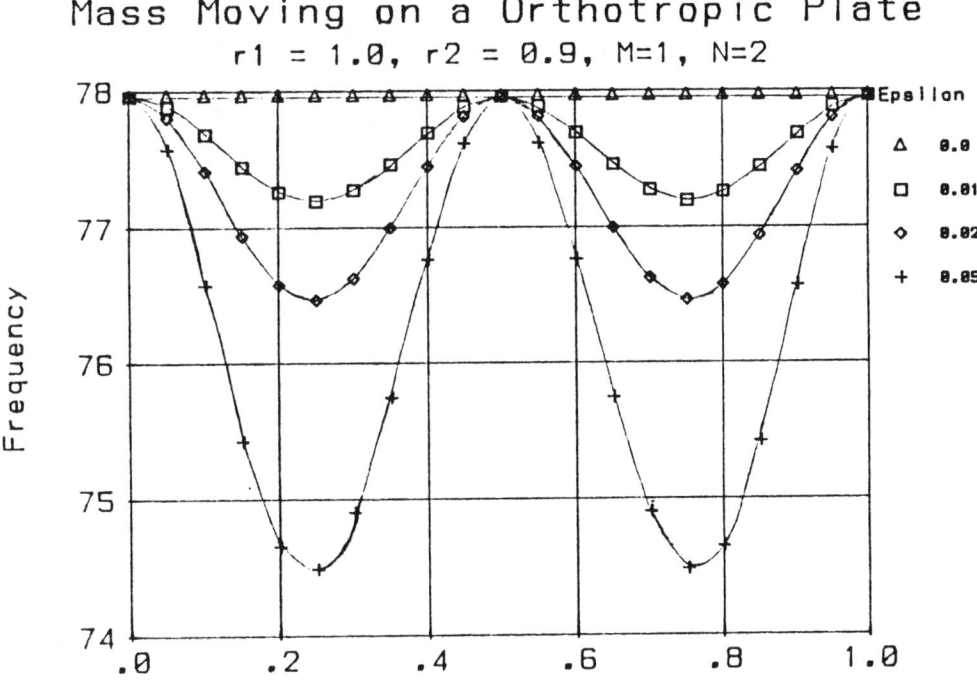

Fig. 2. Nondimensional frequency $\Omega$ vs. position of the moving mass (M=1, N=2)

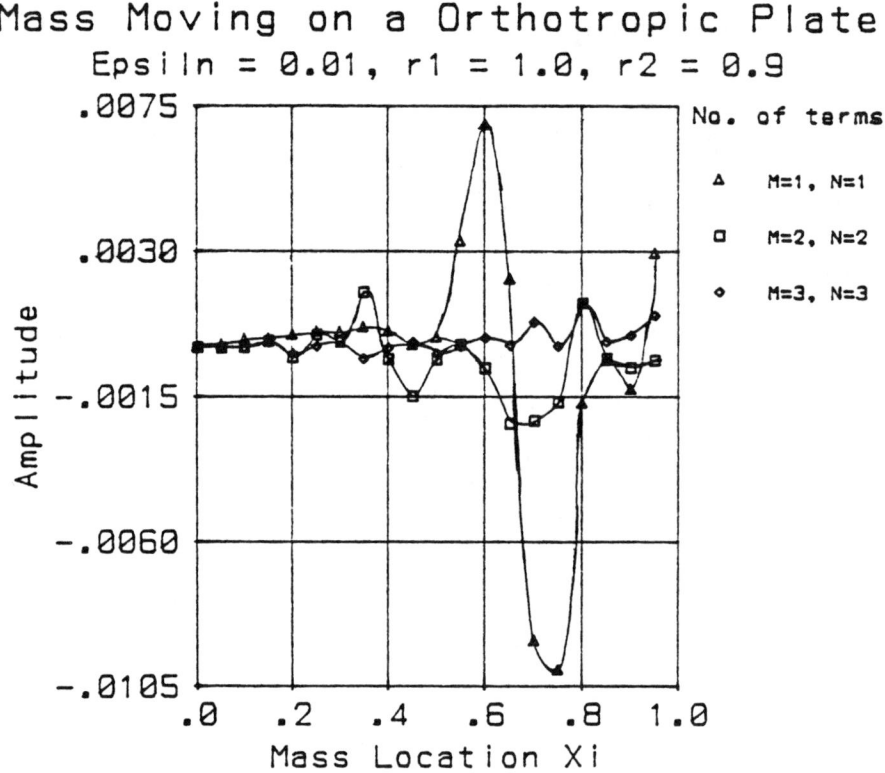

Fig. 3. Convergence of the amplitude-coefficients $a_{mn}$ with increasing mode numbers

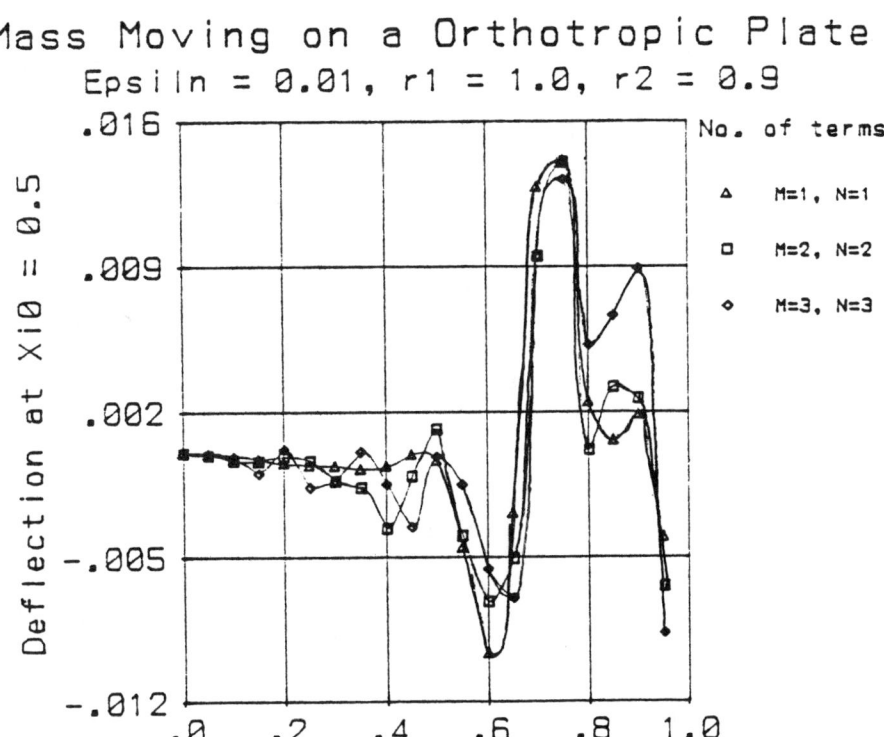

Fig. 4. Response of a point due to moving mass ($\xi_0 = 0.5$, $\eta_0 = 0.25$)

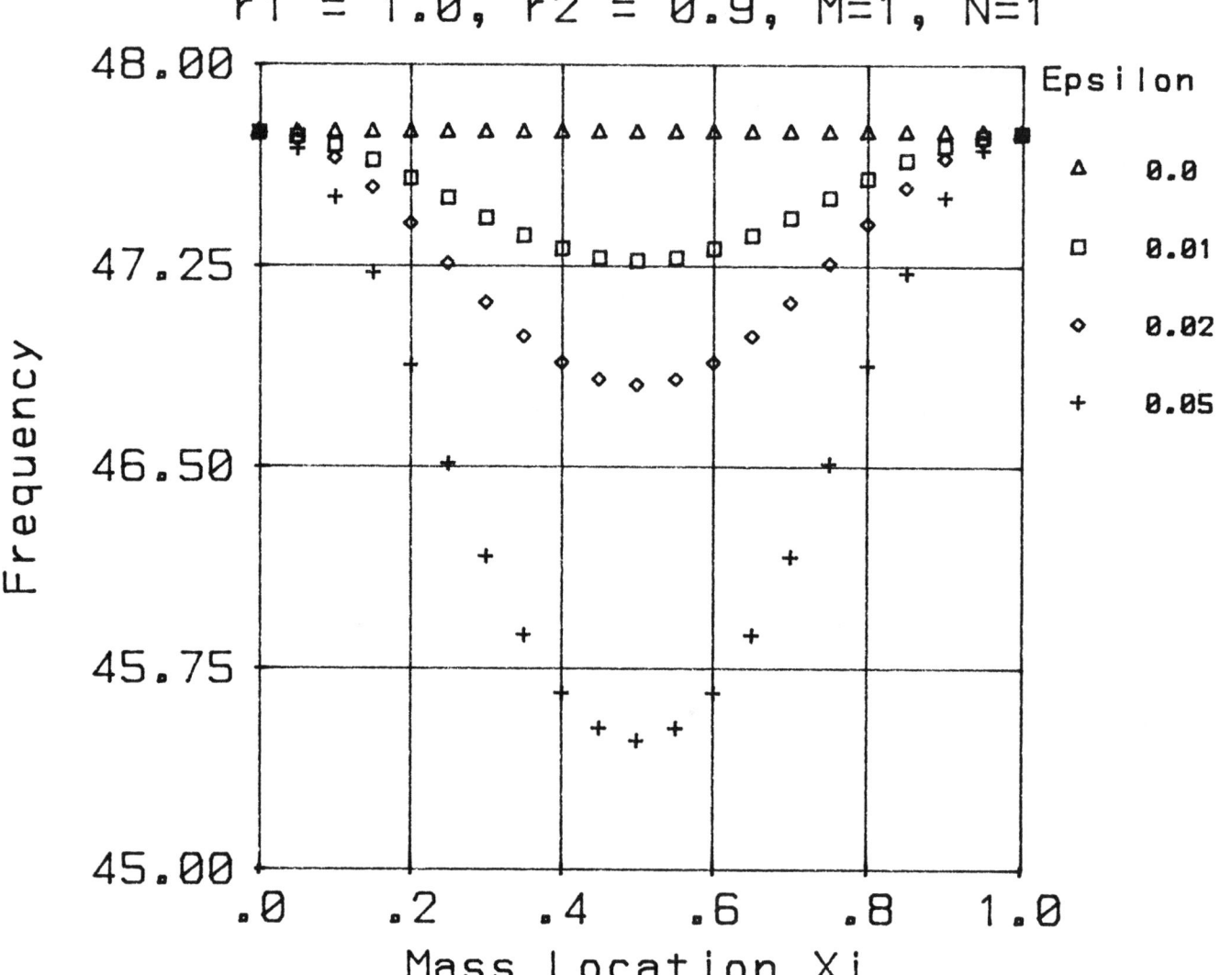

Mass Moving on a Orthotropic Plate
Epsiln = 0.01, r1 = 1.0, r2 = 0.9

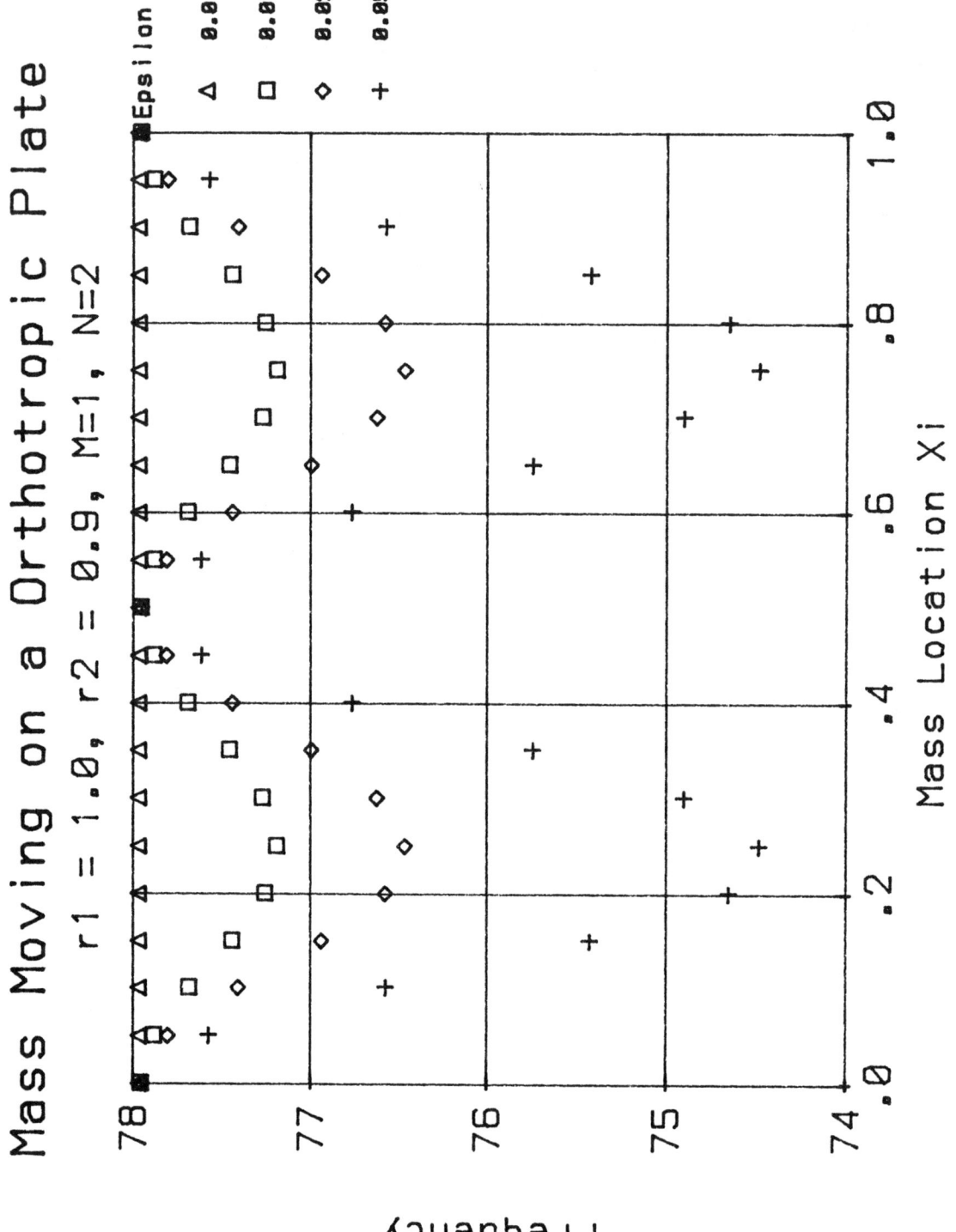

## Mass Moving on a Orthotropic Plate
### Epsiln = 0.01, r1 = 1.0, r2 = 0.9

No. of terms

△  M=1, N=1

□  M=2, N=2

◇  M=3, N=3

IV. Optimal Design

STRUCTURAL TAILORING FOR AIRCRAFT PERFORMANCE

Terrence A. Weisshaar

School of Aeronautics and Astronautics

Purdue University

West Lafayette, Indiana

PREFACE

To prepare a paper for a conference dedicated to the memory of Raymond L. Bisplinghoff it is useful to recall the many accomplishments of the man to be honored. This recollection is aided by a careful reading of a number of significant papers written by Professor Bisplinghoff and by recalling the few occasions, insignificant to him, but highly significant to me, on which I was privileged to meet him. In his 19th Wright Brothers Lecture, Professor Bisplinghoff pointed out the need for integration of aeroelastic considerations into routine design. This call has been largely heeded, aided by rapid advances in high-speed computing and data processing. Still, aeroelasticity remains to many a "black art", a problem area devoid of gain, filled with pitfalls to be avoided. A second interesting item in the Bisplinghoff Lecture of 1955 was a rhetorical question: "How do we attract our best minds to aeronautics?" This latter question is one to be considered carefully by an industry that depends upon new talent and fresh ideas. The re-reading of the 1955 Wright Brothers Lecture provided the opportunity to justify the subject of the present paper. The subject of the present paper, aeroelastic tailoring, is one that has a double purpose, similar to those set down over 30 years ago by Professor Bisplinghoff: to highlight opportunities for aeroelastic synergism; and, to point out opportunities for increased creativity by budding designers. It is to this and to his memory that this paper is dedicated.

INTRODUCTION

Historically the term "performance" has been used primarily to denote the ability of an airplane to fly a mission with a specified amount of fuel and payload. Performance analyses examine different configurations for the mission, how far the aircraft can fly, how high it can fly and how fast it can fly. Aeroelasticity affects some measures of performance in a direct, unquestionable manner. For high-aspect ratio swept wings, static deformation becomes important at high dynamic pressure and may adversely affect lift effectiveness - the ability of a flexible lifting surface to generate lift at a certain incidence to the airstream. Similar problems occur with control effectiveness, the ability of a control system to control the flexible aircraft during commanded maneuvers. There are also secondary problems that arise as a lifting surface is bent and twisted, such as changes in lift and drag distribution. The dynamic response of the aircraft is similarly affected by aeroelastic effects. Flutter, a special and certainly the most fearsome of all special cases of dynamic response, has been known to prevent the achievement of performance objectives. However, none of the aforementioned aeroelastic phenomena is routinely studied in a "Perf I" course. As a result, it is easy to dismiss aeroelasticity as something of a nuisance, to be wrung out of the design by "rigidizing" the aircraft at great expense to the design.

Aeroelasticity does affect the performance of aircraft, without a doubt, so let us retreat one step to the pre-performance, conceptual design phase. For a given mission, why was a certain planform chosen to the exclusion of all others? Why are the wings swept, unswept, delta, high-aspect ratio, and so forth? It is at this stage that one needs to assess the impact of aeroelasticity on performance of a mission, not at the stage when the planform has been defined. Having stated the contention that aeroelasticity affects performance profoundly at the primary level (planform and configuration selection) and at the secondary level (adding "flexibility effects" to classical performance studies and perhaps "unnaturally" restricting airspeeds) let's examine two cases that illustrate this contention. One example, the X-29A research aircraft, has recently flown while the other more recent example, the oblique wing, is only yet in an evolutionary stage. Both are profoundly affected by aeroelastic considerations to the extent that their viability and availability depends upon a successful, creative aeroelastic design.

AEROELASTIC CHARACTERISTICS OF FORWARD SWEPT WING AIRCRAFT

The evolution of the Forward Swept Wing fighter concept (FSW) that culminated in the X-29 research aircraft provides an interesting example of the struggle between modern technology and ancient beliefs (ancient in this case is 40 years). The present author had the good fortune to observe and participate in some of this development, which began as a whimsical conversation in 1972 and culminated in the first flight of a distinctive, beautiful aircraft on December 14, 1984. During that time a number of significant issues related to aeroelasticity were argued and, for the most part, settled. Many of these issues apply to soon-to-be-developed aircraft constructed of composite materials.

It was the contention of Norris J. Krone, Jr. that a laminated sweptforward wing could be constructed in such a way that airload induced bending and twisting would interact favorably to relieve "well-known" undesirable divergence characteristics of such airfoils. Studies by engineers from the Grumman Corporation and Rockwell International predicted several performance advantages, among them [1]:

- Configuration flexibility.
- Significantly higher maneuver L/D.
- Lower trim drag - increased supersonic range.
- Lower stall speed - slower landings.
- Virtually spin proof.
- Better low-speed handling.
- Volume benefits - lower wave-drag.

It was Krone's intention to press for the construction of a manned demonstrator, or, if all else failed, an RPV, as part of his responsibilities at the Defense Advanced Research Projects Agency (DARPA).

The FSW concept was cursed by apparent textbook simplicity. It is easy to illustrate that, due to flexibility, the lift distribution builds on a sweptforward wing and decreases on a sweptback wing - assuming that the angle of attack remains constant. Under conditions of constant lift, the spanwise-center-of-pressure (CP) of the FSW moves outward, due to bending. The converse is true for sweptback wings. (It should be remarked that the reverse trend of CP movement due to sweep is encountered when comparing fore and aft swept wings without flexibility). When bending and torsion are considered, pound-for-pound a conventional FSW is a very shaky concept because of the stiffness that must be added to minimize (if not eliminate) bending deformations. The result is that the above advantages are overwhelmed by increased weight.

Neglecting the fact that Krone's proposal envisioned tailored laminated composites, not metal, as the construction material, a furious battle ensued, since the amount of funding required for a demonstrator was nontrivial in a zero-sum game of budgetting. Figure 1 (taken from Ref. 2) was cited as an example of why money should not be spent developing a FSW. The origin of this figure stems from early studies by Diederich and Budiansky [3] that thoroughly examined the effects of sweep, both fore and aft, on flexible, high-aspect-ratio wings. Krone's results showed something much different (Figure 2, from Ref. 4). His optimized (for least weight) design at different angles of forward and aft sweep showed little necessity for added structural weight to combat divergence. Instead, he recommended changes in thickness distribution and laminate ply orientation as a means of creating a flexible, yet satisfactory, wing structure.

Although aeroelastic tailoring studies had been performed on more conventional planform types (such as the F-16) and had indicated that substantial improvements in aeroelastic stability, lift effectiveness and control effectiveness were possible (at constant weight) Krone's proposal created a stir. A substantial portion of the controversy was due to the lack of widespread, fundamental experience with anisotropic aeroelasticity. What was necessary, in the present author's opinion, was not a sophisticated analytical model such as Krone's, but rather an unsophisticated analytical approach to demonstrate tailoring phenomena.

To focus maximum attention upon the effect of the use of laminated materials to influence static aeroelastic characteristics of lifting surfaces, the simplified approach of Ref. 3 was retained. The result was Reference 5 in which the deformation of a uniform-planform wing was studied. For this study the differential equations of equilibrium of an idealized slender swept wing were formulated. This idealization considered aerodynamic loads acting upon chordwise segments of a beam-like wing, as shown in Fig. 3. The structural model for wing deflection assumes the loads to be carried by a box-beam arrangement in which laminated composite sheets of material form the upper and lower face sheets of the wing box. A reference axis, labeled as the y-axis in Fig. 3, is located equidistant between the front and rear edges of the box beam and lies in the geometric midplane of the box, halfway between the upper and lower surfaces of the cover sheets.

The idealized, laminated box-beam model may be characterized by three stiffness parameters, EI, GJ, and K. These parameters represent the beam bending stiffness, torsional stiffness, and a bending/torsion coupling parameter, respectively. Algebraic expressions for these parameters are available if one assumes a plate-like structural model. In this latter case, the three stiffness parameters closely resemble those developed for laminated plate bending theory (cf. Ref. 6). In any case, EI, GJ, and K are functions of the structural geometry, material properties and the orientation of the various laminae within the cover-sheet laminates.

The wing deflections represent those of a moderate-to-high-aspect-ratio wing under the action of strip theory airloads. The wing deformation consists of an upward bending deflection of the reference axis, $h(y)$, and a nose-up twisting about the y-axis, $\alpha(y)$.

Two aeroelastic parameters appear in the equilibrium equations used to assess the stability of this model. These two parameters, termed a and b (following the symbolic notation used in Ref. 3), are defined in Reference 5 as follows:

$$a = \left(\frac{qcel^2 a_o \cos^2 \Lambda}{GJ}\right)\left(\frac{1-k\tan\Lambda}{1-kg}\right) \qquad (1)$$

$$b = \left(\frac{qcl^3 a_o \cos^2 \Lambda}{EI}\right)\left(\frac{\tan\Lambda-g}{1-kg}\right) \qquad (2)$$

where $k = K/EI$ and $g = K/GJ$. In Eqns. 1 and 2, $a_o$ is the two-dimensional lift-curve slope of an airfoil section perpendicular to the reference axis in Figure 3. This two-dimensional sectional lift-curve slope may be adjusted empirically to reflect the influence of sweep and finite-span effects.

The study of static aeroelastic effects upon metallic wings, first considered in Reference 3, contained parameters similar to those in

Eqns. 1 and 2. The difference between those used in Reference 3 and those in Reference 5 lies in the inclusion of the nondimensional stiffness terms k and g in Eqns. 2 and 3. These terms may be positive or negative. Conventional beam-like structures will have values of K (or k and g) equal to, or nearly equal to, zero. Thus, the influence of aerodynamic sweep appears in the parameter "a" only if cross-coupling is present. Similarly, stiffness cross-coupling is seen to counteract or reinforce wing geometric sweep in the parameter "b" depending upon the sign of g.

With the parameters "a" and "b" as given in Eqns. 1 and 2, together with an equation transformation, the original solution procedure given in Reference 3 was used to examine static aeroelastic characteristics of swept wings. In some cases, figures from Reference 3 can be used for evaluation of composite designs provided that the "advanced" definitions of "a" and "b" are used. This subject, together with further details, is discussed in References 5 and 7. To illustrate the static aeroelastic effects of structural cross-coupling, using this type of model, let us examine several figures from Reference 8. Figure 4 shows the extreme effect that laminate orientation can have upon static aeroelastic divergence. Two forward swept wing angles, $30^o$ and $60^o$, are considered, together with an unswept wing. With bend/twist coupling, induced by shear strain/normal stress coupling within the laminate, the divergence of an unswept wing and a $30^o$ forward swept wing are totally precluded by a small ($10^o$ forward) off-axis ply sweep. For a $60^o$ forward swept wing, dramatic increases are also seen due to built-in anisotropy.

While the increased aeroelastic stability expands the feasible operational envelope, sometimes with reduced weight, advanced composite tailoring also affects sub-critical performance. Figure 5 indicates the potential effect of wing anisotropy upon the spanwise center of pressure movement due to flexibility. Two wing tip-to-chord taper ratios (denoted as $\lambda$ in the figure) at a fixed dynamic pressure are considered. Such center of pressure movement may be deliterious to ride quality, handling qualities or, in extreme cases, overall aircraft stability.

Figure 6 shows an extreme case of center-of-pressure movement for a high-aspect-ratio, forward swept wing operating at a fixed dynamic pressure. A poor design not only leads to a considerable center-of-pressure shift, but also causes divergence. On the other hand, a certain range of ply orientations effectively controls this shift.

Turning to another performance related aspect of aeroelastic design, let us consider lateral control effectiveness, a characteristic important to aircraft maneuverability. Lateral control effectiveness, measured as the steady-state roll rate per unit aileron deflection, can be enhanced considerably by judicious tailoring of laminate geometry. Figure 7 illustrates a measure of aileron effectiveness in the form of flexible-to-rigid ratios of control effectiveness for a laminated composite, $30^o$ forward swept wing at four values of dynamic pressure. As the dynamic pressure is increased, the control effectiveness declines. However, when the laminate fibers are oriented in the forward quadrant, this decline is less rapid. When plies are oriented about $40^o$ forward of the swept reference axis, aileron effectiveness is maximized.

These are but a few of the many examples illustrating performance related concerns of aeroelasticians. Aeroelasticians have always been aware of this impact. However, those in more focused areas of structural design are only now becoming aware of creative uses to which their materials and construction techniques can be applied. The reader is referred to a more complete summary of aeroelastic tailoring applications, up to about 1983, contained in Reference 9.

## OBLIQUE WING AEROELASTIC TAILORING

The recent reappearance of the oblique wing concept offers yet another opportunity for the creative use of composite materials. It has been shown, both theoretically and experimentally, that it is possible to achieve high lift-to-drag ratios for aircraft operating supersonically through the use of a high-aspect-ratio, asymmetrically swept wing [10,11]. Such a configuration has two distinct advantages; first, the engine power required for propulsion is relatively low for high subsonic and supersonic speeds; and secondly, since the wing can be rotated in flight, the landing and takeoff speeds are substantially lower than conventional aircraft. That such an asymmetrically swept wing possesses desirable aerodynamic characteristics in certain speed regimes has been well established. However, the same undesirable experience with static divergence of symmetrically swept forward wings, discussed previously, initially caused a substantial amount of discussion about the aeroelastic stability of oblique winged aircraft (Refs. 12,13,14,15,16).

An important conclusion of these studies was that the inclusion of the freedom of the oblique aircraft to roll as a rigid body during aeroelastic oscillation enhances the stability of the system. In particular, the aeroelastic divergence mode of instability associated with cantilevered swept forward wings disappears and is replaced by a low-frequency, oscillatory instability that occurs at a higher flight speed than does divergence. (This is unlike the freely flying FSW which experiences body freedom flutter at an airspeed <u>lower</u> than its divergence speed). The inclusion of pitch and plunge rigid body freedoms further modifies the low-frequency instability speed, but not to the extent seen with the inclusion of rigid body roll freedom. While the Oblique Wing and X-29 are markedly different designs, they share an element of commonality - the presence of a forward swept wing component that can drive the wing design. (As a historical note, the oblique wing concept predates the X-29. In fact, the oblique wing involvement of the present author led to Krone's initial interest in FSW tailoring.) Therefore, many of the analytical capabilities that have proven so useful to the investigation of the unusual generic features of the X-29 are also adaptive to the oblique wing.

To assess the potential effects of composite materials upon aeroelastic stability of the oblique wing concept, an idealization of a flexible oblique wing, operating in a supersonic flow, was developed. This idealization, shown in Figure 8, assumes the wing to be beam-like with constant geometrical and stiffness properties. An aerodynamic theory similar to piston theory, but with a correction factor applied [17] to make it more acceptable for low supersonic Mach numbers, was used. Equations of motion were developed with assumed deflection functions. For these equations, bending deflection, torsional deflection and roll freedom were permitted, while rigid body pitch and plunge were excluded.

In the jargon of aeroelastic tailoring, "positive cross-coupling" occurs when upward bending is accompanied by nose-up twist. This leads to a wash-in condition and may lower dramatically the divergence speed. Sweptforward metallic wings naturally wash-in under load because of bending and geometric considerations. Thus a wash-out condition, or negative cross-coupling, must be designed into the laminate, as seen previously. On the other hand, sweptback metallic wings naturally wash-out, a condition that results in extremely high divergence speeds, but lowered lift effectiveness.

Since the oblique wing is partly sweptforward, partly sweptback, the most appropriate design strategy to enhance divergence <u>and</u> lift effectiveness is to use a continuous ply arrangement, such as indicated by Figure 9. As a result, cross-coupling is positive on the sweptback wing, but is negative on the sweptforward portion.

Rather than use laminate orientation as a design parameter, as was done for FSW studies, a non-dimensional tailoring parameter was developed to assess "generic" tailoring effects. Consider Eqns. 1 and 2, in which the term 1-kg appears. Strain energy considerations require this term (1-kg) to be greater than zero. As a result kg (a nondimensional product) can take on values only between zero and unity. This nondimensional product may be expressed in terms of a new parameter, denoted as $\psi$ and defined by the following relationship

$$\psi^2 = K^2/EI\ GJ < 1 \tag{3}$$

As a result,

$$-1 < \psi < 1 \tag{4}$$

From Eqn. 4, the limits of the ratio $\psi = \sqrt{EI/GJ}$ enable one to categorize a beam-like structure as highly coupled or lightly coupled, with values near zero categorized as lightly coupled, while absolute values of $\psi$ near unity are associated with highly coupled structures.

The potential effects of advanced composites on oblique wing aeroelastic stability were examined in two stages. In the first stage, the influence of the generic composite parameter $\psi$ upon clamped supersonic sweptforward wing stability was investigated. Then, roll freedom was permitted and the wing aeroelastic stability reanalyzed. Let us turn to some typical results from this study.

Figure 10 shows how the divergence speed of an orthotropic clamped forward swept portion of the oblique wing is affected by sweep angle and the chordwise position of the reference axis (RA for short). For this figure, neither roll freedom nor stiffness coupling is not allowed; the Mach number is 2.5. (Note that with Mach number fixed, an altitude stiffness parameter is probably more appropriate. However, use of the nondimensional velocity shown in Figure 10 is more instructive for present purposes). The wing semi-chord dimension is here denoted as b, while the reference torsional frequency is $\omega_\theta$. The legend RA = 0.1 is used to locate the position of the reference axis with respect to the wing mid-chord. RA equal to zero indicates that the reference axis is exactly at the mid-chord, while RA = 0.1 indicates that the reference axis lies 0.1 semi-chord lengths aft (downstream) of the mid-chord. Negative values of RA are associated with reference axis positions ahead of (upstream) of the mid-chord.

When the reference axis is at the mid-chord (RA = 0) and no stiffness cross-coupling is present, only bending flexibility is important to divergence, even though torsional flexibility has been included in the model. This is so because the aerodynamic center is also at the mid-chord when the flow is supersonic. As a result, there is no static aeroelastic coupling. If RA < 0 then the aerodynamic loads are aft of the reference axis. Favorable (nose-down) aeroelastic twist results. The converse is true for RA > 0. The adverse effect of sweptforward wing bending is seen in Figure 10 since divergence speed decreases as forward sweep increases. Note that Figure 10 depicts two critical speeds, one at low speed, the other at a higher speed. The lower of the two speeds is critical while the upper speed indicates a restabilization of the wing. As forward sweep is decreased, these two divergence speeds merge; divergence disappears from the design as forward sweep is reduced.

The inclusion of freedom to roll substantially modifies the aeroelastic stability of the orthotropic wing, as is indicated in Figure 11. This figure plots nondimensional _flutter_ speed against oblique wing sweep angle. With the root section of each wing portion connected at the fuselage there is now additional inertial and aerodynamic coupling between the two wing portions.

Figure 11 shows the effects of four different reference axis positions on flutter for a typical orthotropic configuration. The instability boundaries in Figure 11 are the result of coupled bend/twist/roll oscillations and occur at slightly higher airspeeds than those shown in Figure 10. The conclusion here is that roll freedom not only changes the form of instability (from static to dynamic), but also modifies the airspeed at which it occurs. Two different modes of instability are visible in Figure 11, as indicated by a cusp in the curves. Both instability modes are low frequency when RA = -0.1; the instabilities occurring at higher sweep angles are also low frequency when RA > -0.1. However, for RA < -0.1, and at the lower sweep angles, the instabilities occur at higher frequencies and are mostly torsion driven. The source of the low frequency instabilities is a lack of stiffness in bending at high dynamic pressures. In short, the flutter speeds at moderate to high sweep in Figure 11 are bending driven and are modified versions of part of the divergence boundaries shown in Figure 10. These results match those found previously for oblique wings constructed of conventional materials.

Now let us consider stiffness cross-coupling and its effects. The inclusion of stiffness cross-coupling will modify the fixed-root divergence behavior considerably, as is indicated in Figure 12. A baseline configuration with RA fixed to be -0.1 was examined in detail for a root-fixed (no roll freedom) case using $\psi$ as a parameter. In addition, because of other constraints, the values of cross-coupling used on each wing reflect continuous ply layups as shown in Figure 9.

A value of $\psi$ greater than zero leads to a destabilizing situation and is so indicated in Figure 12 when $\psi = 0.25$. This value is not unreasonable to expect for a tailored structure. Thus, introduction of this type of cross-coupling into the clamped sweptforward wing provides a detrimental decrease in divergence airspeed, if all other parameters remain the same. On the other hand, with $\psi < 0$ on the sweptforward wing, an increase in static stability is evident.

The inclusion of roll freedom and stiffness cross-coupling leads to stability boundaries indicated in Figure 13. The likely effects of stiffness cross-coupling are more difficult to anticipate when roll-freedom is introduced. First of all, the cusps on three of the stability boundaries indicate "mode-switching" between critical modes. Improvements in stability in one sweep range are accompanied by declines in other sweep ranges. Note particularly, that, in the case considered, a small amount of negative cross-coupling ($\psi = -0.25$) on the sweptforward wing, (and $\psi = 0.25$, on the sweptback wing), is beneficial, but larger amounts ($\psi = -0.50$) are probably not.

Other aeroelastic tailoring issues related to oblique wings remain unresolved. The use of tailoring to enhance lateral control and trim are but a few of the substantial number of aeroelastic problems awaiting seriously study.

The ratio b/a is often referred to as the "sweep parameter" (note that b is now not the semi-chord) in static aeroelastic studies. From Eqns. 1 and 2, this ratio is:

$$b/a = (\frac{L}{e})(\frac{\tan\Lambda - \psi/\sqrt{R}}{1 - \psi\sqrt{R}\tan\Lambda})R \qquad (5)$$

where $R = GJ/EI$ and e is RA times the semi-chord. If $\psi$ is zero, then Eqn. 5 becomes:

$$b/a = (\frac{L}{e})(\tan\Lambda)R \qquad (6)$$

The second quantity in brackets in Eqn. 5 can be thought of as being tangent of the effective aeroelastic sweep, $\Lambda_E$. A value of cross-coupling $\psi_E$ can be found to cause the wing at an actual angle $\Lambda$ to behave aeroelastically as if it were at $\Lambda_E$. The value $\psi_E$ to accomplish this goal is found to be:

$$\psi_E = \frac{\sqrt{R}(\tan\Lambda - \tan\Lambda_E)}{1 - R\tan\Lambda\tan\Lambda_E} \qquad (7)$$

Note that $\psi_E$ must still be bounded and, as a result, it may not be possible to simulate $\Lambda_E$ with cross-coupling.

It is also shown, in Reference 18, that a value of b/a near 3 causes a wing to be aeroisoclinic. As a result, the wing is divergence free and experiences no CP shift. If the value of b/a desired is expressed as the symbol $x_o$, then the value of $\psi$ necessary to accomplish this is

$$\psi_o = \frac{[\sqrt{\overline{R}}\tan\Lambda - x_o(\frac{e}{L})/\sqrt{\overline{R}}]}{[1 - x_o(\frac{e}{L})\tan\Lambda]} \tag{8}$$

CONCLUSION

From this brief discussion, it is seen that aeroelastic optimization with advanced composite materials can lead to favorable improvements in unconventional aircraft aeroelastic stability. As a result, new configurations are available to the designer seeking exceptional performance. One may then ask, "what else?" Most enlightened assessments of the future generally prove to be somewhat off the mark because they are made in the confines of today's needs, not tomorrow's. However, it is safe to say that combinations of efficient, advanced analysis techniques, advanced technology and mission requirements, and new materials will combine to produce configurations for which structures do more than create surface shapes and hold the vehicle together.

In addition, the presence of active controls creates a need for a "friendly" structural configuration that allows for off-design control robustness and damage tolerance. Passive and active control techniques can combine to create vehicles with enhanced ride quality and reduced susceptability to fatigue. Thirty-one years have elapsed since Professor Bisplinghoff's Wright Brothers Lecture. One can only hope to be a witness to the exciting possibilities of the next thirty-one.

REFERENCES

1. N.J. Krone, Jr., "Forward Swept Wing Flight Demonstrator," AIAA Paper No. 80-1882, Anaheim, California, August 1980.

2. R.L. Bisplinghoff, H. Ashley and R.L. Halfman, *Aeroelasticity*, Addison-Wesley, 1955.

3. F.W. Diederich and B. Budiansky, "Divergence of Swept Wings," NACA TN 1680, 1948.

4. N.J. Krone, Jr., "Divergence Elimination with Advanced Composites," AIAA Paper No. 75-1009, Los Angeles, August 1975.

5. T.A. Weisshaar, "Aeroelastic Stability and Performance Characteristics of Aircraft with Advanced Composite Sweptforward Wing Structures," AFFDL-TR-78-116, Wright-Patterson AFB, Ohio, September 1978.

6. S.W. Tsai and H.T. Hahn, *Introduction to Composite Materials*, Technomic Publishing Co., Westport, Conn. 1980.

7. T.A. Weisshaar, "Aeroelastic Tailoring of Forward Swept Wings," *Journal of Aircraft*, Vol. 18, No. 8, August 1981, pp. 669-676.

8. T.J. Hertz, M.H. Shirk, R.H. Ricketts, and T.A. Weisshaar, "Aeroelastic Tailoring with Composites Applied to Forward Swept Wings," AFWAL-TR-81-3043, Wright-Patterson AFB, Ohio, November 1981.

9. M.H. Shirk, T.J. Hertz, T.A. Weisshaar, "Aeroelastic Tailoring-Theory, Practice, Promise," *Journal of Aircraft*, Vol. 23, No. 1, January 1986, pp. 6-18.

10. R.T. Jones, "Reduction of Wave Drag by Anti-symmetric Arrangement of Wings and Bodies," *AIAA Journal*, Vol. 10, No. 2, February 1972, pp. 171-176.

11. R.T. Jones and J.W. Nisbet, "Transonic Transport Wings - Oblique or Swept?," *Astronautics and Aeronautics*, Vol. 12, No. 1, January 1974, pp. 40-47.

12. T.A. Weisshaar and H. Ashley, "Static Aeroelasticity and the Flying Wing," *AIAA Journal of Aircraft*, Vol. 10, No. 10, Oct. 1973, pp. 586-594.

13. T.A. Weisshaar and H. Ashley, "Static Aeroelasticity and the Flying Wing - Revisited," *AIAA Journal of Aircraft*, Vol. 11, No. 11, November 1974, pp. 718-720.

14. T.A. Weisshaar and J.B. Crittenden, "Flutter of Asymetrically Swept Wings," *AIAA Journal*, Vol. 14, August 1976, pp. 993-994.

15. R.T. Jones and J.W. Nisbet, "Aeroelastic Stability and Control of an Oblique Wing," *The Journal of the Royal Aeronautical Society*, Volume 80, August 1976, pp. 365-369.

16. J.B. Crittenden, T.A. Weisshaar, E.H. Johnson and M.J. Rutkowski, "Aeroelastic Stability Characteristics of an Oblique Winged Aircraft," *Journal of Aircraft*, Vol. 15, July 1978, pp. 429-434.

17. M.D. Van Dyke, "A Study of Second-Order Supersonic Flow Theory," NACA Report 1081, 1952.

18. T.A. Weisshaar, "Forward Swept Wing Aeroelasticity," AFFDL-TR-79-3087, Wright-Patterson AFB, Ohio, June 1979.

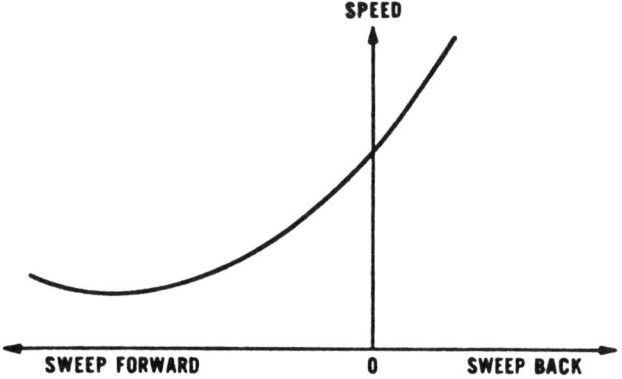

Figure 1 -  A Comparison of wing critical speeds (Ref. 2).

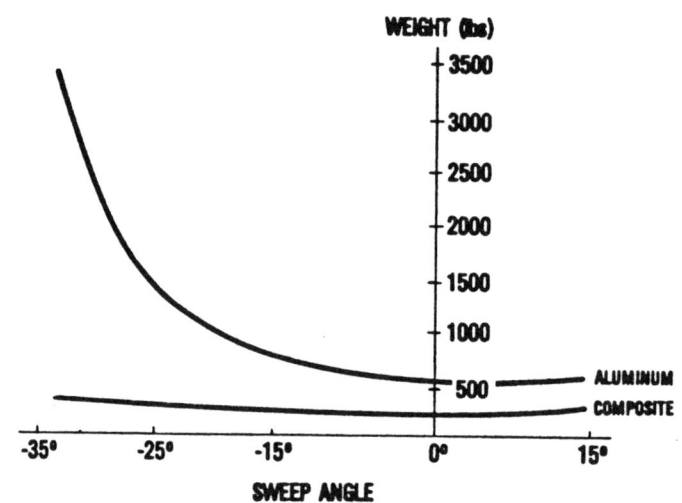

Figure 2 -  Lightweight fighter wing weight versus sweep, comparing optimized composites to conventional materials.

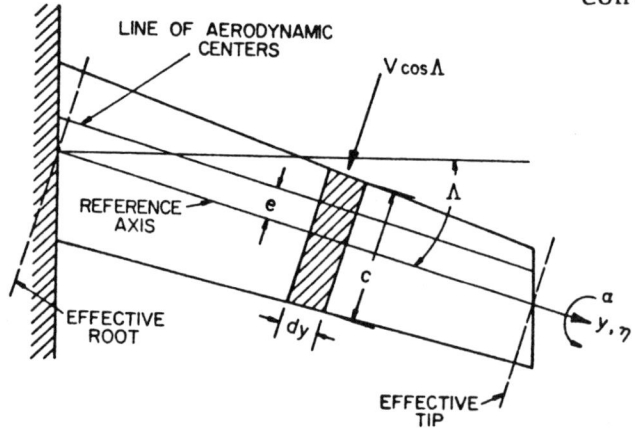

Figure 3 -  Slender swept wing beam idealization showing chordwise strips used to compute aerodynamic loads.

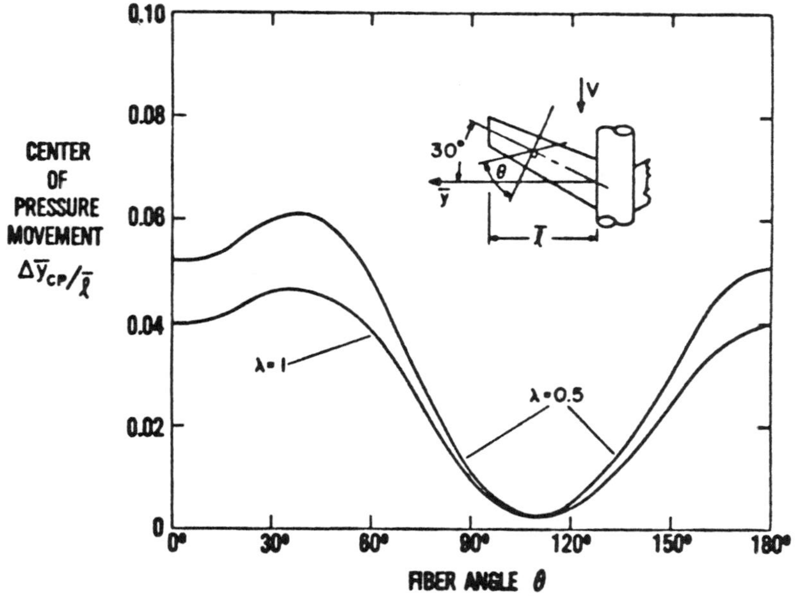

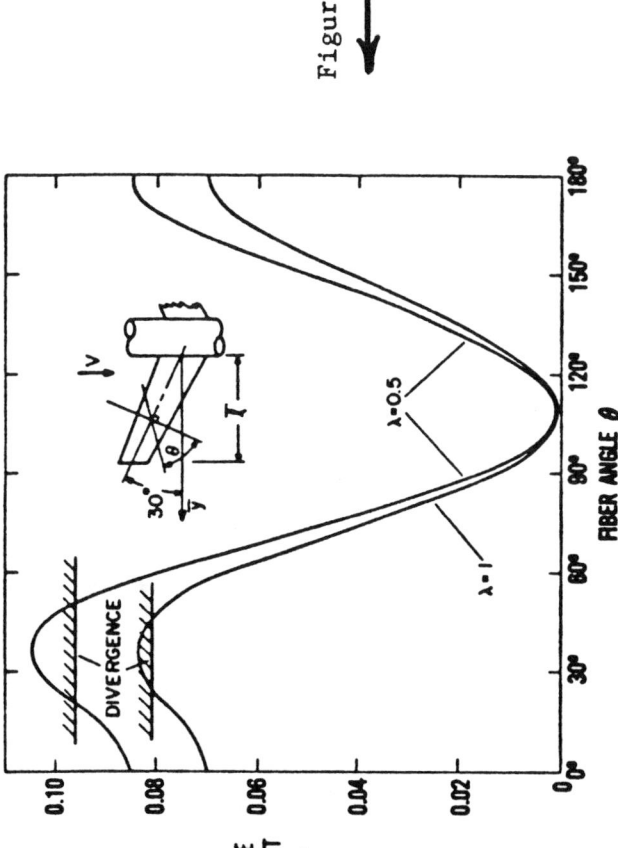

Figure 5 — Normalized change in spanwise CP location versus ply orientation for a 30° sweptforward wing. Dynamic pressure is 20% of divergence dynamic pressure (when $\theta = 90°$). Structural aspect ratio is 3 to 1.

Figure 6 — Normalized change in spanwise CP location versus ply orientation for a 30° sweptforward wing operating at 20% of divergence dynamic pressure (when $\theta = 90°$). Structural aspect ratio is 12.5 to 1.

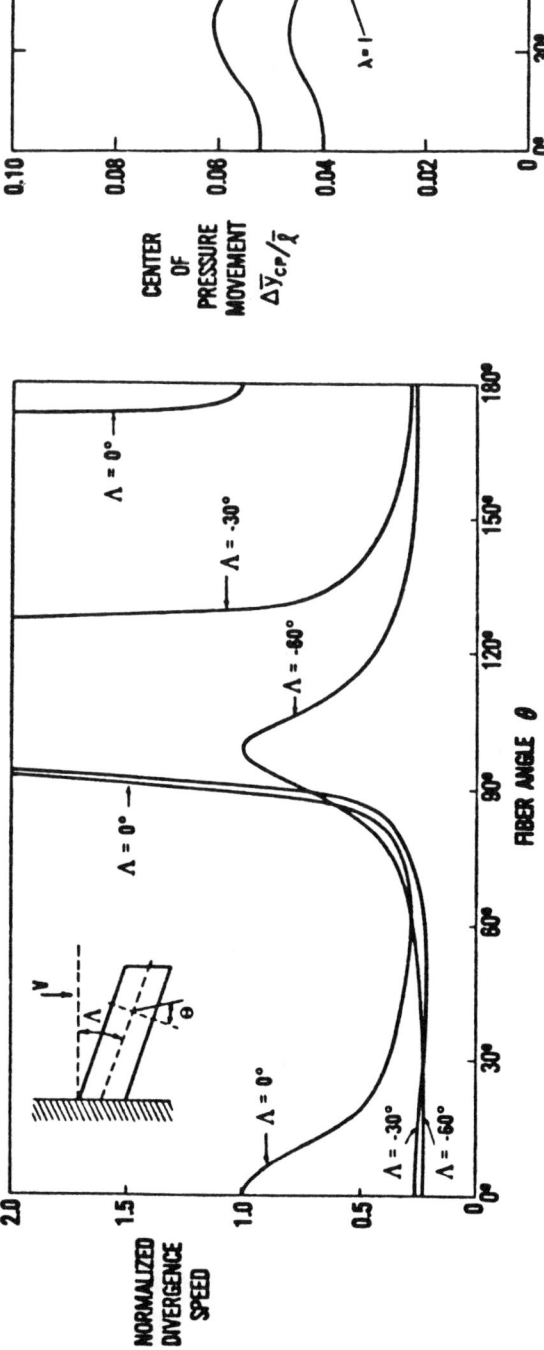

Figure 4 — Normalized divergence speed versus ply orientation (fiber angle). Taper ratio is 1 to 5.

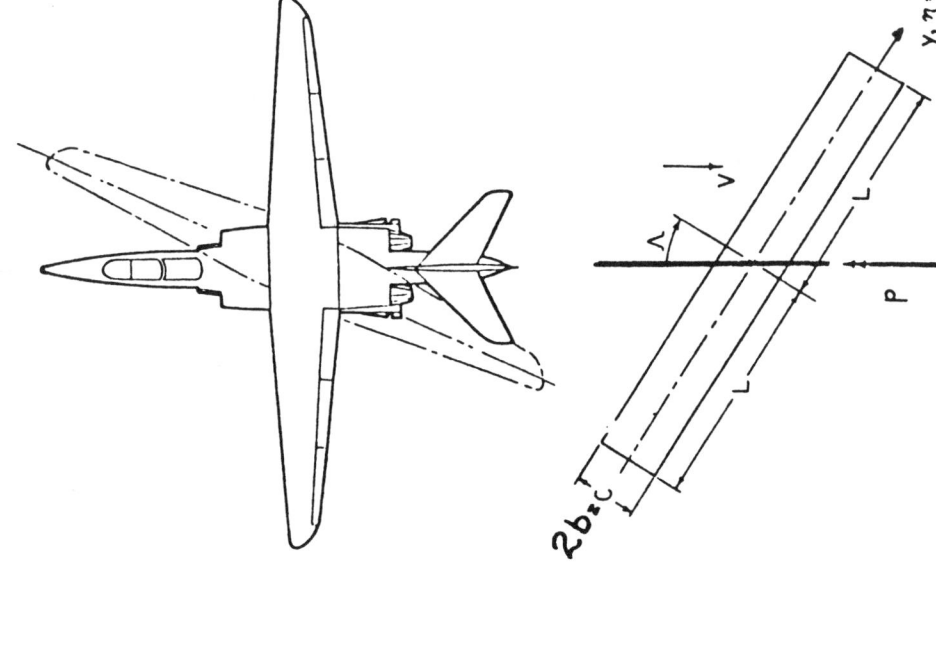

Figure 8 – Proposed oblique wing aircraft configuration, shown with analytical idealization.

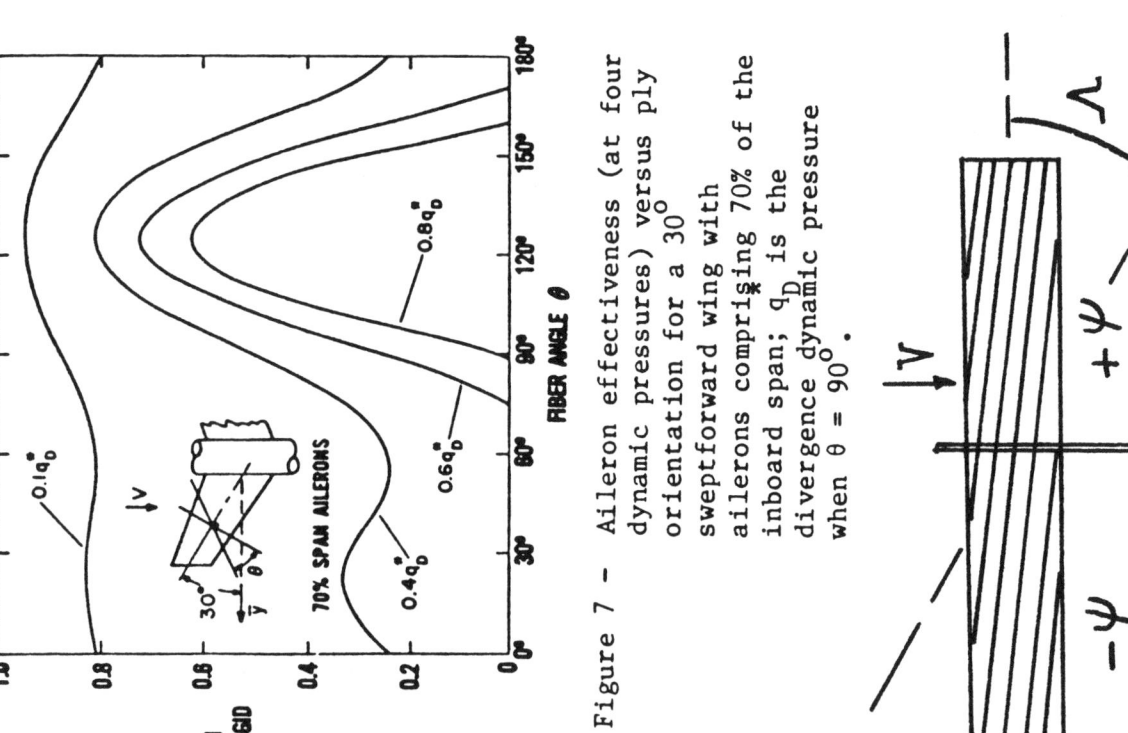

Figure 7 – Aileron effectiveness (at four dynamic pressures) versus ply orientation for a 30° sweptforward wing with ailerons comprising 70% of the inboard span; $q_D$ is the divergence dynamic pressure when $\theta = 90°$.

Figure 9 – Ply arrangement for optimum antisymmetrical performance.

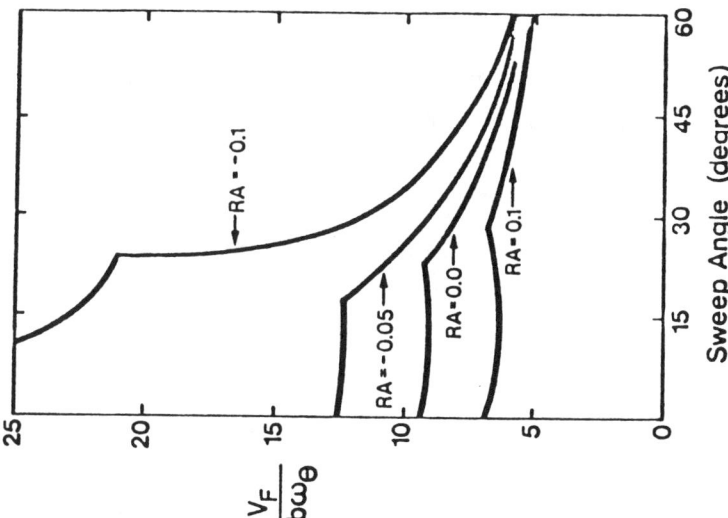

Figure 11 - Nondimensional flutter speed versus sweep angle of a typical oblique wing configuration, free to roll, with the wing characteristics shown in Figure 10.

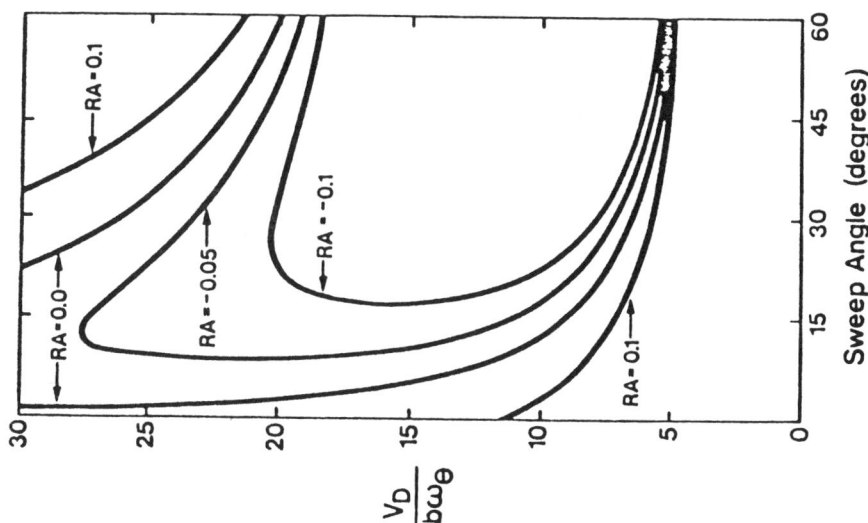

Figure 10 - Wing divergence speed versus sweep forward. Wing is clamped at the root. Aspect ratio (unswept) is 8; no taper. Four reference axis positions of an orthotropic wing are shown with $M = 2.5$.

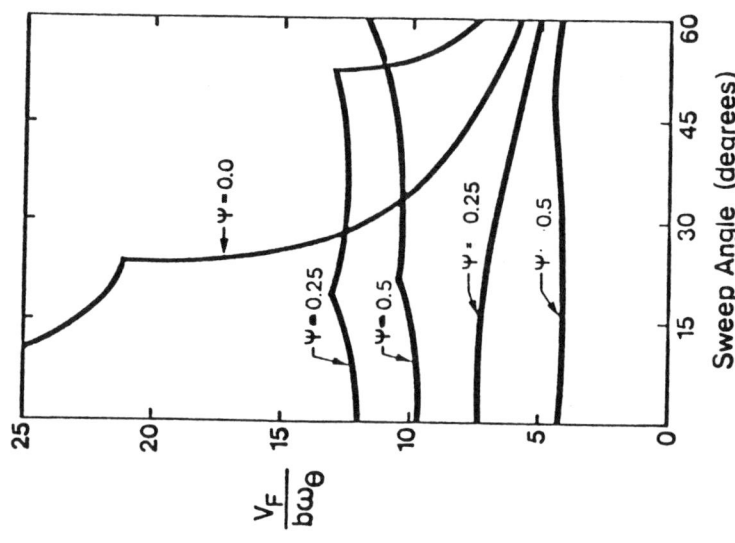

Figure 13 – Nondimensional flutter speed versus sweep angle of a typical oblique wing with roll freedom and wing characteristics as shown in Figure 12. Four values of cross-coupling are shown.

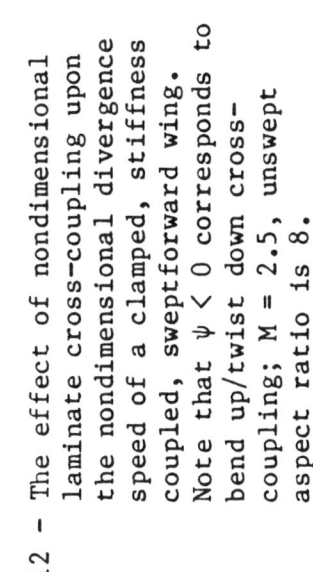

Figure 12 – The effect of nondimensional laminate cross-coupling upon the nondimensional divergence speed of a clamped, stiffness coupled, sweptforward wing. Note that $\psi < 0$ corresponds to bend up/twist down cross-coupling; M = 2.5, unswept aspect ratio is 8.

# STRUCTURAL OPTIMIZATION METHODS FOR INDUSTRIAL APPLICATION

BY

G.N. VANDERPLAATS

Department of Mechanical Engineering
University of California
Santa Barbara, California

## ABSTRACT

The application of numerical optimization techniques to structural design is discussed. The overall design task includes finite element analysis, calculation of the sensitivity of structural responses to the design variables, proper problem formulation, and finally, the actual optimization itself. These various aspects are discussed and examples are offered to demonstrate the present state of the art. It is noted that most major finite element program vendors are in the process of adding some optimization capability to their codes. It is concluded that the use of optimization to solve structural design problems of real complexity is now a reality and promises to lead the way toward fuller automation of the design process.

# INTRODUCTION

Structural optimization has seen major advances in recent years, and today can be considered to be a practical tool for design. This has come about as a result of simultaneous maturing of optimization algorithms, formal approximation techniques, and finite element analysis. Also, the dramatic advances in computer speed now make it possible to do in seconds what, only a few years ago, required hours or even days.

The methodology which now makes industrial use of structural synthesis possible will be reviewed here. The basic ingredients of a structural optimization program include analysis, sensitivity (gradients) and the optimization itself. Included in this is the key issue of problem formulation, and advances here have been instrumental in creating the useful tools now available. These ingredients will be discussed, and the status of encorporating this methodology into large scale commercial programs will be identified. Finally, some examples will be offered to demonstrate the present state of the art in structural synthesis.

# THE BASIC INGREDIENTS

Finite element structural analysis is now commonplace and is considered to be a well established engineeing tool. However, the ability to analyse a proposed design is only a part of the overall design process. Other parts include economic analysis, basic concept formulation, sensitivity of the proposed design to changes in the design parameters, iterative design improvement, and integration of the structural design into the overall system, to list a few.

Structural optimization addresses only some of these aspects. Particularly, this is a useful tool to be used with simple models early in the concept development stage, or with more complex models after the general structure has been defined. In other words, optimizaton is a powerful tool for those aspects of the design process that are, or can be made to be, algorithmic. Optimization provides an efficient means to perform the repetitive computational tasks, and most importantly, is able to deal, simultaneously, with many more variables than is possible with graphical or intuitive methods.

Those ingredients of the design process that are included in the structural optimization task include analysis, sensitivity calculations, problem formulation, and the actual optimization itself. In the following sections, these topics will be briefly reviewed in order to more clearly identify the state of the art. References are offered which provide an extensive review of this technology.

## Analysis

The usual finite element structural analysis problem can be stated as "find the set of displacements, $\underline{u}$, that solves the system of equations"

$$K\underline{u} + C\underline{\dot{u}} + M\underline{\ddot{u}} = F(t) \tag{1}$$

where $\underline{u}$ is considered here to be a function of time. This general form may be added to, or parts deleted, depending on whether static, vibration, dynamic or aeroelastic response is desired. While structural optimization methods can be applied to these general cases, the basic concepts can be best understood by considering only the static case:

$$K\underline{u} = \underline{P} \tag{2}$$

wher $\underline{P}$ is the vector or vectors of applied loads and the mast stiffness matrix, K, is calculated as the sum of elemental stiffness matrices. K is symmetric and sparce and is usually solved by decomposition. Once the nodal displacements, $\underline{u}$, are known, the element level stresses are easily calculated. Because this basic formulation is well understood, further discussion is not necessary except to note that, with few exceptions, modern structural optimization uses this displacement form of the finite element method as the basis for developing the design algorithms.

## Sensitivity Analysis

In the context of structural optimization, sensitivity analysis is taken to mean the rate of change of the structural response with respect to the independent design variables. Here, structural response may be stress, strain, displacment, buckling load factor, vibration eigenvalue, flutter speed, etc., while the design variables are the member sizing variables, joint locations of a discrete structure such as a truss, the shape definition of a continuum structure, or the material properties of a composite structure.

Sensitivity information is valuable in making design decisions whether optimization is used or not. This information provides a direction and amount by which a design variable must be changed in order to reduce weight or reduce some critical stress, as examples. In structural optimization, because many variables will be changed, this information can be used mathematically to find the best way to change them simultaneously. Most modern optimization algorithms require sensitivity information in order to efficiently direct the design process.

In the early years of structural optimizaton, it was generally assumed that because the response quantities are implicit functions of the design variables, the needed sensitivity information must be calculated by finite difference methods. However, noting that sensitivity of stresses is directly calculated from sensitivity of displacements, if displacement sensitivity is available, most of the effort is done. With this in mind, we can implicitly differentiate Equation 2 and re-arrange to give:

$$\frac{\partial \underline{u}}{\partial X_i} = K^{-1}[\frac{\partial \underline{P}}{\partial X_i} - \frac{\partial K}{\partial X_i} \underline{u}] \tag{3}$$

where it is noted that the rate of change of the master stiffness matrix, K, is the sume of the rates of change of the element stiffness matrices with respect to the particular variable. Also, note that once the static analysis has been done, $K^{-1}$ is available in decomposed form so that Equation 3 requires only some additional forward and backward substitutions for its solution.

The use of Equation 3 is referred to as the "direct method," and is commonly used. Once the sensitivity of displacements has been calculated, the sensitivity of stress is recovered from

$$\frac{\partial \sigma}{\partial X_i} = [\frac{\partial}{\partial X_i} \underline{S}]^T u_e + \underline{S}^T \frac{\partial u_e}{\partial X_i} \qquad (4)$$

where $\underline{S}$ is the stress-displacement relationship and $u_e$ is the vector of element nodal displacements.

In many cases, the necessary sensitivity of the element stiffness matrices can be calculated directly. For example, if the design variable is the cross-sectional area of a bar, the element stiffness matrix is the area times a constant (geometric and material) matrix and so its derivative with respect to the member area is just this constant matrix. When considering higher-order elements or geometric design variables, this is not so straight forward and so finite difference methods become useful. However, the finite difference is now only applied to the element stiffness matrix, rather than to the complete system of equations. This is referred to as the "semi-analytical" method and is probably the most direct method of encorporating sensitivity information into an existing finite element program.

A second approach to calculating sensitivity information is referred to as the adjoint or dummy load method [1]. Here, we begin by differentiating the constraint equation as

$$\frac{dg}{dX_i} = \frac{\partial g}{\partial X_i} + \underline{Z}^T \frac{\partial \underline{u}}{\partial X_i} \qquad (5)$$

where $Z_j = \partial g/\partial j$. If the constraint, g, is a stress constraint, $\underline{Z}$ is the stress-displacement relationship. Substituting Equation 4 into Equation 5 and noting symmetry of $K^{-1}$, we have

$$\underline{Z}^T \partial \underline{u}/\partial X_i = \underline{Z}^T K^{-1}[\partial P/\partial X_i - (\partial K/\partial X_i)\underline{u}] \qquad (6)$$

Therefore, this can be evaluated by solving the matrix equation

$$K\underline{Q} = \underline{Z} \qquad (7)$$

and so $\underline{Z}$ just becomes another right-hand side to the original finite element analysis equations. The vector $\underline{Z}$ is often referred to as an adjoint or dummy load.

It may be noted that these two methods are quite similar, but differ in the order of matrix operations. This is true, but depending on the information needed can have a dramatic effect on overall efficiency of the sensitivity calculations. Basically, if the number of responses for which sensitivity is needed is more than the number of design variables times the number of loading conditions, the direct method is preferred. On the other hand, if the needed information is less than this, the adjoint method is preferred [2].

Finally, a third approach, called the material derivative method is useful for such problems as shape optimization of contimuum structures. This method resembles techniques from continuum mechanics where a volume integral is converted to a surface integral. Essentially, the required information is calculated from

$$\frac{\partial \psi}{\partial X_i} = - \int_B \sigma(\underline{u}) \; \varepsilon \; (\underline{\lambda}) \; \underline{n} \cdot \frac{\partial r}{\partial X_i} \; dB \qquad (8)$$

where $\psi$ is the displacement functional for which the derivative is needed. $\sigma(\underline{u})$ are the stresses evaluated from the analysis and $\varepsilon(\lambda)$ are strains calculated from the analysis using adjoint loads. r is the position vector, and n is the normal to the surface. The integration is carried out over the perturbed boundary, B. If gradients of stresses are needed, this requires further computation using the stress-displacement relations for the finite elements being used. The reader is referred to references 3-8 for further details of this methods, a comparison of all three methods, and a general discussion of how the methods fit into overall design/analysis programs.

The choice of which method to use for calculationg sensitivity information depends on the amount of needed information and the specific nature of the problem being solved and the ideal computer program would determine the best method dynamically. In practice, it is likely that the direct, semi-analytical method will be most often used because it most easily fits into the framework of existing finite element programs.

Optimization Algorithms and Software

In finite element analysis, the displacement method is almost universally used today. This is principally because the resulting stiffness matrix is symmetric, positive definite, and quite sparce, as compared to the competing force method in which none of these conditions are met. Thus the displacement method is ideally suited for modern computers which do computations in a serial manner.

On the other hand, optimization methods are as individual as designers themselves. While some algorithms for solving the optimization problem have emerged as methods of choice, there is no clear consensus. The reason for this is that the choice of "search directions," that determine how the design variables are to be changed is not unique. Also, the method chosen is based to a large degree on the designer's philosophy. For example, a conservative designer may choose an interior penalty method since it produces a sequence of improving feasible (acceptable) designs. On the other hand, an unconservative designer may choose sequential linear programming which produces a sequence of improving infeasible (unacceptable) designs. In shrot, the choice of algorithms is often as individual as designers themselves (who are particularly known for their individualism). Nonetheless, there has been an identifiable trend over the years in the choice of methods, brought about in large part by the maturing of the discipline.

In the 1960's, sequential unconstrained minimization techniques (SUMT) were popular. These methods used well developed unconstrained minimization algorithms and included constraints via some form of penalty which accounted for constraints becoming near critical or violated. Also, sequential linear programming methods were popular. These methods first linearized the objective and constraint functions and solved the resulting approximate problem by well known linear programming methods. In each case, the nonlinear design problem was solved by converting it to a form suitable for solution by well known methods.

In the 1970's, methods such as feasible directions, improved SUMT, the Augmented Lagrange Multiplier Method (a form of SUMT) and reduced gradient methods represented an improvement in both efficiency and reliability of the optimization process and indicated a maturing of the field of mathematical programming for nonlinear constrained optimization.

In the 1980's, general optimization algorithms have continued to mature. Perhaps the most popular algorithm at the present time is Sequential Quadratic Programming.  This method approximates the Lagrangian function (the basis for the Kuhn-Tucker necessary conditions for optimality) as a quadratic, but uses only first-order (sensitivity) information.  The constraints are approximated as linear functions.  Also, duality theory has become an important tool for nonlinear constrained optimization.  These techniques have been known for some time in the field of linear programming, but are only recently being applied to nonlinear problems.  Their application to nonlinear problems is presently most appropriate in problems where good "mathematically separable" approximations can be made to the original design problem.

Also, during the 1980's, the earlier methods such as sequential linear programming, the method of feasible directions, sequential unconstrained minimization, and others have continued to mature so that numerous algorithms are still available for the general nonlinear constrained optimization problem, and this maturing process can be expected to continue.  The key point here is that optimization theory itself is fluid and, as the technology advances, the algorithms become increasingly efficient and reliable.

As nonlinear programming methods have matured, numerous computer programs have been developed which can be used by the practitioner.  The intent of this software development is to make the technology available to the practitioner and avoid the need for each designer to develop his/her own software.  This software, which is generally available, is called by such names, and is based on such methods as:

CONMIN [9] Feasible directions
NEWSUMT [10] Newton's method with SUMT
NEWSUMT A [11] Advanced version of NEWSUMT
GRG2 [12] Generalized Reduced Gradient Method
OPT [13] Generalized Reduced Gradient Method plus others
OPDES BYU [14] Variety of methods
IDESIGN [15] Sequential quadratic programming
ADS [16] Variety of methods

This software is, in most cases, available in the public domain or at nominal cost.  Thus, the need for the practitioner to develop optimization software is eliminated unless there is a strong need to develop special purpose algorithms.

Problem Formulation

Perhaps the most dramatic advances in the last ten years have come in the area of problem formulation [17].  It is now recognized that simple coupling of a finite element analysis program, even including sensitivity calculations, is not adequate to provide the efficiency necessary for the design of practical structures.  The principal advances in this regard have been the development of high quality approximations to structural responses which can be used for optimization without the multitude of detailed analyses that was previously required. Coincident with this has been the formalization of such concepts as design variable linking and temporary constraint deletion to further improve design efficiency and realism.

Design variable linking has been used for some time as a means of imposing practical considerations as well as reducing the number of design variables that must be considered in the optimization phase. For example, several finite element dimensions can be controlled by a single design variable as a means of imposing structural symmetry. More importantly, it is now generally agreed that treating (for example) the thickness of each element in the analysis model of an aircraft wing as a design variable is not realistic from manufacturing considerations, and even can introduce significant errors into the analysis model itself. Thus, design variable linking has the practical usefulness of keeping the design process realistic and the theoretical usefulness of reducing the difficulty of the optimization task simply by reducing the design problem size.

The concept of constraint deletion is also nothing more than introducing realism into the automated design process. Experienced engineers seldom consider all design constraints simultaneously since some are easily identified as being non critical (at least at the current point in the design process). However, when optimization has been used, there has been a tendency to routinely include all stress, displacement, frequency, etc. constraints throughout the design process. The disadvantage to this is that, using approximation techniques to be described later, it is necessary to calculate the sensitivity of all constraints to the design variables. Therefore, it is seen that, because the current design stage is only a step toward the optimum, the logical approach is to delete from consideration all constraints that are not currently critical or potentially critical. For example, if a particular stress constraint is far from its limit, it can be ignored for awhile and included later if it becomes near critical.

In addition to the concept of temporary constraint deletion, it is often most efficient to ignore some constraints early in the design process and include them later as the design is refined. Consider for example, the case where stress, displacement, frequency and aeroelastic constraints must be considered in the design. It may be most reasonable to first perform one cycle of the classical fully-stressed design method, even though this design is likely to violate other constraints. Then using this as a starting point, perform one or two design iterations with the displacement constraints included in addition to stress constraints. Following this, add frequency constraints and finally the aeroelastic constraints. The basic concept here is to first solve the easy problem to provide a good initial design for the more complex ones. It is noteworthy that this is what is usually done when optimization is not used, simply because it is most efficient. By ignoring complex constraints early in the design process, we save considerable computational cost by virtue of the fact that they need not be evaluated. Furthermore, as new constraints are added, if they are not critical at the start of the new optimization, only one constraint evaluation is needed to show that. The key idea when using this approach to automated design is that, as the complexity of the constraints considered is increased, all "lower level" constraints are also included. Thus, the optimum solution (assuming it is unique) must be the same as that obtained by including all constraints from the start, but at a much lower cost. While this approach has not yet been widely used, it deserves more attention as practical applications increase because the cost of a single analysis now dominates the design process.

While the concepts just discussed do much to improve the efficiency of the optimization process, the most significant advancement has been the development of approximation concepts themselves. Here it has been recognized that, if the original problem can be approximated in some explicit form, then this approximation can be used for the actual optimization phase. A new approximation can then be created at the proposed design point and the process repeated until it has converged. The most obvious approximation, and one that has now been used for over twenty years is to just linearize the objective and constraint functions using a first-order Taylor series expansion so

$$f(\underline{X}) \simeq f(\underline{X}^0) + \underline{\nabla} f(\underline{X}^0) \cdot \underline{\delta X} \qquad (9)$$

where here F($\underline{X}$) represents any objective or constraint function. The optimization is then solved, without detailed finite element analyses, in this linearized space. This is a linear programming problem and software for its solution is readily available and is efficient.

This approach has some theoretical limitations, particularly related to move limits during the aproximate optimization phase. Nonetheless, experience suggests that it is a reasonable method for general multi-disciplinary applications as well as structural optimization, and the method is presently enjoying renewed interest.

In structural optimization, it is often possible to make very high quality approximations to the response quantities, where the approximations are not linear, but are explicit and are often separable. This allows for solution of a more accurate approximate sub-problem and, where an explicit dual of the problem can be written, duality theory can be used for its efficient solution [18,19].

The basic concept here can be understood by considering the stress in a simple truss element.

$$\sigma = P/A \qquad (10)$$

This is clearly nonlinear in the sizing variable, A. However, if we pick as the design variable X=1/A, this becomes linear in X. Therefore, if a first-order Taylor series approximation is created with respect to X, it will be of very high quality. Indeed, this approximation is precise for statically determinate structures. In the X-space, the objective function now becomes nonlinear (weight is linear in A but nonlinear in X), but is still explicit and easily evaluated, along with its derivatives.

This form of the approximate problem is powerful for bar and membrane structures. For frame structures, a reasonable "intermediate variable" is the reciprocal of the section properties, while for a bending element it is the reciprocal of the cube of the thickness. This straightforward approach is not as simple for general shape optimization problems, but is nonetheless a valuable tool for providing insight here also. For such problems, it has been shown [19] that, using careful problem forumulation, a mixed approach can achieve overall optimization efficiency near that of the more common cases.

The key idea in approximation techniques is to create a high quality approximation to the original finite element based problem. The optimization is then performed on this approximate problem and a new proposed design is produced. The structure is then analysed in detail and the process is repeated until it has converged to a satisfactory solution. Finally, it should be noted that this concept is not limited to structural design. It is a general design methodology that is useful in any discipline as well as in multi-disciplinary optimization.

DESIGN EXAMPLES

In order to demonstrate the power of modern optimization methods to structures of practical size and complexity, two examples are offered here. The first is a large gear housing and the second is the three-dimensional shape optimization of an engine connecting rod.

Gear Housing

The large gear housing shown in Figure 1 was designed for minimum weight [20]. The structure has two planes of symmetry and six loading conditions. The finite element idealization of one fourth of the structure is shown in Figure 2. The analysis and sensitivity analysis was performed using MSC/NASTRAN Version 63. The analysis model consisted of 1623 finite elements and 7239 displacement degrees of freedom. The design model consisted of thirty independent sizing design variables and 5620 nonlinear inequality constraints. The constraints included stress limits under each loading condition and several stringent deformation constraints on the bearing deformations and relative center-to-center movement of the bearings. Some of the deformation constraints were initially violated by 50-90 percent. The initial design was scaled up to provide a near-feasible starting point for optimization.

The iteration history is shown in Figure 3, where one iteration consisted of a detailed finite element analysis and sensitivity analysis of approximately 150 critical or near critical constraints. One iteration required approximately 600 CPU seconds on a Cray 1s supercomputer. Of this, 243 seconds were spent in the analysis, 337 seconds on sensitivity analysis, 18 seconds on sorting operations and 2 seconds solving the approximate optimization sub-problem. The total design time was just under one hour. While this may appear to be expensive, it is highly unlikely that a design near this could be achieved by traditional methods.

Connecting Rod

The automotive connecting rod shown in Figure 4 was designed as a three-diminsional shape optimization problem [21]. The finite element model of one fourth of the structure is made of solid 20-node elements and is shown in Figure 5. The MSC/NASTRAN finite element program was used to perform analysis, while the material derivative approach [3] was used for sensitivity calculations. The analysis model consisted of 105 elements, 928 nodal points and 2126 displacement degrees of freedom. Eight design variables, shown in Figure 4, were used to define the shape of the structure. The objective was to minimize material volume and constraints were imposed on the von Mises stress at each node.

The iteration history is shown in Figure 6, and although the design required 20 iterations, it is clear that a practical optimum was achieved much earlier. This particular example demonstrates that shape sensitivity information can actually be calculated external to the finite element analysis program itself. Thus, while this may not be the desired situation for everyday use of this technology, it is possible to create a rather sophisticated structural optimization program even when the sensitivity capability is not directly available from the finite element analysis program.

## SUMMARY

Structural optimization has progressed dramatically in recent years so that today it can be considered a powerful and practical design tool. While much research remains, the technology has matured to the point that a wide range of realistic design problems can be routinely solved. Most importantly, the major industrial finite element vendors either now have or are encorporating optimization capabilities into their programs. It is clear that it is no longer necessary to be an expert in optimization methods to make good use of these methods. As this technology becomes generally available to a wide user community in the form of commercial quality software, we can anticipate a major expansion in its applications to large scale industrial applications.

## REFERENCES

1. Arora, J.S. and Haug, E.J., "Methods of Design Sensitivity Analysis in Structural Optimization," AIAA Journal, Vol. 17, Sept. 1979, pp. 970-974.

2. Vanderplaats, G.N., "Comment on Methods of Design Sensitivity Analysis in Structural Optimizaion," AIAA Journal, Vol. 18, No. 11, Nov. 1980, pp. 1406-1407.

3. Haug, E.J., Choi, K.K., Hou, J.W., and Yoo, Y.M., "A Variational Method for Shape Optimal Design of Elastic Structures," Optimal Structural Design II, (Ed. R.H. Gallagher) Wiley, New York, 1983.

4. Choi, K.K. and Haug, E.J., "Shape Design Sensitivity Analysis of Elastic Structures," Journal of Structural Mechanics, 11(2), 1983, pp. 231-269.

5. Haug, E.J., Choi, K.K. and Komkov, V., Design Sensitivity Analysis of Structural Systems, Academic Press, 1984.

6. Choi, K.K., "Shape Design Sensitivity Analysis of Displacement and Stress Constraints," Journal of Structural Mechanics (to appear).

7. Yang, R.J. and Botkin, M.E., "The Relationship Between the Variational Approach and the Implicit Differentiation Approach to Shape Design Sensitivities," Proc. 26th AIAA/ASME/ASCE/AHS Structures, Structural Dynamics and Materials Conference, Orlando, FL, April 1985.

8. Vanderplaats, G.N. and Miura, H., "Trends in Structural Optimization: Some Considerations in Using Standard Finite Element Software," Proc. 6th Vehicle Structural Mechanics Conference, Detroit, April 1986.

9. Vanderplaats, G.N., "CONMIN - A FORTRAN Program for Constrained Function Minimization: User's Manual," NASA TM X-62,282, August, 1973.

10. Schmit, L.A. and Miura, H., "A New Structural Analysis/Synthesis Capability - ACCESS1," AIAA Journal, Vol. 14, No. 5, May 1976, pp. 661-671.

11. Haftka, R.T. and Starnes, J.H., Jr., "Application of a Quadratic Extended Interior Penalty Function for Structural Optimization," AIAA Journal, Vol. 14, 1976, pp. 718-724.

12. Lasdon, L., "GRG2 User's Manual," University of Texas, Austin, Texas.

13. OPT Library User's Manuals, K. Ragsdall, University of Misouri, Columbia, Missouri.

14. OPTDES.BYU User's Manual, Design Optimization Laboratory, Brigham Young University, Provo, Utah.

15. Arora, J.S., Thanedar, P.B. and Tseng, C.H., "User's Manual for Program IDESIGN, Version 3.4," Technical Report No. ODL-85-10.

16. Vanderplaats, G.N., Sugimoto, H. and Sprague, C.M., "ADS-1: A New General-Purpose Optimization Program," AIAA Journal, Vol. 22, No. 10, Oct. 1984.

17. Schmit, L.A. and Miura, H., "Approximation Concepts for Efficient Structural Synthesis," NASA CR-2552, May 1976.

18. Schmit, L.A. and Fleury, C., "Structural Synthesis by Combining Approximation Concepts and Dual Methods," AIAA Journal, Vol. 19, Oct. 1980, pp. 1252-1260.

19. Fleury, C. and Braibant, V., "Structural Optimization - A New Dual Method Using Mixed Variables," Proc. AIAA/ASME/ASCE/AHS 25th Structures, Structural Dynamics and Materials Conference, Palm Springs, May 1984.

20. Vanderplaats, G.N., Miura, H. and Chargin, M., "Large Scale Structural Synthesis," Finite Elements in Analysis and Design, Vol. 1, No. 2, August 1985, pp. 117-130.

21. Botkin, M.E., Yang, R.J. and Bennett, J.A., "Shape Optimization of Three-Dimensional Stamped and Solid Automotive Components," General Motors Research Laboratories Report GMR-5168. To appear in Proceedings of the 1985 GMR Symposium on the Optimum Shape.

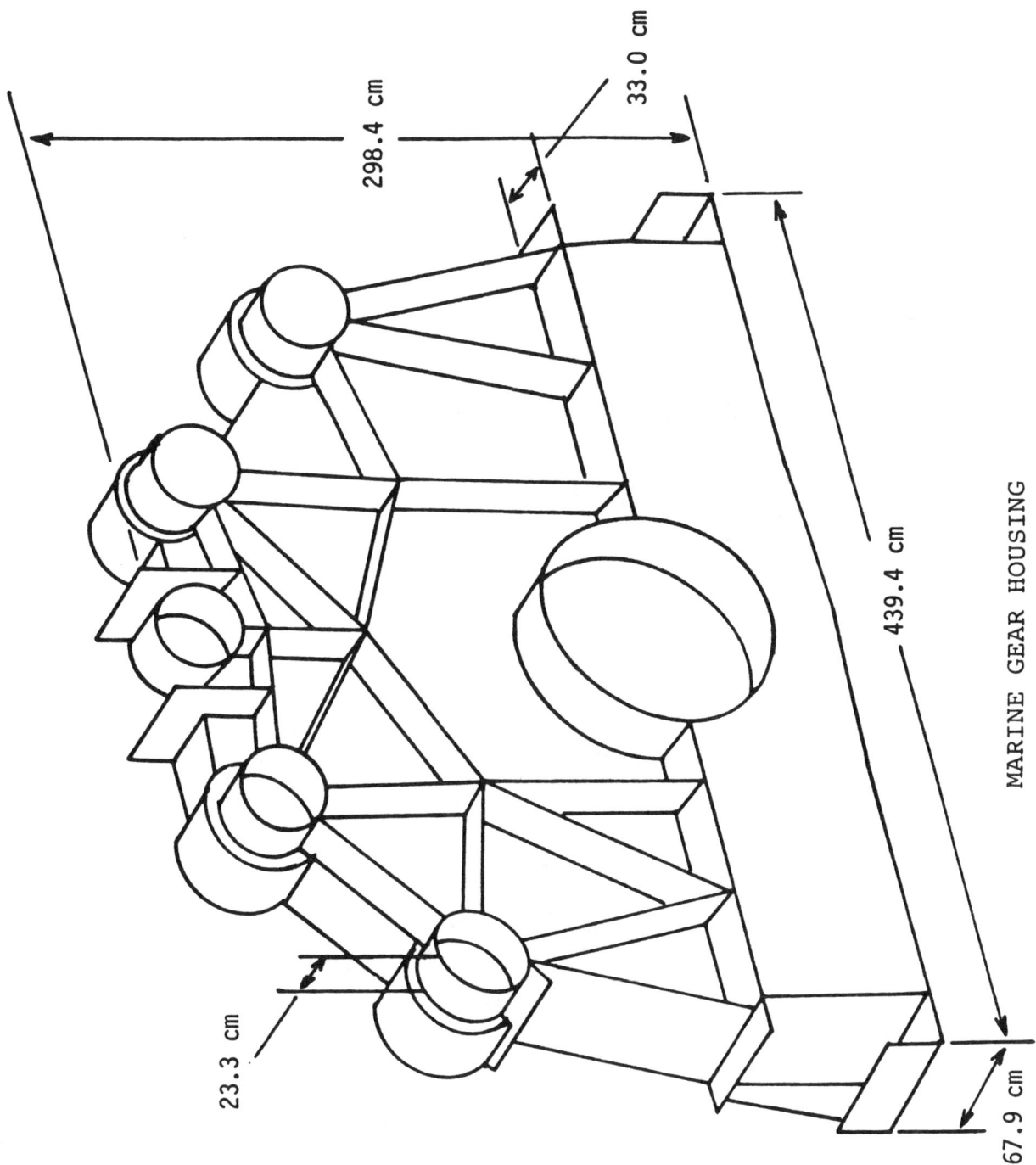

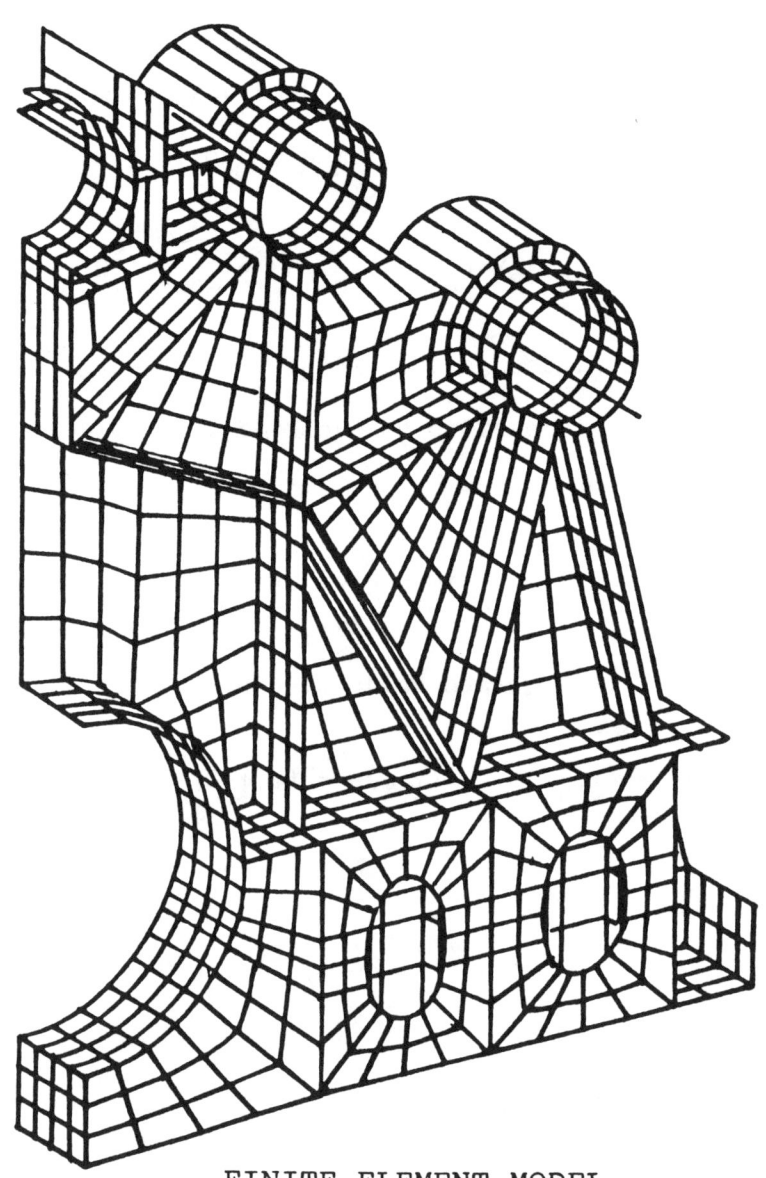

FINITE ELEMENT MODEL

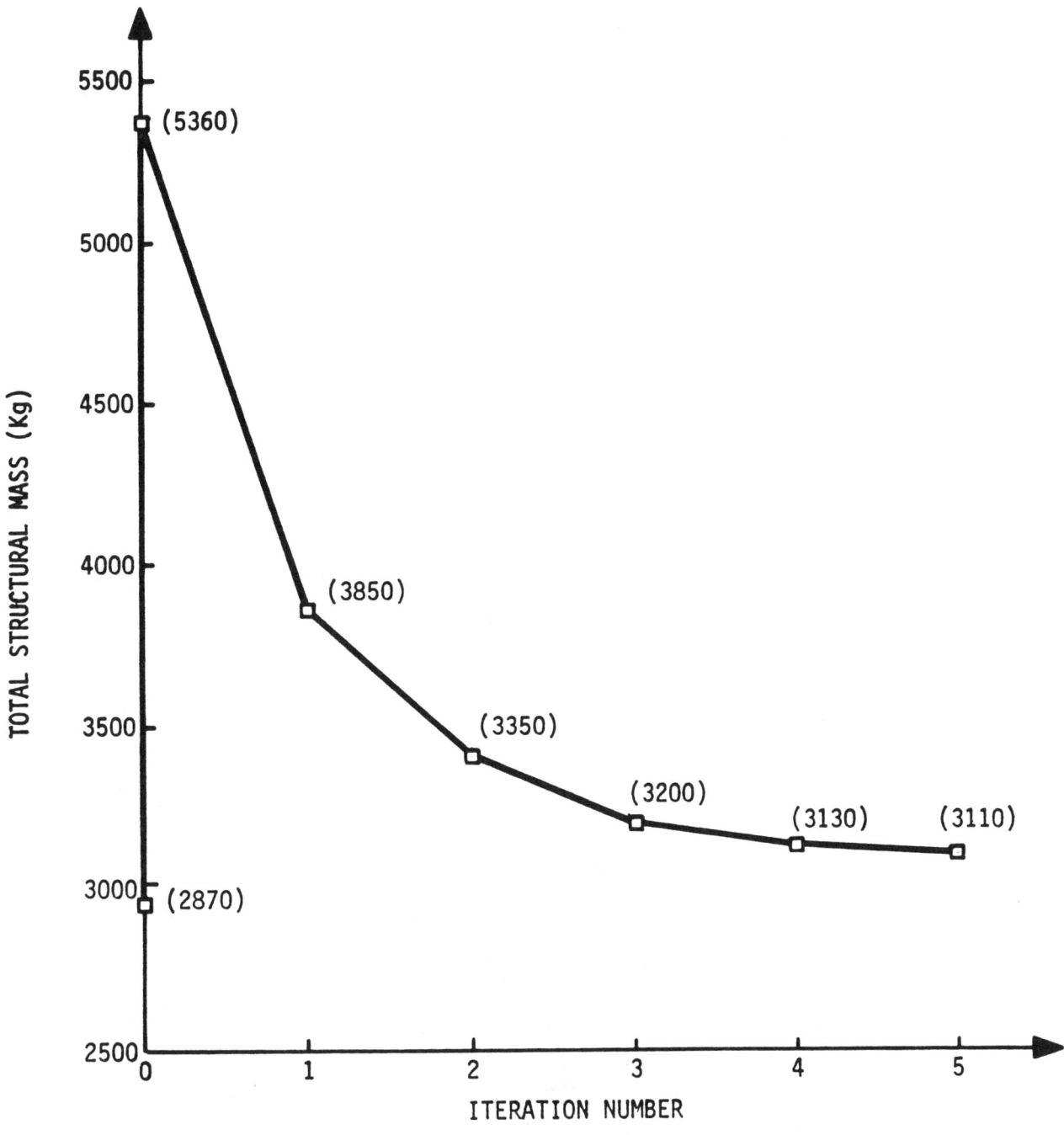

OPTIMIZATION HISTORY

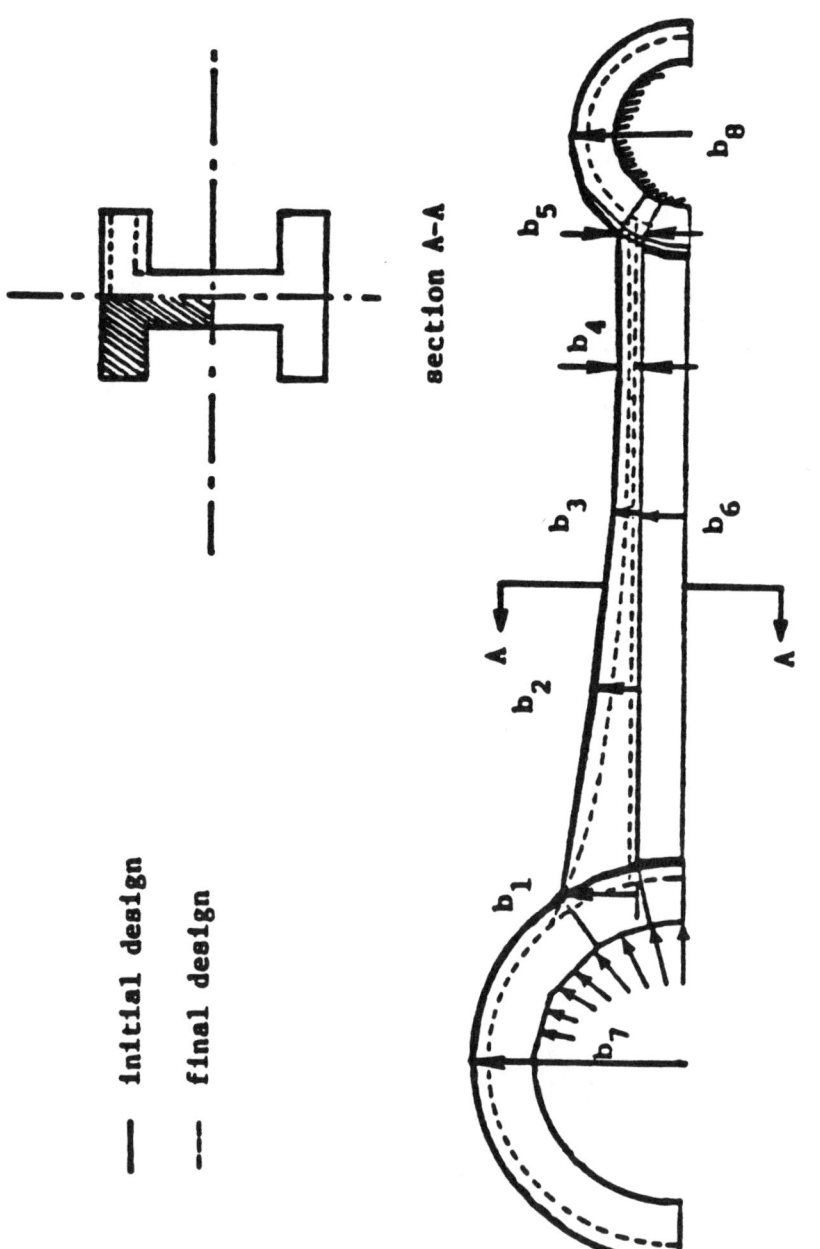

**Generic Model of Engine Connecting Rod**

**Finite Element Mesh of Connecting Rod**

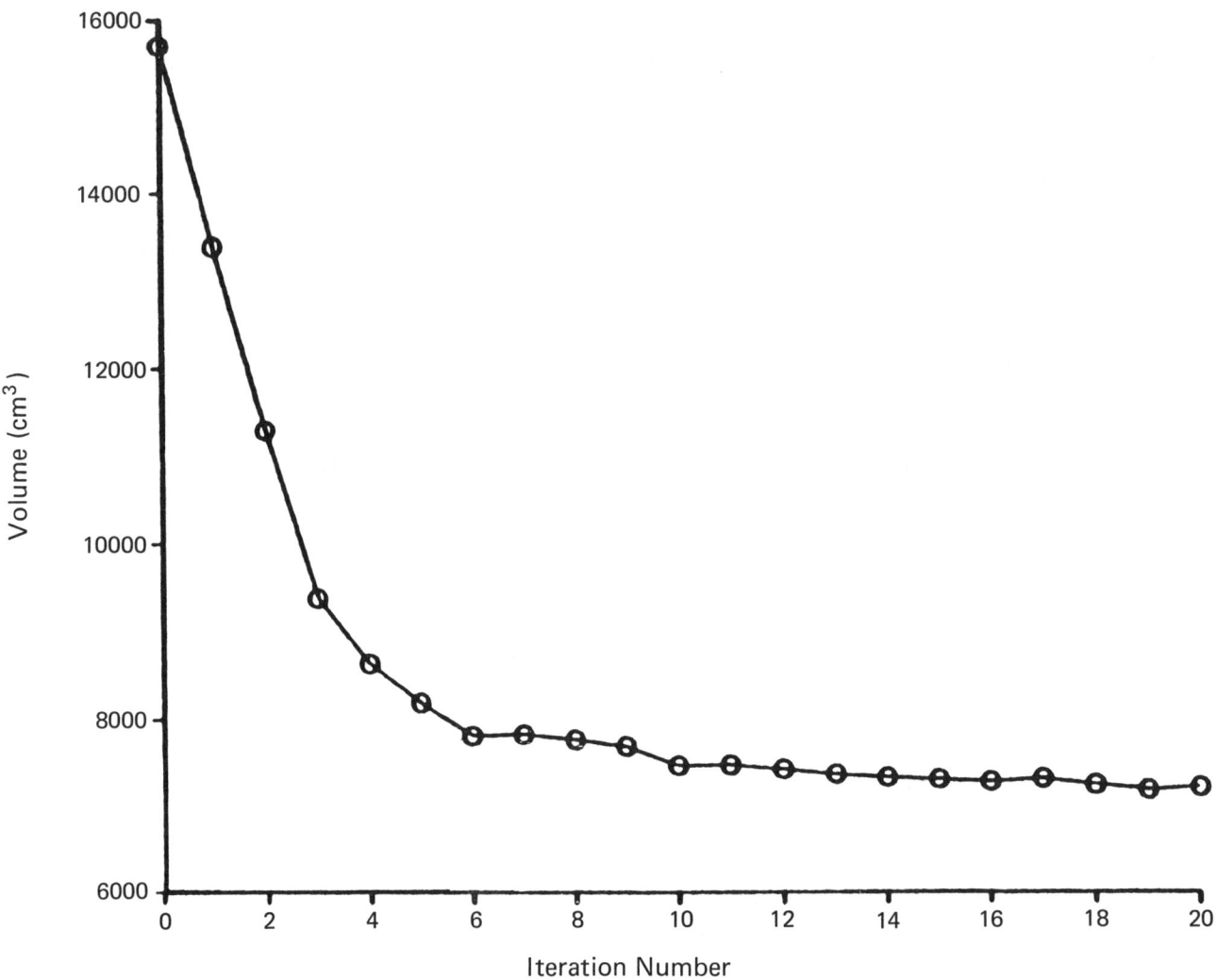

**Design History of Engine Connecting Rod**

CONSTRAINED OPTIMIZATION TECHNIQUES FOR ACTIVE CONTROL
OF AEROELASTIC RESPONSE

Vivekananda Mukhopadhyay
Associate Research Professor
The George Washington University (JIAFS)
NASA Langley Research Center Hampton, Virginia 23665

ABSTRACT

Active control of aeroelastic response is a complex problem in which the designer usually tries to satisfy many design criteria which are often conflicting in nature. To further complicate the design problem, the state space equations describing this type of control problem are usually of high order, involving a large number of states to represent the flexible structure and unsteady aerodynamics. Control laws based on the standard Linear - Quadratic - Gaussian method are of the same high order as the aeroelastic plant and may be difficult to implement in the flight computer. To overcome this disadvantage a new approach was developed for designing low-order optimized robust control laws. In this approach, a nonlinear programming algorithm is used to search for the values of control law design variables that minimize a performance index while satisfying several inequality constraints that describe the design criteria on the stability robustness and responses. The method is applied to a gust load alleviation problem and a stability robustness improvement problem of a drone aircraft.

INTRODUCTION

Active control of aeroelastic response is a complex problem in which the designer usually tries to satisfy many design criteria on the dynamic loads and response, stability robustness etc. which are often conflicting in nature. To further complicate the design problem, the state space equations describing this type of control problem are usually of high order, involving a large number of states to represent the flexible structure and unsteady aerodynamics. Control laws based on the standard Linear - Quadratic - Gaussian (LQG) method are of the same high order as the aeroelastic plant and may be difficult to implement in a flight computer. To overcome this disavantage of the LQG method, a new approach was developed for designing low-order optimal control laws [Ref. 1-4]. In this approach, a nonlinear programming algorithm is used to search for the values of control law variables that minimize a basic performance index while satisfying several inequality constraints that describe the design criteria on the dynamic loads and responses, stability robustness etc. In the usual LQG approach it is often difficult to exert direct control over individual responses and stability margins. In the present design approach the design objectives are incorporated into a set of inequality constraint equations, instead of lumping all of them into a composite performance index J. Through these constraints, the designer has a choice of which design objectives he wishes to directly satisfy, utilizing information on the gradient of the constraints with respect to the design variables. These design constraints may be used for reducing the root mean square (rms) bending moment, torsion moment and shear forces on the flexible wing to a desired value, keeping the control surface rms deflections and rates to within the hardware limitations, lowering dynamic responses [Ref. 2] and improving the stability margins of the multi-input multi-output (MIMO) flight control system [Ref. 3-5]. Another advantage of this method is that the high order state space model of the plant and a stabilizing control law of any prescribed order can be used as the design starting point. Thus the need for reducing the plant model for optimal control law design purposes and the associated spillover problems are eliminated. This method has the disadvantage of requiring the use of nonlinear programming techniques to solve a constrained optimization problem. However, the computational efficiency is significantly improved by obtaining the gradients of the objective function and all the constraints analytically. The method is applied to a gust load alleviation problem and a stability robustness improvement problem of a drone aircraft.

AEROSERVOELASTIC SYSTEM DESCRIPTION

A linear aeroservoelastic system may be described by a set of constant coefficient state-space equations expressed by Eqs. (1) through (6).

Plant
$$\dot{X}_s = F X_s + G_u U \tag{1}$$

$$Y' = H X_s \tag{2a}$$

$$Y_D = H_D X_s \tag{2b}$$

Controller
$$\dot{X}_c = A X_c + B Y \tag{3}$$

$$U' = C X_c \tag{4}$$

Interconnection
$$U = U' + U_{com} \tag{5}$$

$$Y = Y' + V_{com} \tag{6}$$

Equation (1) represents an $N_s$ th order plant having $N_o$ output measurements Y' modeled by Equation (2a) and $N_c$ control inputs U. Equations (3) and (4) represent an M th order feedback controller driven by the output feedback Y. The plant and controller are interconnected by Equations (5) and (6) which include external inputs $U_{com}$ and $V_{com}$. Fictitious white noise processes can be inserted at these two points to improve stability margins at these points.

Development of these equations for a flexible aeroelastic system with unsteady aerodynamic forces are described in several references [Ref. 1, 6, 7]. The plant the matrices F and $G_u$ are composed of structural mass, damping, stiffness, and aerodynamic matrices and are functions of Mach Number and dynamic pressure. The $X_s$ vector is usually composed of modal coordinates, aerodynamic and actuator states. The vector U is the actuator command. In the measurement equation the matrix H contains the modal amplitudes at the sensor locations. The design output matrix $H_D$ contains coefficients for each of the modes that represent quantities such as bending moments, shear forces, torsion moments, and accelerations. The design response $Y_D$ may also contain control surface deflection and rates and other dynamic response vectors. The form of the controller or dynamic compensator is also chosen in state-space form. It is also assumed that the order of the control law is much smaller than the order of the plant.

In order to address the stability margin aspect of the multiloop control system, it is necessary to express the system equations in the input-output form in Laplace domain. A block diagram of the system is shown in Figure 1. The plant and the controller transfer matrices in Laplace domain G(S) and K(S) are defined as

$$Y'(S) = H(IS-F)^{-1} G_u U(S) = G(S) U(S) \tag{7}$$

$$U'(S) = C(IS-A)^{-1} B Y(S) = -K(S) Y(S) \tag{8}$$

The argument S will be dropped in the subsequent chapters for convenience. For the closed loop system one may write

$$\begin{Bmatrix} Y' \\ Y \\ U' \\ U \end{Bmatrix} = \begin{bmatrix} (I+GK)^{-1}G & -(I+GK)^{-1}GK \\ (I+GK)^{-1}G & (I+GK)^{-1} \\ -(I+KG)^{-1}KG & -(I+KG)^{-1}K \\ (I+KG)^{-1} & -(I+KG)^{-1}K \end{bmatrix} \begin{Bmatrix} U_{COM} \\ V_{COM} \end{Bmatrix} \tag{9}$$

In the above transfer-matrix relations, the most fundamental matrices are (I+KG) and (I+GK) which govern the stability margins at the plant input and output denoted by 1 and 2 respectively in Figure 1. A brief review of the stability margins and their relations to the matrix singular values is presented next.

STABILITY ROBUSTNESS REVIEW

The stability robustness of multiloop systems and their relation to the singular value of the return difference matrix are discussed in detail in Refs. 3-5. Singular values of a matrix 'A' are defined as the positive square root of eigenvalues of A*A. The minimum singular value of the return difference matrix at the plant input $\underline{\sigma}(I+KG)$ and the inverse return difference matrix $\underline{\sigma}(I+KG)^{-1}$ are measures of stability margin at the plant input in the following sense. Let a perturbation matrix L whose nominal value is unity be introduced at the plant input denoted by 1 in Figure 1. It can be shown that, if the nominal system is stable, then under certain conditions the stability of the perturbed system is guaranteed if

$$\bar{\sigma}(L^{-1}-I) < \underline{\sigma}(I+KG) \tag{10}$$

or

$$\bar{\sigma}(L-I) < \underline{\sigma}(I+KG^{-1}) \tag{11}$$

over all frequencies $(S=j\omega)$.

The stability robustness can be expressed in terms of gain and phase margins as follows. Consider a specific L matrix which represents simultaneous gain and phase perturbations $k_n$ and $\phi_n$ in every loop, namely

$$L = \text{DIAG}[K_n \exp(j\phi_n)], \quad n = 1,2\ldots N_L, \quad k_n > 0 \tag{12}$$

At nominal condition $k_n=1$ and $\phi_n=0$ for all $N_L$ (i.e. L=I). The stability conditions (10) and (11) can be represented graphically as shown in Figure 2 by the solid line and dashed line, respectively. From the appropriate minimum singular values one can obtain the minimum range of variation of gain $k_n$ and phase $\phi_n$ at the plant input within which the system is guaranteed to be stable. Since inequalities (10) and (11) are conservative conditions one may choose the larger of the ranges. Also the actual gain and phase margins can be higher than those obtained from Figure 2.

If the perturbation is introduced at the plant output denoted by 2 in Figure 1 then the guaranteed stability conditions are similar to the inequalities (10) and (11) except all KG terms in Eqs. (10) and (11) and in Figure 2 are replaced by GK. For a single-input single-output (SISO) system, the terms KG and GK are scalar functions of frequency ω and hence KG=GK for all frequencies. Consequently the stability margins are same at all points in the loop. However, a MIMO system with good stability margin at the plant input may have poor stability margin at the plant output. It is desirable to have margins at both these locations and we should not like small perturbations at either input or output to destabilize the system. In Ref. 1 a so-called fictitious noise adjustment procedure was described to improve stability margins at the plant input or output. This technique is very useful in arriving at robust LQG controllers for a MIMO feedback control system. The robust controllers obtained by this technique can be improved by the present constrained optimization scheme and is described next. A detailed description of this scheme without any constraints may be found in Ref. 1, where full and reduced order controller design for a flutter suppression problem was presented.

OPTIMIZATION SCHEME

A simplified block diagram of the optimization scheme is shown in Figure 3. The basic objective of the control law synthesis scheme is to find the values of the matrices A, B and C in Equations (3) and (4) which represent a controller or a dynamic compensator for a stable closed loop system, such that a performance index is minimized and a set of inequality constraints are satisfied. The performance index and constraints are defined next. It is assumed that all external inputs $U_{com}$ and $V_{com}$ are stochastic white noise processes and all responses are defined by their RMS values. The uncorrelated noise intensity matrices of $U_{com}$ and $V_{com}$ are denoted by $R_u$ and $R_v$ respectively.

PERFORMANCE INDEX

The performance index is in the standard LQG form defined by a weighted sum of the steady state RMS values of the closed loop plant and controller output

$$J = E[Y'^T Q_1 Y' + U'^T Q_2 U'] \tag{13}$$

which can be computed by solving a steady state Lyapunov equation of order $(N_s+M)$. The gradients of the performance index can be computed by solving an adjoint Lyapunov equation of the same order. The details are available in Ref. 1. In order to exert control over individual responses one may treat most of the responses as constraints instead of lumping their weighted values into a single performance index.

Constraints

In the constrained minimization approach the performance index J is minimized by changing the design variables p subject to the inequality constraints

$$g_i(p) \leq 0 \quad , \quad i = 1, 2 \ldots N_g \tag{14}$$

$g_i(p)$ may represent a constraint on the minimum singular value $\underline{\sigma}(I+KG)$ or $\underline{\sigma}(I+GK)$, or on maximum RMS response.

Constraints on Singular Value

Let us assume that the minimum singular value $\underline{\sigma}(I+KG)$ of a stable system is $\underline{\sigma}_M$ as shown in Figure 4. It is desired to increase the minimum singular value to a higher level denoted by the line $\underline{\sigma}_D$. Two types of lower bounds are imposed in this paper. The type A denotes a constant $\underline{\sigma}_D$. The type B denotes a frequency dependent $\underline{\sigma}_D(\omega)$ where the singular values are required to attenuate to unity after a specific break frequency. The selection of an achievable $\underline{\sigma}_D(\omega)$ is based on engineering experience and judgement depending on the control power and response limitations since control power may be wasted in designing a compensator to satisfy a tight bound with high $\underline{\sigma}_D(\omega)$. Moreover, because of the conservative nature of the singular value measures, inability to obtain such a compensator does not necessarily imply a robustness problem.

The constraints function $g_i(p)$ is basically a cumulative measure of the vertical distance between $\underline{\sigma}_D(\omega)$ and $\underline{\sigma}(I+KG)$ or $\underline{\sigma}(I+GK)$ as shown in Figure 4 and is defined as

$$g_1(p) = \frac{1}{N} \sum_{n=1}^{N} [\max\{0, [\underline{\sigma}_D(\omega_n) - \underline{\sigma}(I+KG(j\omega_n,p))]\}]^2 \tag{15}$$

$$g_2(p) = \frac{1}{N} \sum_{n=1}^{N} [\max\{0, [\underline{\sigma}_D(\omega_n) - \underline{\sigma}(I+GK(j\omega_n,p))]\}]^2 \tag{16}$$

The summation is taken over a large number of frequency points $\omega_n$ and the spacing of the frequency points in a frequency range are left to the designer depending on the specific problem. The objective is to minimize (preferably reduce to zero) the shaded area below the $\underline{\sigma}_D(\omega)$ line by satisfying the inequality constraint of Eq. (14). The expressions for the singular values and their gradients with respect to the design variables are obtained analytically as described in Ref. 3. Constraints based on inequality relation (11) were not used since they are complementary to inequality relation (10) and would have added unnecessary computational burden without tangible improvement in results.

Constraint on RMS Response

In the constrained design approach, instead of lumping all the responses in the performance index J, the designer has the choice of choosing individual responses as a set of inequality constraints

$$g_i(p) = (Y_D/Y_{Dmax})_i^2 - 1 \leq 0, \quad i = 2 \text{ or } 3 \ldots N_g \tag{17}$$

where $Y_D$ is the RMS design response and $Y_{Dmax}$ is the maximum allowable RMS value. $N_g$ is the total number of inequality constraints. The response constraints and their gradients with respect to the design variables are obtained analytically by solving a set of Lyapunov equations.

A general purpose optimization software [Ref. 8] which employs the method of usable-feasible directions is used to search for the controller design variables which minimize J subject to $g_i(p) \leq 0$. The method uses the performance index and constraint gradient information to determine a parameter move direction and a scalar multiplier in the usable-feasible direction in order to satisfy the constraints. During the linear search in the usable-feasible direction the eigenvalues are monitored to prevent the system from becoming unstable. The method also employs a relaxation technique where only the constraints close to the $g_i = 0$ boundary are activated in each iteration and seems to work well when the nominal system does not satisfy the constraints.

EXAMPLE PROBLEM

To illustrate the control law synthesis procedure, it is applied to a gust load alleviation problem and a stability robustness improvement problem of an unmanned drone aircraft used in NASA's Drone for Aerodynamic and Structural Testing (DAST) program [Ref. 9].

Gust Load Alleviation Example

The basic features of this design problem is illustrated in Figure 5. The objective is to design a gust load alleviation system. Both an outboard control surface and a horizontal stabilizer are used as the control surfaces. Two accelerometers, one in the wing tip and one in the fuselage near the airplane center-of-gravity, are used as feedback sensors. A 30th order model was generated that included two rigid body modes, three flexible modes, and actuator dynamics. A 2nd order filter that approximates the Dryden gust spectrum and driven by white noise was added to give a 32nd order system. The order of the controller was selected as four. Design reponses were selected as wing root bending moment, wing root shear, outboard bending moment and torsion, accelerations, and control surface rates and deflections. The performance index was selected as outboard control input. Constraints were formulated to decrease wing root bending moment and shear by 40% below their open loop values while not increasing outboard bending moment and torsion above their open loop values.

First an optimal linear quadratic (LQ) full state feedback control law was designed with $Q_1$ and $Q_2$ selected using Bryson's inverse square rule. Then an optimal Kalman state estimator was designed. The control law and state estimator combination is called full order (32nd order) LQG controller. Next a 4th order controller was obtained by truncating full order LQG controller retaining only the pitch, pitch rate, first flexible mode displacement and velocity. The structure of this reduced order controller is analogous to the LQG controller. The constrained optimization was then applied using the B and C matrices of this 4th order controller as the design variables. A matrix A is now a linear function of B and C as in a LQG controller. Figure 7 shows the variation of performance index (J), wing-root-bending-moment (WRBM), wing-root-shear (WRS) and several of the design variables during five iterations of the constrained optimization. The performance index, WRBM and WRS are normalized by the corresponding values with the full order LQG controller. In five iterations the WRBM and WRS were reduced to within 2 percent of the LQG value without increasing the wing-outboard-bending moment (WOBM) and wing-outboard torsion (WOT). Note that during fourth and fifth iteration, WOBM and WOS are reduced at the cost of increased aileron control activity. Some comparative results are shown in Figure 6. This bar chart compares the wing loadings and control surface activity using full-state feedback (LQ), full-order output feedback (LQG), and the present approach (4th order controller) respectively. In the first row of figures, the loadings are normalized by the corresponding open loop values, shown by the first unit bar. The second bar is the LQ value which is the ideal case since all states are assumed measured. The third bar is the LQG value. The last bar is the result using the present approach. Both the LQG and 4th order controller reduce the WRBM and WRS approximately 40%. The WOBM does not change appreciably. The wing outboard torsion (WOT) is reduced by about 30% using the present approach.

The second row of figures compares the closed loop stabilizer and outboard control surface activities. The 4th order controller always requires higher control surface activity. The maximum stabilizer deflection is about half the allowable limit of 0.157 degrees. The stabilizer's rate and those of the outboard control are well within allowable limits.

Stability Robustness Improvement Example

In this example, the synthesis method is applied to the rigid drone aircraft with a lateral attitude control system. A block diagram of the system is shown in Figure 8. The 6th order plant state vector is defined as

$$X_s = [\beta \; \dot{\phi} \; \dot{\psi} \; \phi \; \delta_1/20 \; \delta_2/20]^T \tag{18}$$

The plant matrices F, $G_u$ and H as defined in Eqs. (1) and (2) were presented in Refs. 3, 5. The eigenvalues of the nominal open loop system are -0.03701, 0.1889 $\pm$ j1.058, -3.25, -20.0 and -20.0. The unstable complex eigenvalue represents the Dutch Roll mode. The plant input position 1 is defined at the entry point to the elevon and rudder actuators denoted by $U_1$ and $U_2$ in degrees unit. The plant output position 2 is defined at the roll rate and yaw rate sensor outputs denoted by $Y_1$ and $Y_2$ in degrees/second unit.

Design of Full Order LQG Controller

In this section, the objective is to demonstrate the ability to shape the minimum singular value by adjusting the noise intensity matrices and illustrate them by the singular value plots at the plant input and output presented in Figures 9 and 10 for a full order LQG controller. The noise intensity matrices used in the design are shown beside each plot and in Table 1 and are designated as Design No. 1 to 6. In all these designs, the weighting matrices $Q_1 = I$ and $Q_2 = 0.5I$. The full order LQG controllers are obtained by solving the standard Kalman filter and controller Riccati equation to obtain filter gain matrix B and controller gain matrix C, respectively. The matrix A is given by $A=(F-BH+G_uC)$. In design No. 1 to 3, $R_u = I$ and $R_v$ takes the values $100.0I$, $I$ and $0.01I$ respectively. The resulting singular values are plotted in Figure 9. In Design No. 1 with the high value of $R_v$ the minimum $\underline{\sigma}(I+KG)=0.15$ and $\underline{\sigma}(I+GK)=0.83$. Most of the stability robustness is at the plant output. The singular values also have rapid attenuation at higher frequencies. In Design No. 2 with $R_v=I$ the minimum $\underline{\sigma}(I+KG)=0.2$ and $\underline{\sigma}(I+GK)=0.5$. When $R_v$ is lowered to $0.01I$ in Design 3 most of the stability robustness is achieved at the plant input. There is substantial loss of attenuation at higher frequencies. Note that the relative values of $R_u$ and $R_v$ is the main factor in these noise adjustment procedures. In Design No. 4, shown in Figure 10, the stability margin at the input is increased further using $R_u = 1000I$ and $R_v=I$. The effect of unequal noise intensity in each loop is investigated in Design No. 5 which uses $R_u=\text{Diag.}[1\ 100]$ and $R_v=\text{Diag}[1\ 10]$ imposing more uncertainty in the rudder channel. The minimum singular value plot shown in Figure 10 indicate that the minimum $\underline{\sigma}(I+KG)$ is 0.33 and minimum $\underline{\sigma}(I+GK)$ is 0.25 and both have good high frequency attenuation. This design is used as the starting point for testing the present constrained optimization design procedure to improve stability robustness at both the plant input and output. In these designs the weighting matrices $Q_1$ and $Q_2$ are the same as before. The noise intensity matrices $R_u$ and $R_v$ are set to zero. The system only contains a unit RMS white noise at the elevon actuator input. All the elements of the matrices B and C are chosen as design variables. The minimum desired singular value $\underline{\sigma}_D=0.45$ is chosen as constraints on both $\underline{\sigma}(I+KG)$ and $\underline{\sigma}(I+GK)$ at the plant input and output respectively. Type 'A' constraint is used in Design Nos. 6 and 8 and Type 'B' constraint is used in Design No. 9. Figure 4 shows the exact shape of the constraint $\underline{\sigma}_D(\omega)$ imposed on the singular values $\underline{\sigma}(I+KG)$ and $\underline{\sigma}(I+GK)$. The frequency ranges from 0.1 to 100 radians/second beyond which both KG and GK attenuate to zero for all designs. All high frequency uncertainties are assumed to be above 100 radians/second. In the cumulative constraints of Eqs. (15) and (16), the frequency points are chosen uniformly as 50 divisions per decade. In all the design cases the optimization cycle was stopped after five iterations since more iterations usually resulted only in marginal improvements.

In Design No. 6, the full order controller obtained in Design 5 was optimized. After five iterations, the minimum singular values are reshaped as shown in Figure 10. The minimum $\underline{\sigma}(I+KG)$ and $\underline{\sigma}(I+GK)$ are increased from 0.33 and 0.25 to 0.4 and 0.38, respectively. Thus the present constrained optimization procedure is able to improve stability robustness at both input and output. However, the improvement is at the cost of some loss of high frequency attenuation at the input. The transient response plots (not presented here) indicate adequate damping of the Dutch roll, heading and roll modes for a drone aircraft.

Design of Reduced Order Controller

The next three designs designated as Design No. 7 to 9 in Table 1 represent results of an attempt to obtain reduced order controllers starting from design No. 6 using truncation and then constrained optimization technique. First, a third order controller is obtained from controller No. 6, retaining only the second, third and sixth states which correspond to the roll rate, yaw rate and rudder actuator states. In Table 1, this is designated as Design No. 7. The fact that such a truncated controller can stabilize the system is a manifestation of stability robustness of Design No. 6. The corresponding minimum singular value plots of (I+KG) and (I+GK) are shown in Figure 11. The minimum $\underline{\sigma}(I+KG)$ and

$\underline{\sigma}(I+GK)$ are 0.17 and 0.1, respectively. To improve its stability robustness, the present constrained optimization procedure is applied with Type 'A' and Type 'B' constraints and are designated as Design No. 8 and 9, respectively, in Table 1. The B and C matrices in Eqs. (3) and (4) are the design variables. After five iterations the minimum singular values $\underline{\sigma}(I+KG)$ and $\underline{\sigma}(I+GK)$ are reshaped as shown in Figure 11. This represents a moderate improvement at the plant input and output. The improvement is at the cost of some loss of high frequency attenuation.

The numerical examples indicate that an increase in the minimum singular value at the plant input is always accompanied by a decrease in the minimum singular value at the plant output and vice versa. Using the present procedure, the minimum singular values can be increased at both plant input and output but only to a limited extent. This is due to the intrinsic relations between $\underline{\sigma}(I+KG)$ and $\underline{\sigma}(I+GK)$.

RMS Response

The RMS response of the system for a unit RMS noise at the plant input $U_1$ was computed using covariance analysis. The RMS side slip $\beta$, roll rate $\dot{\phi}$ and yaw rate $\dot{\psi}$, elevon and rudder deflections $\delta_1$, $\delta_2$ are presented in Table 1. The general trend in the design No. 1 to 6 is a progressive reduction of RMS responses of $\beta$, $\dot{\phi}$, and $\dot{\psi}$ while $\delta_1$ and $\delta_2$ remain more or less same. The Design No. 6 and 8 have very low side slip response. In the Design Nos. 7 to 9 the controller order reduction through truncation and reoptimization did not result in significant increase in responses compared to Design No. 6. The design software has the provision of treating each of these responses as individual constraints instead of lumping them in the performance index. The variation of the normalized performance index and constraints on singular values at the plant input and output is shown in Figure 12 for the Design No. 6. The RMS responses and weighting matrices indicate that U and Y roughly contribute equally to the performance index J. After the first iteration $g_2$ is nearly satisfied with slight increase in J. After the second iteration $g_1$ is also reduced at the cost of increased J. During iteration four the algorithm attempts to minimize the performance index J by violating the second constraint $g_2$ slightly to reach a compromised solution. The convergence pattern for Design No. 8 and 9 are similar.

CONCLUSIONS

The paper describes a design methodology which uses constrained optimization techniques to reduce random gust loading and improve the stability margins of a multiloop system at both the plant input and output, while minimizing a performance index. The capabilities of the method are demonstrated using a two-input two-output system representing a drone aircraft. Both full order and reduced order controllers were designed for gust load alleviation and stability robustness improvement at both input and output. The examples show that the constrained optimization procedure can be used in conjunction with the noise adjustment procedure to synthesize reduced order controller and to improve the stability margins at the plant input and plant output.

## REFERENCES

1. Mukhopadhyay, V., Newsom, J. R., and Abel, I., "A Method for Obtaining Reduced Order Control Laws for High Order Systems Using Optimization Techniques," NASA TP-1876, August 1981.

2. Newsom, J. R. and Mukhopadhyay, V., "Application of Constrained Optimization to Active Control of Aeroelastic Response," NASA TM-83150, June 1981.

3. Newsom, J. R. and Mukhopadhyay, V., "A Multiloop Robust Controller Design Study using Singular Value Gradients," Journal of Guidance Control and Dynamics, Vol. 4, No. 8, July-August 1985, pp. 514-519.

4. Mukhopadhyay, V., "Stability Robustness Improvement Using Constrained Optimization Techniques," AIAA Paper No. 85-1931, August 1985.

5. Mukhopadhyay, V. and Newsom, J. R., "A Multiloop System Stability Margin Study Using Matrix Singular Values,' Journal of Guidance Control and Dynamics, Vol. 7, No. 5, Sept.-Oct. 1984, pp. 582-587.

6. Abel, I., "An Analytical Technique for Predicting the Characteristics of a Flexible Wing Equipped With an Active Flutter Suppression System and Comparison With Wind Tunnel Data," NASA TP-1376, February 1979.

7. Mahesh, J. K., Stone, C. R., Garrard, W. L., and Hausman, P. D., "Active Flutter Control for Flexible Vehicles," NASA CR-159160, Vol. 1, November 1979.

8. Vanderplatts, G. D., "CONMIN - A Fortran Program for Constraint Function Minimization. User manual," NASA TM-X-62282, 1973.

9. Murrow, H. N. and Eckstrom, C. V., "Drone for Aerodynamic and Structural Testing (DAST) - A Status Report," Journal of Aircraft, Vol. 16, No. 8, August 1979, pp. 521-526.

Table - 1  Summary of Design Parameters and RMS Responses

| Design No. | Input Noise Intensity $R_u$ | Sensor Noise Intensity $R_v$ | Order of Controller | Design Procedure* | $\underline{\sigma}_{min}(I+KG)$ | $\underline{\sigma}_{min}(I+GK)$ | RMS RESPONSE TO UNIT RMS NOISE AT $U_1$ | | | | |
|---|---|---|---|---|---|---|---|---|---|---|---|
| | | | | | | | Side Slip $\beta$ | Roll Rate $\dot\phi$ | Yaw Rate $\dot\psi$ | Elevon Defl. $\delta_1$ | Rudder Defl. $\delta_2$ |
| 1 | I | 100I | 6 | LQG | 0.15 | 0.83 | 0.27 | 2.69 | 0.59 | 3.37 | 0.36 |
| 2 | I | I | 6 | LQG | 0.20 | 0.50 | 0.11 | 1.68 | 0.21 | 3.43 | 0.25 |
| 3 | I | 0.01I | 6 | LQG | 0.53 | 0.10 | 0.11 | 1.30 | 0.24 | 3.26 | 0.59 |
| 4 | 1000I | I | 6 | LQG | 0.68 | 0.05 | 0.09 | 1.19 | 0.21 | 3.14 | 0.53 |
| 5 | $\begin{bmatrix}1 & \\ & 100\end{bmatrix}$ | $\begin{bmatrix}1 & \\ & 10\end{bmatrix}$ | 6 | LQG | 0.33 | 0.25 | 0.07 | 1.24 | 0.20 | 3.24 | 0.45 |
| 6 | $\begin{bmatrix}0 & \\ & 0\end{bmatrix}$ | $\begin{bmatrix}0 & \\ & 0\end{bmatrix}$ | 6 | O(A) | 0.40 | 0.38 | 0.02 | 1.27 | 0.13 | 3.31 | 0.22 |
| 7 | $\begin{bmatrix}0 & \\ & 0\end{bmatrix}$ | $\begin{bmatrix}0 & \\ & 0\end{bmatrix}$ | 3 | T | 0.17 | 0.10 | 0.16 | 1.56 | 0.19 | 3.26 | 0.36 |
| 8 | $\begin{bmatrix}0 & \\ & 0\end{bmatrix}$ | $\begin{bmatrix}0 & \\ & 0\end{bmatrix}$ | 3 | O(A) | 0.22 | 0.40 | 0.03 | 1.48 | 0.14 | 3.59 | 0.71 |
| 9 | $\begin{bmatrix}0 & \\ & 0\end{bmatrix}$ | $\begin{bmatrix}0 & \\ & 0\end{bmatrix}$ | 3 | O(B) | 0.26 | 0.22 | 0.08 | 1.46 | 0.14 | 3.25 | 0.12 |

\* O(A) - Optimization (Type A Constraint)
  O(B) - Optimization (Type B Constraint)
  T   - Truncation of Design No. 6

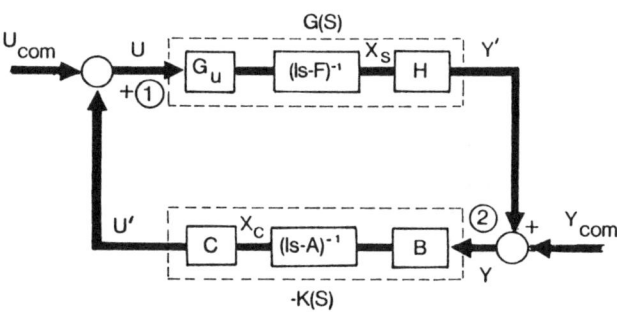

Figure 1. Block diagram of a multiloop feedback control system

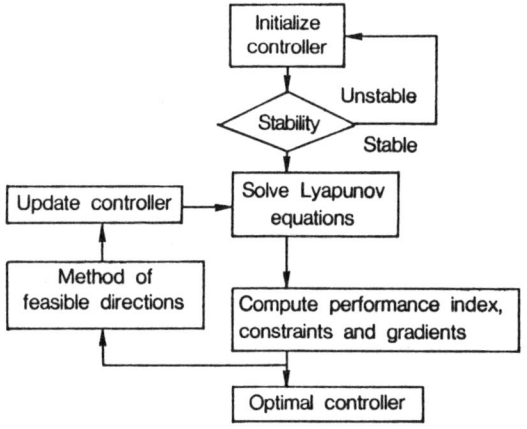

Figure 3. Simplified block diagram of the optimization scheme.

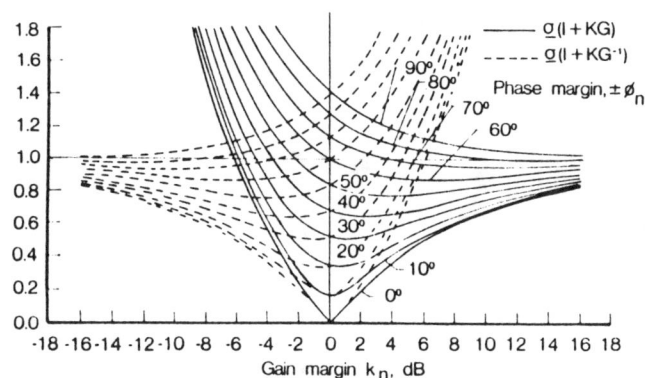

Figure 2. Universal diagram for gain-phase margin evaluation

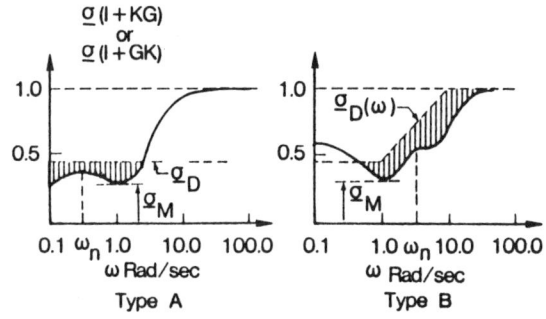

Figure 4. Geometric description of two types of cumulative constraints on singular values $\underline{\sigma}(I+KG)$ and $\underline{\sigma}(I+GK)$

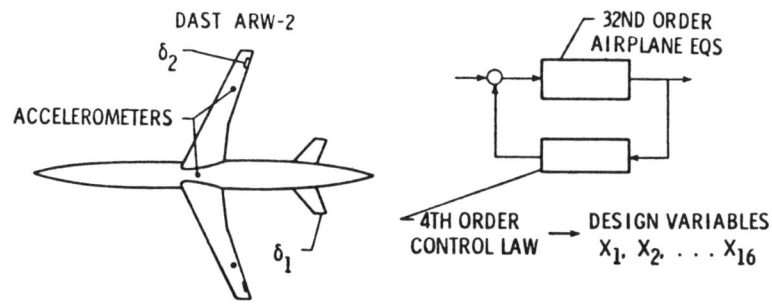

| PHYSICAL QUANTITIES | DESIGN OBJECTIVES WHAT WE WANT TO DO | HOW WE DO IT |
|---|---|---|
| ROOT BENDING MOMENT | 40% REDUCTION | CONSTRAINT |
| ROOT SHEAR | 40% REDUCTION | CONSTRAINT |
| OUTBOARD BENDING MOMENT | NO INCREASE | CONSTRAINT |
| OUTBOARD TORSION | NO INCREASE | CONSTRAINT |
| $\delta_1$ DISPLACEMENT | WITHIN MAX LIMITS | CONSTRAINT |
| $\delta_1$ RATE | WITHIN MAX LIMITS | CONSTRAINT |
| $\delta_2$ DISPLACEMENT | WITHIN MAX LIMITS | OBJECTIVE FUNCTION |
| $\delta_2$ RATE | WITHIN MAX LIMITS | CONSTRAINT |

Figure 5. Design objectives for gust load alleviation example.

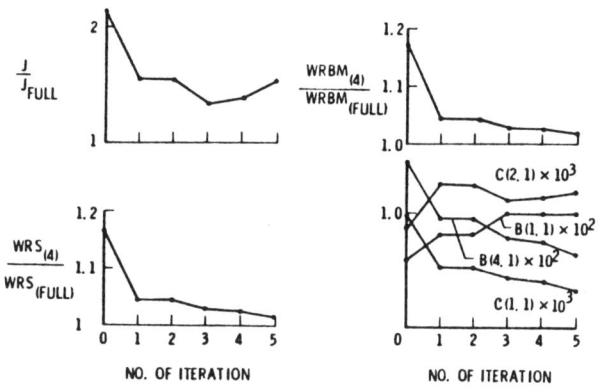

Figure 6. Comparison of root mean square responses (gust load alleviation example).

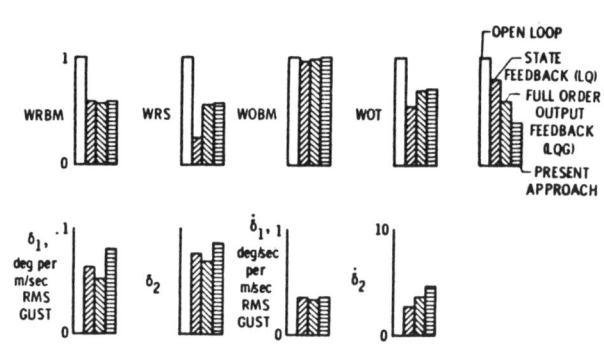

Figure 7. Optimization sequence using present approach.

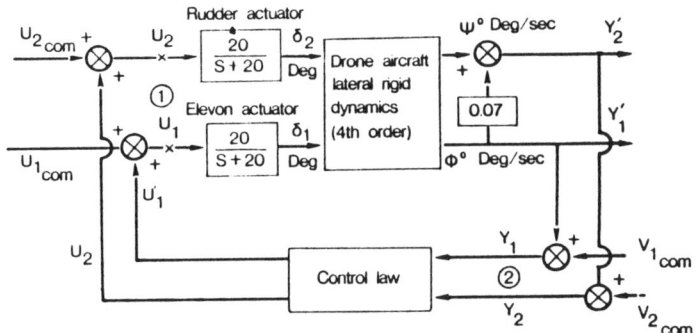

Figure 8. Block diagram for stability margin improvement example.

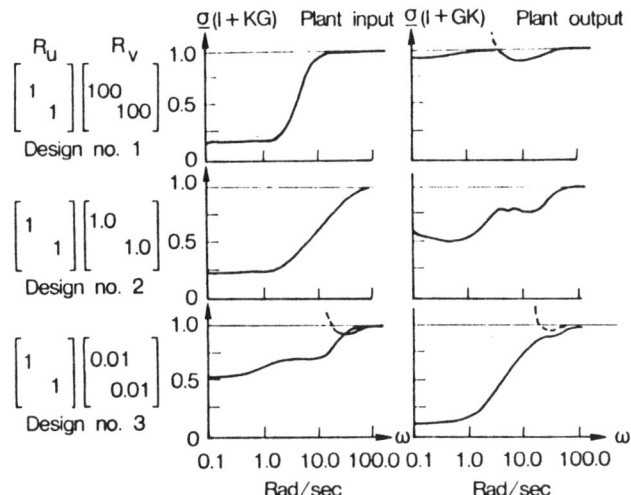

Figure 9. Singular value shaping by noise adjustment (Full order controller).

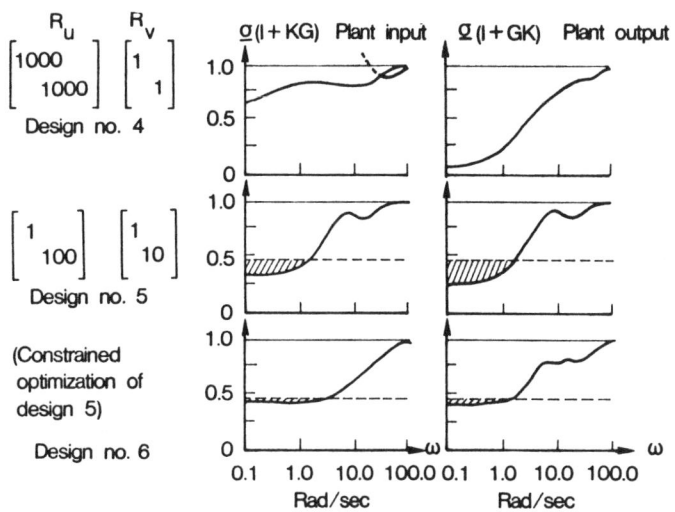

Figure 10. Singlar value shaping by noise adjustment and constrained optimization (Full order controller).

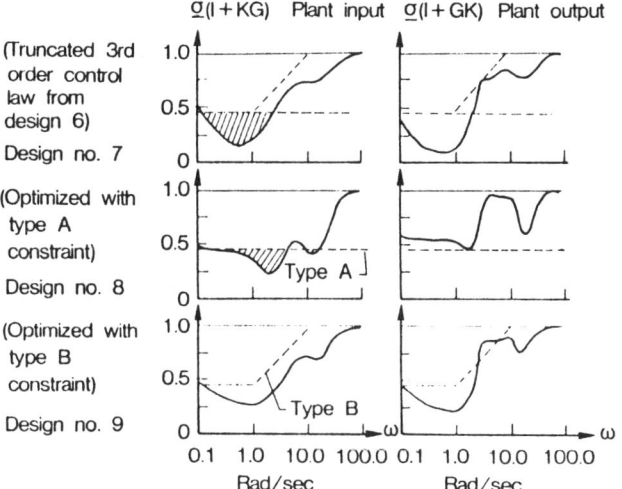

Figure 11. Singular value shaping by constrained optimization (Third order controller).

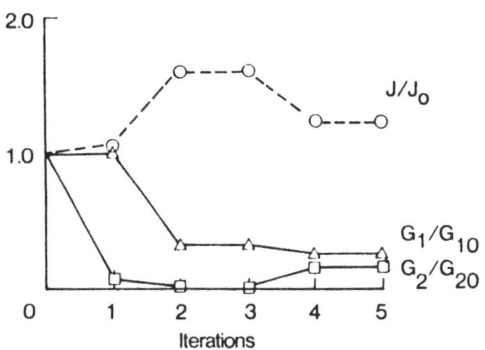

Figure 12. Normalized performance index and constraints variation with iteraton.

# FIRST AND SECOND-ORDER SENSITIVITY ANALYSIS
# OF LINEAR AND NON-LINEAR STRUCTURES

Raphael T. Haftka
Department of Aerospace and Ocean Engineering
Virginia Polytechnic Institute and State University
Blacksburg, VA, USA 24060

Zenon Mróz
Institute of Fundamental Technological Research
U. Swieto-Krzyska 21
00-049 Warsaw, Poland

## ABSTRACT

The paper employs the principle of virtual work to derive sensitivity derivatives of structural response with respect to stiffness parameters using both direct and adjoint approaches. The computations required are based on additional load conditions characterized by imposed initial strains, body forces or surface tractions. As such they are equally applicable to numerical or analytical solution techniques. The relative efficiency of various approaches for calculating first and second derivatives is assessed. It is shown that for the evaluation of second derivatives the most efficient approach is one which makes use of both the first order sensitivities and adjoint vectors. Two example problems are used for demonstrating the various approaches.

## INTRODUCTION

Whenever physical response is calculated from a mathematical model there also is an interest in the sensitivity derivatives of that response with respect to parameters of the problem. This sensitivity information may be used to assess the effect of uncertainties in the mathematical model, to predict the change in response due to change in parameters, and to optimize a system with the aid of mathematical optimization techniques. It is no wonder, then that methods of sensitivity analysis have been explored in various fields of science and engineering such as electronics and control (e.g., [1]), physical chemistry (e.g., [2], physiology (e.g., [3]), thermodynamics (e.g., [4]) and aerodynamics (e.g., [5]).

While the methods applied in these various disciplines sometimes have their idiosyncracies, most of them fall into three broad categories. The simplest approach is finite difference calculation of sensitivities based on successive perturbation of parameter values followed by reanalysis of the system. This approach is often inefficient or may have accuracy problems, and in such cases analytical approaches are more appropriate. Direct analytical approaches are based on differentiating the equations which describe the system with respect to the desired parameters and the solution of these sensitivity equations. Adjoint methods define an adjoint physical system whose solution permits rapid evaluation of the desired sensitivities. Both the direct and adjoint methods generate linear equations even if the original system is governed by nonlinear equations.

In structural mechanics interest has been focused mostly on the sensitivity of local or global structural response to the variation of structural stiffness parameters. For linear problems the adjoint approach has been known as the dummy load method. Haug and Arora and their coworkers generalized the adjoint approach for the calculation of first order sensitivities for both discrete and distributed parameter structural systems (e.g., [6] - [8]). Dems and Mróz [9] and [10] treated first and second order sensitivities by a general variational approach which is applicable to both discrete and distributed parameter linear systems, and that approach has been generalized to nonlinear systems in [11]. Finally, Haug [12] and Haftka [13] presented adjoint formulations for second order sensitivities of discrete systems.

The present paper follows the variational formulation of [9] - [11] in presenting the various alternatives for computing first-and second-order sensitivities of structural systems. The emphasis in the comparison is on the efficiency of the various alternatives when applied to structural systems where a single analysis is computationally costly. The present analysis is limited to structures with fixed configuration, small deformations and nonlinear material properties. However, the efficiency considerations probably have a wider range of applicability. The variational formulation permits applications to both analytical and numerical solution techniques. One analytical example, and one finite-element example are used for demonstration.

FIRST-ORDER SENSITIVITY ANALYSIS

Consider a structure subjected to body forces $\underline{f}$, surface tractions $\underline{T}^o$ on a part of its surface $S_T$ and prescribed displacements $\underline{u}^o$ on the remaining portion $S_u$. Consider a general nonlinear stress-strain relation

$$\underline{\sigma} = \underline{s}(\underline{\varepsilon}, \underline{b}) \tag{1}$$

where $\underline{\sigma}$ denotes the stress tensor, $\underline{\varepsilon}$ is the strain tensor and $\underline{b}$ is a vector of stiffness parameters. When the elastic strain energy, $U$, is used to specify Eq. (1), we have

$$\underline{\sigma} = \frac{\partial U(\underline{\varepsilon}, \underline{b})}{\partial \underline{\varepsilon}} = \underline{s}(\underline{\varepsilon}, \underline{b}) \tag{2}$$

To simplify our presentation, assume that the vector $\underline{b}$ has only two components $p$ and $q$. Thus

$$\underline{\sigma} = \underline{s}(\underline{\varepsilon}, p, q) \tag{3}$$

The equations of equilibrium can be expressed through the principle of virtual work as

$$\int \underset{\sim}{\sigma} \cdot \delta \underset{\sim}{\varepsilon} \, dV = \int \underset{\sim}{T}^\circ \cdot \delta \underset{\sim}{u} \, dS_T + \int \underset{\sim}{f} \cdot \delta \underset{\sim}{u} \, dV + \int \underset{\sim}{T} \cdot \delta \underset{\sim}{u}^\circ \, dS_u \qquad (4)$$

where a dot between two tensors or vectors denotes their scalar product, and $\delta\underset{\sim}{\varepsilon}$ and $\delta\underset{\sim}{u}$ are kinematically admissible strain and displacement fields. Equations (3) and (4) are complemented by the strain-displacement relation

$$\underset{\sim}{\varepsilon} = B\underset{\sim}{u} \qquad (5)$$

where B is a linear operator. Equations (3) - (5) can be used to obtain the differential equilibrium equations or be solved directly by a Galerkin-type formulation.

Similarly as in [9, 10] we consider the problem of evaluating the first and second order sensitivity derivatives of a functional G of the form

$$G = \int \psi(\underset{\sim}{\sigma}, p, q) \, dV + \int h(\underset{\sim}{u}, p, q) \, dV + \int F(\underset{\sim}{T}) \, dS_u + \int g(\underset{\sim}{u}) \, dS_T \qquad (6)$$

In the following we use the following notation for partial and total derivatives

$$\frac{\partial \psi}{\partial p} \equiv \psi,_p \qquad \frac{d\psi}{dp} = \psi_p \equiv \psi,_p + \psi,_{\underset{\sim}{\sigma}} \underset{\sim}{\sigma}_p \qquad (7)$$

First derivatives - direct approach

The first derivative $G_p$ is obtained by differentiating Eq. (6)

$$G_p = \int (\psi,_p + \psi,_{\underset{\sim}{\sigma}} \cdot \underset{\sim}{\sigma}_p) \, dV + \int (h,_p + h,_{\underset{\sim}{u}} \cdot \underset{\sim}{u}_p) \, dV$$

$$+ \int F,_{\underset{\sim}{T}} \cdot \underset{\sim}{T}_p \, dS_u + \int g,_{\underset{\sim}{u}} \cdot \underset{\sim}{u}_p \, dS_T \qquad (8)$$

To obtain the first-order sensitivity fields $\underset{\sim}{u}_p$, $\underset{\sim}{\sigma}_p$ and $\underset{\sim}{T}_p$ we differentiate the equilibrium equations, Eq. (4), with respect to p

$$\int \underset{\sim}{\sigma}_p \cdot \delta \underset{\sim}{\varepsilon} \, dV = \int \underset{\sim}{T}_p \cdot \delta \underset{\sim}{u}^\circ \, dS_u \qquad (9)$$

where $\underset{\sim}{\sigma}_p$ is obtained from (5), namely

$$\sigma_{\sim p} = s,_\varepsilon \cdot \varepsilon_{\sim p} + s,_{\sim p} \tag{10}$$

Note that for a nonlinear material, $s,_\varepsilon$ is a tangential stiffness matrix and the relation between $\sigma_{\sim p}$ and $\varepsilon_{\sim p}$ is linear, whereas the second term of (10) can be regarded as an initial stress within the body. From (7) it follows that

$$\varepsilon_{\sim p} = Bu_{\sim p} \tag{11}$$

Substituting (10) into (9), and assuming $\delta u^o = 0$ on $S_u$, we obtain

$$\int \delta\varepsilon \cdot s,_\varepsilon \cdot \varepsilon_{\sim p} dV = -\int s,_{\sim p} \cdot \delta\varepsilon dV \tag{12}$$

Eq. (12) may be regarded as the equilibrium condition of a linear elastic structure with a stiffness matrix $D = s,_\varepsilon$, and initial stress $s,_{\sim p}$. The stress, strain and displacement fields for that structure are the sensitivity fields $\sigma_{\sim p}, \varepsilon_{\sim p}$ and $u_{\sim p}$. Let us note that when the stress-strain relation (2) are generated by an elastic potential U

$$D = s,_\varepsilon = \frac{\partial^2 U}{\partial\varepsilon \partial\varepsilon} \tag{13}$$

and the tangent stiffness matrix is symmetric. Moreover, when U is a quadratic function of strain, then D is a constant stiffness matrix of a linear elastic body.

The equilibrium equation, (12), may be solved with a zero traction boundary condition on $S_T$ and a zero prescribed displacement field on $S_u$ to obtain $\sigma_{\sim p}$, $\varepsilon_{\sim p}$ and $u_{\sim p}$. Then Eq. (8) is used to evaluate $G_p$. The process has to be repeated for evaluating the sensitivity derivatives with respect to q or any other parameter.

First derivatives - adjoint structure approach

Consider now the alternative method of defining an adjoint structure with an associated adjoint solution $\sigma^a$, $\varepsilon^a$ and $u^a$. Following [9, 10] the adjoint structure is linear with a stress-strain relation of the form

$$\sigma^a = s,_\varepsilon \cdot (\varepsilon^a - \varepsilon^{ia}) \tag{14}$$

where $s,_\varepsilon$ is evaluated at the equilibrium solution $\sigma, \varepsilon$, of the primary structure. It obeys the linear strain-displacement relation

$$\underset{\sim}{\varepsilon}^a = B\underset{\sim}{u}^a \tag{15}$$

is subjected to body forces

$$\underset{\sim}{f}^a = \underset{\sim}{h},_u \quad \text{within } V \tag{16}$$

surface tractions and displacements

$$\underset{\sim}{T}^{ao} = \underset{\sim}{g},_u \text{ on } S_T, \quad \underset{\sim}{u}^{ao} = -\underset{\sim}{F},_T \text{ on } S_u \tag{17}$$

and an imposed field of initial strains

$$\underset{\sim}{\varepsilon}^{ia} = \underset{\sim}{\psi},_\sigma \quad \text{within } V \tag{18}$$

The equations of equilibrium of the adjoint structure are expressed through the virtual work equation

$$\int \underset{\sim}{\sigma}^a \cdot \delta\underset{\sim}{\varepsilon} \, dV = \int \underset{\sim}{g},_u \cdot \delta\underset{\sim}{u} \, dS_T + \int \underset{\sim}{h},_u \cdot \delta\underset{\sim}{u} \, dV + \int \underset{\sim}{T}^a \cdot \delta\underset{\sim}{u}^o \, dS_u \tag{19}$$

Setting $\delta\underset{\sim}{\varepsilon} = \underset{\sim}{\varepsilon}_p$ and $\delta\underset{\sim}{u} = \underset{\sim}{u}_p$ in (19) we obtain

$$\int \underset{\sim}{\sigma}^a \cdot \underset{\sim}{\varepsilon}_p \, dV = \int \underset{\sim}{g},_u \cdot \underset{\sim}{u}_p \, dS_T + \int \underset{\sim}{h},_u \cdot \underset{\sim}{u}_p \, dV \tag{20}$$

But in view of (10) and (14)

$$\int \underset{\sim}{\sigma}^a \cdot \underset{\sim}{\varepsilon}_p \, dV = \int \underset{\sim}{s},_\varepsilon \cdot (\underset{\sim}{\varepsilon}^a - \underset{\sim}{\varepsilon}^{ia}) \cdot \underset{\sim}{\varepsilon}_p \, dV$$

$$= \int (\underset{\sim}{\varepsilon}^a - \underset{\sim}{\varepsilon}^{ia}) \cdot (\underset{\sim}{\sigma}_p - \underset{\sim}{s},_p) \, dV \tag{21}$$

and since by virtue of (9) we have

$$\int \underset{\sim}{\sigma}_p \cdot \underset{\sim}{\varepsilon}^a \, dV = -\int \underset{\sim}{T}_p \cdot \underset{\sim}{F},_T \, dS_u \tag{22}$$

Eq. (20) can be written in the form

$$\int (\underset{\sim}{\psi},_\sigma - \underset{\sim}{\varepsilon}^a) \cdot \underset{\sim}{s},_p \, dV = \int \underset{\sim}{g},_u \cdot \underset{\sim}{u}_p \, dS_T + \int \underset{\sim}{h},_u \cdot \underset{\sim}{u}_p \, dV$$

$$+ \int \psi,_\sigma \cdot \sigma,_p dV + \int T,_p \cdot F,_T dS_u \qquad (23)$$

Using (21) and (23), the derivative $G_p$ given by (8) can be expressed as

$$G_p = \int (\psi,_p + \psi,_\sigma \cdot s,_p - \varepsilon^a \cdot s,_p + h,_p) dV \qquad (24)$$

that is explicitly in terms of derivatives of the integrands of (6) with respect to p and the fields $s,_p$ and $\varepsilon^a$. A similar expression is obtained for the derivative $G_q$. It is thus seen that when there are n stiffness parameters $b_i$, all the derivatives $G_{b_i}$ may be evaluated by using the primary and adjoint structure solutions, that is two solutions. The adjoint approach is, therefore, more economical than the direct approach which requires n+1 solutions. However, when there are m functionals and n stiffness parameters, then the direct approach would still require n+1 solutions whereas the adjoint approach would need m+1 solutions in order to generate the derivatives $G_{b_i}$. Thus, the choice between the two approaches depends on the ratio of m to n as well as the relative difficulty of obtaining adjoint solutions versus sensitivity solutions.

SECOND-ORDER SENSITIVITY ANALYSIS

The second derivative $G_{pq}$ is obtained by differentiating (8) with respect to q, that is

$$G_{pq} = \int (\psi,_{pq} + \psi,_{p\sigma} \cdot \sigma,_q + \psi,_{q\sigma} \cdot \sigma,_p + \sigma,_p \cdot \psi,_{\sigma\sigma} \cdot \sigma,_q + \psi,_\sigma \cdot \sigma,_{pq}) dV$$

$$+ \int (h,_{pq} + h,_{pu} \cdot u,_q + h,_{qu} \cdot u,_p + u,_p \cdot h,_{uu} \cdot u,_q + h,_u \cdot u,_{pq}) dV$$

$$+ \int (T,_p \cdot F,_{TT} \cdot T,_q + F,_T \cdot T,_{pq}) dS_u + \int (u,_p \cdot g,_{uu} \cdot u,_q + g,_u \cdot u,_{pq}) dS_T \qquad (25)$$

Let us now discuss the three methods of evaluating the derivative $G_{pq}$.

Second derivatives - direct approach

After calculating the first-order derivatives from Eqs. (10) - (12) we differentiate them with respect to q to obtain

$$\int \sigma,_{pq} \cdot \delta\varepsilon \, dV = \int T,_{pq} \cdot \delta u^\circ \, dS_u \qquad (26)$$

$$\sigma_{pq} = s,_\varepsilon \cdot \varepsilon_{pq} + \varepsilon_p \cdot s,_{\varepsilon\varepsilon} \cdot \varepsilon_q + s,_{p\varepsilon}\varepsilon_q + s,_{q\varepsilon}\varepsilon_p + s,_{pq} \tag{27}$$

$$\varepsilon_{pq} = Bu,_{pq} \tag{28}$$

The last four terms on the right-hand side of (27) are initial stress terms, whereas $\sigma_{pq}$ and $\varepsilon_{pq}$ are linearly related by the tangent stiffness matrix $D = s,_\varepsilon$. The nature of the problem is, therefore, the same as for the first-order sensitivity analysis, that is an initial stress problem for a linear material with zero surface tractions on $S_T$ and zero prescribed displacements on $S_u$.

The systems of equations is to be solved for $\sigma_{pq}$, $\varepsilon_{pq}$ and $u_{pq}$ and then Eq. (25) may be used to evaluate $G_{pq}$. However, when the number of stiffness parameters increases the direct method becomes quite expensive as there are $n(n+1)/2$ second derivatives and each requires an independent solution. Counting also the required primary solution and first-order sensitivities the total number of solutions is $(n+1)(n+2)/2$.

<u>Second derivatives - mixed approach</u>

An alternative method which will be called the 'mixed approach' uses the first order sensitivity solution together with the adjoint structure solution.

Setting $\delta\varepsilon = \varepsilon_{pq}$, $\delta u = u_{pq}$ in (19) and noting that $u_{pq} = 0$ on $S_u$, we obtain

$$\int \sigma^a \cdot \varepsilon_{pq} dV = \int g,_u \cdot u_{pq} dS_T + \int h,_u \cdot u_{pq} dV \tag{29}$$

However, from (14) and (27) it follows that

$$\int \sigma^a \cdot \varepsilon_{pq} dU = \int (\varepsilon^a - \varepsilon^{ia}) \cdot s,_\varepsilon \cdot \varepsilon_{pq} dV =$$

$$\int (\varepsilon^a - \varepsilon^{ia}) \cdot (\sigma_{pq} - \varepsilon_p \cdot s,_{\varepsilon\varepsilon} \cdot \varepsilon_q - s,_{pq} - s,_{\varepsilon p} \cdot \varepsilon_q - s,_{\varepsilon q} \cdot \varepsilon_p) dV \tag{30}$$

On the other hand, setting $\delta\varepsilon = \varepsilon^a$, $\delta u^o = -F,_T$ in Eq. (26) we obtain

$$\int \sigma_{pq} \cdot \varepsilon^a dV = - \int T_{pq} \cdot F,_T dS_u \tag{31}$$

so that Eq. (30) becomes

$$- \int (\varepsilon^a - \varepsilon^{ia}) \cdot (\varepsilon_p \cdot s,_{\varepsilon\varepsilon} \cdot \varepsilon_q + s,_{pq} + s,_{\varepsilon p} \cdot \varepsilon_q + s,_{\varepsilon q} \cdot \varepsilon_p) dV$$

$$= \int g,_u \cdot u_{pq} dS_T + \int h,_u u_{pq} dV + \int T_{pq} \cdot F,_T dS_u + \int \psi,_\sigma \cdot \sigma_{pq} dV \qquad (32)$$

The second derivative $G_{pq}$ may, therefore, be expressed as

$$G_{pq} = \int (\psi,_{pq} + \psi,_{p\sigma} \cdot \sigma_q + \psi,_{q\sigma} \cdot \sigma_p + \sigma_p \cdot \psi,_{\sigma\sigma} \cdot \sigma_q) dV$$

$$+ \int (h,_{pq} + h,_{pu} \cdot u_q + h,_{qu} \cdot u_p + u_p \cdot h,_{uu} \cdot u_q) dV$$

$$+ \int T_p \cdot F,_{TT} \cdot T_q dS_u + \int u_p \cdot g,_{uu} \cdot u_q dS_T$$

$$- \int (\varepsilon^a - \psi,_\sigma) \cdot (\varepsilon_p \cdot s,_{\varepsilon\varepsilon} \cdot \varepsilon_q + s,_{pq} + s,_{\varepsilon p} \cdot \varepsilon_q + s,_{\varepsilon q} \cdot \varepsilon_p) dV \qquad (33)$$

That is in terms of solutions of the primary and adjoint structures, the first order sensitivities and the derivatives of the integrands of G when we have n stiffness parameters and m functionals the mixed approach require n+m+1 solutions which is more efficient than the (n+1)(n+2)/2 solutions required by the direct approach unless n<<m.

## Second derivatives - other approaches

Other methods to the calculation of second derivatives were presented in [10, 12, 14]. These methods all follow the same basic approach which for our problem would start by differentiating Eq. (24) with respect to q.

$$G_{pq} = \int (\psi,_{pq} + \psi,_{p\sigma} \cdot \sigma_q + \psi,_{q\sigma} \cdot s,_p + \sigma_q \cdot \psi,_{\sigma\sigma} \cdot s,_p + \psi,_\sigma \cdot s,_{pq}$$

$$-\varepsilon^a_q \cdot s,_p - \varepsilon^a \cdot s,_{pq} + h,_{pq} + h,_{p\sigma} \cdot \sigma_q) dV \qquad (34)$$

Eq. (34) requires the evaluation of the first order sensitivities as well as the first order sensitivities of the adjoint solution (which may be obtained by differentiating Eqs. (14), (15) and (19)). For m functionals and n stiffness parameters the total number of solutions is 1+n+m+nm = (n+1)(m+1) which is always greater than the mixed approach.

## EXAMPLE - BAR SUBJECT TO APPLIED TEMPERATURE

The bar shown in Figure 1 is clamped between two rigid walls and subject to a linearly varying temperature field $T = T_o(x/L)$. The beam material obeys the nonlinear constitutive law

$$\varepsilon = \alpha T + t + aT_o t^2 \qquad (a1)$$

where t is a normalized stress $t = \sigma/E$, a is a constant, and $\alpha$ is the coefficient of thermal expansion. The bar is assumed to have a uniform cross section. We want to calculate the average displacement $u_{av}$ and its derivatives with respect to $\alpha$ and $T_o$

$$u_{av} = (1/L) \int_0^L u\, dx \tag{a2}$$

## Nominal solution

Assuming small strains the equation for the displacement is

$$u,_x = \alpha T_o x/L + t + aT_o t^2 \tag{a3}$$

$$u(0) = u(L) = 0$$

so that

$$u = \alpha T_o x^2/2L + (t + aT_o t^2)x \tag{a4}$$

where t is found from the condition that $u(L) = 0$.

$$\alpha T_o + 2(t + aT_o t^2) = 0 \tag{a5}$$

and using (a5)

$$u = 0.5\alpha T_o (x^2/L - x) \tag{a6}$$

## Direct approach for sensitivities

For our example Eq. (a6) provides an analytical solution for the displacement which can be differentiated for the derivatives of u. Since in general such analytical solution are not available we differentiate instead the differential equation for u, Eq. (a3). Differentiating with respect to $\alpha$

$$u_{\alpha,x} = T_o x/L + t_\alpha(1 + 2aT_o t)$$

$$u_\alpha(0) = u_\alpha(L) = 0 \tag{a7}$$

so that

$$u_\alpha = T_o x^2/2L + t_\alpha(1 + 2aT_o t)x \tag{a8}$$

$$t_\alpha = -0.5 T_o/(1 + 2aT_o t) \tag{a9}$$

and then

$$u_\alpha = 0.5T_o(x^2/L-x) \tag{a10}$$

Similarly we obtain

$$u_T = \alpha x^2/2L + [t_T(1+2aT_o t) + at^2]x \tag{a11}$$

$$t_T = -(0.5\alpha + at^2)/(1+2aT_o t) \tag{a12}$$

or

$$u_T = 0.5\alpha(x^2L-x) \tag{a13}$$

The second derivatives are treated similarly. For example, differentiating Eq. (a7) with respect to $T_o$ and integrating

$$u_{\alpha T} = x^2/2L + [t_{\alpha T}(1+2aT_o t) + 2att_\alpha + 2aT_o t_\alpha t_T]x \tag{a14}$$

$$t_{\alpha T} = -(0.5 + 2att_\alpha + 2aT_o t_\alpha t_T)/(1+2aT_o t) \tag{a15}$$

so that

$$u_{\alpha T} = 0.5(x^2/L-x) \tag{a16}$$

These displacement derivatives are then used to evaluate the derivatives of $u_{av}$

$$(u_{av})_{,\alpha} = (1/L)\int_0^L u_\alpha dx = -T_o L/12 \tag{a17}$$

$$(u_{av})_{,T} = (1/L)\int_0^L u_T dx = -\alpha L/12 \tag{a18}$$

$$(u_{av})_{,\alpha T} = (1/L)\int_0^L u_{\alpha T} dx = -L/12 \tag{a19}$$

## Adjoint and mixed approaches

The adjoint structure obeys a linear stress-strain relation

$$\varepsilon^a = t^a(1+2aT_o t) \tag{a20}$$

and is subject to body forces, $f^a = h,_u = 1/L$. The equation of equilibrium is

$$t^a,_x = -f^a/E = -1/EL \tag{a21}$$

so that

$$t^a = -x/EL + c \tag{a22}$$

The strain displacement relation and boundary conditions are

$$u^a{}_{,x} = \varepsilon^a = t^a(1+2aT_o t)$$

$$u^a(0) = u^a(L) = 0 \tag{a23}$$

Substituting from Eq. (a22) and integrating

$$u^a = (1+2aT_o t)(-x^2/2EL+cx) \tag{a24}$$

$$c = 1/2E \tag{a25}$$

We can now evaluate $(u_{av})_{,\alpha}$ and $(u_{av})_{,T}$ by using Eq. (24)

$$(u_{av})_{,\alpha} = \int_0^L -\varepsilon^a{}_{,s,\alpha} dx \tag{a26}$$

Because the stress-strain relationship is not given in the form $\sigma = s(\varepsilon,\alpha, T_o)$ we need to evaluate $s_{,\alpha}$ indirectly by differentiating Eq. (a1)

$$\varepsilon_\alpha = T + t_\alpha(1+2aT_o t) \tag{a27}$$

But also

$$t_\alpha = \sigma_\alpha/E = s_{,\alpha}/E + s_{,\varepsilon}\varepsilon_\alpha/E \tag{a28}$$

substituting from (a28) to (a27) we get

$$s_{,\varepsilon} = E/(1+2aT_o t) \tag{a29}$$

$$s_{,\alpha} = -ET_o(x/L)/(1+2aT_o t) \tag{a30}$$

substituting from Eqs. (a22), (a23), (a25) and (a30) into Eq. (a26)

$$(u_{av})_{,\alpha} = \int_0^L T_o(x/L)(0.5-x/L)dx = -T_o L/12 \tag{a31}$$

which agrees with the result obtained from the direct result, Eq. (a13). Similarly,

$$(u_{av}),_T = \int_0^L -\varepsilon^a s,_T dx \tag{a32}$$

To calculate $s,_T$ we note that

$$\varepsilon_T = \alpha(x/L) + t_T(1+2aT_o t) + at^2 \tag{a33}$$

and

$$t_T = \sigma_T/E = s,_T/E + s,_\varepsilon \varepsilon_T/E \tag{a34}$$

so that

$$s,_T = -E[\alpha x/L + at^2]/(1+2atT_o) \tag{a35}$$

Substituting into Eq. (a32) we obtain

$$(u_{av}),_T = \int_0^L (\alpha x/L + at^2(0.5-x/L)dx = \frac{-\alpha L}{12} \tag{a36}$$

For the second derivative $(u_{av}),_{\alpha T}$ we need to use Eq. (33)

$$(u_{av}),_{\alpha T} = -\int_0^L \varepsilon^a(\varepsilon_\alpha s,_{\varepsilon\varepsilon}\varepsilon_T + s,_{\alpha T} + s,_{\varepsilon\alpha}\varepsilon_T + s,_{\varepsilon T}\varepsilon_\alpha)dx \tag{a37}$$

The second derivatives of $s$ in Eq. (a37) are obtained by differentiating Eq. (a27) with respect to $T_o$

$$\varepsilon_{\alpha T} = (x/L) + t_{\alpha T}(1+2aT_o t) + 2at_\alpha(t+T_o t_T) \tag{a38}$$

and noting that

$$t_{\alpha T} = \sigma_{\alpha T}/E = (s,_{\alpha T} + s,_{\varepsilon\alpha}\varepsilon_T + s,_{\varepsilon T}\varepsilon_\alpha$$
$$+ s,_{\varepsilon\varepsilon}\varepsilon_\alpha\varepsilon_T + s,_\varepsilon \varepsilon_{\alpha T})/E \tag{a39}$$

Substituting from Eqs. (a24), (a34) and (a39) into Eq. (a38) we obtain

$$s,_{\alpha T} = -\frac{E(x/L) + 2as,_\alpha(s,_T T_o/E + t)}{1+2aT_o t} \tag{a40}$$

$$s,_{\varepsilon T} = -2a(t+s,_T T_o/E)s,_\varepsilon/(1+2aT_o t) \tag{a41}$$

$$s,_{\varepsilon\alpha} = -2aT_o s,_\alpha s,_\varepsilon/E(1+2aT_o t) \tag{a42}$$

$$s,_{\varepsilon\varepsilon} = \frac{-2as^2,_\varepsilon T_o/E}{1+2aT_o t} \tag{a43}$$

substituting into Eq. (a37) we obtain

$$(u_{av}),_{\alpha T} = -L/12 \tag{a44}$$

as from Eq. (a19).

EXAMPLE - TRUSS STRUCTURES

As an example of the finite-element implementation of the procedure consider the application to linear analysis of truss structures. In the example the derivatives of the average stress in the i-th element are considered. That is first derivatives of the functional

$$G = \frac{1}{V_i} \int_{e_i} \sigma dV \tag{b1}$$

with respect to Young's modulus of the j-th element, $E_j$, are calculated as well as second derivatives with respect to $E_j$ and $E_k$.

(a) <u>Direct approach, first derivatives</u>

Eq. (8) specializes to

$$G_p = \int (\underset{\sim}{\psi},_\sigma \cdot \underset{\sim}{\sigma}_p) dV \tag{b2}$$

where the parameter p represents $E_j$, and the function $\psi$ corresponding to Eq. (b1) is

$$\psi = \frac{1}{V_i} \underset{\sim}{\sigma} \chi_i \tag{b3}$$

where $\chi_i$ is the characteristic function of the i-th element (Equal to 1 in the element, zero elsewhere). The stress field $\underset{\sim}{\sigma}_p$ satisfies Eq. (10) which is based on a stress-strain relationship $\underset{\sim}{\sigma} = \underset{\sim}{s}(\underset{\sim}{\varepsilon})$. For the present example

$$\sigma = E\varepsilon \tag{b4}$$

and, therefore, Eq. (10) becomes

$$\sigma_p = E\varepsilon_p + \chi_j \varepsilon = E(\varepsilon_p + \chi_j \varepsilon/E) \tag{b5}$$

That is the sensitivity field may be obtained by imposing an initial strain of magnitude $\varepsilon_j/E$ on the j-th element, and $G_p = (\sigma_p)_i$. To implement in a finite element program the following steps are required.

   (i)   Extract strain in j-th element

   (ii)  Divide strain by $E_j$ to obtain imposed initial strain field

   (iii) Apply initial strain field, and solve for resultant stress field $\sigma_p$.

(b) <u>Adjoint method, first derivative</u>

According to Eq. (18) the adjoint field is due to an initial strain of magnitude $\psi,_\sigma = \frac{1}{V_i}$ (see Eq. 3) in the i-th element. Eq. (24) then reduces to

$$G_p = \int (\psi,_\sigma - \underset{\sim}{\varepsilon}^a) \cdot \underset{\sim}{s},_p dV = \frac{1}{V_i} \int_{e_i} x_j \varepsilon dV - \int \underset{\sim}{\varepsilon}^a \underset{\sim}{s},_p dV \tag{b6}$$

The second term in Eq. (b6) represents the work done by the initial stress of the sensitivity field on the adjoint strain field. This also is equal to the work done by the nodal forces corresponding to the initial sensitivity stresses by the adjoint displacement. Accordingly, to implement in a finite-element program, the following steps should be taken.

   (i)   Impose initial strains of magnitude $1/V_i$ on i-th element

   (ii)  Solve for adjoint displacement field

   (iii) Calculate nodal forces corresponding to $\sigma_p$ as in direct method

   (iv)  Calculate negative of cross work.

   (v)   Add strain in element i when i=j.

(c) <u>Direct approach, second derivatives</u>

First we calculate the sensitivity field with respect to $E_k$ by the direct approach as before. Eq. (27) becomes

$$\sigma_{pq} = E\varepsilon_{pq} + x_p \varepsilon_q + x_q \varepsilon_p \tag{b7}$$

where p represents $E_j$ and q represents $E_k$. Equation (b7) means that we apply an initial stress of magnitude $\varepsilon_q$ in the j-th element plus an initial stress of magnitude $\varepsilon_p$ in the k-th element. So steps are:

(i) Extract $\frac{\partial \varepsilon}{\partial E_j}$ in k-th element and $\frac{\partial \varepsilon}{\partial E_k}$ in j-th element from direct solution for first derivatives

(ii) Apply as initial strain field in these elements

(iii) Solve for 2nd derivative field

(d) <u>Mixed approach, second derivatives</u>

Equation (33) becomes

$$G_{pq} = -\int (\underset{\sim}{\varepsilon}^a - \underset{\sim}{\psi},_\sigma) \cdot (\underset{\sim}{s},_{\varepsilon p} \cdot \underset{\sim}{\varepsilon}_q + \underset{\sim}{s},_{\varepsilon q} \cdot \underset{\sim}{\varepsilon}_p) dV$$
$$= -\int \varepsilon^a (\chi_p \varepsilon_q + \chi_q \varepsilon_p) dV + \frac{1}{V_i} \int_{e_i} (\chi_p \varepsilon_q + \chi_q \varepsilon_p) dV \qquad (b8)$$

the first term is the work of the strain field of the adjoint field on the initial stress field used for the direct sensitivity calculation. So to implement this term in a finite element program we take the negative of the scalar product of the displacement of the adjoint field with the forces due to initial stresses in Eq. (b7). To that we add the strain in the i-th element if i=j or i=k and twice that strain if i=j=k.

CONCLUDING REMARKS

The principle of virtual work was used to derive sensitivity derivatives of structural response with respect to stiffness parameters. The relative efficiency of various approaches for calculating first and second derivatives was assessed. It was shown that for the evaluation of second derivatives the most efficient approach is one which makes use of both the first order sensitivities and adjoint vectors. Two example problems were used for demonstrating the various approaches.

ACKNOWLEDGEMENT

This work was supported in part by NASA grant NAG-1-224.

REFERENCES

1. Tomovic, R. and Vukobratovic, M., <u>General Sensitivity Theory</u>, American Elsevier, 1972.

2. Tilden, J. W. and Seinfeld, J. H., "Sensitivity Analysis of Mathematical Model for Photochemical Air Pollution," <u>Atmospheric Environment</u>, Vol. 16, No. 6, pp. 1357-1364, 1982.

3. Leonard, J. I., "The Application of Sensitivity Analysis to Models of Large Scale Physiological Systems," NASA CR-160288, October 1974.

4. Irwin, C. L. and O'Brien, T. J., "Sensitivity Analysis of Thermodynamics Calculations," U.S. Dept. of Energy Report DOE/METC/82-53, 1982.

5. Dwyer, H. A., Peterson, T. and Brewer, J., "Sensitivity Analysis Applied to Boundary Layer Flow," Proc. 5th International Conference on Numerical Methods in Fluid Dynamics, June 28-July 2, 1976.

6. Haug, E. J. and Arora, J. S., "Design Sensitivity Analysis of Elastic Mechanical Systems," Computer Methods in Applied Mechanics and Engineering, Vol. 15, pp. 35-62, 1978.

7. Haug, E. J. and Rousselet, B., "Design Sensitivity Analysis in Structural Mechanics, I: Static Response," Journal of Structural Mechanics, Vol. 8, pp. 17-41, 1980.

8. Haug, E. J., Komkov, V. and Choi, K. K., Design Sensitivity Analysis of Structural Systems, Academic Press, 1984.

9. Dems, K. and Mróz, Z., "Variational Approach by Means of Adjoint Systems for Structural Optimization and Sensitivity Analysis, Part I, Variation of Material Parameters within Fixed Domain," International Journal of Solids and Structures, Vol. 19, pp. 677-692, 1983.

10. Dems, K. and Mróz, Z., "Variational Approach to First- and Second-Order Sensitivity Analysis of Elastic Structures," International Journal for Numerical Methods in Engineering, 1984 (in press).

11. Szefer, G., Mróz, Z. and Dembowicz, L., "Variational Approach to Sensitivity Analysis in Non-linear Elasticity," submitted for publication to Archives of Mechanics.

12. Haug, E. J., "Second-Order Design Sensitivity Analysis of Structural Systems," AIAA Journal, Vol. 19, pp. 1087-1088, 1981.

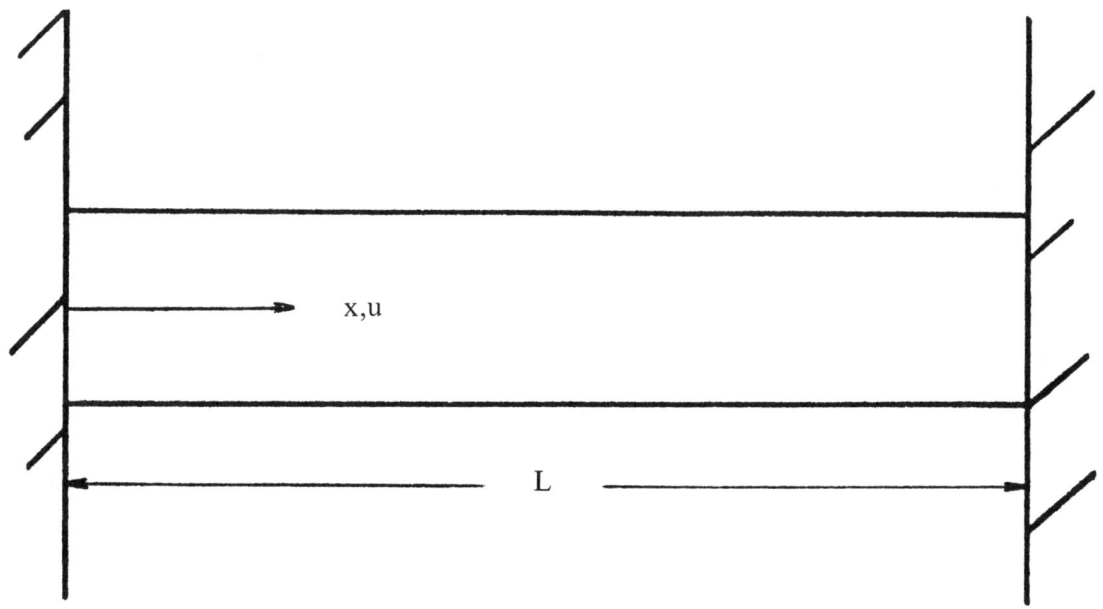

Figure 1: Geometry of bar example

A SURVEY OF METHODS AND PROBLEMS IN
AEROELASTIC OPTIMIZATION


Prabhat Hajela
Department of Engineering Sciences
University of Florida
Gainesville, Florida 32611


ABSTRACT

The applications of optimization methods in the aeroelastic design environment are reviewed. Extension of the structural optimization problem to include aeroelastic constraints, is assessed for static divergence, flutter and gust response problems. Applications in the area of fixed-wing and rotary-wing aeroelasticity are considered. In addition to the structural optimization problem, the state-of-the-art in active control technology is presented with a special emphasis on the use of formal optimization methods.

INTRODUCTION

The past decade has witnessed the emergence of automated structural synthesis as a discipline in its own right. Improvements in analysis techniques and rapidly improving capabilities for digital computing have contributed to this trend. The early inspirations of Maxwell [1] and Michell [2] are constant reminders of how far this discipline has progressed. Although variational methods of optimization continue to provide challenging mathematical problems [3], the practical aspects of large-scale structural synthesis have been addressed by two distinct approaches. These are referred to in literature as the optimality criteria methods [4], and the mathematical nonlinear programming approaches [5,6]. Of the two, the latter is recognized as the more flexible and general technique.

The significance of formal optimization methods in structural synthesis governed by aeroelastic constraints, cannot be emphasized enough, especially in view of the many counter intuitive solutions that have resulted from such applications. A review of the state-of-the-art assessment in the field of aeroelastic optimization was conducted by Stroud [7]. That review was largely confined to the implementation of flutter constraints and focussed on the significant drawbacks in those approaches. The present paper seeks to complement that survey rather than to replace it. The solution techniques described in the paper are dwelt upon in passing and not repeated in their entirety. However, new developments or improvements in those techniques are discussed. Additionally, the present paper extends the survey to include structural optimization under nondeterministic loads (gust loading) and applications in the area of rotary-wing aeroelasticity. The use of optimization methods in the optimal configuration of control systems for aeroelastic load alleviation, and suppression of aeroelastic instabilities is also examined.

THE STRUCTURAL OPTIMIZATION PROBLEM

The mathematical statement of the minimum weight problem can be written as follows:

$$\text{Minimize} \quad W(\bar{d}) \tag{1}$$

$$\text{Subject to} \quad g_j(\bar{d}, t) \leq 0 \qquad j = 1, 2, \ldots m \tag{2}$$

$$h_k(\bar{d}, t) = 0 \qquad k = 1, 2, \ldots p \tag{3}$$

$$d_i^L \leq d_i \leq d_i^u \qquad i = 1, 2, \ldots n \tag{4}$$

Here $W(\bar{d})$ is the objective function and is most typically the structural weight; $\bar{d}$ is an n-dimensional vector of design variables and usually represent structural member dimensions. Response constraints $h_k$ and $g_j$ are categorized as equality and inequality constraints, respectively, and are used to specify bounds on response quantities such as displacements, stresses, natural frequencies and elastic or aeroelastic stability boundaries. The presence of time 't' in the response constraints introduces an added difficulty in the implementation, and is addressed in a later section. The components of the design variable vector $d_i$, are also subject to user prescribed lower and upper bounds, $d_i^L$ and $d_i^u$, respectively. Application of a nonlinear programming algorithm to solve the above problem is fraught with hazards of inordinately large computational times in the presence of a large number of design variables and constraints. These problems are further compounded in applications where objective and constraint function evaluations are computationally burdensome. Repetitive analysis, as required in numerical optimization methods, often render such strategies intractable. Approximation concepts such as design variable linking, efficient treatment of constraints, and strategies that circumvent exact analysis in each iteration of the design process, have been proposed as a solution to these drawbacks [8,9]. A detailed enumeration of these methods is beyond the scope of the present paper.

STRUCTURAL DESIGN WITH AEROELASTIC CONSTRAINTS

A review of technical literature for structural synthesis problems with aeroelastic constraints, identifies two distinct areas of interest. Researchers approaching the structural optimization problem from a variational methods standpoint, have studied a class of one or two dimensional structures that are constrained to have divergence or flutter speeds above prescribed limits. The second group has primarily dwelt on implementing mathematical programming methods for the design of more practical built-up lifting surface assemblies. Of the former group, the problem of panel flutter has been exhaustively studied by several researchers and is reviewed here.

## Panel Flutter

A commonly used approach has been to obtain the minimum weight thickness distribution while holding the flutter speed constant and equal to that of a uniform reference panel. Finite-element studies of this problem have been reported with varying degrees of success [10,11]. The continuous problem solutions denote the bulk of published work in this area. McIntosh et. al. [12] and Weisshaar [13] used indirect techniques to obtain solutions to the problem. The finite-span panel flutter problem has been successfully attempted by Van Keuren and Eastep [14], using a Galeskin approach. Pierson and his co-workers [15,16,17] have studied the continuous, infinite span problem, from a direct optimal control approach with success. Results for both solid and sandwich panels, with prescribed upper and lower bounds on thickness are reported.

The equation of motion for an infinite span panel which is exposed to supersonic air flow on one side and considering the aerodynamic forces predicted by linearized strip theory, can be written as

$$\frac{\partial^2}{\partial x^2}\left[ D \frac{\partial^2 w}{\partial x^2} \right] + N_{xx} \frac{\partial^2 w}{\partial x^2} + \frac{2q}{\sqrt{M^2 - 1}} \frac{\partial w}{\partial x} + \frac{2q(M^2 - 2)}{U(M^2 - 1)^{3/2}} \frac{\partial w}{\partial t} + m \frac{\partial^2 w}{\partial t^2} = 0 \quad (5)$$

where 'w' is the transverse deflection, 'x' is the lengthwise co-ordinate, 'D' is the panel bending rigidity, '$N_{xx}$' is the edge traction, 'q' is the dynamic pressure, 'M' is the Mach number and 'U' is the free stream velocity. Assuming simple harmonic behavior for the displacement w, the fourth order equation can be represented by four first-order ordinary differential equations, by an appropriate selection of the state variables. The optimization problem can be posed as the minimization of an integral objective function subject to nonlinear differential constraints and prescribed boundary conditions and nonlinear terminal state constraints. Pierson and co-workers have successfully used a gradient projection method that first seeks the terminal state constraints and then executes a search along this boundary to the constrained optimum.

Interest in this problem has not waned as is evidenced by a recent survey of this problem by Librescu and Reiner [18]. Interesting new studies that include axisymmetric circular cylindrical shells and rectangular orthotropic flat panels are included in that survey.

## Flutter and Divergence of Built-up Lifting Surfaces

The first flutter synthesis effort can be traced to a 1969 publication of Turner [19]. Several studies involving both the optimality criteria methods and the nonlinear programming approaches have resulted since then [20,21,22]. Two production level codes TSO (Aeroelastic Tailoring and Structural Optimization) [23], and FASTOP (Flutter and Strength Optimization Procedures) [24], are representative of more recent advances. With the availability of efficient discrete analysis programs and significant improvements in the performance of mathematical programming based optimization algorithms, automated structural design with flutter constraints is not a difficult problem at a conceptual level. Flutter constraints are imposed by requiring that instability in any of the aeroelastic modes appear at a flight speed greater than or equal to a specified value. Since flutter constraints are imposed in addition to static strength requirements, there are two solution strategies that have been presented. The first approach is one of sequential design in which a strength design is first obtained by neglecting the flutter constraint. The strength design then constitutes a lower bound for a redesign cycle with only a flutter constraint. Due to a change in the load path which may result in a flutter redesign, this cycle is repeated till convergence. The other approach involves a search in the design space which includes all constraints, including the flutter constraint. Methods of implementing the flutter constraint and providing analytical gradients of these constraints for both the nonlinear programming and optimality criteria methods are summarized by Stroud [7]. Amongst the former, the implementation of the flutter constraint by the V-g method has enjoyed widespread application. The flutter equation can be written in a linearized form as follows.

$$(\Omega [\bar{K}] - [\bar{M}] - [\bar{A}])\{z\} = 0 \qquad (6)$$

where $[\bar{K}]$, $[\bar{M}]$ and $[\bar{A}]$ are generalized stiffness, mass and aerodynamics matrices, respectively. The parameter $\Omega$ is the complex eigenvalue parameter expressed in terms of the circular frequency '$\omega$' and a measure of the net damping of the structure '$g$' as follows

$$\Omega = \frac{1 + ig}{\omega^2} \qquad (7)$$

The generalized aerodynamics matrix $[\bar{A}]$ depends upon a reduced frequency parameter '$k$' defined in terms of a velocity '$V$' and a reference semichord '$b$' as follows

$$k = \frac{\omega b}{V} \qquad (8)$$

The matrix is in general complex, and hence equation 6 would yield complex eigenvalues and eigenvectors. The flutter speed calculation is done for an operating regime by choosing a set of reduced frequency parameters '$k$' of practical interest. For each of these values, the complex eigenvalues $\Omega$ are computed and for which a corresponding set of V-g points can be obtained. The crossover of the $g = 0$ line in any mode indicates a flutter condition for that mode at the corresponding flight condition. An implementation of this constraint in the context of a feasible-usable search direction algorithm for optimization, is described in Reference 25. In particular, remedial measures to ensure function and gradient continuity are suggested in that publication. Although the ease of analysis renders this approach attractive, there are serious drawbacks that suggest the search for an alternative method. These drawbacks can be summarized as follows:

a) The V-g plots are dependent upon a somewhat ambiguous reduced frequency parameter k, corresponding to which the V-g pairs are computed. Even though the V-g pairs for two successive values of k may be located in the feasible region of the design space, an intermediate value of '$k$' may result in an infeasibility that goes undetected in the design.
b) In the V-g approach, instability is expressed in terms of an artificial damping parameter '$g$' which is physically meaningful for only $g = 0$. Hence no safety margins can be prescribed.

c) A third problem that has been traced in the V-g solutions is a lack of uniqueness in the damping parameter at a given flight speed. This can result in convergence problems in the optimization algorithm.

A root locus based analysis procedure for flutter optimization has been suggested [26], that circumvents the above problems. In the Laplace s-domain, the aeroelastic stability equation is written as

$$(\bar{s}^2 [\bar{M}] + [\bar{K}] - q[\bar{A}(\bar{s})]) X(\bar{s}) = 0 \qquad (9)$$

where $[\bar{M}]$, $[\bar{K}]$ and $[\bar{A}]$ are the generalized mass, stiffness and aerodynamic matrices and $X(\bar{s})$ is vector of modal amplitudes. For various values of the dynamic pressure q, the complex roots $\bar{s}$ are obtained from the characteristic equation. The system would be unstable for that value of 'q' for which one of the roots has a positive real part. A constraint can thus be formulated by choosing a set of dynamic pressures corresponding to flight speeds of interest, computing the roots of the characteristic equation and requiring that the real part be negative. Furthermore, since the root locus approach permits computation of the viscous damping in each mode, constraints can also be imposed to bound the available viscous damping in each mode. Details on the implementation are proposed in the reference.

In both the V-g method and the s-plane formulations, the number of design constraints would be prohibitively large if treated in a conventional manner. A novel approach of folding these constraints into an efficient single representative constraint is discussed in the cited references.

The re-emergence of the forward swept wing concept with the advent of advanced filamentary composites, has introduced the need for including a constraint on the static divergence speed. The problem can be posed in a conventional form, requiring the flight regime of interest to be free of flutter and divergence instabilities. An alternative formulation of the problem would maximize both the flutter and divergence speeds. This approach results in a multiobjective problem and can be addressed by concepts of Pareto optimality.

More recently, efforts have been galvanized to develop an industry-standard program for the automated synthesis of aerospace structures [27]. This program, currently under development on a grant from the U.S. Air Force, will include static strength, thermal loads, dynamic loads and aeroelastic considerations as the design options. Response constraint sensitivities and a range of optimization options are also expected to be available.

## The Gust Response Problem

In addition to constraints on aeroelastic stability, it is imperative to establish optimization methods for sizing airframe structures that are subjected to random flight loads. The random vibration environment introduces the need for selecting a statistical process that best describes the random loads and permits computation of the dynamic response parameters of interest. Measurements of air turbulence samples indicate both a temporal and a spatial variation. A rapid penetration of the gust field justifies freezing the gust in time [28]. One and two dimensional spatial variations have been considered in the literature.

A survey of the available methods for optimum sizing under nondeterministic loads is available in [29]. A conclusion of that survey indicates that it is not only critical to select a statistical process for the gust but the formulation of constraints should be such that they be computationally viable and minimize the conservativeness in the design.

The system response can be computed in both the time and the frequency domain. Recent results of gust response constrained optimization in the time domain, using an interior penalty function type optimization algorithm, have been published by Rao [30]. However, computation of the system response in the frequency domain is more elegant wherein the turbulence field is typically characterized by the gust velocity power spectral density (PSD) distribution. The PSD function is representative of the variation of the mean square values of the gust velocities with frequency. It is used in conjunction with the response admittance functions to compute the linear response of the system to the gust loading. The availability of root mean square (rms) values of individual loads such as shear, bending moment and torsion at various points in the structure does not contain important phase information that is necessary to obtain the combined effect of these loads. A constant probability of combination criterion has been suggested to circumvent this problem. A normal probability distribution density function for two variables, x and y, is given as

$$p(x,y) = \frac{1}{2\pi\sigma_x\sigma_y\sqrt{1-\rho_{xy}^2}} \exp\left[-\frac{1}{2(1-\rho_{xy}^2)}\left[\left(\frac{x-\mu_x}{\sigma_x}\right)^2 - 2\rho_{xy}\left(\frac{x-\mu_x}{\sigma_x}\right)\left(\frac{y-\mu_y}{\sigma_y}\right) + \left(\frac{y-\mu_y}{\sigma_y}\right)^2\right]\right] \quad (10)$$

where $\sigma_x$ and $\sigma_y$ are the rms values of x and y; $\mu_x$ and $\mu_y$ are the mean values of x and y, respectively, and $\rho_{xy}$ is a quantity referred to as the correlation coefficient.

Equation 10 represents ellipses in planes parallel to the x - y plane and an infinite number of load conditions with an equal probability of occurrence are obtained as points on the boundary of the ellipse.

The quantity $\rho_{xy}$ is defined as

$$\rho_{xy} = \frac{1}{\overline{A}_x \overline{A}_y} \int_0^\infty \phi_w(\omega) \left[H_{x_{real}}(\omega) * H_{y_{real}}(\omega) + H_{x_{imag}}(\omega) * H_{y_{imag}}(\omega)\right] d\omega \quad (11)$$

where, $\phi_w(\omega)$ is the gust PSD; $H_x$ and $H_y$ are the frequency response functions for the load quantities x and y; $\bar{A}_x$ and $\bar{A}_y$ are the ratios of the design rms loads $\sigma_x$ and $\sigma_y$ to the design rms gust intensity $\sigma_w$, respectively.

The failure modes that must be accounted for in the optimal design for such nondeterministic loads include a single excursion failure and a fatigue failure. Constraints for these failure modes are formulated and implemented in a recent publication [31]. The prediction of fatigue failure by a classical Palmgren Miner approach is considered inadequate in view of more recent developments in fracture mechanics. The previous reference also includes constraint formulation and implementation in the fracture mechanics environment.

Another publication [32] attempts to integrate these developments into a research oriented design and synthesis package. ISAC [33] is a system of programs developed at NASA Langley Research Center, primarily for integrating aeroelastic design with active control technology. An optimization capability that includes gust response constraints has been added to this system at a preliminary level and is expected to be augmented with full analytical gradient capability.

## Applications in Rotorcraft Aeroelasticity

The formulation and solution of rotorcraft aeroelastic stability and dynamic response has matured significantly in the last decade. Several research publications and monographs pertaining to the discipline have appeared over this period. A recent survey [34], examines the effects of geometric nonlinearities and dynamic stall on the aeroelastic stability and dynamic response of the vehicle. Stability of isolated rotor blades in hover and forward flight and coupled rotor/fuselage dynamics are addressed in context of structural, inertial and aerodynamic operators. The development of such a significant body of literature clearly indicates the need to complement the analysis tools with an automated synthesis capability. Two problem areas can be identified where the analysis capability is developed enough to warrant optimization applications. The simpler of the two pertains to the isolated blade aeroelastic stability - coupled flap-lag and coupled flap-lag-torsional instabilities. The second is the coupled rotor-fuselage aeromechanical stability. Recent studies [35] seem to indicate that the blade stability problem appears to be less significant from a structural optimization standpoint than the problem of vibration reduction by a suitable modification of stiffness and mass properties.

Since helicopter rotor blades are constructed from modern composite materials, a significant potential exists to tailor the stiffness and mass properties to obtain favorable dynamic characteristics. Very early efforts [36] with nonoptimal metallic rotors achieved vibration reduction by significant weight increases to alter the spanwise and chordwise mass distributions. More recent publications recognize the potential of using formal optimization towards this end.

Taylor [37] attempts the problem of vibration reduction in forward flight by a concept he refers to as "modal shaping". The modal shaping parameter is a measure of the blade modal vibration susceptibility. The process involves densensitizing a normal mode to the aerodynamic excitation. No formal optimization algorithm was used for this task. Another study due to Peters et. al. [38]

describes the use of formal optimization methods in the design of helicopter blades for a desired placement of the blade natural frequencies to avoid resonance from the system forcing frequencies. The task of selecting windows for frequency placement is a difficult one. Additionally, the axial stiffness in the blades due to the centrifugal effect dominates the values of the lower mode frequencies, making it difficult to change them significantly by redistributing the mass and stiffness characteristics.

Two recent studies pose the structural synthesis problem as one of minimizing the vertical hub shears. Bennett [39] achieves vibration reduction resulting in a 20 - 40% reduction in the vertical hub shears. Friedman and Shantakumaran [40] minimize the hub shears and hub rolling moments subject to constraints of frequency placement and aeroelastic stability boundaries. The aeroelastic stability margins are specified for hover conditions with the assumption that forward flight alleviates the aeroelastic instability problem.

The advantages of incorporating optimization in helicopter design tools are now clearly recognized and is evident in a survey by Miura [41]. Yet another treatment of the problem, and, in particular, the identification of the immediate problems of concern is available in a publication of Sutton and Bennett [42]. The extension of the structural optimization problem to the control-coupled aeroelastic problem holds significant promise and is addressed in a survey [43].

## ACTIVE CONTROLS IN AEROELASTIC APPLICATIONS

Increasing stringent specifications on the performance of military aircraft have contributed significantly to the emergence of a new discipline of research referred to as aeroservoelasticity. Elements of control theory are incorporated into a conventional aeroelastic synthesis procedure. The procedure enhances aerodynamic efficiency and offers a reduction in the structural weight. The system configured in this approach has an extended envelope of operation, in which structural intregrity and system performance are assured by the successful implementation of an active control system. A survey of this discipline including an assessment of the analytical tools was conducted by Newsom et al. [44].

The field of aeroservoelasticity encompasses the more conventional disciplines of structural dynamics, the interaction with steady/unsteady aerodynamic loads, and a coupling of methods of control theory. The equations of motion for a lifting surface can be expressed in modal coordinates as follows:

$$M_i \ddot{q}_i(t) + K_i q_i(t) = Q_i(t) \qquad i = 1,2 \ldots n \qquad (12)$$

Here, $M_i$, $K_i$ and $Q_i$ are the generalized mass, stiffness and aerodynamic force, respectively, and $q_i$ is the generalized co-ordinate. The aerodynamic force term comprises of contributions due to lifting surface motion, control surface deflections and atmospheric turbulence. Equation 12 can be rewritten as

$$M_i \ddot{q}_i(t) + \omega_i^2 M_i q_i(t) + \sum_{j=1}^{n} Q_{ij} q_i(t) = - Q_{i\delta} \delta(t) - Q_{i w_g} W_g(t) \qquad (13)$$

where 'n' is the number of generalized co-ordinates; $\delta(t)$ is the control surface deflection and $W_g(t)$ is the gust velocity. The aerodynamic force matrices can be represented as functions of the reduced frequency 'k' or alternatively in terms of the Laplace s variable, using Pade' approximant techniques [45]. Newsom [46] reformulates the second order differential Equation 13, into a state space form as follows

$$\{\dot{x}\} = [A]\{x\} + [B]\{u\} + [H]\{\eta\} \tag{14}$$

There [A] is the system dynamics matrix; [B] is the control matrix; [H] is the gust distribution matrix; [x] is the state vector; [u] is the vector of control surface deflections and its time derivates; $\{\eta\}$ is a vector of gust velocity and its derivatives. The system aeroelastic stability, the effects of control surface deflection, and atmospheric turbulence are encompassed in Equation 14.

The application of optimization theory in the synthesis of such aeroservoelastic systems is fairly obvious. The state space representation of Equation 14 allows for the implementation of concepts of optimal control theory that are suited for the synthesis of MIMO (multi-input, multi-output) systems. The state vector in the above formulation consists of the system modes and its time derivatives, which can be superposed to obtain the system response. The control input must be able to influence each of these state variables for the system to be controllable. The control system is synthesized by minimizing a time integral quadratic function that is dependent on the state variables and the control input

$$J = \int_0^\infty (x^T Q x + u^T R u) dt \tag{15}$$

There Q and R are weighting matrices on the state and control variables, and are specified by the designer. The application of optimality conditions results in the optimal control u,

$$u = - Gx \tag{16}$$

$$G = R^{-1} B^T P \tag{17}$$

where G is the gain matrix and P is the steady state solution of the matrix Ricatti equation

$$-\dot{P} = PA + A^T P - R^{-1} B^T P + Q \tag{18}$$

A similar procedure is used by Chipman et al. [47] in the active control of aeroelastic divergence of forward swept wings. The aerodynamic destiffening of the first bending mode for such a structure causes it to couple with the short-period mode and results in a low-frequency dynamic instability. The dynamic and static stabilities are controlled by synthesizing a control system in a manner similar to the one described above.

The accurate representation of the aeroelastic system response requires a large number of state variables. A linear quadratic Gaussian performance index is generally minimized to obtain a control law that has the same order as the plant (full-state). Mukhopadhyay et al. [48] assert that such high-order controllers are sensitive to modeling errors and are often too complex to implement. They synthesize a reduced order control law by minimizing a weighted sum of mean-square steady-state system responses and control input commands. A conjugate-gradient algorithm is used for the minimization and the gradient information is obtained analytically as a solution of the Lyapunov equations [49]. Expressed mathematically, the control system dynamics are written as,

$$\dot{x}_c = A x_c + B y \tag{19}$$

$$u = C x_c \tag{20}$$

where $x_c$ is the control state vector and is of a lower order than the plant model. Select elements of matrices A, B, and C are specified as the design variables selected to minimize the performance index. The authors apply the technique in the synthesis of an active flutter suppression control. A 25th-order and 65th-order plant were controlled successfully by a fourth-order controller. Furthermore, comparisons of the reduced order controllers with full order controllers showed only a minor deterioration in the performance index.

In an application of numerical optimization methods to improve the stability robustness of MIMO feedback control systems, Mukhopadhyay [50] minimizes a standard LQG performance index with constraints on the singular value at the plant input or output. The lateral attitude control system of a drone aircraft is used as a test problem and both full-state and reduced-state controllers are synthesized.

Another study that addresses the problem of dual structural-control optimization for very large flexible space structures [51] involves both optimal control and standard nonlinear programming approaches in the synthesis procedure.

CONCLUDING REMARKS

The use of optimization methods in aeroelastic applications will continue to expand as more effective analysis tools are developed. Such applications are necessary from the standpoint of design automation. Furthermore, optimization studies often yield counter-intuitive results which become invaluable as the complexity of the problem develops. The present paper has examined the state-of-the-art developments in aeroelastic optimization and has focused on problems of current interest. Applications in rotorcraft aeroelasticity and coupled structural-aeroelastic-control optimizaion show potential for promising benefits.

## REFERENCES

1. Maxwell, C., Scientific Papers, Vol. 2, 1869.
2. Michell, A. G. M., "The Limits of Economy of Material in Frame Structures", Philosophical Magazine, Series 6, Vol. 8, 1904.
3. Chung, K., et al., "Shape Optimization of Elastic Structures to Meet Local Criteria and Constraints", presented at the 22nd SES Annual Technical Meeting, October 7-9, 1985.
4. Berke, L., and Khot, N. S., "Use of Optimality Criteria Methods for Large Scale Systems", AGARD LS No. 70, Structural Optimizaion, October 1974, pp. 1-29.
5. Ragsdell, K. M., "The Utility of Nonlinear Programming Methods for Engineering Design", 11th ONR Naval Structural Mechanics Symposium, Tuscon, Arizona, October 19-22, 1981.
6. Schmit, L. A., Jr., "Structural Synthesis - Its Genesin and Development", AIAA Journal, Vol. 19, No. 10, October 1981.
7. Stroud, W. J., "Automated Structural Design with Aeroelastic Constraints: A Review and Assessment of the State-of-the-Art", Structural Optimization Symposium, AMD, Vol. 7, 1974.
8. Schmit, L. A., Jr., and Miura, H., "Approximation Concepts for Efficient Structural Synthesis", NASA CR-2552, March 1976.
9. Hajela, P., and Sobieski, J. E., "The Controlled Growth Method - A Tool for Structural Optimization", AIAA Journal, Vol. 20, No. 10, October 1982.
10. Turner, M. J., "Optimization of Structures to Satisfy Flutter Requirements", AIAA Journal, Vol. 7, No. 5, 1969.
11. Weisshaar, T. A., "Panel Flutter Optimization - A Refined Finite Element Approach", International Journal of Numerical Methods for Engineering, Vol. 10, No. 1, 1976.
12. McIntosh, S. C., Jr., Weisshaar, T. A., Ashley, H., "Progress in Aeroelastic Optimization - Analytical Versus Numerical Approaches", SUDAAR No. 383, Stanford University, Stanford, California, 1969.
13. Weisshaar, T. A., "Aeroelastic Optimization of a Panel in High Mach Number Supersonic Flow", Journal of Aircraft, 9, 1972.
14. Van Keuren, J. M., Jr., and Eastep, F. E., "Use of Galerkins Method for Minimum Weight Panels with Dynamic Constraints", International Journal of Numerical Methods in Engineering, Vol. 10, No. 1, 1976.
15. Pierson, B. L., "Application of a Gradient Projection Optimal Control Method to a Class of Panel Flutter Optimization Problems", ERI-73186, Iowa State University, Ames, Iowa, 1973.
16. Pierson, B. L., "Aeroelastic Panel Optimizaion with Aerodynamic Damping", AIAA Journal, Vol. 13, No. 4, 1975.
17. Pierson, B. L., and Hajela, P., "Optimal Aeroelastic Design of an Unsymmetrically Supported Panel", Journal of Structural Mechanics, 8(3), 1980.
18. Librescu, L., and Reiner, I., "Recent Results of the Weight Minimization of Panels with a Flutter Speed Constraint", Second International Symposium on Aeroelasticity and Structural Dynamics, Aachen, April 1-3, 1985.
19. Turner, M. J., "Optimization of Structures to Satisfy Flutter Requirements", AIAA Journal, Vol. 7, No. 5, 1969.
20. Rudisill, C. S., and Bhatia, K. G., "Optimization of Complex Structures to Satisfy Flutter Requirements", AIAA Journal, Vol. 9, No. 8, 1971.
21. Greene, W. H., and Sobieski, J. E., "Minimum Mass Sizing of a Large Low Aspect Ratio Airframe for Flutter Free Performance", NASA TM-81818, May 1980.

22. O'Connell, R. F., Hassig, H. J., and Radovich, N. A., "Study of Flutter Related Computational Procedures for Minimum Weight Structural Sizing of Advanced Aircraft", NASA CR-2607, March 1976.
23. Wilkinson, K., et al., "An Automated Procedure for Flutter and Strength Analysis and Optimization of Aerospace Vehicles", Volumes I and II, AFFDL-TR-75-137, December 1975.
24. Lynch, R. W., et al., "Aeroelastic Tailoring of Advanced Composite Structures for Military Aircraft", Volumes I - III, AFFDL-TR-76-100, April 1977.
25. Hajela, P., "Techniques in Optimum Structural Synthesis with Static and Dynamic Constraints", Ph.D. Thesis, Stanford University, July 1982.
26. Hajela, P., "A Root Locus Based Flutter Synthesis Procedure", Journal of Aircraft, 21(6), January 1984.
27. Johnson, E. H., and Venkayya, V. B., "Progress Report on the Automated Strength-Aeroelastic Design of Aerospace Structures Program", NASA CP-2327, Vol. 1, April 1984.
28. Turner, E. W., "An Exposition on Aircraft Response to Atmospheric Turbulance Using Power Spectral Density Analysis Techniques", AFFDL-TR-76-162, 1977.
29. Hajela, P., "Comments on Gust Response Constrained Optimization", NASA CP-2327, Vol. 2, April 1984.
30. Rao, S. S., " Optimization of Airplane Wing Structures Under Gust Loads", Computers and Structures, Vol. 21, No. 4, 1985.
31. Hajela, P., and Lamb, A., "Automated Structural Synthesis for Nondeterministic Loads", to be published in Computer Methods in Applied Mechanics and Engineering, 1986.
32. Hajela, P., "Optimal Airframe Synthesis for Gust Loads", NASA CR-178047, Feb. 1986.
33. Peele, E. L., and Adams, W. M., Jr., "A Digital Program for Calculating the Interaction Between Flexible Structures, Unsteady Aerodynamics, and Active Controls", NASA TM 80040, January 1979.
34. Friedmann, P. P., "Formulation and Solution of Rotary-Wing Aeroelastic Stability and Response Problems", presented at the Eighth European Rotorcraft Forum, Paper 3-2, August 31 - September 3, 1982.
35. Bielawa, R. L., "Techniques for Stability Analysis and Design Optimization with Dynamic Constraints of Nonconservative Linear Systems", presented at the 12th AIAA/ASME SDM Conference, Anaheim, California, April 1971.
36. Hirsch, H., Dutton, R. E., and Rasumoff, A., "Effects of Spanwise and Chordwise Mass Distribution on Rotor Blade Cyclic Stresses", Journal of AHS, Vol. 1, No. 2, 1956.
37. Taylor, R. B., "Helicopter Vibration Reduction by Rotor Blade Modal Shaping", Proceedings of the 38th Annual Forum of AHS, May 1982, Anaheim, California.
38. Peters, D. A., Ko, T., Korn, A., and Rossow, M. P., "Design of Helicopter Rotor Blades for Desired Placement of Natural Frequencies", Proceedings of the 39th Annual Forum of AHS, 1983, St. Louis, Missouri.
39. Bennett, R. L., "Optimum Structural Design", Proceedings of the 38th Annual Forum of the AHS, May 1982, Anaheim, California.
40. Friedmann, P. P., and Shanthakumaran, P., "Aeroelastic Tailoring of Rotor Blades for Vibration Reduction in Forward Flight", presented at the 24th AIAA/ASME/ASCE/AHS SDM Conference, Lake Tahoe, Neveda, May 1983, AIAA-83-0916-CP.
41. Miura, H., "Overview: Applications of Numerical Optimization Methods to Helicopter Design Problems", NASA CP-2327, Part 2, April 1984.

42. Sutton, L. R., and Bennett, R. L., "Aeroelastic/Aerodynamic Optimization of High Speed Helicopter/Compound Rotor", NASA CP-2327, Part 2, April 1984.
43. Johnson, W., "Self-Tuning Regulators for Multicyclic Control of Helicopter Vibration", NASA TP-1996, March 1982.
44. Newsom, J. R., et al., "Active Controls: A Look at Analytical Methods and Associated Tools", NASA TM 86269, July 1984.
45. Vepa, R., "On the Use of Pade Approximants to Represent Unsteady Aerodynamic Loads for Arbitrary Small Motions of Wings", AIAA Paper No. 76-17, 1976.
46. Newsom, J. R., "Active Flutter Suppression Synthesis Using Optimal Control Theory", M. S. Thesis, George Washington University, May 1978.
47. Chipman, R. R., Zislin, A. M., and Waters, C., "Active Control of Aeroelastic Divergence", AIAA Paper No. 82-0684, 1982.
48. Mukhopadhyay, V., Newsom, J. R., and Abel, I., "A Method for Obtaining Reduced-Order Control Laws for High-Order Systems Using Optimization Techniques", NASA TP-1876, August 1981.
49. Kwakernaak, M., and Sivan, R., Linear Optimal Control Systems, John Wiley and Sons, Inc., 1972.
50. Mukhopadhyay, V., "Stability Robustness Improvement Using Constrained Optimization Techniques", AIAA Paper No. 85-1931-CP, August 1985.
51. Messac, A., and Turner, J., "Dual Structural-Control Optimization of Large Space Structures", NASA CP-2327, Part 2, April 1984.